KING TYRANT

KING TYRANT

A Natural History of *Tyrannosaurus rex*

MARK P. WITTON

Requests for permission to reproduce material from this work
should be sent to permissions@press.princeton.edu
Published by Princeton University Press
41 William Street, Princeton, New Jersey 08540
99 Banbury Road, Oxford OX2 6JX
press.princeton.edu

ISBN 978-0-691-24558-4
ISBN (e-book) 978-0-691-24559-1

British Library Cataloging-in-Publication Data is available

Editorial: Robert Kirk and Megan Mendonça
Production Editorial: Kathleen Cioffi
Text Design: D & N Publishing, Wiltshire, UK
Jacket Design: Ben Higgins
Production: Steven Sears
Publicity: Matthew Taylor and Caitlyn Robson
Copyeditor: Dana Henricks
Jacket illustration by Mark P. Witton

This book has been composed in Neue Kabel (main text, captions and smaller headings) and Bodega Serif (main headings)

Printed in Italy

10 9 8 7 6 5 4 3 2 1

For Georgia, Earth's most formidable biped for the last 66 million years

... although the world has gone on to produce saints and philosophers and poets, what has it ever done more impressive than the tyrannosaurus?

—Heywood Broun, *New York Tribune*, November 22, 1920

CONTENTS

Acknowledgments   ix

1. **Dinosaur Superstar**   1
   It's All in the Name   1
   Unearthing *T. rex*: From Cretaceous to Celebrity   6
   The Highs and Lows of Fame   36

2. **What, in Actuality, Is a *T. rex*?**   43
   The Long Road to *Tyrannosaurus*   43
   The Tyrannosaur Family Tree   53
   Tyrannosaurus rex, the Species   65

3. **Inside and Out**   83
   The Essence of Form: Bones and Muscles   83
   A (Partial) Internal Tour   105
   Clothing King Tyrants: The Life Appearance of *T. rex*   111

4. **Breathing Life into Bones**   131
   Running Hot, Living Large   131
   The Smarts, the Senses, the Voice   151
   Cometh the Mechatyrants   162

5. **Lands of the Tyrants**   181
   King Tyrants in Time and Space   181
   Where the Rexes Roamed   188
   Meeting the Neighbors   194

CONTENTS

6. **Life, Food, Love, Death** — **217**

The King Tyrant Social Scene — 217

Tyrannosaurus rex: Dinosaur Hunter — 228

Love, Death, and *T. rex* — 241

7. **Death of a Species, End of an Era** — **257**

Approaching the Finish Line — 257

Goodbye, *T. rex* — 266

Afterlife — 274

References — 280

Image Credits — 302

Index — 303

ACKNOWLEDGMENTS

WORKING ON THE *King Tyrant* project has been an excuse to synthesize some of the most exciting research on one of the most comprehensively understood species from deep time. In doing so, it has been difficult not to be awed by the mass of knowledge that, specimen by specimen and project by project, has accumulated about *Tyrannosaurus*. This book stands on the shoulders of countless field teams, researchers, authors, and artists who have made *T. rex* the poster child of dinosaur research that it is today.

A number of individuals helped bring this book into existence by providing access to specimens, giving useful advice or feedback, assisting with the never-ending search for literature, being soundboards for discussions of dinosaur biology, or simply providing encouragement. Thanks to Kirstin Brink, Vicky Coules, John Conway, Thomas Cullen, Charles Deeming, Dougal Dixon, David Evans, Richard Fallon, Richard Hing, Thomas Holtz, David Hone, Hilary Ketchum, Derek Larson, Susie Maidment, Jordan Mallon, Chris Manias, Darren Naish, Pat O'Connor, Elsa Panciroli, Eric Snively, and Andy Wood. Julianne Kiely deserves special mention for preparing comprehensive advice on Cretaceous paleobotany to guide my illustrations. Kevin Padian and an anonymous reader provided helpful comments on my manuscript, and Jon Davies kindly allowed me to use his privately commissioned art in chapter 6. Dana Henricks has done a terrific job with copyediting my text. If your name should be included on this list but wasn't, please accept my apology and demand however many beverages is appropriate recompense at our next meeting.

The staff at Princeton University Press have been very supportive of this project from the moment my pitch arrived in their inbox—I cannot recall a more enthusiastic publisher response to a book proposal. Creating *King Tyrant* was supported by the very kind people who follow and sponsor my work at Patreon.com, without whom I would struggle to write and illustrate books. I hope the finished product lives up to their expectations.

It would be remiss not to mention my parents, who see less and less of me as these book projects get more and more involved, but provide plenty of encouragement from afar. I have a particularly early, vague memory of being carried by my dad alongside the British Museum (Natural History)'s now long disassembled mounted *"Dynamosaurus"* at some point in my very early childhood. So, in a way, this is all their fault.

A few personal thanks are due to people who directly experienced the production of *King Tyrant*. Simon Hardy and Suzie Clark endured an enormous amount of *Tyrannosaurus* exposure on the first leg of the five museum Tyrannoversary Tour™ around the United States. Their company made an amazing but exhausting trip all the better. Simon, along with Richard Hing and Julianne Kiely, also provided proofing and readability checks on the *King Tyrant* manuscript.

And what of Georgia Witton-Maclean, six-time book widow, one time Disacknowledgment (see Witton 2013) and provider of photographs superior to my own? As usual, her contributions were minor, merely accompanying me to museums at home and abroad, helping to shift (very heavy) king tyrant bones around museum stores, tolerating a lot of talk about oversized dinosaur predators, *and* demonstrating enormous patience during the eleventh-hour push to get *King Tyrant* into a publishable format. So, you know: a 6–7 out of 10 performance.

Finally, and most importantly, I'd like to acknowledge myself for having the restraint not to use the innumerable puns that could, nay, *begged* to be included in a book about *Tyrannosaurus rex*. It was only with the greatest resolve that I suppressed section headers like *"Tyrannosoreus"* (pathologies); *"Tyrannosaurus wrecks"* (predatory behavior); *"Tyrannosaurus pecs"* (arm musculature); *"Tymannosaurus"* (distinguishing male specimens); *"Die, rannosaurus!"* (extinction); and, of course, *"Tyrannosaurus tex"* (Texan tyrannosaur fossils), among many more. Readers are encouraged to annotate their books with these and other hilarious *T. rex* puns to enhance their enjoyment, while I congratulate myself for being such a professional.

DINOSAUR SUPERSTAR

IT'S A CLICHÉ to say that some topics need no introduction, but this is undoubtedly true for *Tyrannosaurus rex*, the most famous dinosaur on the planet. *T. rex* fossils were first unearthed over a century ago, but their popularity does not solely reflect raw osteological charisma. King tyrants are so beloved because establishing them as the "king of dinosaurs" was critical to the individuals and institutions that shaped their early public impression; a history of nepotism, pop-culture exploitation, and the monetary value of their fossils has only raised their popularity further. These contribute to a unique *Tyrannosaurus* subculture that sits, sometimes uncomfortably, on the boundary between science and sensationalism.

IT'S ALL IN THE NAME

UNLIKE THE COMMON names we apply to plants and animals, the scientific names of organisms are universal, standardized labels that apply across cultures and history. Geographic differences dictate whether we refer to "brown bears" or "grizzly bears," just as we now use the term "leopard seals" rather than the nineteenth-century phrase "sea leopards." Scientists, however, will refer to these animals as *Ursus arctos* and *Hydrurga leptonyx* regardless of where they are from or what language they speak. Scientific names can honor people or make reference to the geography of a species, but they can also be descriptive, often drawing attention to characteristics of anatomy or behavior. *Ursus arctos* means "bear bear" (its name merely being the Latin and Greek names for "bear," implying it is the archetype of the bear group) and *Hydrurga leptonyx* is "thin-clawed water-worker." These are both fine, robust scientific names, but they are not especially exciting

or inspirational. Fossil organisms are subjected to the same naming conventions, and many also have descriptive and functional denominations. Famous extinct reptiles such as *Diplodocus carnegii*, *Allosaurus fragilis* and *Triceratops horridus* have straightforward etymologies: respectively, "Carnegie's double-beam" (a reference to bones in the tail); "fragile different lizard" (referring to once-unique vertebral anatomy); and "rugose, three-horned face" (which needs no explanation for *Triceratops*).

There are, however, some scientific names that defy these conventions. They are a special type that eschew description and function for evocativeness and character, instilling their namesakes with qualities that command attention and respect. Often, we don't even need to translate these words to feel their weight and excitement. Among extinct species, some of the best known dinosaurs have such names: *Brontosaurus excelsus*, the

FIGURE 1.1

Tyrannosaurus rex. Kingly in name and perhaps in nature, these dinosaurs were notable in their time for being among the largest terrestrial carnivores to have ever existed. Today, they have a second role, as the most popular and studied of all the non-avian dinosaurs.

"noble thunder lizard," and *Velociraptor mongoliensis*, "Mongolia's swift thief."

But the pinnacle of such titles must be *Tyrannosaurus rex*, "tyrant lizard king," a name assigned in 1905 to a partial skeleton of a large theropod, or predatory dinosaur, found in Montana a few years prior (fig. 1.1). Formed from Greek (*Tyrannosaurus*) and Latin (*rex*), even those of us with a no grasp of ancient languages grasp that this animal was no mere tyrannical reptile, but their veritable *king*. Like all the best scientific names, *Tyrannosaurus rex* is also easily articulated by even the most leaden tongue and it abbreviates perfectly to the slick, dangerous, and eminently marketable "*T. rex*." Even when this is corrupted to *T. Rex* (with a nonscientific capital *R*; species names are always written with lowercase letters) or hyphenated to *T-Rex*, the most offensive variant to scientific writing standards, it leaves a commanding presence in a block of text. It's hard to argue that *Tyrannosaurus rex* is not one of the most arresting and memorable scientific names ever applied to an animal, living or extinct.

This iconic designation was coined by the American paleontologist Henry Fairfield Osborn for appropriate reasons. Working from fossils recently acquired from Montana by the American Museum of Natural History, Osborn announced that the size of *T. rex* "greatly exceed[ed] that of any carnivorous land animal hitherto described" (Osborn 1905, p. 259) and warranted a label with appropriate gravitas. Osborn could not have known how prudent it was to name these fossils for their immensity because, over a century after their discovery, *Tyrannosaurus* remains a strong contender for the largest terrestrial carnivore in history. Similarly, while the kingly title emphasizes king tyrant proportions, it is also apt for the cultural status of *Tyrannosaurus* as the world's most famous dinosaur. Countless starring appearances in books, TV series, films, magazine articles, museum exhibits, theme park rides, and more have made *T. rex* a bona fide paleontological celebrity. Their faces—either in the form of their now-iconic skulls, or restored with flesh to roar at us with bared teeth—emblazon no end of paleontological media and products aimed at any and all demographics, within which other fossil species are frequently sidelined. Nowadays, *Tyrannosaurus* can sustain intense cultural attention all by themselves, with entire documentaries, books, and touring exhibits constructed purely around king tyrants (fig. 1.2).

Without a doubt, *Tyrannosaurus* has become the leading face of not only dinosaurs, but prehistoric life in general, and this interest is expressed in ways beyond TV ratings, book sales, and museum footfall. Academic interest in *Tyrannosaurus* is proportional to their public profile, and their remains are actively sought by museums and universities. We now have dozens of excellent *T. rex* specimens, and they are regularly topics of scientific research papers. So popular are *T. rex* as a research subject that we probably know more about this 66-million-year-old dinosaur species than we do many living species. Demand for their specimens is also unprecedented among fossil collectors, and private sales of their remains now attract bids of tens of millions of dollars. *Tyrannosaurus* are such sought-after and enjoyed animals that they have formed their own subculture within paleontology, an accolade that few other fossil species can boast.

The question of why dinosaurs, and *Tyrannosaurus* in general, are so well liked is a complex one, with their natural appeal pitched against our own efforts at marketing and exploitation. The history of *Tyrannosaurus* popularization certainly includes much of the latter, with professional rivalries, institutional nepotism, influences from popular culture, and a surprisingly recent ascent to superstardom shaping *T. rex* as we see it in public today. We would be remiss, however, to deny that visiting one of the many mounted *T. rex* skeletons

FIGURE 1.2

King tyrants attract so much attention that their skeletons, both real and replica, tour the planet like rock stars. Alongside museums and universities, their remains can end up in unexpected places, from airports and theme parks to gothic thirteenth-century British cathedrals.

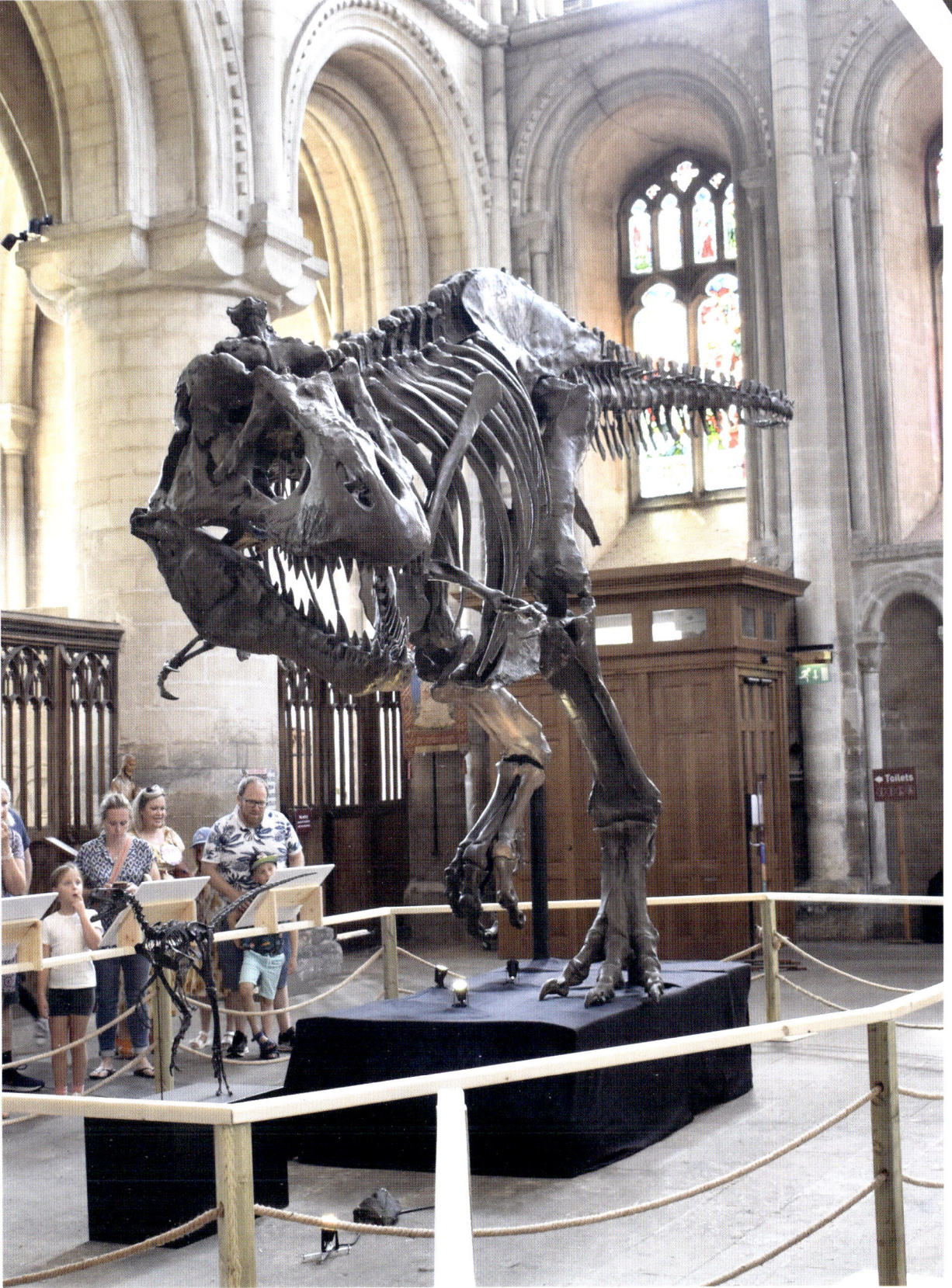

in museums around the world is a special experience. It's true that most institutions with mounted *T. rex* skeletons have replicas rather than real fossils (especially outside of the United States) but techniques for casting ancient bones are so sophisticated that these copies are indistinguishable from the originals to all but expert eyes. Even for those uninterested in fossils, it's hard to argue that *Tyrannosaurus* are not awesome animals. The size of even a modestly sized adult is remarkable. They stand at 11–12 m long and over 3 m tall at the hips, and 1.4 m long jaws tower over our heads as we gaze up at even a calmly posed skeleton. Their famous proportions are also, if we try to momentarily forget how familiar they have become, bizarre to behold: a clashing combination of giant head, huge teeth, tiny arms, and long legs. Embodying the characteristics of an elephant, bird, and crocodile, there is nothing like *Tyrannosaurus* in modern times. We sometimes have to remind ourselves that the owners of these skeletons were denizens of the same Earth that we now inhabit, albeit millions of years later. For all their imposing size and strangeness, they were not fantasy or imaginary creatures.

This effect we feel when looking at *Tyrannosaurus* skeletons is their raw osteological charisma. They are natural centerpieces that draw us into museum spaces, generating excitement and intrigue that often overshadows neighboring displays. They are not the only species to do this, of course: many animal skeletons are undeniably exciting objects that draw us to them, and not all of them are giant dinosaurs. *T. rex* has an appeal that is now self-perpetuating, however, fueled by decades of celebrity status and our interest in their maximized embodiment of predatory lifestyles. We intuitively appreciate that, if flesh and movement were restored to their bones, we would be easy prey. Is there something fundamentally, primally *humbling* about gazing at a set of once-living giant jaws lined with huge, spike-like teeth? Next to *T. rex*, we finally understand how a mouse views a cat. But we are also challenged to understand how *Tyrannosaurus* worked. For while *T. rex* bones are recognizable in basic form, their size and proportions are beyond anything in our experience.

As with many dinosaurs, we are forced to conclude that they were some sort of *superanimal*; a species that pushed evolutionary forces to their known limits to become bigger, and perhaps stronger, than anything that walks the Earth today.

This notion may provide one of the main factors in our continued interest in *Tyrannosaurus* and, perhaps, in dinosaurs in general. Their grand sizes and unusual anatomies once saw them regarded as a race of lumbering, dimwitted reptiles with twisted genomes and few evolutionary prospects. From the late twentieth century onward, however, this view was overturned and dinosaurs were reinterpreted as active, intelligent, and behaviorally complex species. This was an awakening, one could argue, to the realization that dinosaurs were species that not only attained many of the biological accomplishments we see in living species but, in some cases, exceeded them (fig. 1.3). Their gigantic sizes, exaggerated horns, spikes, and armor, as well as unusual, outlandish body plans were not mere aesthetic features, but anatomical and biomechanical marvels of physiology and biology. Dinosaurs are to zoology and anatomy what the Grand Canyon or the Himalayas are to geology. Yes, there are other canyons and mountains in nature, but no, you won't find more awesome examples of them. This does not, of course, make dinosaurs somehow "better" than other modern or prehistoric species, nor did they exceed every biological feat we might choose to measure them by. But they were remarkable for their propensity for magnificence, and the aspects of their biology that allowed them to attain their celebrated body forms and biomechanical regimes are quite different to those that govern our own mammalian evolution.

In being large, powerful, and predatory, *Tyrannosaurus* is one of the ultimate examples of dinosaurs at their most spectacular. They are reminders of the awesome formative power of natural selection and the possibilities of evolutionary processes. As curious, intelligent beings, how could we not want to admire and learn more about them? How rapidly did they grow, how fast could they move, and how strong were their

FIGURE 1.3

The size of *Tyrannosaurus* eclipses that of all the largest terrestrial mammalian predators, recent (polar bears) and extinct (*Andrewsarchus*). Their magnitude and remarkable appearance surely contribute to their popularity, although other factors have shaped their fame as well.

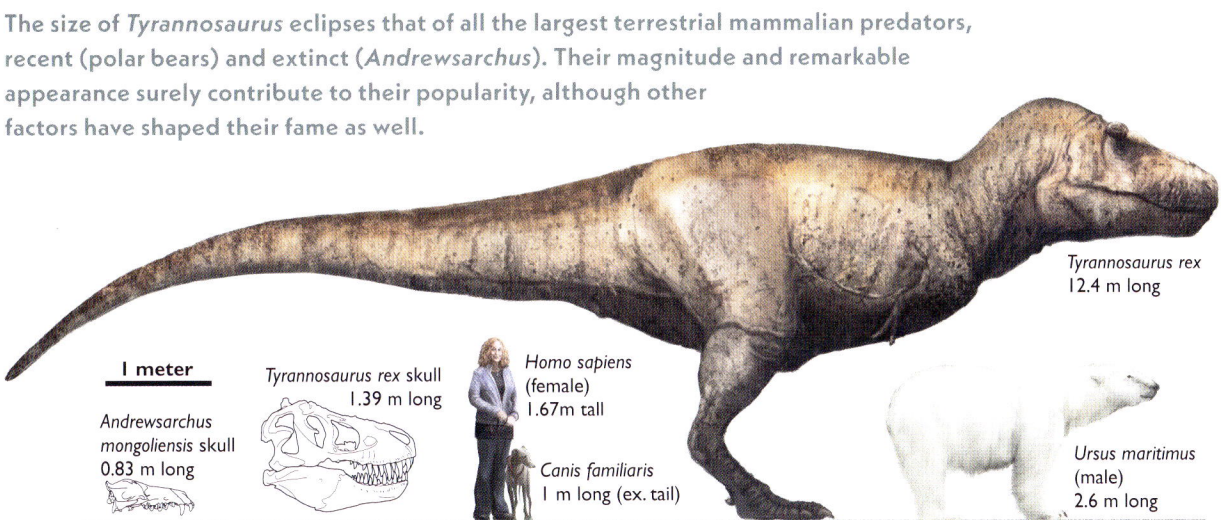

Tyrannosaurus rex
12.4 m long

I meter

Tyrannosaurus rex skull
1.39 m long

Homo sapiens
(female)
1.67m tall

*Andrewsarchus
mongoliensis* skull
0.83 m long

Canis familiaris
I m long (ex. tail)

Ursus maritimus
(male)
2.6 m long

bites? And how, as artists, could we not want to visualize what they might have looked like? As with classic artworks depicting mythological figures, drawing *T. rex* is an attempt to capture something beyond human experience. The enormous number of artworks of *T. rex* produced over the last century shows how satisfying and exciting king tyrants are as art subjects, and accurately rendering their anatomy—their "T"-shaped skulls with their pinched snouts and distinct ornaments, their high-waisted, barrel chests, their surprisingly slender hips and shoulders—can make for some of our most impressive, memorable scenes of deep time.

It's here, comparing *T. rex* to mythological figures and declaring them "superanimals" that we have already verged dangerously close to another facet of *Tyrannosaurus* interpretation: a tendency to exaggerate, overhype, and romanticize their biology and cultural standing. As with any celebrity with an enthusiastic following, our intense interest in *Tyrannosaurus* has spawned an imagined and hyperbolic version of *T. rex* that has little bearing on reality—"reality" in this case being their fossils and the science that aims to understand them. Some of what we see of *Tyrannosaurus* in popular culture are exaggerations of facts known about the real animals, but many aspects are effectively

fiction, a result of unrestrained molding of their image and character by a media exploiting their appeal. Even basics of their proportions, recorded as facts by fossil remains, are distorted by mainstream portrayals, such that the visions of adult *Tyrannosaurus* depicted in toys, films, and television programs show a relatively lithe, long-legged monster with a lipless mouth and hooded eyes. These interpretations bear only passing resemblance to the robust, surely lipped, stout-limbed, deep-bodied, and wide-headed creature known from fossils. They are perhaps what popular culture *wants T. rex* to have been, not what they actually were. In the same vein, portrayals of *Tyrannosaurus* in both fiction and nonfiction frequently follow agendas, casting them in stereotyped roles as sinister villains or action heroes, cuddly pets or terrifying creatures, slow, lazy scavengers or fast, regal predators. Because *Tyrannosaurus* was a real animal, and products featuring them often have a vestige of educational or scientific content, such visualizations bleed into the general conversation about this most popular dinosaur, confusing what paleontologists have determined from *T. rex* fossils with ideas and concepts introduced by filmmakers and toy manufacturers.

This book lies at this intersection between popularized *Tyrannosaurus* and our sober, scientific interpretations of this animal. Stripped of its hype, *T. rex* remains a fascinating and unusual species worthy of our attention and investigation. We know enormous amounts about their paleobiology and the world they lived in thanks to their fossils receiving a wealth of scientific study. There is much to be said on where and how they lived, how they functioned physiologically, and how they operated as gigantic, multi-tonne predators. And around this, we know that *Tyrannosaurus* were among the last of the dinosaurs, a genuine victim of the mass extinction that ended the Age of Reptiles. Understanding *T. rex* for what they were, and not what we want them to be, reveals a far more enlightening and fascinating species than any of their portrayals in fiction.

UNEARTHING *T. REX*: FROM CRETACEOUS TO CELEBRITY

UNLIKE THE RESEARCH history of some fossil species, the discovery, naming, and reconstruction of *Tyrannosaurus* was a relatively straightforward sequence of events. This reflected, in part, early twentieth-century dinosaur science being of sufficient maturity that new remains did not bamboozle and confuse scientists in the way that they had in previous decades. Equally, the first fossils identified as *Tyrannosaurus* were well-preserved partial skeletons that, as early as 1905, gave clear indications of their size, body shape, and posture (Osborn 1905). What makes the research history of *Tyrannosaurus* interesting, therefore, is not the story of how they were found and interpreted, but the blurred boundary between the development of *T. rex* as a scientific entity and public celebrity. It is easy to detach these aspects for most fossil animals, there being obvious distinctions between the objective, technical studies of paleontologists and the drive to interpret and promote that science to the public. It is far more difficult to tease these strands apart for *Tyrannosaurus*, however, because their research history runs parallel to their role as a prominent, even the preeminent entity in the popularization of dinosaur paleontology. Indeed, the very discovery of *Tyrannosaurus* was driven by a need to broadcast spectacular dinosaur science, and, over the next century, popular culture became as much of a driving force for our learning about king tyrants as scientific necessity. *Tyrannosaurus* research has always sat at a sometimes uncomfortable interface between science and sensationalism, and between the needs of certain individuals and the general public.

1900–1920s: THE DISCOVERY OF KING TYRANTS

No discovery of a fossil species can ever be said to be premeditated, but there was an element of strategy in the locating and unearthing of the first *Tyrannosaurus* bones in 1902. Along with dozens of other extinct species, *T. rex* was discovered through museum-driven campaigns to unearth spectacular fossils from the American West at the turn of the twentieth century (Rieppel 2019). The late 1900s had seen the foundation of several new American museums that were financed by rich private benefactors, mostly industrialists flush with new money from burgeoning transport, building, and agricultural industries (Rieppel 2019). Large philanthropic donations had financed grand buildings to archive and display objects of American heritage and achievement, but, in contrast to traditional museums where research needs were prioritized over public edification, these new institutions considered public outreach of equal concern to academic pursuits. This attitude appreciated, however, that museum visitors required enthrallment and entertainment as much as education, for which some degree of spectacle or showmanship was needed. The appeal of creating grand displays was not lost on the financiers of these institutions either, these exhibits being physical monuments to their wealth, influence, and importance (Rieppel 2019).

Partly fueled by the allure of its own dramatic landscapes, natural history was a popular topic in the United

States in the late nineteenth century. The fantastic animals and fossils of this region provided vast scope for the sort of large, impressive museum exhibits sought by museum directors. Vertebrate paleontology was a particularly obvious discipline that could deliver on these requirements, such that several museums began acquiring scores of fossil bones that would fill their collections and display rooms. It was as the curator of the newly formed Department of Vertebrate Paleontology at the American Museum of Natural History (AMNH) that Henry Fairfield Osborn entered this story in 1891 (fig. 1.4A). Osborn was one of the most important figures in the twentieth-century history of vertebrate paleontology, his tenure at the AMNH setting the stage for generations of research. His legacy, however, is tarnished by atrocious views on race, the implementation of which shaped both the museum and twentieth-century science (Regal 2018). Born into wealth and high social standing thanks to his father's presidency of the Illinois Central Railroad, Osborn was expected to occupy a position of comparable influence and power in whatever

career he pursued. Directing his attention to paleontology, Osborn decreed that his Department of Vertebrate Paleontology would be the most important and influential in the country, and set about hiring teams of fossil hunters to make this a reality.

Ambitious and overbearing even to his closest allies, Osborn also had personal grievances to settle. Like all American vertebrate paleontologists working at the turn of the twentieth century, Osborn was operating in the shadow of the first Great Dinosaur Rush, a period popularly known as the "Bone Wars" (Brinkman 2010). This infamous feud saw Edward Drinker Cope, of the Academy of Natural Sciences of Philadelphia, and Othniel Charles Marsh, of the Peabody Museum of Natural History, Yale, engaged in a ruthless competition to find, describe, and name the most fossil vertebrate species during the late nineteenth century. Their intense rivalry lasted for decades, even into the final years of their lives, and spilled over into the next generation of American paleontologists. Included in this fallout was Osborn, who was mentored by Cope in the 1870s.

FIGURE 1.4

The individuals who discovered and named *T. rex*. (A) Henry Fairfield Osborn, c. 1909, then curator of the Department of Vertebrate Paleontology at the American Museum of Natural History and the eventual president of the same institution; (B) famed fossil hunter Barnum Brown, 1914, on fieldwork in Montana. Under Osborn's orders, Brown's fossil hunts in Montana uncovered the fossils that Osborn would name *Tyrannosaurus rex*.

DINOSAUR SUPERSTAR

The two men would become friends, and, eventually, Cope relied on Osborn's support both academically and financially. Allying himself with Cope saw Osborn naturally view Marsh as a rival. His professional crosshairs were aimed at not only amassing a world-class collection of fossils that would shame that compiled by Marsh at Yale, but also on outshining Marsh's respected research program on extinct vertebrates (Brinkman 2010). It was not enough to merely outperform Marsh, however; Osborn also sought to undermine his legacy by revising and refuting the bulk of his academic output (Brinkman 2010). Through the 1890s Osborn began sending field crews to fossil sites frequented by Marsh's expeditions to recover remains that would achieve these goals. Osborn's early focus was on mammals, but the rocks his teams scoured were of Mesozoic age: that is, the dinosaur-bearing sediments of 250–66 million years ago. Mammal remains were few, but AMNH field crews were stepping over find after amazing find of dinosaur skeletons in Jurassic-age (200–145 million years ago) strata. From 1897 onward, Osborn turned his department's attention to dinosaurs.

In doing so, Osborn joined a second Dinosaur Rush, his teams competing with those from Pittsburgh's Carnegie Museum and the Field Columbia Museum in Chicago (renamed the Field Museum of Natural History in 1905) to find ever more spectacular dinosaur skeletons. The prize each museum sought was a complete Jurassic sauropod (long-necked) dinosaur that could be reconstructed as an exhibition centerpiece. Mounted dinosaur skeletons were still novelties at this time, and no one had attempted to assemble the skeleton of a giant like *Brontosaurus* or *Diplodocus* before, at least for public display. Osborn felt that such spectacles would be a major draw to the AMNH (Rieppel 2019) and would also impress museum patrons (Brinkman 2010). Not everyone agreed this was a worthwhile pursuit, however. Researchers felt that such a mount would restrict access to important specimens, and in any case, earlier, outdated attempts at restoring prehistoric animals had become scientifically embarrassing and misleading to the public: What was to stop the same fate from

befalling these new efforts (Brinkman 2010; Nieuwland 2019; Rieppel 2019; Witton and Michel 2022)? Undeterred, Osborn pushed ahead, feeling that such displays were the most effective means of communicating the size and characteristics of extinct animals. He even persuaded the AMNH directors to build a new hall specifically for dinosaurs in 1903, having full faith that his field teams would find complete skeletons to fill it. When these proved elusive, Osborn ordered his fossil hunters to prioritize searching for anatomies that would complement incomplete remains already filling AMNH collections, allowing for the creation of complete, if chimeric skeletons (Brinkman 2010).

By 1905, the race to reconstruct a giant dinosaur was over, and first place had gone to Osborn and the AMNH. Although a complete sauropod had remained elusive, combining several specimens and sculpting missing elements saw the AMNH assemble and mount a complete skeleton of the iconic sauropod *Brontosaurus excelsus*, the first giant dinosaur assembled for museum exhibition. This proved to be a sensation with the public, their *Brontosaurus* quickly establishing itself as one of the most visited parts of the museum. It was aided by innovative displays that featured reconstructions of the animals as in life alongside the fossils, each adding to the spectacle as much as public education. Away from the public galleries, Osborn's efforts to turn his Department of Vertebrate Paleontology into a world-leading institution had also been an inescapable success. The AMNH became a giant among paleontological institutions and established a pioneering legacy for the modernization and democratization of museums and knowledge. Osborn, meanwhile, attained a status of supreme influence within the world of vertebrate paleontology.

While the AMNH busied itself with displaying dinosaur fossils, however, Osborn's fossil hunters were running out of luck. Their prospecting efforts were finding increasingly slim returns from the Jurassic rocks that had hitherto supplied the museum with their most noteworthy finds (Brinkman 2010). Although far from exhausted of fossil bones, Osborn's teams felt they had taken the best material from their most productive sites and suggested that new localities may bear novel, more interesting specimens. Osborn accordingly dispatched his best fossil hunters to search for new field sites in the early 1900s. In 1902 he sent one of his top fossil hunters, Barnum Brown (fig. 1.4B), to Cretaceous (145–66 million years ago) localities in Montana to seek new dinosaurs. These rocks, from the final days of the Mesozoic Era, were geologically younger than the Jurassic sediments that had yielded *Brontosaurus* and promised new specimens, as well as new attractions, for the museum. Brown had worked with Osborn since the early 1890s and was renowned for his ability to find significant vertebrate remains even at sites already picked clean by other prospectors. Often described as an eccentric and colorful individual, Brown's personality sometimes clashed with Osborn's overbearing qualities, but the two enjoyed a long and successful professional relationship. Brown remained loyal to the AMNH for his entire paleontological career.

The 1902 Montana expedition had been organized following a tip-off from Dr. W. T. Hornaday, director of the Bronx Zoological Garden, who had found large fossil bones thought to belong to the famous horned dinosaur *Triceratops* in a location hitherto unknown for its fossils (Brown 1915). Their destination was close to the town of Jordan, which even today only just earns the title of "town": the 2020 census gives a population of just over 350 people. Twenty-six miles north of Jordan flows Hell Creek, a tributary of the Missouri River, a body of water that has shaped Cretaceous clays and sandstones into dramatic badlands. These were described by Brown in his memoirs as

> a variety of variegated sculpted cliffs and domes intersected by deep canyons with scattered pine trees and pockets of junipers; while on the hillsides in the broader valleys, lines of cottonwoods mark the stream courses. That fall when the cottonwood leaves had turned yellow it [sic] was such a striking scene my saddle horse stopped to look. (Brown, quoted in Dingus and Norell 2010, pp. 251–252)

Following Hornaday's tip-off, Brown and his team headed straight to Hell Creek Canyon. On their first day of prospecting, "before the cook's call for dinner" (Brown 1915, p. 322), they found large sandstone blocks containing large bones. Tracking them to their source, they found dinosaur remains encased in an unusually hard blue sandstone located on a steep slope. Situated on an awkward hill and entombed in a rock that yielded little to even the strongest strike with a pick, excavating this animal would be challenging. Even entombed in sediment, however, Brown realized that his team had found something special. He wrote to Osborn, "Quarry No. 1 contains the femur, pubes, [partial] humerus, three vertebrae and two undetermined bones of a large Carnivorous Dinosaur not described by Marsh ... I have never seen anything like it from the Cretaceous" (Dingus and Norell 2010, p. 82). He went on to justify the time and expense it would take to extract the bones, which required new equipment, more horses, and even dynamite to remove the overburden. Working through a severe case of gout, Brown supervised his team to procure several large blocks of fossil-filled sandstone, the largest of which weighed almost two tonnes and required six horses to transport it to the nearest road. Unable to secure the entire specimen in one season, Brown and his team returned to finish the excavation in 1905. They eventually left a hole in the Montana landscape 9 m long, 6 m wide and almost 8 m deep.

This giant carnivorous dinosaur was exciting, not the least because it represented something entirely new. Although his vertebrate paleontology program was undoubtedly a success, Osborn had largely failed to discredit Marsh academically, and, adding insult to injury, his museum was filled with reminders of Marsh's successes in North American dinosaur studies. The AMNH Hall of Reptiles was filled with species named by Marsh, including *Stegosaurus*, *Allosaurus*, as well as, of course, *Brontosaurus* (Preston 1986; Davey 2019). This new predator thus presented an opportunity for Osborn to put his own name to a spectacular new dinosaur, but he would have to wait for the opportunity. Removing the newly found bones from their granite-hard matrix proved no easier than extracting them in the field. Brown later described how the largest block, which contained the pelvis, had to be broken into pieces to allow for more delicate work, after which the bones could be reassembled (Dingus and Norrell 2010). More material had been recovered than was originally estimated, and it took two years to remove all the fossils. When finished, the inventory included much of the skull, many vertebrae, some gastralia (belly ribs), an arm bone, the entire pelvis, and the major leg bones. Before the preparatory work was concluded, however, the AMNH team learned that another giant Cretaceous dinosaur similar to their own had been acquired by a rival institution, the Carnegie Museum. Found in Wyoming during 1902 by former AMNH fossil hunter Olaf A. Peterson, this specimen contained a similar inventory of skull material, limb bones, and vertebrae (fig. 1.5D). In the ongoing competition for prestige and scholarly significance among America's natural history museums, this presented an unacceptable risk to Osborn's Department of Vertebrate Paleontology. Might Peterson and the Carnegie Museum announce and name their new predator while the AMNH team toiled away at preparing their own example?

Osborn was sufficiently concerned to write a short paper on the AMNH's specimen before either field or preparatory work was concluded (Breithaupt et al. 2008). This seven-page, 1905 paper had little information to convey about the fossil and could only give the roughest outline of its size and appearance. A simple and schematic skeletal reconstruction by W. D. Matthew showed the size of the specimen relative to a human figure and was the only illustration of its bones (fig.

1.6A). Even so, Osborn's short, premature paper was sufficient to give the still-emerging specimen a name: *Tyrannosaurus rex* (Osborn 1905). As noted above, this name reflected the specimen's size, but such a grandiose, self-important label was surely also chosen for its publicity potential. Osborn's Department of Vertebrate Paleontology had not only won the race for mounting a giant sauropod, but was now credited with the discovery and ownership of the literal king of tyrant dinosaurs. Naturally, their specimen was destined for display in the Hall of Fossil Reptiles as soon as it could be fully prepared.

Brief as it was, Osborn's 1905 paper did not only name *Tyrannosaurus*. It also described material of another carnivorous dinosaur discovered during an AMNH expedition to northeastern Wyoming in 1900, also led by Brown. The specimen was of a similar size to their first *T. rex*, although the bones were more fragile. Although also not yet fully extracted from their sedimentary matrix, Osborn was able to provide more details about these than the *Tyrannosaurus* fossils, including an excellent illustration of its lower jaw (fig. 1.5C). The specimen would ultimately prove to comprise both lower jaws, some skull bones, a number of vertebrae (including a complete neck), a partial pelvis, and one limb bone. It also seemed to have several bony scutes (that is, bones embedded in skin), seemingly indicating an armored hide in life. It was chiefly the latter that prompted Osborn to think these remains may belong to a genus distinct from *Tyrannosaurus*, for which he created a title of similar grandeur: *Dynamosaurus imperiosus*, the "imperious power lizard." This name was not as elegant as *Tyrannosaurus rex*, but its similar energy would work well

FIGURE 1.5

Historically important *Tyrannosaurus* specimens. (A) Possibly the oldest scientific record of a *Tyrannosaurus* fossil, a drawing of a Denver Formation *T. rex* tooth by Edward Berthoud, figured in an 1872 letter to Othniel Marsh; (B) vertebrae named *Manospondylus gigas* by Cope, 1892, eventually identified as *T. rex* by Osborn; (C) part of Osborn's other tyrannosaur specimen, the lower jaw of *Dynamosaurus imperiosus*; (D) lower jaw of a partial *T. rex* skeleton discovered by the Carnegie Museum in 1902, the existence of which prompted Osborn to rush out the description and name of *Tyrannosaurus rex*. (A) After Carpenter and Young (2002); (B) from Cope (1892); (C) from Osborn (1905).

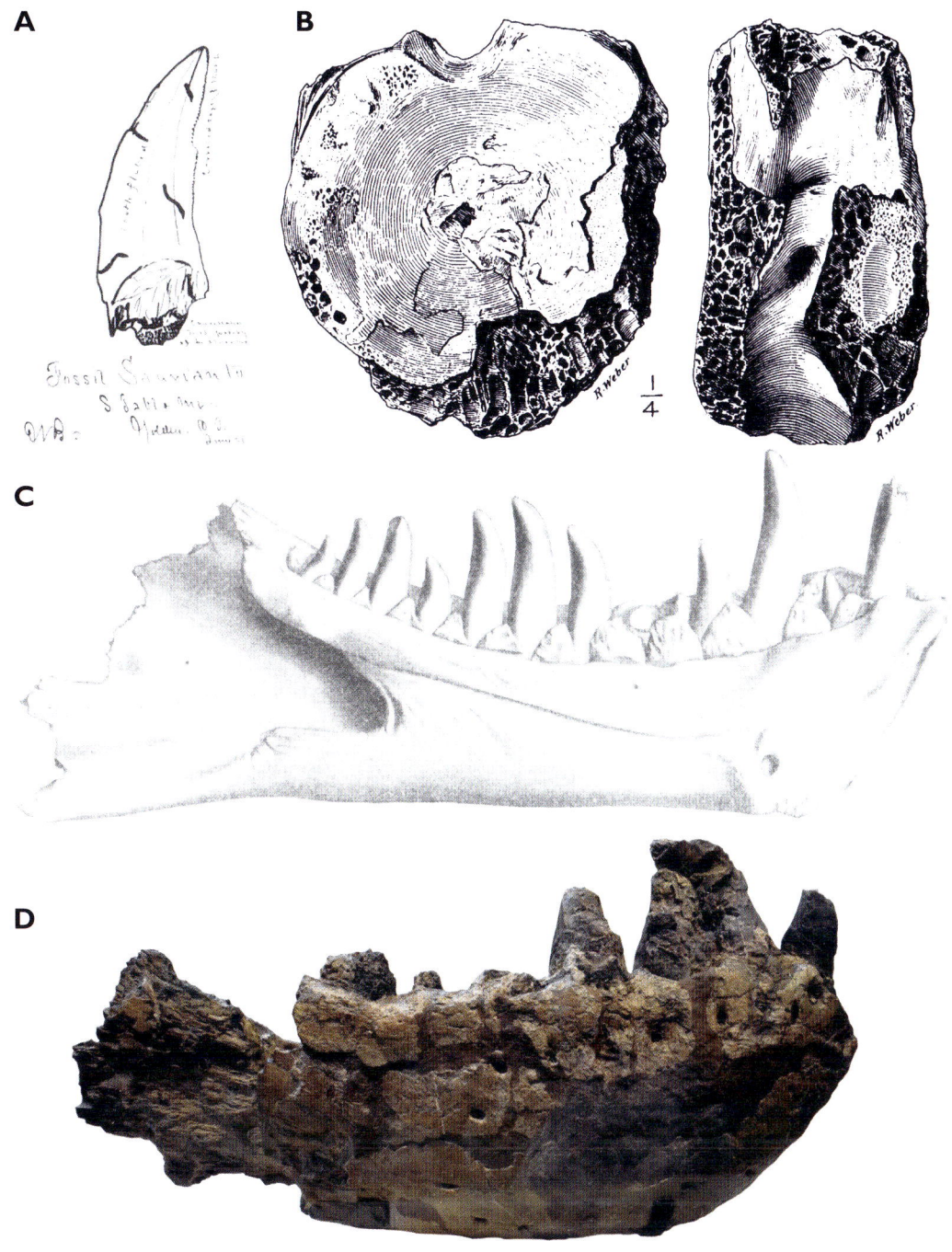

A

B

C

D

DINOSAUR SUPERSTAR

FIGURE 1.6

(A) The first effort at reconstructing *Tyrannosaurus* from 1905, before the AMNH's original specimen had even been prepared fully; (B) a second attempt from one year later, 1906, executed with superior, though not yet complete, knowledge of *Tyrannosaurus* anatomy. (A) Illustration by W. D. Matthew, from Osborn (1905); (B) by L. M. Sterling, from Osborn (1906).

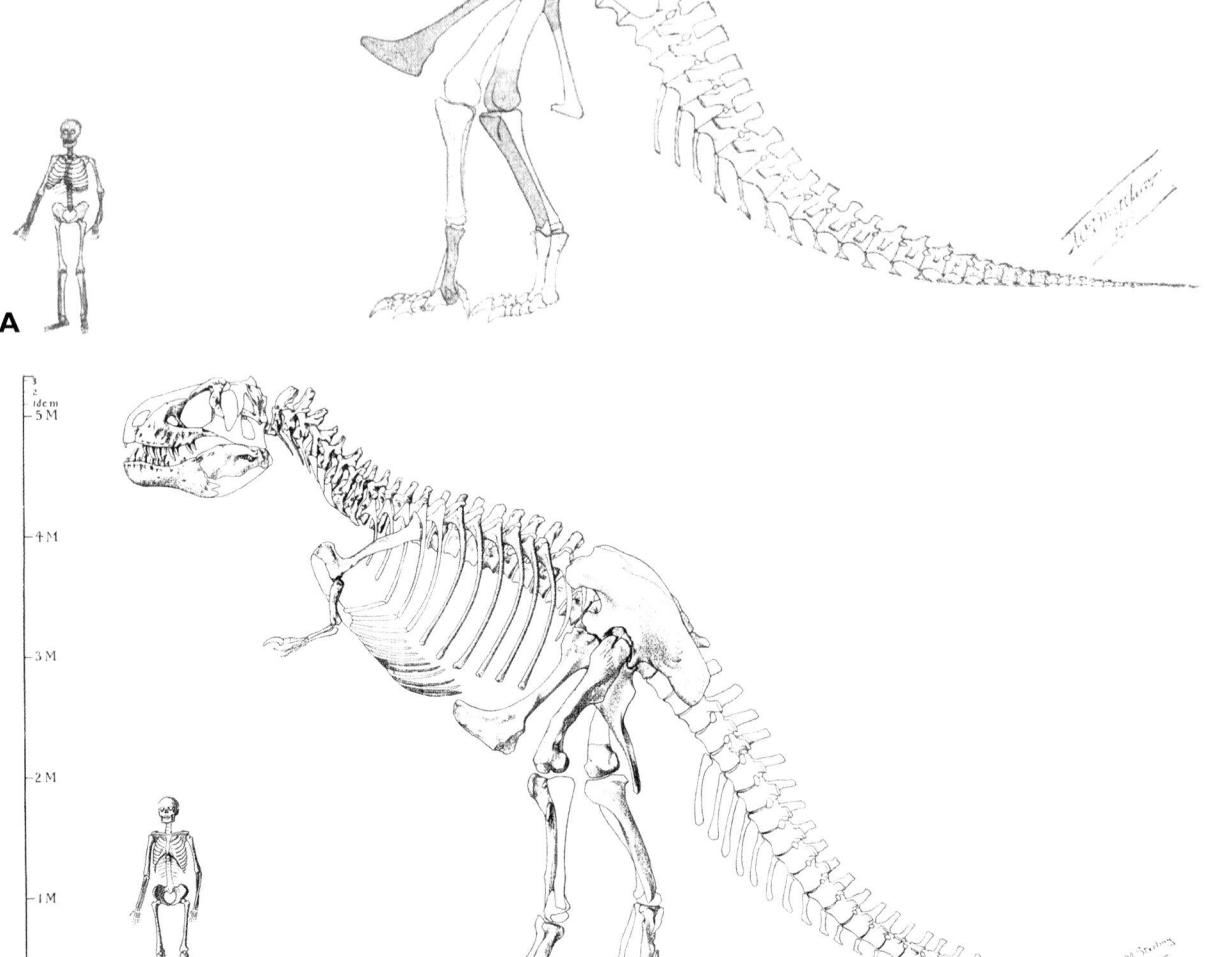

in AMNH publicity. Of course, *Dynamosaurus* was also not truly ready for description either; one has to wonder if Osborn was simply loading his bases to secure his department's legacy against research competition. Rivalry for dinosaur remains in Wyoming and Montana was already strong (Dingus and Norell 2010) and as museum departments competed for scientific and cultural kudos, it was better for Osborn's competitors to find further examples of *Tyrannosaurus* or *Dynamosaurus* than to establish their own genera.

To Osborn's credit, his lackluster first paper was followed in 1906 by a more in-depth effort that described the by then fully prepared *Tyrannosaurus* and *Dynamosaurus* specimens (fig. 1.7). Now able to see the specimens in full, and augmented with additional hindlimb material from a smaller third king tyrant specimen found in 1905, Osborn realized that his identification of two distinct species was in error. *Tyrannosaurus rex*, as the first-named and thus senior title, took priority. This 1906 paper was the first to accurately characterize *Tyrannosaurus* as a long-legged, large-headed carnivore with enormous teeth and a powerfully muscled pelvis, as well as a species underendowed in arm anatomy. The original *T. rex* specimen had included both a shoulder girdle and an upper arm bone, but the latter was so small compared to the rest of the animal that Osborn was uncertain it was from the same individual. The recovery of other dinosaurs, including other arm bones, with the original *T. rex* specimen had only confused matters further (Osborn 1905). By 1906, however, Osborn realized that the small *T. rex* arm was genuine, although the iconic two-fingered hand was not yet apparent. Based on other large dinosaur carnivores, Osborn assumed that they had three fingers. The armored scutes remained vexing, however, with Osborn noting that the presence of other dinosaurs in the quarry confused their association with *Tyrannosaurus*. Uncertain of their significance, he no longer considered them a diagnosing feature of the species.

With several prepared *Tyrannosaurus* specimens to work with, 1906 marked the first time *T. rex* anatomy could be visualized in relative fullness. Osborn had

a new skeletal illustration prepared for his paper by AMNH artist and sculptor Lindsey Morris Sterling, who produced a much more detailed and accurate piece than had been possible just a year before (fig. 1.6B). From this, a life reconstruction was worked up by the premiere paleoartist of this time, AMNH employee Charles Knight (fig. 1.8A). Knight's work was based on Sterling's osteological reconstruction, so they shared the same excusable errors: a face too short and tall, three-fingered hands, and overly raptorial feet. While both artworks would quickly be dated by new science, Knight's painting established the classic paleoart composition of *Tyrannosaurus* facing off against *Triceratops*, a group of these horned dinosaurs set somewhat back in the painting from the central king tyrant. This artwork was created for publicizing *Tyrannosaurus* and appeared in media coverage about the animal around the world. It was also present alongside *Tyrannosaurus* bones displayed in the Hall of Fossil Reptiles from December 30, 1906. Although this first *T. rex* mount featured just the legs, hips, and a reconstructed skull, it proved a major hit with the public and received generous coverage in newspapers (Dingus and Norell 2010). As impressive as this was, however, Osborn wanted a full king tyrant for his Hall of Reptiles. He devised a further expedition with Brown to return to Montana in the 1908 field season for more specimens (Dingus and Norell 2010).

Once again, Brown's uncanny ability to locate excellent fossils did not disappoint. After several weeks of limited success, he identified a promising series of tail vertebrae in Big Dry Creek, eastern Montana on July 1. Within two weeks, he had ascertained that these led to a near complete and excellently preserved series of vertebrae, the pelvic girdle, lower jaw, and skull, and that they unquestionably represented *Tyrannosaurus* (Dingus and Norell 2010). This was exactly what he and Osborn had hoped for, and cause for additional celebration was their easy procurement from the encasing rock. Unlike Brown's 1902 *T. rex*, this specimen was surrounded by relatively soft sandstone and easily excavated, although the sheer size of the fossil bones, still encased in matrix and wrapped in plaster for additional protection, complicated their delivery to New York. It was not until mid-October that this new *T. rex*, the most complete specimen yet, arrived at the AMNH. Although lacking limb bones, the new *Tyrannosaurus*, the famous American Museum of Natural History specimen no. 5027, was essentially identical in size to the 1902 find, allowing AMNH staff to combine the limb bones from the original example and visualize, for the first time, the full anatomy of a king tyrant (figs. 1.9–10). Only a minority of bones, including the hand, remained unknown.

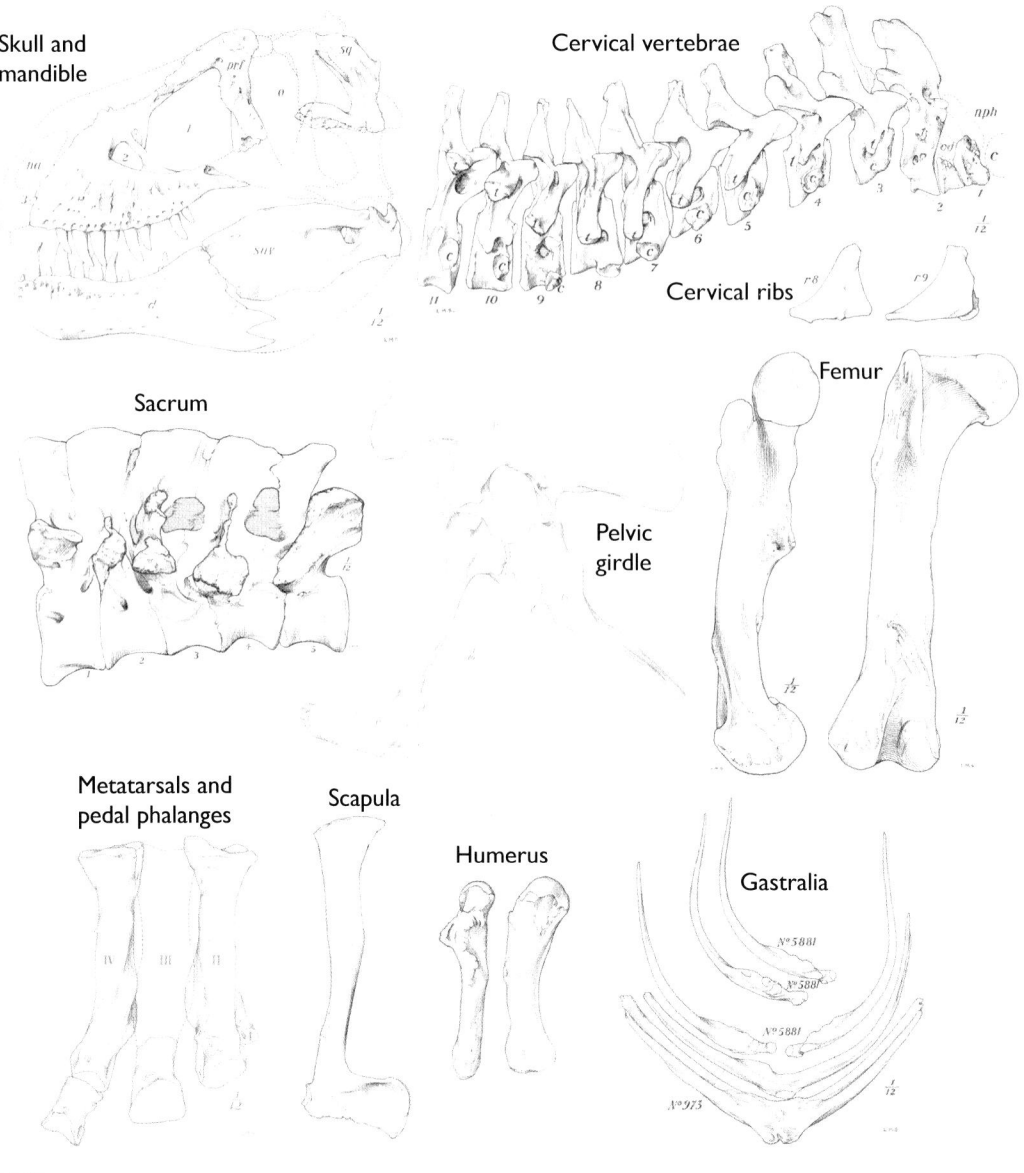

Skull and mandible

Cervical vertebrae

Cervical ribs

Sacrum

Pelvic girdle

Femur

Metatarsals and pedal phalanges

Scapula

Humerus

Gastralia

FIGURE 1.7

The original, holotype specimens of *Tyrannosaurus rex* as depicted in Osborn's 1906 description. Images not to scale.

FIGURE 1.8

Charles Knight's iconic and enormously influential paintings of *Tyrannosaurus*. (A) From 1906, the first published life restoration of a king tyrant. Based on an early, short-faced, and three-fingered reconstruction of *Tyrannosaurus* that was superseded by superior anatomical knowledge within a few years, Knight's *T. rex* is nevertheless an imposing presence. The juxtaposition of potential *Triceratops* prey established one of the most pervasive themes in paleontological restoration and featured more directly in (B), Knight's c. 1927 Field Museum mural depicting *Tyrannosaurus* confronting *Triceratops*. The low posture of the foremost *T. rex* looks "correct" to our modern eyes but reflects the forward advancement of the animal, not a depiction of a habitual pose. A kangaroo-like, upright king tyrant, more typical of the era, occurs in the background.

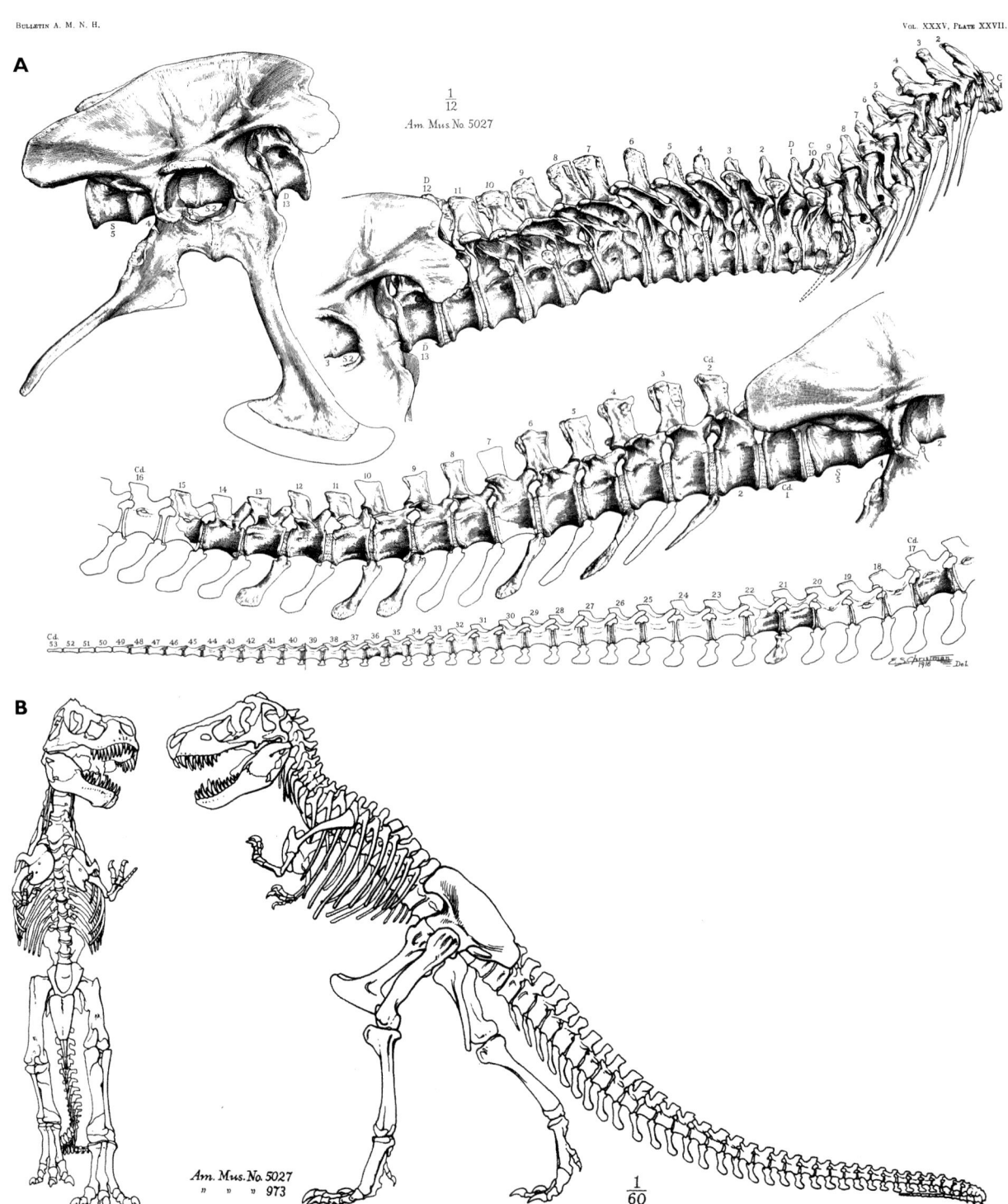

BULLETIN A. M. N. H.

VOL. XXXV, PLATE XXVII.

A

$\frac{1}{12}$

Am. Mus. No. 5027

B

Am. Mus. No. 5027

973

$\frac{1}{60}$

FIGURE 1.9

(A) The wonderfully rendered vertebrae and pelvis of American Museum of Natural History specimen 5027; (B) line art of the same specimen, combined with elements of the original *T. rex* material, reconstructed into the iconic skeletal mount that occupied the Dinosaur Hall in the AMNH for several decades. Illustrations by E. L. Christman, from Osborn (1917a).

Now equipped with knowledge of almost the entire *Tyrannosaurus* skeleton, Osborn wrote a further two papers documenting their osteology. These works completed the set of primary academic references for this species that would stand for almost a century (Osborn 1912, 1917a). Osborn's analysis of the vertebrae permitted greater insight into *T. rex* taxonomy and allowed him to come to terms with some historic details. It was now apparent that Brown's various discoveries—even his 1900 "*Dynamosaurus*"—were not the earliest find of *T. rex.* In fact, Osborn's late colleague Edward Drinker Cope had found two weathered king tyrant vertebrae in 1892 and named them *Manospondylus gigas* ("giant porous vertebra") (fig. 1.5B). Although this older name took priority over *Tyrannosaurus,* Osborn sensibly regarded Cope's specimen (which by then comprised only a single vertebra, the other having been lost) as too fragmentary to justify having a scientific name, and it thus could not supersede *Tyrannosaurus.*

Apparently unbeknownst to Osborn was that other *Tyrannosaurus* fossils had also been discovered during the nineteenth century. A *T. rex* tooth, found in 1874,

is probably the first documented king tyrant find (fig. 1.5A) but it stemmed from Colorado's Denver Formation rather than Montana. A historic irony saw the discoverer of the tooth, Arthur Lakes, send it to Yale for analysis by Cope and Osborn's rival, Othniel Marsh (Carpenter and Young 2002; Breithaupt et al. 2008). Marsh did not publish details of Lakes's tooth nor several other king tyrant bones he was presented with in the early 1890s (Dalman 2013), but he did describe and name *Tyrannosaurus* hindlimb material unearthed from the Lance Formation of Wyoming in the 1880s. Assuming that they represented a giant form of the ostrich dinosaur *Ornithomimus,* he named them *Ornithomimus grandis,* "giant bird mimic" (Marsh 1890). In doing so, he scooped Osborn's perception of *T. rex* by several years, writing that "this gigantic carnivore was one of the most destructive enemies of the herbivorous Ceratopsidae" (Marsh 1896, p. 206). Marsh's work on *T. rex* quickly fell into obscurity amid his countless other projects and the "*Ornithomimus grandis*" specimen went missing just decades after it was named (Osborn 1917a). Osborn was aware of "*O. grandis*" but dismissed it as a

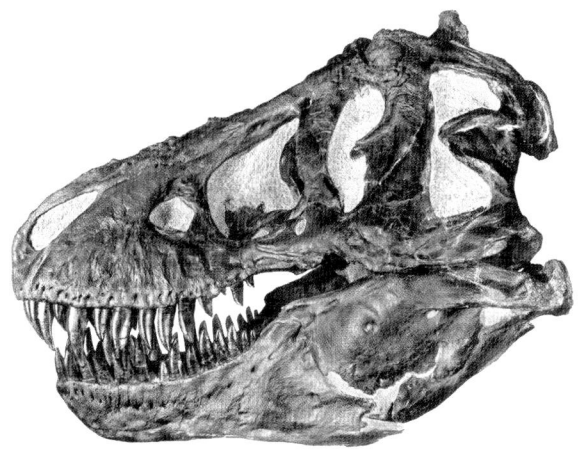

FIGURE 1.10

1912 photograph of the skull of American Museum of Natural History specimen 5027, the first completely known *Tyrannosaurus* skull and the fossil that informed depictions of *T. rex* for the better part of a century. After Osborn (1912).

representative of "*Deinodon*" (Osborn 1917a), a problematic dinosaur only known for certain from teeth. The irony here is inescapable: although history has largely forgotten the event, Osborn's professional nemesis had actually beaten him to naming and characterizing the world's most famous dinosaur. Even with *T. rex*, the ultimate Osborn dinosaur, he struggled to fully escape Marsh's shadow.

Matters of older names and misplaced fossils concerned the past, however, and AMNH 5027 represented the future of *T. rex* science. Its discovery made *T. rex* a potential museum centerpiece as it filled the anatomical gaps left by other AMNH *Tyrannosaurus* specimens, giving museum staff all they needed to replace the legs, pelvis, and skull then on display with a fully reconstructed skeleton. A new mount was being planned as early as 1913 when Osborn wrote about the museum's ambitious plans to display not one, but *two* tyrannosaurs arguing over the remains of another dinosaur. Replica bones would have provided a second skeleton as well as replaced some of the heavier individual elements, such as the skulls, which could not be mounted on account of their weight. A scale model was made of this dramatic display (fig.1.11), but it ultimately proved unfeasible to realize it at life size, there simply not being enough space in the Hall of Fossil Reptiles to accommodate two *T. rex* as well as the other exhibits. The mount of a single individual, reconstructed at over 14 m long, was no less impressive however, and the AMNH unveiled this to the public in December 1915 to much media interest and fanfare (fig. 1.9B).

Restored with the sloping back and dragging tail typical of dinosaur depictions in the early twentieth century, the AMNH *T. rex* skeleton would go on

to become one of the most iconic and recognizable mounted dinosaur fossils in history. Described by Brown (1915) as one of the finest mounted skeletons yet achieved, hindsight would show that only a few anatomical errors were made. Chiefly, the tail was overly long, the hands had one finger too many (Newman 1970), and the shafts of the feet were too broad. Few contemporary critics had issues with how Osborn and the AMNH visualized their king tyrant. Among the only doubters was Oliver P. Hay who, in 1910, argued that the sauropod dinosaur *Diplodocus* held its limbs in a lizard-like sprawl rather than, as scientific consensus held, beneath the body (Hay 1910). Hay's

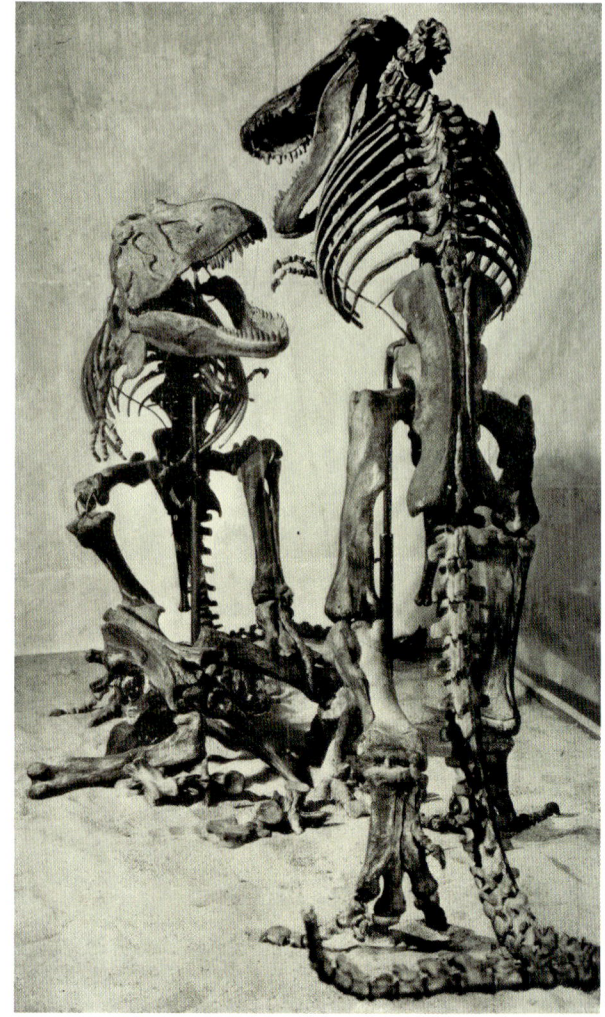

FIGURE 1.11

Scale model of dueling *Tyrannosaurus* by E. S. Christman. This model was commissioned by the AMNH as part of an aborted plan to erect two combative *T. rex* in their dinosaur hall. From Osborn (1913).

paper is often remembered for its inclusion of artwork by Miss Mary Mason Mitchell showing two sprawling, belly-dragging *Diplodocus* (fig. 1.12A), but he argued for similar conclusions about the large theropods *Allosaurus* and *Tyrannosaurus* in his text (fig. 1.12B). He suggested that the configuration of the *T. rex* hindlimb betrayed a "straddling" gait where the legs could not be put in front of one another without considerable effort, concluding that large theropods must have walked somewhat like penguins. All available data on dinosaur hindlimbs show that Hay was mistaken, and his ideas on dinosaur limb posture had little impact on depictions of *T. rex* or other species.

For the next fifty years, *T. rex* were imagined just as Osborn and the AMNH presented them, their display being filmed and photographed countless times and forming the basis for an avalanche of *Tyrannosaurus* artwork. The legacy of this reconstruction continues today in forming the basis for the *Jurassic Park* franchise logo, an accolade that might make the historic AMNH *T. rex* mount the most famous, or at least the most reproduced, dinosaur exhibit in history. *Tyrannosaurus*

can be seen as the ultimate success for Osborn and the AMNH's efforts to establish themselves as giants in vertebrate paleontology and public engagement. In less than two decades, they had not only taken full ownership of America's biggest predatory dinosaur, but turned it into an institutional flagship.

1920s–1940: MANUFACTURING A CELEBRITY

Academic activity around *Tyrannosaurus* slowed from the late 1910s as global events, including world wars and economic crashes, diverted attention and funds away from certain scientific pursuits (N. Larson 2008). But while scientists struggled to find time for *T. rex*, their cultural cachet only increased (fig. 1.13). Newspapers had been running photos of mounted specimens and Knight's artwork accompanied by hyperbolic quotes from Osborn even before the whole animal was on display. Entire pages of newspapers were devoted to AMNH *Tyrannosaurus* imagery, and their accompa-

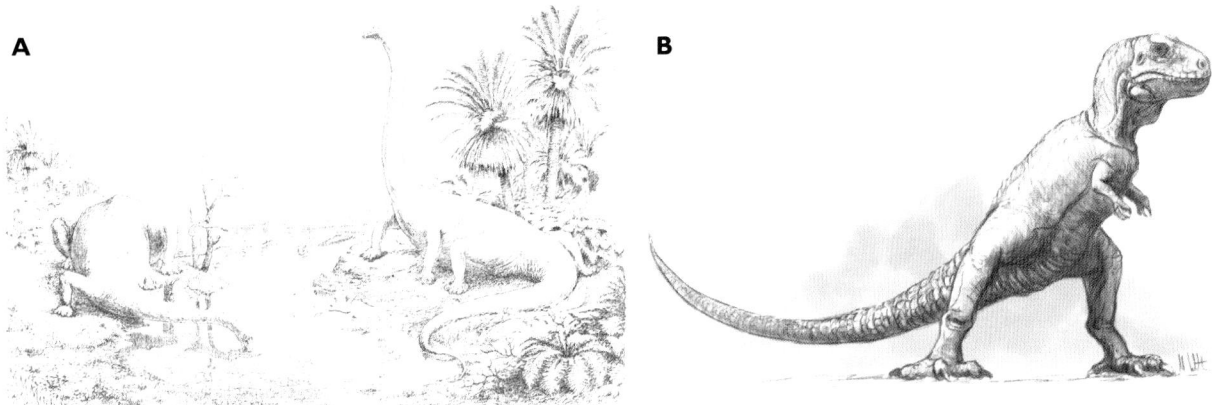

A **B**

FIGURE 1.12

(A) Mary Mason Mitchell's famous illustration of sprawling *Diplodocus*, reflecting Oliver P. Hay's hypothesis that dinosaurs sprawled like lizards (from Hay 1910). Best known for discussing the limb postures of the long-necked sauropods, Hay also considered, but never commissioned an illustration of, sprawling poses for *T. rex*. A modern interpretation of a Hay-esque sprawling *Tyrannosaurus* is shown in (B).

nying text demonstrates that the core components of *Tyrannosaurus* popularity had already been decided by the press. They were *the largest and most ferocious predatory dinosaurs*; the *culmination of predatory dinosaur evolution*; and among *the last of the dinosaurs*. *T. rex* was, in other words, not just another dinosaur species, but the pinnacle of dinosaur evolution.

The narrative established around *T. rex* was, in many respects, a dinosaurian reflection of Western values in the early twentieth century. Western nations widely viewed themselves as the end-product and pinnacle of mammalian evolution as well as the masters of the natural world and less-wealthy countries. In much the same way, *Tyrannosaurus* was the final form and master of dinosaurian development. But, in alignment with then-fashionable principles of orthogenic evolution (the idea that life becomes more sophisticated and complex over time), *Tyrannosaurus* was a dark, savage imitation of humanity. This early period of popularizing *Tyrannosaurus* assigned them the role of vicious monsters, "the last word in reptilian frightfulness" as described by H. G. Wells in his *A Short History of the*

World (Wells 1922). *T. rex* may have been the rulers of the Cretaceous, but they were never allowed the regalness of something like a lion or eagle. Rather, they were brutish, unthinking creatures that subdued prey by sheer physicality. Authors emphasized the size of their teeth and foot claws and focused much attention on their relatively small brains. Encapsulating such contemporary thoughts in one of the first popular books devoted to dinosaurs, W. E. Swinton (1934) wrote:

> *Tyrannosaurus* could not have been other than a clumsy and awkward giant battling against other equally cumbrous forms. Not a cunning or highly-brained creature, it would be guided largely by instinct ... the mechanical limits of the body were all against sudden leaping movements, swift pursuit, or the battle of wits that characterizes mammalian contests ... the contests of the Cretaceous world ... would seem to have the stiffness of amateur activity rather than the smoothness of the professional even though the feud was real and terrible. (Swinton 1934, p. 70)

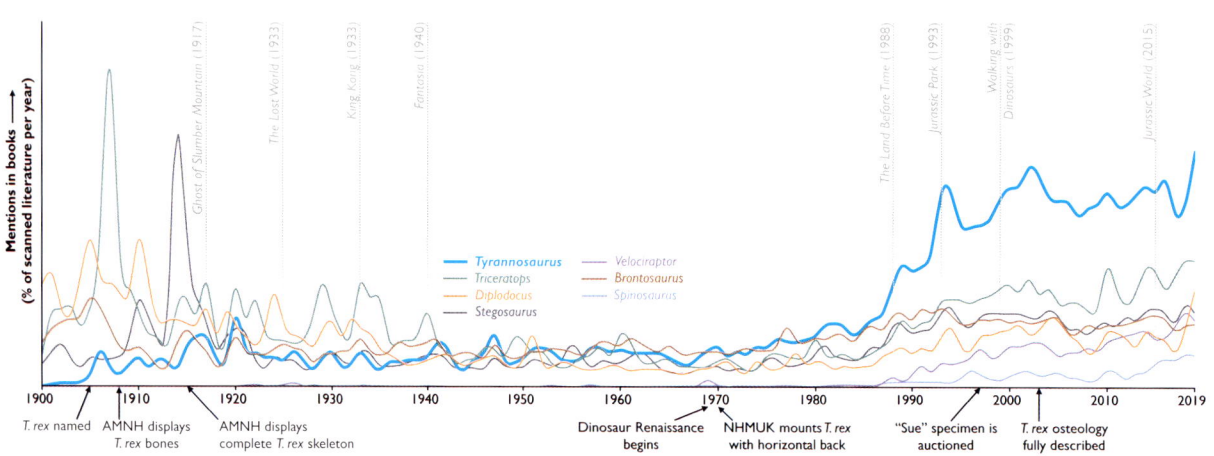

FIGURE 1.13

Mentions of *Tyrannosaurus* in books, newspapers, and other media compared to other famous dinosaurs (*Diplodocus*, *Brontosaurus*, *Stegosaurus*, *Triceratops*, *Velociraptor* and *Spinosaurus*), plotted against the release years of several major *Tyrannosaurs*-starring movies and documentaries. Data derived from Google Books Ngram Viewer (search conducted within "English 2019" corpus, smoothing set to 0).

This autonomous savagery was turned against human-ity—especially children—in dinosaur text and visuals, something the AMNH press department was not above: a particularly noteworthy 1919 AMNH advertisement portrayed *T. rex* as the once-living embodiment of fairy-tale dragons (fig. 1.14). Crowning *Tyrannosaurus* as the most developed and masterful of dinosaurs was thus not entirely complimentary: they were cruel, dim-wit-ted, and ultimately failed rulers of a clumsy reptilian world, and they only succeeded because they embod-ied the most terrible qualities of their age.

Unsurprisingly absent from any of this promotional material was the fact that, within a decade of discovery, *T. rex* was challenged for the title of biggest predatory dinosaur. New Cretaceous fossils unearthed in Egypt re-vealed two other gigantic and remarkable dinosaur car-nivores: *Spinosaurus aegyptiacus*, named in 1915, and *Carcharodontosaurus saharicus*, named in 1931 (Noth-durft and Smith 2002). *Carcharodontosaurus* was the most directly comparable to *Tyrannosaurus* in possess-ing a large and deep-snouted skull lined with formida-ble teeth, while *Spinosaurus* was more enigmatic and unusual, not the least because it bore an enormous bony torso sail. Both were described and named by the famed German paleontologist Ernst Stromer and, even though he was working from fragmentary remains, their enormous sizes were inescapable. Stromer reasoned skull lengths of 1.2 m or more for each and reconstruct-ed an upright, relatively short-tailed *Spinosaurus* as ex-ceeding 12 m in body length (Stomer 1936).

It was not only *Spinosaurus* and *Carcharodontosaurus* that remained absent from the popularized paleonto-logical canon of the early twentieth century. Following their discovery, the 12 m long *Allosaurus*-like predators "*Saurophagus*" (known as *Saurophaganax maximus* to-day: see Ray 1941; Chure 1995) and *Acrocanthosaurus atokensis* (Stovall and Langston 1950) also made little impression. Such acts expose the careful and deliber-ate publicity efforts driving the popularity of *T. rex*. It would be naive to ignore the multitude of factors that likely influenced the popularity of these new discover-ies, but it's also clear that Osborn and the AMNH were not interested in revising their promotional efforts to

FIGURE 1.14

The American Press could not get enough of *T. rex*, but when free promotion in the form of king tyrant news was in short supply, the AMNH ran advertisements. This example from the *New York Tribune*, May 11, 1919 (p. 4), is a particularly egregious early example of the "monsterization" that has plagued *Tyrannosaurus* since their discovery. The illustration at the top of the image, inferred to be *T. rex*, is actually based on a Charles Knight image of *Allosaurus*.

promote *Tyrannosaurus* as *one of several* giant dinosaur predators: it served their interests more if *T. rex* remained *the* giant dinosaur predator. With such an influential institution holding this steadfast attitude, it's not surprising that other giant carnivorous dinosaurs were unable to obtain a foothold in public consciousness. This was perhaps made easier by none of the fossils of the *T. rex* challengers being anywhere near as complete as those of *Tyrannosaurus*, leading to few attempts to restore them in artwork or museum displays. An infamous exception was *Spinosaurus*, the fossils of which were displayed in Munich until an Allied bomb destroyed them in 1944, along with the original bones of *Carcharodontosaurus* (Nothdurft and Smith 2002; J. B. Smith, et al. 2006).

In being portrayed as the biggest, baddest dinosaur by museums, books, and periodicals, it was only natural that *Tyrannosaurus* would play the same roles in the emerging industry of motion pictures. Dinosaur popularity and films are closely linked because cinematic portrayals spread interpretations of prehistoric life far wider than most books, articles, or television shows (e.g., Tattersdill and Witton 2025). The rapid adoption of *T. rex* into dinosaur cinema was undoubtedly a major boost to their popularity but, once again, this was no accident: institutional nepotism ensured that *Tyrannosaurus* would be stars. Hollywood producers turned to the AMNH and its paleontologists, including Barnum Brown, as consultants for the biggest dinosaur films of the early twentieth century (Goldner and Turner 1976; Culhane 1987). Thus, *Tyrannosaurus*, inevitably, received high billing. These early cinematic *T. rex* were invariably based on Charles Knight works, mostly referencing his original 1906 painting or his even more influential c. 1927 *Tyrannosaurus* and *Triceratops* mural for the Field Museum (fig. 1.8B; Milner 2012).

Among major film releases, *Tyrannosaurus* made their cinematic debut at the end of Willis O'Brien's 1917 *The Ghost of Slumber Mountain*, battling a *Triceratops* before briefly pursuing the film's hero. A pioneer of stop-motion animation, O'Brien would animate *Tyrannosaurus* again in Harry O. Hoyt's 1925 silent movie *The Lost World* (fig. 1.15). Although not appearing in Arthur Conan Doyle's source novel, *T. rex* featured heavily in this picture, rampaging across a South American plateau and dispatching a variety of dinosaurs and pterosaurs. Included among these victims of *Tyrannosaurus* was Marsh's best-known predatory dinosaur, *Allosaurus*, a (probably unintentional) symbolic moment for the Osborn-Marsh feud. *T. rex* becomes a recurrent threat throughout the film and a catalyst for several major plot events, principally with a fight against a *Brontosaurus* that allows for the sauropod's capture and transportation to London, where it runs amok. *The Lost World T. rex* are only unsettled by a huge volcanic eruption, throughout which they nevertheless continue to attack and devour other animals.

Shortly after *The Lost World*, *Tyrannosaurs* would appear again in Ernest B. Schoedsack and Merian C. Cooper's 1933 *King Kong* where, in a classic scene from the Golden Age of Hollywood, one wrestles the famous giant ape for a chance to eat Fay Wray's screaming Ann Darrow. It is a testament to O'Brien's animation and choreography that, in a film overflowing with iconic prehistoric animal imagery, this short sequence stands out as one of the most memorable. The *Tyrannosaurus* in this film is an especially obvious recreation of Charles Knight's 1906 painting, down to the large scales on its hide, crocodylian-like tail and overly deep, abbreviated skull. It is in this film that *Tyrannosaurus* vocalized for the first time in millions of years, although it snarls and screams rather than, as we've come to expect from subsequent depictions, roars and bellows. Perhaps unsettling, high-pitched squeals were thought better suited to a cold-blooded, unsympathetic reptilian antagonist, while the noble, lion-derived roars were reserved for the film's gorilla antihero? Whatever the reason, this choice at least allowed for clear vocal distinction between the *T. rex* and the baritone ape.

Tyrannosaurus terrorized their fellow animals again in Disney's 1940 *Fantasia*. An animated tour through the evolution of life set to Igor Stravinsky's *The Rite of Spring* proceeds from stellar dust and microbes to a zenith of a rich (if anachronistic) and mostly peaceful

FIGURE 1.15

T. rex go to the movies. Among the first major on-screen appearances for king tyrants was Harry O. Hoyt's 1925 *The Lost World*, in which *Tyrannosaurus* is shown rampaging across a South American plateau, dispatching a number of other dinosaurs. Despite only appearing in the second act, scenes of *T. rex* wrecking London dominated publicity for the film, eclipsing the *Brontosaurus* that actually devastates the city in the film's finale.

reptile-dominated ecosystem. This peace is shattered when, toward the end of the sequence, amid flashing lightning, driving rain, and a musical crescendo, *Tyrannosaurus* arrives. Consistently framed from a low angle so as to give a sense of scale and deliberately avoiding the anthropomorphism that characterized Disney animation (Culhane 1987), *Fantasia*'s cinematography encapsulates the ferocious persona built around *T. rex* more than any film before it. Animals of all kinds flee this king tyrant, and only the slow *Stegosaurus* cannot escape. Forced into a dramatic

showdown with the tyrannosaur, it quickly loses. Unlike its cousin in *King Kong*, the *Fantasia Tyrannosaurus* has no voice, the sequence being—as with all of *Fantasia*—a score-driven silent movie, but its roars are choreographed to the lowest, most threatening notes of Stravinsky's composition.

That *Tyrannosaurus* would appear in *Fantasia* was an idea woven into the very fabric of *The Rite of Spring*. In a 1938 development meeting, *Fantasia*'s conductor and major collaborator Leopold Stokowski mused, "If we could put ... the most terrific and terrifying of the animals fighting and eating each other, people would gasp," to which Walt Disney replied, "We could have a battle and build it to a grand climax. It is the fight for life" (Culhane 1987, p. 108). Some discussion took place over which species the *T. rex* would battle, with *Stegosaurus* eventually winning over *Triceratops* because of the animation potential of a spiked tail (Culhane 1987). The use of *T. rex* was never questioned, however: no other carnivorous dinosaur known to the public in the 1930s could have embodied the grim side of natural history at such scale. Like the 1925 *The Lost World*, *The Rite of Spring* concludes by demonstrating that the only threats to *Tyrannosaurus* were planetary. The sequence ends with a severe drought, geological upheaval and floods that weaken, starve, and specifically destroy the bony remains of *T. rex*. This placed *Tyrannosaurus* on such a lofty pedestal that *Fantasia* feels like an endpoint for this first phase of *T. rex* mania, before general and widespread disinterest in dinosaurs temporarily reduced their public impact for the following decades.

1940s–1960s: DINOBORES

The Second World War was very impactful on American paleontology (Wilson 1990). Resources necessary to get into the field, including tires and fuel, were reserved for the war effort, and many American paleontologists joined the armed forces. As the effects of this and other global events of the early twentieth century waned, vertebrate paleontology entered a new phase no longer dominated so forcibly by the United States. Instead, more cachet was found in exploring untapped parts of the planet for diverse fossil faunas, and researchers from other nations joined the conversation about life in Earth's past (Romer 1959; Buffetaut 1987; Wilson 1990; Colbert et al. 2012). Amid this, there was seemingly little drive to return to western North America for more *T. rex* bones. Few *Tyrannosaurus* discoveries of note occurred in the mid-twentieth century (N. Larson 2008), although some finds of eventual importance were made (e.g., Gilmore 1946a).

Dinosaur popularity, indeed, was lessening among both the public and scientists at this time. The excitement that had accompanied the unveiling of the giant sauropods, the great carnivores, and other charismatic species was now over, and dinosaurs were increasingly viewed as a spectacular but ultimately unimportant part of Earth's history. They were slow, unintelligent, and cold-blooded reptiles that distracted attention from more important topics, such as mammal and human evolution. One has to wonder if this declining interest in dinosaurs factored into the sale of the first, less-complete AMNH *Tyrannosaurus* skeleton to the Carnegie Museum in 1941 (fig. 1.16). Traded for other fossils and $7,000 (plus $108 for transportation), this sale was ostensibly instigated as a safeguarding measure against the threat of war spreading to US soil; Norell et al. (1995) report that this story is hard to authenticate from historic records. The Carnegie had their new *T. rex* skeleton on display by 1942 and set it against a dramatic, towering mural painted by Ottmar von Fuehrer in 1950.

Dinosaurs remained a museum draw among the public of this time, even as their novelty wore away (Gould 1995). But with little in the way of new science to respond to, their reconstructions remained static, changing little from their depictions in the early twentieth century. Among the noteworthy paleoartworks of *Tyrannosaurus* from this interval were those of Rudolph Zallinger (Volpe 2007), Zdeněk Burian (Müller et al. 2023) and his contemporaries Neave Parker and Maurice Wilson (Lescaze 2017). While iconic in their own right, their scientific advancements over those of Knight and earlier artists were incremental, where present at all.

That even *Tyrannosaurus* struggled to spark major interest in the mid-twentieth century is evidenced by their lessened role in feature films. The most famous dinosaur (or dinosaur-inspired) movies of this era focused on other predatory species. *One Million Years BC* (1966) featured *Allosaurus* and *Ceratosaurus*, while others created dinosaur chimeras to suit their narratives. *Valley of Gwangi* (1969) blended *Tyrannosaurus* with *Allosaurus* to create the titular Gwangi, a large but

FIGURE 1.16

Although the AMNH lacked the space to mount a *Tyrannosaurus* confrontation, the modern, excellent Carnegie Museum dinosaur hall has mounted a reconstruction of the original *T. rex* skeleton (left) with a replica of "Peck's rex" (right) in a similar composition to that outlined by Osborn (1913). A giant pterosaur skeleton, *Quetzalcoatlus northropi*, soars overhead.

generic carnivorous dinosaur (Harryhausen and Dalton 2010). Meanwhile, one of the most famous dinosaur-inspired creatures in film history resulted from crossing *T. rex* with *Iguanodon* and *Stegosaurus* to forge the 1954 *Godzilla* (Designing Godzilla 2006). A "real" *Tyrannosaurus* appeared in the now-obscure 1960 picture *Dinosaurus!* in the form of a stop-motion figurine and conventional puppet, and another ostensibly cameoed in a remake of *The Lost World* of the same year. Neither a high point for dinosaur cinema or animal welfare, the 1960 *Lost World* dressed living reptiles as "prehistoric" animals and filmed them fighting one another, visibly

drawing blood. At one point, these living animals were dropped from great heights to simulate falls from their lost plateau home. A tokay gecko, meant to be a baby *T. rex*, appears at the end of the film.

Not all reverence for *Tyrannosaurus* had been lost in the mid-twentieth century, but it was more evident away from the silver screen. The seeds of tyrant-mania sown by Osborn and Brown clearly influenced the famous science fiction author Ray Bradbury who, in 1952, centered the short time-travel story *A Sound of Thunder* around the idea of hunting and killing a living *Tyrannosaurus* (fig. 1.17). Bradbury leaned into the voracious

DINOSAUR SUPERSTAR

FIGURE 1.17

A scene from Ray Bradbury's 1952 *A Sound of Thunder*, where time-traveling big-game hunters encounter a *Tyrannosaurus*. Bradbury's story is a rare high point for *T. rex* in popular culture of the mid-twentieth century, a period in which king tyrants were often overlooked in favor of other dinosaurs, real and fictional.

ideal of *T. rex*, describing it as "the Tyrant Lizard, the most incredible monster in history" and establishes that humanity's need to dominate nature goes against all considerations of safety and even the very stability of Earth history. Primarily a tale of hubris, the *Sound of Thunder Tyrannosaurus* is an elaborate MacGuffin that could be replaced with any number of historic entities, but Bradbury's fondness for dinosaurs means his *T. rex* was given a grandeur and energy that was ahead of its time. He wrote of *Tyrannosaurus* as moving on "great oiled, resilient, striding legs" that were each "a piston, a thousand pounds of white bone, sunk in thick ropes of muscle, sheathed over in a gleam of pebbled skin like the mail of a terrible warrior." When moving, Bradbury's *T. rex* "ran with a gliding ballet step, far too poised and balanced for its ten tons." Even the famously small arms were given their due: "from the great breathing cage of the upper body those two delicate arms dangled out front, arms with hands which might pick up and examine men like toys." And, on the formidable head and teeth: "a ton of sculptured stone, lifted easily upon the sky. Its mouth gaped, exposing a fence of teeth like daggers. Its eyes rolled, ostrich eggs, empty of all expression save hunger." (All quotes from Bradbury 2003, pp. 68–69). Much contrast exists between this graceful, agile, and even regal tyrannosaur and the lumbering dinosaur carnivores of mid-20th-century popular culture.

Bradbury would play on this perception of *Tyrannosaurus* a few years later in a 1962 short story simply titled *Tyrannosaurus rex* (originally titled *The Prehistoric Producer*), a satirical tale of a Hollywood special effects artist who mockingly alters a *T. rex* stop-motion puppet to resemble a reviled movie producer. Here, the qualities that make *Tyrannosaurus* a monster were projected unfavorably onto a person but, in a twist, the mocked individual is of such low character that they are happy for any recognition that such comparisons would bring. Bradbury's contrasting views on *Tyrannosaurus* represent early insights into our own complex relationship with king tyrants; animals we have imbued with various and sometimes conflicting characteristics that we equally admire and dislike.

1960s–1980s: A WITNESS TO REVOLUTION

With midcentury popular and academic interest in dinosaurs in a slump, a spark was needed to reignite our passion for these great reptiles. This arrived in the late 1960s when researchers reappraised the concept of dinosaurs as a sorry group of slow, dim-witted, and cold-blooded evolutionary dead-ends, revolutionizing dinosaur paleobiology almost beyond recognition over the next few decades. Influential researchers such as Armand de Ricqlès, Robert Bakker, John Ostrom, Peter Galton, Jacques Gauthier, and Luis Alvarez demonstrated that dinosaur anatomy and physiology were more like those of birds and mammals than modern reptiles, that dinosaurs were powerful, terrestrially adept creatures that needn't be confined to swamps, that dinosaur extinction was likely cataclysmic and unexpected rather than a foregone conclusion of their biology, and that birds were incontrovertibly members of the predatory dinosaur lineage (e.g., Ostrom 1969, 1974; de Ricqles 1974; Bakker 1975; Desmond 1975; see Benton 2000 and Naish 2021 for retrospectives). This "Dinosaur Renaissance" laid the foundation of modern dinosaur science and established the view of dinosaurs that we largely follow today.

Although predatory dinosaurs were instrumental to Dinosaur Renaissance science, *Tyrannosaurus* was not crucial to this reinvention. Their birdier cousins, such as *Deinonychus* and *Archaeopteryx*, were far more informative and influential in this regard (e.g., Ostrom 1969, 1974), and, indeed, dedicated research on *Tyrannosaurus* remained rare at this time. But *T. rex* did contribute to one influential study of this era, demonstrating that large bipedal (two-legged) dinosaurs carried themselves with horizontal backbones rather than standing tall like kangaroos. Scholars and artists had tentatively experimented with this concept in dinosaur restorations for a number of decades (e.g., Heilmann 1927; Augusta and Burian 1958; Colbert 1965) but little dedicated investigation of dinosaur postures had been performed. Barney W. Newman's 1970 work on *Tyrannosaurus,* catalyzed by the efforts of the British Museum (Natural History) (now the Natural History Museum) to mount a composite *Tyrannosaurus* skeleton in their dinosaur gallery, was one of the first major contributions to this discussion. Basing his observations in part on the *"Dynamosaurus"* remains, purchased by the British Museum from the AMNH in 1960, he argued that a bird-like pose with a horizontal spine, "S"-shaped neck, and counterbalancing tail was a better interpretation of *T. rex* anatomy than the classic tail-dragging pose. Furthermore, Newman noted that traditional *T. rex* tail reconstructions had far more vertebrae than those of other tyrannosaurids and suggested that their

length had been overestimated by a substantial 4 m. A more cynical perspective of Newman's reposed *T. rex* was that the museum ceiling was too low to mount their king tyrant in a traditional pose, and that his technical rationalization was an effort to cover this fact (see Moody and Naish 2010 for discussion and references). Whatever the rationale, Newman's reposing and tail adjustments represent the first, and so far only, major revision to the appearance of *Tyrannosaurus.* His work, along with that of other researchers arguing for reposed bipedal dinosaurs, went on to be highly influential (e.g., Russell 1970; Galton 1970), even if his ideas of *T. rex* locomotion—a waddling, short-striding gait—left little impression (fig. 1.18).

FIGURE 1.18

The modern *Tyrannosaurus* mount of American Museum of Natural History specimen 5027, reposed from its iconic, sloping-back guise in the early 1990s. Although not complete, it remains one of the best preserved and most studied *T. rex* specimens.

Perhaps helped by coverage of the Dinosaur Renaissance, public interest in dinosaurs began to pick up in the 1970s (fig. 1.13). This was aided by a wave of revolutionary paleoart swapping out lumbering, dim-witted creatures for new sporty and muscular interpretations. Two particularly influential figures in this transformation were American dinosaur researchers and artists Robert Bakker and Gregory S. Paul, both of whom also published highly influential popular books on dinosaurs in the 1980s that continued Newman's transformation of *Tyrannosaurus* into their post-Renaissance guise. Bakker was one of the architects of the Dinosaur Renaissance and had argued for mammal- or bird-like energetics, postures, and ecologies in dinosaurs throughout the 1970s (e.g., Bakker 1972, 1975). His seminal 1986 book *The Dinosaur Heresies* gave space to apply these principles to *T. rex*, which Bakker envisaged as a strong-biting predator that could run at over 40 mph (64 kph).

Paul, who studied under Bakker, took this further. A dinosaur artist who also contributes to debates on dinosaur appearance, taxonomy, locomotion, and physiology, Paul has written extensively about *Tyrannosaurus*. His 1988 book *Predatory Dinosaurs of the World* summarized his early views on these animals, crafting a vision that, as expressed in his dynamic art, could not be more contrasting to that of earlier decades. His *T. rex* was lithe, muscular, bird-like and dangerous, as fast as a race horse and capable of delivering crippling, meter-long bites into anything it could catch. This was the *T. rex* on steroids promised by the Dinosaur Renaissance, but, despite being a faster, smarter, and potentially more formidable animal than ever imagined, this new *Tyrannosaurus* was not presented as a monster. Rather, it was viewed as awe-inspiring. As Paul (1988a) wrote, *Tyrannosaurus* "is *the* theropod. ... Its place as the greatest of known land predators is secure—no other giant consists of such complete skeletons, is bigger, or as powerful" (p. 344). This reverential tone set the stage for the next stage of *T. rex* history: unquestioned dominance of dinosaur popular culture and the subject of intense research and commercial interests.

1990s–PRESENT: KING OF THE DINOSAURS

The generation of dinosaur scientists that followed the original "Renaissance" period considered *Tyrannosaurus* a more remarkable and important animal than any previous one. The latest 1980s and 1990s saw a dramatic uptick in the discovery and scientific study of *Tyrannosaurus* fossils, marking the first period of significant academic interest in T. rex since the early twentieth century (N. Larson 2008). *Tyrannosaurus* moved from being a datapoint in wider dinosaur studies to a specific research topic of their own, often being the principle subjects of pioneering investigative methods. Analytical avenues included all aspects of their biology, from raw anatomy and the variation within their fossils (e.g., Carpenter 1990; Molnar 1991), to functional morphology and ecology (Farlow et al. 1995; Erickson et al. 1996; Erickson and Olsen 1996), as well as their relationships to other tyrant dinosaurs (e.g., Carpenter 1992; Holtz 1994; Carr 1999). Scientific understanding of king tyrants quickly grew to a point where the results could fill a whole book: this tome was Jack Horner and Dom Lessem's 1994 *The Complete T. rex*, the first major synthesis of *T. rex* research.

Academic curiosity in *T. rex* only grew as the 1990s became the twenty-first century. Here, *T. rex* studies become too numerous to succinctly summarize. They included the first thorough description of the entirety of the *Tyrannosaurus* skeleton by Christopher A. Brochu in 2003, along with numerous insights into *T. rex* biomechanics (e.g., Hutchinson and Garcia 2002; Meers 2002; Rayfield 2004; Snively et al. 2006; Snively and Russell 2007a, b), the remarkable, if controversial, recovery of still-spongy cell material from a *T. rex* limb bone (Schweitzer et al. 2005a) and detailed, seminal investigations of *Tyrannosaurus* growth (e.g., Erickson et al. 2004; Horner and Padian 2004; Carr 2020). So rapidly was research progressing that the discovery of *Tyrannosaurus* was celebrated on its centenary with a conference at the Black Hills Institute in South Dakota.

DINOSAUR SUPERSTAR

Two multiauthor collections of scientific papers resulted, filled with new data and insights on *Tyrannosaurus* and tyrannosaurids: *Tyrannosaurus rex, the tyrant king* (2008, edited by Peter Larson and Kenneth Carpenter) and *Tyrannosaurid Paleobiology* (2013, edited by J. Michael Parrish, Ralph E. Molnar, Phillip J. Currie, and Eva B. Koppelhus). Shortly after, another book, David W. E. Hone's *The Tyrannosaur Chronicles* (2016), brought the public up to date on the science of *T. rex* and kin. This unparalleled interest and willingness to interrogate *T. rex* specimens for data, and the number of specimens unearthed by tyrannophilic paleontologists, was transforming *Tyrannosaurus* into a "model organism": the dinosaur equivalent of the *Drosophila melanogaster* fruit fly or laboratory mouse. This approach means *Tyrannosaurus* science is now so mature that we can say something intelligent about just about every aspect of their biology, a statement that cannot be made for the overwhelming majority of fossil species—or a great number of living ones.

It was not only researchers that found *T. rex* alluring during the 1990s. Mentions of *Tyrannosaurus* in written media had been climbing since the Dinosaur Renaissance began, and, by the 1980s, *Tyrannosaurus* was on course to being the most mentioned dinosaur species in books, magazines, and other publications (fig. 1.13). A decade later, in the 1990s, *T. rex* was mentioned twice as often than other well-known dinosaur species, a status it retains today. Historians have yet to investigate the factors that saw *Tyrannosaurus* rise to the top of the dinosaur canon so many decades after their discovery, but one possibility is that, for all the reengagement with dinosaurs spurred on by Renaissance science, no new giant dinosaur predators had been found or heavily promoted in the preceding decades. Stars of the Dinosaur Renaissance, like the raptorial *Deinonychus* and the large-brained *Stenonychosaurus*, had brought human-sized danger and (perceived) greater intelligence to predatory dinosaurs, but they lacked the otherworldly size and commanding presence so synonymous with dinosaurian reptiles. *T. rex* still fully delivered these qualities, however, and a late twentieth-century surge of interest in science fiction and fantasy media gave them a wider stage to demonstrate their prowess. Blockbuster cinema and child-focused cartoons, often with tie-in merchandise, made wide use of *T. rex* as their dinosaurian villain and, in their new counterbalanced, sleek, and muscular form, king tyrants embodied dinosaurian terror more than ever. Indeed, they remained vicious savages even in universes populated by intelligent talking dinosaurs, such as 1988's animated Don Bluth film *The Land Before Time*.

The 1990s saw the blooming of another product of the 1980s dinosaur craze: the 1993 film adaptation of *Jurassic Park* (Shay and Duncan 1993). First conceived as a film script in the early 1980s by author Michael Crichton, he rewrote this screenplay as a novel that became an immediate bestseller upon its 1990 publication. Filmmakers were so desperate to create an adaptation that they pitched to obtain the movie rights before the book was finished (Shay and Duncan 1993). This led, of course, to the arrival of a movie *Tyrannosaurus* that, to this day, defines this animal in the public eye more than any other: the *Jurassic Park T. rex* (fig. 1.19).

The *Jurassic Park* movie, of course, needs little introduction to anyone reading a book about *T. rex*. Directed by Steven Spielberg and featuring revolutionary special effects, *Jurassic Park* and its sequels are probably the most influential dinosaur products of all time. Closely based on Crichton's novel (although notably lacking its second, juvenile *Tyrannosaurus* character), *Jurassic Park*'s story of resurrecting dinosaurs for a futuristic zoo recalibrated public perception of dinosaurs by showing them as active, intelligent and bird-like creatures that could be cute, awesome, and terrifying in equal measure. A deliberate effort by the filmmakers sought to "find the animal within the dinosaur as opposed to the monster in the dinosaur" (Shay and Duncan 1993, p. 14) and this, along with creating dinosaur effects largely in line with the science of the early 1990s, made the movie as much a hit with scientists as with critics and the public.

As one of the most important dinosaurs in the picture, engineering an impressive *Tyrannosaurus* effect

FIGURE 1.19

No depiction of *Tyrannosaurus* has been more iconic or influential, at least among the public, than the version created for *Jurassic Park*. This film-accurate recreation from Florida's Universal Studios Islands of Adventure theme park shows some of the liberties taken with king tyrant anatomy, such as the larger eyes and teeth, the peculiar jawline, and the hood-like structure over the eye.

was instrumental to the success of *Jurassic Park*. Pioneering mechanical and digital techniques would bring it to life, ultimately representing a landmark moment in the history of cinema (Shay and Duncan 1993). The *T. rex* was heavily teased to audiences in promotional material (not the least in skeletal form in the film's logo, which, as noted above, borrows the silhouette of the famous 1917 AMNH mount) but deliberately held back for the first half of the film. Its eventual nocturnal, rain-soaked arrival is a moment seared into the brains of many moviegoers: this was a deafeningly loud, enormous, and powerful animal that was equal parts breathtaking and terrifying. The sequence where it attacks a vehicle in an effort to reach two children is rightly regarded as an iconic piece of cinema: a mix of spectacular special effects, terrific performances, and edge-of-your-seat Spielbergian tension. Further scenes featured the incredible feats attributed to a post–Dinosaur Renaissance tyrannosaur: chasing a speeding jeep and ambushing other dinosaurs like a stealthy, fleet-footed bird. By the time the credits rolled, *Jurassic Park* had banished the lumbering, upright *Tyrannosaurus* of previous decades from popular consciousness. The film undeniably owes much of its effectiveness to *T. rex*: no other dinosaur could have stirred up so much pre-release hype and anticipation among moviegoers, and then so satisfyingly delivered on this promise.

Thirty years on, dinosaur pop culture is still in the Age of *Jurassic Park*. The film and its five (at the time of writing) sequels have inspired a generation of dinosaur fans and scientists, and they remain the entry point of public conversations about dinosaurs, but the long-term legacy of *Jurassic Park* is mixed. The success and influence of the series means that its fictional dinosaurs now eclipse many paleontological realities, especially so because the *Jurassic* franchise has proven reluctant to update its dinosaurs to reflect the post-1993 *Jurassic Park* generation of paleontological science. This has contributed to a stagnation of popular dinosaur portrayals and slowed mainstream acceptance of the reality of feathered dinosaurs, the *Jurassic* films now being infamous among scientists for denying *Velociraptor* and related species the

bird-like plumage they actually had (Turner et al. 2007).

The success of the series has also cemented *Jurassic Park* design idiosyncrasies into the public conscience as the actual life appearance of certain species. *Tyrannosaurus* has especially suffered from this. As impressive as the *Jurassic Park T. rex* is, a number of creative decisions distinguish it from true *Tyrannosaurus*, including its larger, more prominent teeth, the "hoods" above the eyes formed from sharp (rather than blunt) bony ridges, a strongly undulating jaw shape, generally slender build and relatively shallow pelvic region. It's certainly no crime that these creative liberties were made when crafting this film—*Jurassic Park* was, after all, not claiming to be a documentary—but most *T. rex* toys, films, videogames, and artwork now owe some or all of their design decisions to this movie concept, not to fossils. For scientists, outreach is frustrated when these artistic choices are taken as fact and well-established tenets of paleontology are treated skeptically. Some of the topics covered later in this book, such as the well-evidenced probabilities of *T. rex* having protofeathers and lips, or even the realities of their generous torso proportions and blunted facial ornaments, prove needlessly controversial when presented to the public because a three-decades-old movie design looms so large in popular perception.

Since *Jurassic Park*, dinosaur popularity has been consistently high, and *Tyrannosaurus* have attained a different level of pop-culture penetration to their most famous relatives. A steady trickle of dinosaur films, documentaries, and other media invariably feature *T. rex*, and, almost as often, they ensure that king tyrants are the most obviously promoted element of their products. Only risk taking, perhaps almost contrarian media ventures substitute *Tyrannosaurus* with another large predator as their headliner (e.g., 2000's *Jurassic Park 3* and 2011's BBC documentary *Planet Dinosaur*, both of which cast *Spinosaurus* as their headliner species). None of these efforts have had any long-term impact on the dominance of *T. rex*, nor do these substitute carnivores have anywhere near as much cultural impact (see *Spinosaurus* and *Velociraptor* curves in Figure

1.13). Indeed, *Tyrannosaurus* are so deeply embedded in popular culture that they have become the standard yardstick against which all dinosaur news is measured. Is a new discovery older or younger than *T. rex*? Bigger or smaller? Closely or distantly related? And most importantly, is it deadlier?

With *Tyrannosaurus* saturated into every pore of dinosaur media, a case can be made that the hype around *T. rex* has become more problematic than helpful, and not only because the interesting reality of the animals themselves has become, at times, of secondary importance. Nowhere is this more evident than what could be considered the latest significant development in the human history of *Tyrannosaurus*: their newfound status as trophy fossil specimens on the commercial fossil market. The sale and trade of fossils is, today, a controversial topic, but it was not always. Fossils have been collected and sold between individuals and institutions for hundreds of years, this being the mechanism by which many important specimens became accessible to science or traded around the world. Several famous figures in paleontological history, such as Mary Anning, Barnum Brown, and Charles Sternberg, were trusted freelance fossil collectors who found and excavated important fossil specimens for scientific research, sometimes in direct employment of museums.

Few researchers would argue, however, that recent developments surrounding *Tyrannosaurus* fossils are ideal. Since the 1990s, the value of tyrannosaur fossils has become so significant that the pursuit of specimens troubles scientific and even legal considerations, intensifying debates around the sale of important fossil specimens. The 1997 sale of the especially large and complete *T. rex* specimen known as "Sue" (fig. 1.20) was a watershed moment in this ongoing drama (P. Larson and Donnan 2002; Jones 2020). A dispute between "Sue's" discoverer, the company that excavated and prepared its bones, and the US government resulted in the specimen being seized by the FBI. Protracted court proceedings followed, which, eventually, saw the owner of the land that the specimen was found in, Maurice Williams, take ownership. "Sue" was put up for auction

in 1997, and rumors circulated that the specimen would go for a record price. The auction was over in less than ten minutes, and "Sue's" new owners—a mix of corporate and private individuals working with the Field Museum, Chicago—had won with a $7.6 million bid, the most ever paid for a fossil specimen at that time. The final cost, factoring auction fees, was over $8 million.

From that moment, anyone with a decent dinosaur skeleton, even one that had not been excavated yet, was in possession of a potential major financial asset. Since the late 1990s auction houses around the world have been selling dinosaur skeletons for hundreds of thousands or even millions of dollars, and financial magazines, including *Forbes*, have ranked fossils as sound investments (Rohleder 2001). A 2019 article in the *The Observer* newspaper reported that the going rate for a *Triceratops* skull was now $170,000 to $400,000, while *Diplodocus* skeletons could retail for $570,000 to $1.1 million (Helmore 2019). In 2024, the highest price ever paid for a fossil, $44.6 million, was paid for a *Stegosaurus* nicknamed "Apex" (Nicholls 2024). Such prices have spread to non-dinosaurs too, with remains of fossil primates now also commanding million-dollar valuations (Kjærgaard 2012).

Even as other species become commodified, frequent sales and constant high market values suggest that *T. rex* remains the fossil to own. Even partial skeletons without any scientific authentication will command high bids. So long as a specimen is from the right rocks and looks *T. rex*–like, it can sell for a seven-figure sum. In 2019, an alleged juvenile *T. rex* nicknamed "Baby Bob" was put up for sale on the auction website eBay for $2.95 million, following another probable juvenile *Tyrannosaurus* fossil (part of the so-called "dueling dinosaurs" specimen, a tyrannosaurid and horned dinosaur preserved alongside one another) put on sale for a reserve of $6 million in 2013, which it did not meet (Pantuso 2019). These values pale against the 2020, $31.8 million sale of the *Tyrannosaurus* known as "Stan," a particularly famous specimen that exists as casts in museums around the world and is widely commented on in tyrannosaur research. "Stan" broke the 1997 record

DINOSAUR SUPERSTAR

FIGURE 1.20

"Sue" on display at Chicago's Field Museum. Discovered in 1990, "Sue" remains one of the largest, and certainly the most completely known, *T. rex*. The purchase of this specimen by the Field Museum rocked the paleontological world by making dinosaur specimens, and especially *Tyrannosaurus*, big business. Sold for over $8 million dollars in 1997, the asking price for less complete examples is now tens of millions of dollars.

for the sale of "Sue" as the world's most expensive fossil several times over (Vogel 2020), but was quickly eclipsed by the sale of "Apex" the *Stegosaurus* in 2024. It's difficult not to wonder what, if stegosaurs are now worth over $40 million, the next near-complete king tyrant will sell for. With dinosaur fossil prices leaping up by tens of millions of dollars so rapidly, how long will it be before substantial *Tyrannosaurus* specimens command $100 million values?

The increased acquisition of *T. rex* specimens in the last thirty years has been driven, at least in part, by these staggering commercial interests, and their high price tags is forcing museums and researchers to either watch, powerlessly, as scientifically important specimens enter private ownership or else to rely on philanthropy to reach market values (Vogel 2020). The dollar value of *T. rex* has had a measurable impact on the availability of their fossils for research and public access. In 2021, tyrannosaur expert Thomas Carr predicted that 47 out of 105 notable *T. rex* specimens were privately owned, representing about half of our total global inventory (Carr 2021). The fate of these specimens is varied. Some will eventually be donated or sold to museums (as has happened with the "dueling dinosaurs," now at the North Carolina Museum of Natural Sciences), others will remain privately owned but publicly displayed, such as the skull of a complete *Tyrannosaurus* nicknamed "Samson" (bought in 2009 for c. $5 million, now located in the lobby of a Californian software company [Reynolds 2018]). Others simply disappear into private collections, a risk that hangs over any *T. rex* not accessioned into a museum. For scientists, such fossils are not worth researching, even if they are temporarily accessible. Science is dependent on reproducibility of observations and the private ownership of specimens cannot guarantee this: unlike a museum, a private owner can retire their dinosaurs from researcher access whenever they choose. This is no mere hypothetical scenario, either: some private *T. rex* specimens, including the aforementioned specimen known as "Stan," are well described and widely interpreted in scientific literature (e.g., P. Larson 2008a), but are inaccessible at the time of writing.

As *Tyrannosaurus* fossils have become multimillion-dollar business assets, the stakes for those involved in their collection and sale have risen higher and higher. Selling *Tyrannosaurus* bones is not illegal under United States law as they belong to whoever owns the land surface rights where they occur. Their sale is permitted, therefore, so long as appropriate permits and permissions have been obtained. Even so, battles over the ownership of *T. rex* fossils and even trademarked specimens have led to legal threats, lawsuits, government interventions and even prison terms. The legislative wrangling over "Sue" is the most famous example of tyrannosaur ownership court action (P. Larson and Donnon 2002), but legal intervention was also required to address ownership of the "dueling dinosaurs" (Cornwall 2020). The sale of "Stan" was forced by the need to liquidate assets within the company that owned it, following a legal dispute between corporate directors (Greshko 2020). Some companies have even started to obtain legal protection for *Tyrannosaurus* skeletons and their nicknames (perhaps better viewed, in this context, as brand names?), resulting in lawsuits filed for unauthorized copying of *Tyrannosaurus* bones when reconstructing less-complete specimens (Associated Press 2012; Santangelo Law Offices 2015; Begum 2022; Neate 2022). Concerns over such commoditization of fossils saw the Society of Vertebrate Paleontology, the world's leading organization dedicated to the study of vertebrate fossils, work with the US government to establish the Paleontological Resources Preservation Act during the 1990s and 2000s to protect fossils on federal land from unauthorized commercial collecting.

Lawsuits and government intervention have also taken place outside of the United States. The close evolutionary relationship between *Tyrannosaurus rex* and the Mongolian *Tarbosaurus bataar*—termed *Tyrannosaurus bataar* by a few researchers (see chapter 2)—has dragged *Tarbosaurus* specimens into demand as an alternative and typically cheaper tyrannical display piece for the super-rich. Efforts to subvert Mongolian collecting laws have resulted in some *Tarbosaurus* remains being looted from Mongolia to be sold in the West, resulting in dramatic interventions from governments, prison time for criminal fossil dealers, and repatriation of specimens to their home nation (Reuters 2015; Williams 2018).

Tyrannosaurus specimens are not unique for being viewed as commodities first, scientific specimens second, but they are exceptional for their value, legal dramas, and the personal risks undertaken to obtain them. The theatrics associated with finding, excavating, selling, and suing others over these remains have spawned their own subgenre of books (e.g., Larson and Donnan 2002; Williams 2018), TV shows (e.g., the Discovery Channel TV series *Dino Hunters*) and films (*Dinosaur 13*, 2014). The court case surrounding "Sue" has been used to teach law students the vagaries of contract defense (Cherry 2005) and is cited as precedent for the ownership of fossils in American law. It's difficult to believe that any other extinct species could spawn their own line of "reality-based" media, or become a legal teaching aid.

The desirability of *T. rex* fossils among society's wealthiest can be traced back to the century of promotion that they have received from academic, educational, and entertainment venues. If any fossils were going to join art, classic cars, and expensive houses as sought-after commodities for the super-rich, could it be anything other than *Tyrannosaurus*? This competition for ownership, both literal and symbolic, exemplifies the conflict between parties interested in king tyrants.

T. rex are a research topic, a museum attraction, a status symbol, a movie monster, and a real, if extinct, part of nature. Many of these statuses directly oppose one another and see individuals at odds about the cultural importance of *T. rex*. In the midst of this jostling throng of researchers, fans, media companies, and commercial fossil dealers, we might examine our current relationship, good and bad, with king tyrants.

THE HIGHS AND LOWS OF FAME

A CENTURY OF promotion and hyperbole has set *Tyrannosaurus* as a figurehead for Dinosauria; a species that encapsulates everything we enjoy about this diverse group and their hundreds of millions of years of evolutionary history. With decades of historic hindsight to consider, we might ask ourselves about the pros and cons of this relationship: Is *Tyrannosaurus* a valuable icon and ambassador for science in our increasingly busy, media-driven lives? Or is it a drag: an over-researched, over-exposed, and over-hyped bore that is long overdue retirement from the public eye? Any in-depth look at *Tyrannosaurus* subculture shows that the attention we pay to *Tyrannosaurus* has both positive and negative impacts, and also that what benefits some parties interested in *T. rex* rarely satisfies all others.

Tyrannosaurus are not merely the most discussed dinosaur on the planet (fig. 1.13): they are cultural icons with a level of recognition afforded to celebrities, clothing labels, and beloved media franchises. Wherever you are in the world, there's a good chance that people will know what a *Tyrannosaurus* is, and, despite their vastly different visual interpretations (below), their two-fingered hands, and large, toothy jaws make them instantly recognizable. This gives king tyrants a quality that can sell consumer products ranging from movies to Big Macs (Mitchell 1998), as well as a potent agency for communicating science. Whether broadcasting a specific message about dinosaur paleobiology or representing science more generally, a *Tyrannosaurus* connection

garners great interest from the general public, almost like a celebrity endorsement. This can have measurable benefits, such as bringing additional footfall and revenue to museums and educational events, as well as helping scientific studies and projects gain promotional traction. *Tyrannosaurus* act as ambassadors for science, encouraging immersion in paleontology, geology, and beyond, and are potent tools for enlightening people about the natural world.

For all this, however, mainstream interest in *Tyrannosaurus* is demonstrably superficial, a fact aptly demonstrated by our limited consideration of their appearance. At the core of the scientific concept that we call "*Tyrannosaurus rex*" is a series of fossil bones that, however else we want to imagine this animal, constrain many aspects of their body shape. We do not know exactly what *T. rex* looked like (chapter 3), but there are some fundamental "rights" and "wrongs," and also many "better" and "worse" ways to visualize them. Outside of specialist artworks however, our king tyrant depictions pay little attention to these data: any two-legged dinosaur can be labeled *T. rex* so long as it has two fingers, a large head, and sharp teeth. Some artistic modifications to *Tyrannosaurus* occur for understandable stylistic reasons, such as the softening and anthropomorphizing of their features to appeal to very young children (epitomized, of course, by TV shows like *Barney & Friends*), but we often present "monsterized" versions, with bigger, more conspicuous teeth, a shallower belly, and exaggerated cranial scales or horns, as realistic takes on their anatomy (fig. 1.21). As noted by Paul (1988a), the bulk of our *T. rex* art misses the real, distinctive shapes recorded in their fossils, often making them more generic and cartoonish.

This laissez-faire concern for *Tyrannosaurus* form is exposed further by studies finding that few people can visualize *Tyrannosaurus* even half accurately (Ross et al. 2013). Groups of pupils and students asked to draw *T. rex* show that our collective mental image of *Tyrannosaurus* owes more to their long-defunct, slope-backed guise than their "modern" (now fifty-year-old!) horizontal-backed form (Ross et al. 2013). This finding is alarming for those of us involved in communicating dinosaur science, suggesting that the general public have not advanced their understanding of dinosaurs in the last half century. The abundance of outdated depictions of king tyrants and other dinosaurs across popular culture, and especially in dinosaur media aimed at children, is thought to play some role in this outcome. From

FIGURE 1.21

Depictions of *Tyrannosaurus rex* in popular culture have rarely borne any resemblance to the real species since their discovery in the early 1900s. Depicted as a sluggish, formless beast for much of the twentieth century, this has been replaced in modern times by cartoonish, garish caricatures like that shown here; a version based on influential film depictions and stereotypes about dinosaur appearance and behavior rather than on fossil data.

DINOSAUR SUPERSTAR

plush toys to chicken nuggets, young minds first meet *T. rex* as they were imagined a century ago, and few of us maintain childhood interests in dinosaurs long enough to understand that these old interpretations have been replaced with newer ones (Ross et al. 2013). Perhaps this explains why *T. rex* occupies a permanent and yet malleable place in pop culture: they are mainstays of many formative childhood years, but few of us learn much about them beyond the most superficial information. Outside of groups with special interests in paleontology, *Tyrannosaurus* perhaps becomes a label applied to large,

predatory reptiles—an ill-defined "brand" of extinct animal—rather than a specific extinct organism (fig. 1.22).

It's not only visualizations of *T. rex* that fall into this schism between vague concept and reality. A century of *T. rex* popularization has whittled down their biology to a few core facts: they were the largest and most powerful predatory dinosaur, had bone-crushing bites, and were "the last of their kind." Combined with a surprisingly narrow range of depicted behaviors in film and television, *Tyrannosaurus* has become a major victim of stereotyping and cliché; a perpetually roaring,

FIGURE 1.22

This spectacular arrangement of *T. rex* and *Triceratops* at the Los Angeles Natural History Museum makes great impressions on the public, and yet the information taken from such exhibits, it seems, is limited: most people having very poor knowledge of even basic aspects of dinosaur appearance and biology, even for *Tyrannosaurus*. What does this imply about king tyrant celebrity and the status of dinosaurs as ambassadors of science?

violent animal always looking for the next meal. Like historical figures canonized by popular culture, this portrait—what we might call the popular "character" of *Tyrannosaurus*—blurs the line between reality (i.e., what we know from fossils) and fiction (a blend of their most popular portrayals). The result is that the complexities and nuances of their actual existences are frequently overlooked or dismissed.

Among the greatest character tropes now associated with *T. rex* stems from their monarchial name: the ubiquitous idea that they "ruled" or "dominated" North American dinosaurs. What we actually mean by prehistoric species "ruling" or "dominating" one another has been questioned by mammal palaeontologist Elsa Panciroli (2021), as has the related concept of "evolutionary successes." The animals described as "dominant" or "successful" in popular media are invariably exciting, large-bodied predatory species like *Tyrannosaurus*, a qualification that overshadows measures that might give us a more objective insight into the "success" of a lineage, such as geographic range or evolutionary duration. The latter can be just as great, if not longer, for non-predatory, non-giant animal species like the mammals and invertebrates that lived alongside dinosaurs, but these groups are rarely portrayed as "successes" by the popular press (Panciroli 2021). That measures of evolutionary "success" owe more to a hyperbolic, stereotyped telling of the history of life rather than objective measures of evolution or environmental impact is aptly demonstrated among non-tyrannosaurid dinosaur predators. By the usual measures of "success," the tyrannosaurids (the group to which *T. rex* belongs; see chapter 2) were less "successful" or "dominant" than another theropod group, the carcharodontosaurids, which were globally distributed, had a longer evolutionary range, and likely surpassed even *T. rex* in body sizes (fig. 1.23, also see chapter 2). And yet, carcharodontosaurids remain relatively unknown to the public because they are not part of the narrative that has, by rote, become established for dinosaurs and wider prehistory.

Despite this superficiality, the "character" of *Tyrannosaurus* seems important to our adoration of king tyrants,

and efforts to revise or counter stereotypes associated with *T. rex* can face resistance. In outreach, scientists frequently have to explain why interpretations based on actual fossils are so different from better-known sensationalized versions, often to be met with skepticism from people who cling to fantasy versions of king tyrants. This scenario is especially prevalent when scientific interpretations remove perceived "awesomeness" by suggesting, for instance, that adult *Tyrannosaurus* could not run fast, or that their famous teeth were hidden behind lips. Such reactions forget that our interpretations of *T. rex* are hypotheses—testable inferences deduced from experimentation and analysis of evidence—and thus prone to change with new data and ideas. Science demands a vision of *Tyrannosaurus* that grows and morphs to fit new data, while their popular counterparts are only sluggishly, seemingly begrudgingly modernized.

Acceptance of new *Tyrannosaurus* science is not assisted by bizarre assignments of political or social meaning to certain scientific developments. Depictions that deliberately depart from the familiar, monstrous view of *T. rex* have been labeled as pandering to political correctness by tabloid newspapers (e.g., McPhee 2022), or are branded disappointing or emasculating (e.g., the 2004 BBC documentary "T. rex: *Warrior or Wimp?*"; 2007 *Times* article: "A Slow, Clumsy Beast: How *T. rex* Lost His Crown"; 2009 *Gizmodo* piece "T Rex's [*sic*] Hunting Habits Disappoint Fans of Carnage"). Paleontological author Riley Black wrote at length on such matters in her 2022 *Slate* article "Give *T. rex* a Rest!," noting that *Tyrannosaurus* culture "inevitably bears a masculine gloss, thanks to so many decades of fearsome pop culture portrayals and the fact that it's almost always men who step forward as the prime interpreters of *T. rex* biology." We undeniably add to this macho portrayal, even in scientific publications, by referring to "the tyrant king" as a singular masculine individual rather than, as is more apt for an animal species, "king tyrants," equivalent to king cobras, king vultures, or king crabs. Online responses to Black's article from female paleontologists show the consequences of this masculine bias, noting that experienced, knowledge-able, and highly qualified female scientists avoid discussing *Tyrannosaurus* in public because certain men dislike being informed about their favorite dinosaur by non-male presenters. Outright dismissal or even abuse can follow. This politicization and toxification of science communication is surely the nadir of *Tyrannosaurus* hype, and a situation we can only push to change as soon as possible.

Academics are not immune to the hype of *Tyrannosaurus*, either. As noted above, *T. rex* is an intensely studied animal; so much so, in fact, that we might ask whether they are *over*-studied (Black 2022). This case has to be made against the counterargument that *T. rex* has become a "model organism" for dinosaurs, the likes of which are subjected to investigation precisely because we have so much data to contextualize new results. In the background of ongoing *T. rex* examination, however, are countless extinct organisms that have never been subjected to even elementary paleobiological investigations. Working on a popular fossil species like *Tyrannosaurus* can, of course, help to garner critical research funding and media attention, the likes of which can be essential to modern academic careers. Moreover, research involving king tyrants is often not exclusively focused on them: we learn much about other fossil organisms through studies that may have been funded because of a loose connection with *T. rex*. Such nuances are inevitably lost in news coverage of this science, however, which generally overemphasize any mention of *Tyrannosaurus* in dinosaur news stories or even crowbar king tyrants into news where they have no relevance. On those occasions when *T. rex* is the focus of a news story, the state of their science is often exaggerated and misleading. Contrary viewpoints are presented as sides in fierce arguments and debates, and unlikely "fringe" ideas are framed as overturning better-established hypotheses (e.g., the so-called "predator-scavenger debate" [chapter 6], or "the *Nanotyrannus* debate" [chapter 2]).

Whatever their rationale, our academic and media preoccupation with *Tyrannosaurus* creates an inflated sense of their importance to paleontological science and wider culture. In turn, this contributes to the private sales of *Tyrannosaurus* specimens and the debates about commercialization of fossils. This discussion has its own issues with hype as the strong feeling generated by high-profile fossil auctions, the likes of which sell *Tyrannosaurus* and similar-grade specimens, overshadows the nonproblematic, noncontroversial fossil sales that take place around the globe daily (Hippensteel and Condliffe 2013). Acknowledging this, however, does not exonerate fossil dealers from some of the tactics used to inflate *Tyrannosaurus* prices at auction, which include the application of dubious taxonomic labels to give specimens more prestige and value (e.g., listing specimens as the doubtful genus "*Nanotyrannus*"), or attaching paleobiological interpretations beyond those established by science (identifying specimens as male; see chapter 6). Another method, the sale of chimeric or very fragmentary skeletons with substantial amounts of replicated elements, is also now common, even as the transformation of *T. rex* bones into intellectual property makes selling replicated skeletons a risky business in itself: the auction house Christies rescinded a *Tyrannosaurus* specimen from sale in 2022 following unauthorized use of trademarked specimens in its reconstruction (Begum 2022; Neate 2022).

Hyping the sale of *T. rex* is not only unduly emptying the pockets of *T. rex* patrons; it is also having a measurable impact on the accessibility of *Tyrannosaurus* fossils to scientists and the public. As noted above, almost 50 substantial *T. rex* fossils, representing half our global inventory, have entered private ownership since the 1990s, setting a pace of acquisition that outstrips that of museums by roughly 17 percent (Carr 2021). That same figure, 17 percent, also accounts for the number of commercially-excavated *Tyrannosaurus* specimens that have found their way to museums, leaving most outside of the public trust (Carr 2021). If these rates remain steady, privately owned *Tyrannosaurus* will outnumber those accessioned to museums in the next few decades. Among them are rare and especially significant juvenile and subadult specimens that, as will be

FIGURE 1.23

The giant, North African theropod *Carcharodontosaurus saharicus* attacks a rebbachisaurid sauropod. Carcharodontosaurids were, by usual measures of evolutionary "success," a bigger deal than tyrannosaurids: this has not translated into greater cultural penetration for the group, however, perhaps owing to narrative stereotypes in our public discussion of dinosaur biology and evolution.

explored in later chapters, could tell us much about *T. rex* biology. As private *T. rex* ownership increases, concerns are growing not only about the immediate loss to science and our collective natural heritage, but also its longer-term implications. *Tyrannosaurus* is relatively abundant for a large dinosaur, and their fossils occur over a wide geographic area, but there are not infinite numbers of king tyrant fossils awaiting discovery. As a nonrenewable resource, continued collecting will, eventually, exhaust their supply. When that happens, will the bulk of *T. rex* specimens be in museums, or in private hands?

We could easily turn the rest of this book into a discourse purely on the curious culture that has developed around *T. rex*. This, however, is not our goal. Instead, our focus is on peeling back the sensationalism, politics, and controversy surrounding *Tyrannosaurus* to

understand them as real, if long extinct, animals. Echoing the process a paleontologist may take to studying a *T. rex* specimen, we will learn where king tyrants fit into reptile evolution, describe their anatomy, and then, with this data in hand, make interpretations about their lifestyles. We will deconstruct hype and controversy where we find it, and put common assumptions about *T. rex* biology under the microscope to determine what, if anything, current science says about those topics. This myth-busting approach will reveal not only the sometimes genuinely amazing biology of *T. rex*, but also the wealth of human effort that has unraveled the details of their existence. Whatever opinion we have on the human history of *Tyrannosaurus*, the fact we know so much about them, and their capacity to shine a light on the biology of other dinosaurs, is something to celebrate.

WHAT, IN ACTUALITY, IS A *T. REX?*

2

THE QUESTION ABOVE has a seemingly obvious answer: everyone knows that *Tyrannosaurus* were enormous reptilian carnivores that pursued other large dinosaurs with great jaws and tiny arms. But to scientists looking to understand *T. rex* evolution, this assessment is far from sufficient. To understand king tyrants in detail, we must first learn what sort of dinosaurs they represented, where they fit in reptile evolution, and what, exactly, defines them as a species.

THE LONG ROAD TO *TYRANNOSAURUS*

T. REX DID not spring, fully formed, into existence from some primordial ether. They were the outcome of a long, complex evolutionary process, and their famous body plan was constructed not from entirely novel features, but by modifying traits that developed deep within their reptilian past. Indeed, if we consider king tyrant anatomy in the most general sense—as gigantic, large-headed, robust-bodied terrestrial reptiles adapted for predating other large animals (fig. 2.1)—we can appreciate that they were just one of several reptile groups to attain this form during the Mesozoic Era. The repeated appearance of these similar ecomorphs (that is, the expression between anatomical form and ecological role) is a consequence of evolutionary convergence; incidences where distantly related species develop similar adaptations to "solve" similar evolutionary challenges. Large, big-headed carnivorous reptiles began developing in the Early Triassic Period, some 180 million years before *Tyrannosaurus* evolved (fig. 2.2). Introducing some of these convergent species allows us to explore the deeper evolutionary pathways that ultimately led to the development of *T. rex*.

Considerable scientific effort goes into classifying evolutionary relationships between organisms, living and extinct. Doing so reveals how their anatomy was shaped through the process of natural selection, what extinctions their lineage endured, and what opportunities allowed them to diversify. Our means to classify organisms have, in the last few decades, diverged from the traditional means of discussing evolution and biological classification still taught in most schools, so appreciating where *Tyrannosaurus* fit in the tree of life might be aided by a brief primer on these methods. Many of us learn that organisms are categorized using a system established by Carl Linnaeus in the eighteenth century where species are placed in ranks: kingdoms, orders, families, and genera. These ranks have been all but abandoned by contemporary scientists in favor of modeling evolutionary trees, known as phylogenies, where organisms are placed in clades: groups of

FIGURE 2.1

The reconstructed holotype of *Tyrannosaurus rex*, as displayed at the Carnegie Museum, Pittsburgh. Holotypes are pivotal specimens in organismal classification as they define the characteristics of a species and are the remains to which a scientific name will be forever attached. No matter what other decisions occur in *T. rex* classification, this specimen, Carnegie Museum no. 9380, will always define the king tyrant species.

species with shared ancestry united by common anatomies. Clades are named and nested within one another, but are not ranked. After all, if life on Earth has evolved through a continuous, unbroken chain of natural selection, hierarchies are arbitrary and scientifically meaningless. Instead, we simply refer to taxonomic units: clades and "taxa" (or, singular, "taxon"), the latter being a name given to any named evolutionary unit, be it a single species or a clade of any size. "*Tyrannosaurus rex*" and "*Homo sapiens*" are taxa, and so are groups like "Tyrannosauridae" and "Hominidae." Evolutionary relationships between taxa are frequently expressed via branching diagrams known as phylogenetic trees.

The only holdovers of older Linnaean schemes are genera and species, the scientific binomials applied to discrete populations of organisms. Species are, generally speaking, the most fundamental taxonomic units, a population of organisms that are distinguishable in some way from all others. A genus may contain one or more species, and they are knowingly artificial; a taxonomic practicality that ensures consistency with hundreds of years of biological classification, as well as a means to more easily produce new taxonomic names. Each organism requires a unique name and, as we know from the names we give ourselves, a binomial system such as first name and surname, or genus and species, offers far greater variety than a single word.

ARCHOSAURIFORMES AND THE FIRST LARGE, GIANT-HEADED TERRESTRIAL PREDATORS: ERYTHROSUCHIDS

We might start our march toward the evolution of *Tyrannosaurus* by meeting the earliest successes in achieving a *T. rex*-like ecomorph: the erythrosuchids (figs. 2.2, 2.3A). These peculiar-looking reptiles were a widely distributed group of Early-Middle Triassic carnivores that resembled large lizards with enormous, deep heads (see Ezcurra et al. 2013 for an overview of this group).

As members of the clade known as Archosauriformes, erythrosuchids were more closely related to archosaurs, the group that includes crocodylians and dinosaurs (the latter, of course, including birds), than they were to lizards and snakes. They were, nevertheless, a relatively archaic lineage only distantly related to any animals alive today (fig. 2.2). Archosauriformes were a major radiation of reptiles in the early Mesozoic Era. They are anatomically united by a unique hole in their snouts (the antorbital fenestra; see chapter 3), deeply rooted teeth set into sockets within their jaws, and appear to have been more energetic than other reptiles. They diversified across Triassic ecosystems emptied of their previous occupiers, our own distant mammalian relatives, by the mass extinction at the end of the Permian period 252 million years ago. This Triassic diversification of the archosaur line gave rise to many famous reptile types: the strange tanystropheids, the squat, beak-faced rhynchosaurs, the extremely diverse crocodylian lineage, the flying pterosaurs, and the dinosaurs. Following this rapid explosion in reptile form, the Triassic saw these lineages competing for the same niches within terrestrial ecosystems, from which dinosaurs emerged with the greatest stakes in herbivorous and carnivorous roles.

Erythrosuchids did not significantly overlap in geological time with dinosaurs as their 11-million-year evolutionary run, which spanned 251–242 million years ago, was largely concluded before dinosaurs appeared. Their Early and Middle Triassic habitats were still recovering from the Permian mass extinction and they lived among faunas and floras distinct from those that would develop later in the Mesozoic. Within this timeframe, erythrosuchids operated as predatory species that, in some cases, were arch-carnivores: giants reaching 5 m long. Scientific exploration of erythrosuchid ecology is still in its infancy, but they were evidently powerful animals equipped with strong bites. What they ate, and where they sourced their food is still uncertain, but most recent opinion has favored terrestrial carnivory over prey sourced in semi-aquatic roles (Ezcurra et al. 2013).

Our limited understanding of erythrosuchid paleoecology, and the fact that their bodies resembled

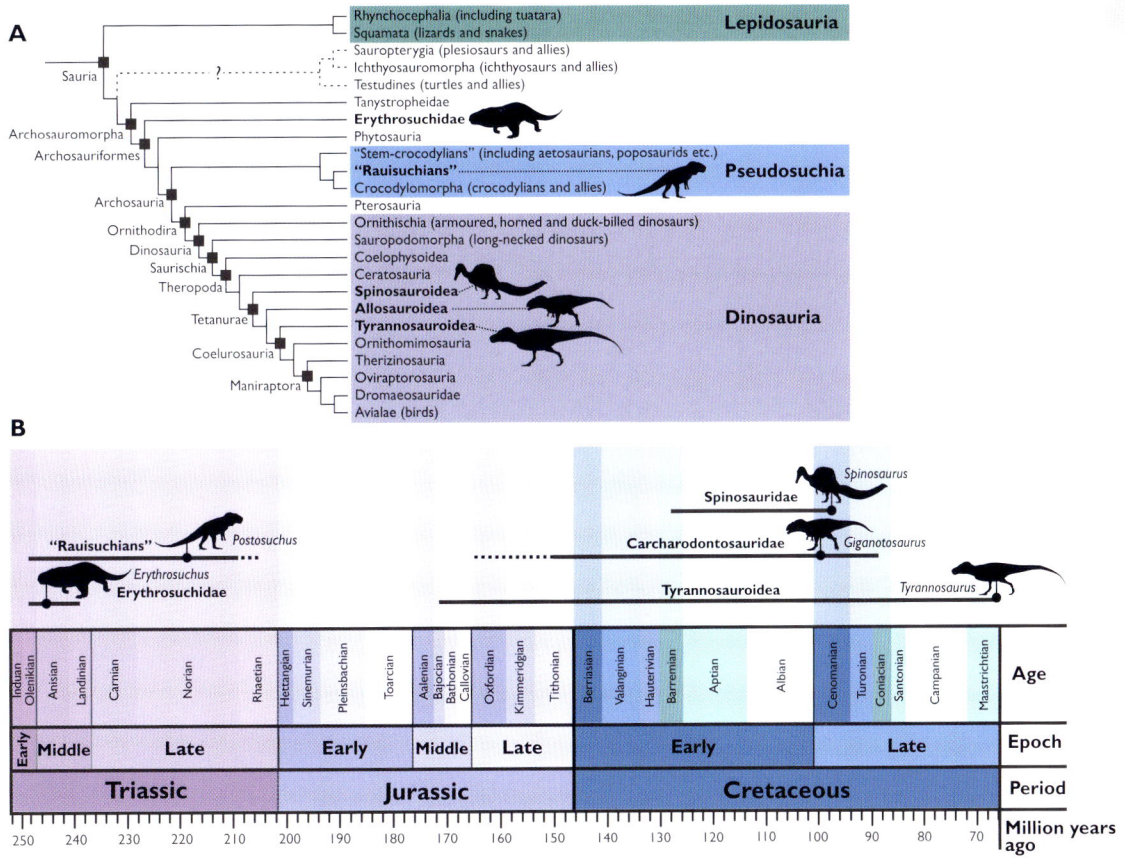

FIGURE 2.2

Tyrannosaurus in a broad evolutionary context. (A) Summary of reptile evolution; silhouettes and bold text denote clades discussed in this chapter; (B) temporal distribution of discussed groups and species across the Mesozoic Era.

those of lizards or crocodylians mean that it is their enormous, well-built heads that make them primarily tyrannosaur-like. Despite their evolutionary distance, the fundamental similarities between tyrannosaur and erythrosuchid skulls are impressive, with shared solutions found for strengthening the neck-skull joint, increasing bite forces, and reinforcing the skull against stresses incurred during predation. Like *T. rex*, the posterior faces of erythrosuchid skulls were expanded to accommodate larger neck musculature, and their neck vertebrae were proportionally large. Their skulls also bore large spaces for jaw muscles, facilitating a strong bite. Broad skull bones and an associated reduction of skull openings likely increased the strength of their crania, although some vacuities, likely related

to weight-saving, air-filled sinus tissues, remained. These same features underpin the morphology of the *Tyrannosaurus* head and neck (see chapters 3 and 4) and show that the genetic and functional potential for large-headed predators originated far outside of the tyrannosaur, or even dinosaur lines. Indeed, archosaurs appear to have been particularly well adapted to developing oversized heads, these appearing in several unrelated groups: erythrosuchids, predatory dinosaurs, horned dinosaurs, and the flying pterosaurs. In this sense, we can see that one of the most critical parts of *T. rex* anatomy, their enormous, powerful skulls, were not an evolutionary novelty, but pulled from evolutionary resources developed deep within their archosauriform ancestry.

FIGURE 2.3

Long before *Tyrannosaurus*, non-dinosaurian Triassic reptiles were experimenting with large-headed predatory body plans. These included (A) the 5–6 m long erythrosuchid *Erythrosuchus africanus* from South Africa; and (B) the superficially dinosaur-like *Postosuchus kirkpatricki*. More closely related to crocodylians than dinosaurs, *Postosuchus* is so convergent with tyrannosaurs that some initial interpretations posited that tyrant dinosaurs evolved from similar animals.

TWO-LEGGED, ROBUST-SKULLED, SHORT-ARMED CROCODILE RELATIVES: THE PREDATORY "RAUISUCHIANS"

Moving closer to *Tyrannosaurus* in reptile phylogeny introduces us to animals which, superficially speaking, looked a lot like predatory dinosaurs. These Triassic animals stood upright on two legs, had short arms and large skulls equipped with flesh-cutting teeth, but closer inspection reveals several obvious differences with dinosaurs. Their feet were short and compact rather than bird-like, and they were plantigrade: that is, they placed their ankle on the floor when walking and standing. Their legs were shorter and held in a more upright, columnar fashion, and some bore armor on their backs. Their necks were short and lacked the "S" shape so common to dinosaur predators, and their skulls were more massive. These were not dinosaurs at all, but certain members of "Rauisuchia," a grade of reptiles from the crocodylian evolutionary line (figs. 2.2, 2.3B).

Modern crocodylians are often regarded as dinosaur-like animals or even living dinosaurs but, despite sharing a common ancestor, dinosaurs and crocodylians are not closely related. They are both part of Archosauria, the archosauriform group that includes crocodylians and their ancestors in one major clade, and the flying pterosaurs and the dinosaurs (including birds) in another. As a group, archosaurs elaborated on features already common to archosauriforms by further lightening their skulls and developing powerful hindlimb muscles, the latter of which allowed for deviation from the quadrupedal gaits of other reptiles. The crocodylian lineage is known as Pseudosuchia, a once-diverse group that, during the Triassic, were major competitors for the niches dinosaurs would eventually occupy. They included species that were radically different to living crocodylians, including gracile taxa that could run fast, others that adapted to eat plants, and even types that walked on two legs. The "rauisuchians" were some of the most dinosaur-like of all pseudosuchians, such that their fossils are often confused for one another (Nesbitt et al. 2013a). It's widely appreciated that "rauisuchian" evolutionary relationships are unresolved and, presently, this group acts as a taxonomic "wastebasket" for Triassic pseudosuchians that do not fit into other, better-defined lineages (Gower 2000; Nesbitt et al. 2013a). "Rauisuchia" is thus not considered a natural group but a collection of species with disparate sizes, lifestyles, and methods of locomotion, some of which may not be closely related to one another. In time, further study will tease out their evolutionary affinities.

Most "rauisuchians" were carnivores and among their diverse forms were the first truly large terrestrial reptilian predators, animals that reached or exceeded 7 m long. These giants included species like *Postosuchus kirkpatricki* (fig. 2.3B), which were among the first powerful, two-legged reptilian predators to adopt a *Tyrannosaurus*-like ecomorphology. Equipped with a stout, reinforced skull and large teeth adapted for tearing flesh (Weinbaum 2011), *Postosuchus* were arch-predators of the southern United States in the Late Triassic. Comparisons with this species and *Tyrannosaurus* are not merely casual as the original describer of *Postosuchus*, Sankar Chatterjee, felt that this Triassic carnivore was so *T. rex*–like that the two surely had close evolutionary affinities (Chatterjee 1985). Citing shared features of their skulls, hips, and limbs, Chatterjee felt that tyrannosaurs were not dinosaurs at all, but actually late-surviving descendants of *Postosuchus*-grade pseudosuchians (this idea, it's perhaps needless to say, has not endured). Whether *Postosuchus* was fully bipedal or not has been debated, but most authors have concluded that their short, slender arms would be of little use in terrestrial progression, as well as noting features optimizing two-limbed locomotion in their hips, spines, and legs (e.g., Chatterjee 1985; Weinbaum 2011; Nesbitt et al. 2013a).

As convergent on *T. rex* as *Postosuchus* was, comparing their anatomy highlights some important features that distinguish the *Tyrannosaurus* body plan from that of even the most formidable bipedal "rauisuchians." Chiefly, although the *Postosuchus* skeleton was lightly built compared to their close relatives (Chatterjee 1985), terrestrial pseudosuchians lacked adaptations to develop extremely large body sizes and fast locomotion (Nesbitt et al. 2013a). Their upright limbs and bipedal poses may have conferred some speed advantages over equivalently sized pseudosuchians with four, relatively sprawled limbs, but "rauisuchians" had relatively short, plantigrade legs that were less optimized for running than those of predatory dinosaurs. Their long bodies likely negatively impacted their agility at speed, too. It is thought that these properties limited "rauisuchian" predatory habits to ambushing prey rather than pursuing it over long distances, which was probably an option for at least some carnivorous dinosaurs. Additionally, they may have been relatively heavy compared to theropods, as pseudosuchians lack unambiguous evidence of the neck and torso air sacs that lightened dinosaur bodies (Butler et al. 2012; Weinbaum 2011). As with erythrosuchids and all other archosaur-line reptiles, bony signatures of air sacs were present in "rauisuchian" skulls (e.g., Weinbaum 2011) and likely contributed to reducing the mass of the head, but the lack of weight-saving features elsewhere in their bodies may have precluded the attainment of *T. rex*–grade sizes and further impacted locomotion. We might view *Postosuchus*-like "rauisuchians" as demonstrating how close reptiles could get to a *Tyrannosaurus* ecomorph without developing those key anatomical and physiological properties that made dinosaurs well suited to developing gigantic, relatively fast predators.

DINOSAURS, THEROPODA, AND THE OTHER KING-SIZED DINOSAUR PREDATORS

One of the greatest archosaurian lineages were the dinosaurs, a major clade that is perhaps the most familiar group in the evolutionary address of *Tyrannosaurus rex*. Mesozoic dinosaurs were unique among terrestrial animals for their capacity to combine gigantic body sizes with particularly efficient terrestrial locomotion, with many adaptations to these traits distinguishing them from their pseudosuchian archosaur relatives. Historically, dinosaurs were characterized by features related to strengthening their skeletons and moving quickly, including reinforced pelves, hindlimbs adapted for upright postures with a specialized ball-and-socket joint between the femur and hip, and a hinge-like ankle. Learning more about the early origins of dinosaurs has smeared some of these features across other archosaurs, however (e.g., Brusatte et al. 2010a; Nesbitt 2011; Nesbitt et al. 2013b), and dinosaurs are today distinguished by minutiae of their skulls, vertebrae, and especially their limbs, the latter of which are also hallmarks of powerful locomotion (Coombs 1978; Nesbitt 2011; Cuff et al. 2022).

It is thought that this fast, low-cost movement was of particular utility to the first dinosaurs, which were predatory in habit (Benton 2004). In this respect, *Tyrannosaurus* and their carnivorous relatives represent

continuations and refinements of the original dinosaur ecology, eschewing the evolutionary opportunities taken by omnivorous and herbivorous species elsewhere on the dinosaur tree. Indeed, as relatively long-legged bipedal reptiles with grasping hands, the first dinosaurs already had the fundamental foundation of the tyrannosaur body plan in place, and abundant evidence suggests that they already had fast metabolic rates (i.e., that they were endothermic, or "warm-blooded" animals with fast metabolisms, like birds and mammals, not ectothermic, or "cold-blooded," like living reptiles; see chapter 4). Although there is no fossil data indicating that the first dinosaurs had protofeathers (the fluffy skin covering that would eventually give rise to true feathers), such features are predicted in evolutionary models thanks to the presence of skin fibers in pterosaurs and several dinosaur groups (e.g., Campione et al. 2020; Benton 2021). Fluffy tyrannosauroids (see below) may have also inherited these features from early dinosaur ancestors (Xu et al. 2004, 2012).

Within Dinosauria are three well-established lineages: the ornithischians, or beaked herbivores, the sauropodomorphs, the long-necked dinosaurs and kin, and the theropods, exclusively two-legged, mostly predatory species. How these groups are related to each other has been largely unquestioned among researchers for the last century, with the saurischian group ("lizard-hipped" dinosaurs, with a forward-projecting pubic bone making for a triradiate pelvis; see chapter 3 for more on dinosaur anatomy) containing the sauropodomorphs and theropods, while Ornithischia ("bird-hipped," with a backward-angled pubic bone) stand independently. This long-standing model has been challenged through new studies of early dinosaur anatomy, however, as some features uniquely shared between theropods and ornithischians may indicate a closer affinity than previously appreciated. In this model, the theropod and ornithischian group is known as Ornithoscelida (Baron et al. 2017). Although the conventional view of dinosaur evolution has generally been upheld in subsequent study, the Ornithoscelida hypothesis and other revisions to the base of the dinosaur tree have shown that traditional interpretations were not as concrete as previously thought. Discussions about the arrangement of the base of the dinosaur tree are ongoing (e.g., Langer et al. 2017; Müller and Garcia 2020; Norman et al. 2022).

Happily, this rocking of the dinosaur evolutionary boat does not affect our discussion as the composition of Theropoda, the predatory dinosaur line, is largely noncontroversial. Theropods first appeared in the Carnian stage of the Late Triassic Period (approximately 237–227 million years ago) and occupied most Mesozoic terrestrial predatory niches from the Early Jurassic onward (Brusatte et al. 2010a). A tremendously diverse group, they developed into gigantic and tiny forms during the Mesozoic, as well as fast runners and slow lumberers, carnivores, herbivores, and omnivores, and had skin covered in everything from scales to downy-fluff to fully developed, avian-grade feathers. In some other respects, theropods were evolutionarily conservative, never abandoning the two-limbed gaits of their ancestors, mostly retaining flesh-eating habits (at least in the Mesozoic), and never experimenting with extensive body armor. The origin of birds within Theropoda makes this the only dinosaur group to survive to modern times, as well as the most speciose of the three main dinosaur lineages. Living birds comprise something like ten thousand species, dwarfing the totality of Mesozoic dinosaurs known from fossils.

Amid this diversity arose several superficially tyrannosaur-like animals: species that combined large body size with big heads, short forelimbs, and rapid locomotion (fig. 2.2). Mesozoic theropods seem predisposed to developing large sizes because this was achieved by a variety of Cretaceous theropod types, including the noncarnivorous ornithomimosaurs (ostrich dinosaurs) and the bird–like oviraptorosaurs. Among the predatory lineages, however, the most tyrannosaur-convergent were members of the Allosauroidea and Spinosauridae: two diverse, geographically widespread Jurassic and Cretaceous lineages that evolved *T. rex*–grade stature and carnivorous capacity. These are the only non-tyrannosaurid species known that challenge *Tyrannosaurus*

for the title of biggest terrestrial predator among Dinosauria or, indeed, any animals.

The interrelationships of large carnivorous theropods are somewhat uncertain (see below), but we can confidently assume that these gigantic species evolved independently of one another. One branch of theropod evolution gave rise to Spinosauridae, a group of long-snouted, low-skulled Cretaceous theropods that may (or may not) be allied with the famous Jurassic predators *Megalosaurus* and *Torvosaurus*. They ranged widely through the Mesozoic, spreading across Europe, Asia, Africa, and South America during the Early and mid-Cretaceous. Among the last and most spectacular of their kind was *Spinosaurus aegyptiacus*, a North African, Late Cretaceous (probably Cenomanian, 100–94 million years ago; Figure 2.2B; Ibrahim et al. 2020a) species famous for their tall, bony sails. This species needs little introduction to dinosaur aficionados as it has enjoyed a recent surge in popularity thanks to high-profile research (Ibrahim et al. 2014, 2020b) and a starring role in 2001's *Jurassic Park III*. Famously, the original *Spinosaurus* fossils were destroyed by Allied bombing during the Second World War, leaving this species recorded only through photographs and illustrations until relatively recently when, from 1990s onward, new fossils were finally unearthed in various North African nations (e.g., Dal Sasso et al. 2005; Ibrahim et al. 2014, 2020b).

The most significant of these new finds is a partial skeleton which has recast *Spinosaurus* as a short-legged theropod with a heightened, laterally compressed tail (Ibrahim et al. 2014, 2020b). They seem to have stood on four toes thanks to an enlarged first toe, or "hallux," that, unusually for a theropod, could reach the ground. How anatomically unusual *Spinosaurus* were remains controversial, and these matters may not be resolved until a more complete skeleton is found. Partial remains of key skeletal elements suggest that they bore the large, three-fingered hands and long, low snouts typical of all spinosaurids, but attempts to create whole-body reconstructions differ in subtle but important aspects of proportion and size, especially leg length (Evers et al. 2015; Paul 2016; Henderson 2018; Hartman 2020;

Sereno et al. 2022). Considerations of body proportions play into debates about virtually every interpretation of this animal, from their basic taxonomy, gait, and locomotive prowess to diet and lifestyle (e.g., Ibrahim et al. 2014, 2020b; Evers et al. 2015; Hone and Holtz 2017, 2021; Sereno et al. 2022). Taxonomic issues are perhaps the most crucial debates facing *Spinosaurus* at present because, currently, no consensus exists on how many spinosaurid species lived across North Africa and existing *Spinosaurus* reconstructions lean heavily on compiling specimens from across African countries. Are we are reconstructing the anatomy and lifestyle of one species, or creating a chimeric "Frankenstein" spinosaurid by combining several?

Uncertainty aside, *Spinosaurus* was clearly an aberrant species unlike any other giant theropod currently known. A wealth of evidence shows that spinosaurids obtained much of their food from aquatic settings (e.g., Ibrahim et al. 2014, 2020b; Hone and Holtz 2017, 2021), and *Spinosaurus* represents the furthest specialization of this evolutionary pathway. Some researchers have proposed that they adopted a pursuit-swimming predatory lifestyle (Ibrahim et al. 2014, 2020b; Fabbri et al. 2022), while others argue that they lived, as is generally assumed for other spinosaurids, somewhat like a mix of grizzly bear and heron: generalist foragers that waded into lakes and rivers to apprehend swimming prey (Hone and Holtz 2017, 2021; Henderson 2018; Sereno et al. 2022). Currently, the "wader" hypothesis seems more likely because several studies have shown that the *Spinosaurus* body plan was ill-suited to pursuing prey underwater (Henderson 2018; Hone and Holtz 2021; Sereno et al. 2022, see chapter 4 for more on theropod swimming) and, in this model, we can imagine *Spinosaurus* grabbing aquatic animals with their low, slender skulls and strongly clawed arms. *Spinosaurus* was further distinguished among predatory giants by their investment in enormous sails that ran along their backs and tails, structures that were almost certainly sociosexual display organs (Hone and Holtz 2021). These not only differentiate *Spinosaurus* from *T. rex*, which has reduced ornamentation (chapter 3) but

other theropods in general. *Spinosaurus* was the peacock of the giant predatory dinosaurs (fig. 2.4).

In being such an unusual theropod, *Spinosaurus* has little else in common with *Tyrannosaurus* other than raw size. Since their discovery in the early twentieth century, *Spinosaurus* has been regarded as a giant that rivaled or exceeded *T. rex* in body length. Like all dinosaurs, *Spinosaurus* has evidence of mass-reducing air sacs within their bodies and necks that, along with other features, allowed them to grow very large. Our most substantial, but still far from entire, *Spinosaurus* skeleton perhaps measured nearly 11 m long when complete, not accounting for the natural curvature of the animal when standing. This individual may have massed around 4 tonnes (Ibrahim et al. 2020b), and scaling it to the largest *Spinosaurus* fossils implies a body length greater than that of the biggest *Tyrannosaurus*: 14–15 m. Estimated body masses were modest for such a long creature, however; just 7.4–9.5 tonnes (Ibrahim et al. 2020b; Sereno et al. 2022). This indicates that *Spinosaurus* may have been longer than the c. 12 m–long *T. rex*, but large king tyrants likely exceeded 8 or 9 tonnes, making them proportionally heavier (see chapter 4 for discussion of *T. rex* body mass). This assessment comes with the caveat that, without a complete, or even near complete *Spinosaurus* specimen, errors in our scaling assumptions, length estimates and mass predictions are near certain (Therrien and Henderson 2007; Persons et al. 2020). We need to learn a lot more about *Spinosaurus* before determining how it truly ranked in the Battle of the Biggest Theropods.

With *Spinosaurus* differing from *Tyrannosaurus* in virtually all attributes other than size, the group that converged most with giant tyrannosaurids are the carcharodontosaurids: a grand Cretaceous lineage of predatory dinosaurs that includes famous species like *Acrocanthosaurus atokensis*, *Giganotosaurus carolinii* (fig. 2.5) and *Carcharodontosaurus saharicus* (fig. 1.23). These theropods were probably the closest any group came to developing tyrannosaurine characteristics outside of the tyrannosaur line itself, and they are superficially similar in many attributes: large heads, small arms, and strongly

FIGURE 2.4

Longer than *Tyrannosaurus* but perhaps not as heavy, *Spinosaurus aegyptiacus* were among the most unusual of all theropods. A predator of aquatic prey, their anatomy contrasts markedly with that of king tyrants, not the least for the development of ostentatious sails on their torsos and tails. *T. rex*, in contrast, reduced their bony ornaments. In terms of fashion at least, there's no question who wins the battle between *Spinosaurus* and *Tyrannosaurus*.

FIGURE 2.5

The giant carcharodontosaurid *Giganotosaurus carolinii*. Large Cretaceous carcharodontosaurids such as this and *Carcharodontosaurus* are perhaps the only theropods that rivaled *Tyrannosaurus* in size and predatory scale.

built, massive bodies. In detail, however, they are quite different. Carcharodontosaurids lacked the elongate, narrow feet of tyrannosaurs, as well as their barrel-shaped chests and two-fingered hands. Their bite strengths were weaker, and their teeth better suited to tearing flesh than puncturing bone. In many respects, they resemble especially large versions of their Jurassic allosauroid ancestors, epitomized by the famous *Allosaurus*, and we might consider the giant Cretaceous carcharodontosaurids an archaic grade of predatory dinosaurs inflated to a new size class rather than, as with tyrannosaurs, a new kind of large-bodied predatory form (see below).

This is not, however, to imply that carcharodontosaurids should be regarded as somehow inferior to the tyrannosaur line. As noted in chapter 1, carcharodontosaurids were among the most widespread and longest-surviving groups of large theropods with a distribution across the Americas, Africa, Europe, and Asia, as well as an evolutionary history spanning much of the Cretaceous (fig. 2.2; Candeiro et al. 2018). They disappeared from the fossil record approximately 83 million years ago, and it was only after this that the tyrannosaur lineage, apparently in quick succession, developed large, robust forms (Brusatte et al. 2010b; Brusatte and Carr 2016). A patchy mid–Cretaceous dinosaur record obscures the specifics of this transition, but current data might indicate that the evolution of large tyrannosaurs was suppressed by the existence of giant carcharodontosaurids. Furthermore, the likes of the North African *Carcharodontosaurus* and South American *Giganotosaurus* were among the largest terrestrial predatory animals to ever live, equaling or surpassing *T. rex* in length and mass. Carcharodontosaurids were undoubtedly important theropods that shaped the development of Mesozoic terrestrial ecosystems.

As with *Spinosaurus*, working out exactly how big carcharodontosaurids could get remains challenging because their largest known skeletal remains are too incomplete for precise size estimates. The first-found specimen of *Giganotosaurus* has a thigh bone length surpassing that of the largest *T. rex* by 4 cm (Coria and Salgado 1995) and an estimated skull length of over 1.6 m (Canale et al.

2022), a value similar to that predicted for *Carcharodontosaurus* skulls (Sereno et al. 1996). These estimates are about 20 cm longer than the biggest complete *Tyrannosaurus* skull on record. Another *Giganotosaurus* specimen, an isolated fragment of lower jaw, hints at a 6.5–8 percent larger animal or, alternatively, a considerably more robust individual than the original specimen (Calvo 1998; Hartman 2013). If representing the former, this would equate to a skull surpassing 1.7 m long, almost 25 percent bigger than that of *T. rex*. How this translates to body length can only be coarsely estimated, but the 12 m attained by the biggest *T. rex* seems like a minimum value for such giants: 13 m or more seems plausible. Based on thigh bone proportions, they may have been heavier than *T. rex* as well (Persons et al. 2020). Attempts to compare the masses of big carcharodontosaurids and *Tyrannosaurus* have found that the robust chests of *T. rex* (chapter 3) make them relatively heavy for their size, but the greater stature of giant carcharodontosaurids may have out-massed *T. rex*, despite their more gracile construction (Hartman 2013; Paul 2016).

The above discussion should not be considered definitive, however. Our specimens of large carcharodontosaurids are even less complete than those of *Spinosaurus*, so any comparison between them and *Tyrannosaurus* is correspondingly imprecise. If we must force a winner in the contest for largest theropod, however, probability does not favor king tyrants. *T. rex* is a relatively well-sampled species and it has, thus far, presented a fairly consistent adult body length of 11–12.5 m. Large carcharodontosaurids, in contrast, are known by just a handful of bones that already hint at animals equaling or surpassing *T. rex* in magnitude. We may have lucked out by finding unusually big, exceptional carcharodontosaurids right away, but it's more statistically likely that these first discoveries represent averagely sized carcharodontosaurid adults, and that larger ones await excavation. It remains true that larger *Tyrannosaurus* may be discovered as well, but probability favors carcharodontosaurids having more surprises in store for us. Time, and further fieldwork, will reveal how accurate these predictions are.

THE TYRANNOSAUR FAMILY TREE

APPRECIATING HOW *TYRANNOSAURUS* arose from other dinosaurs, and distinguishing their true ancestry over mere convergent development of body plans, requires that we begin a more forensic examination of tyrannosaur evolution. In our discussion above we have already pinned down a broad phylogenetic address for *T. rex*: they belong to the reptile clades of Archosauriformes, Archosauria, Dinosauria, and Theropoda (fig. 2.2A). We have also mentioned that Mesozoic theropods were a particularly diverse and speciose bunch, and it is within this group that we should begin to narrow our focus. Theropod evolution has drawn particular interest among dinosaur researchers, and the broad structure of their phylogeny is now well established (e.g., Gauthier 1986; Holtz 1994, 2000; Rauhut 2003; Carrano & Sampson 2008; Carrano et al. 2012). Plenty of controversies and disagreements still exist over theropod interrelationships, but we can outline their evolution and place lineages, such as the *Tyrannosaurus* line, with some degree of confidence (fig. 2.2A).

FINDING THE RIGHT BRANCH: TYRANNOSAUROIDEA

The basics of Mesozoic theropod evolution can be broadly understood with knowledge of a few major clades. Their first radiation were the Neotheropoda, which included coelophysoids and the incredibly diverse, long-ranging ceratosaurs (within which we find the eponymous *Ceratosaurus*, the frequently bizarre abelisaurs, and even some fast-running herbivores—see Delcourt 2018). The tetanurans were the next major clade, the first offshoots of which gave rise to three famous carnivorous lineages: megalosaurids, spinosaurids and allosauroids. As discussed above, the spinosaurids and allosauroids contain some of the largest theropod species known, but grand size developed independently several times within this clade and these giants were

not closely related. This convergence on body size has only been appreciated relatively recently thanks to the advent of computerized means to identify animal clades. In the early and mid-twentieth century, in contrast, all large predatory theropods, including *Tyrannosaurus*, *Allosaurus*, *Spinosaurus*, and *Megalosaurus* were contained in a group known as "Carnosauria," a loose assortment of large and giant theropods that bore few shared anatomies. Today, we realize that *Allosaurus*, *Tyrannosaurus*, *Megalosaurus*, and so on only resemble one another in superficial features related to their great body sizes, and, in detail, their anatomy is too different to suggest close common ancestry. A clade known as Carnosauria survives in some phylogenetic schemes, but it is different in composition to its historic forebear, and no longer has any bearing on the *Tyrannosaurus* lineage (Rauhut 2003; Padian et al. 1999; Rauhut and Pol 2019).

The most diverse tetanuran clade was Coelurosauria, a group that, among other features, is marked by possessing the only incontrovertibly feathered (and protofeathered) theropods (Campione et al. 2020; Benton 2021). The coelurosaurs comprise the ornithomimosaurs, the ostrich dinosaurs, and their sister group, the Maniraptora. Maniraptorans were an extremely disparate group, including the giant, herbivorous therizinosaurs and the tiny insectivorous alvarezsaurs, as well as the Penneraptora, a collection of bird-like dinosaurs that eventually gave rise to avians themselves. The beaked oviraptorosaurs, omnivorous troodontids, and predatory dromaeosaurs (the group that houses *Velociraptor* and *Deinonychus*) are generally considered the closest non-avian dinosaur relatives of modern birds and their Mesozoic ancestors.

Within this scheme, *Tyrannosaurus* are placed at the base of Coelurosauria. This position makes them the closest relatives of birds among all of the giant predatory theropods, but they were not closely related to birds in any real sense. Their particular branch of coelurosaur evolution is the Tyrannosauroidea, a group of

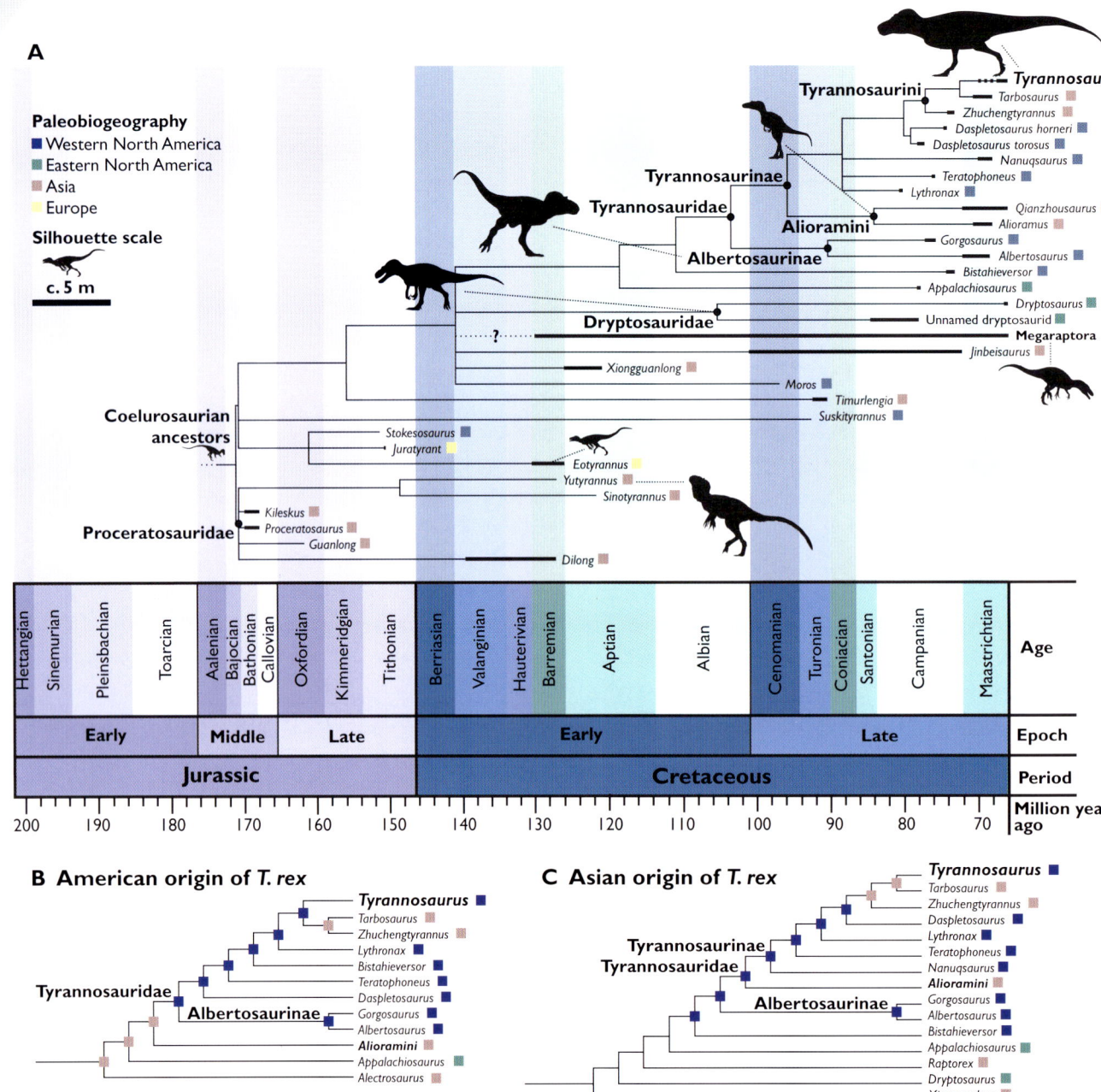

FIGURE 2.6

Tyrannosauroid evolution. (A) Chronogram (a time-calibrated phylogenetic tree) of Tyrannosauroidea with paleobiogeography indicated by colored boxes; (B–C) competing phylogenies of Tyrannosauroidea with different implications for the geographic origins of *Tyrannosaurus*. (A) Topology adapted from Brownstein (2021); (B) after Loewen et al. (2013); (C) after Brusatte and Carr (2016).

theropods that first appeared 170–165 million years ago, during the Middle Jurassic (fig. 2.6; Brusatte et al. 2010b; Brusatte and Carr 2016; Brownstein 2021; Naish and Cau 2022). Apparently restricted to the northern continents (though read on), they can be distinguished from other theropods by a number of features (Holtz 2004): a tall, blunted premaxillary bone (the bone at the front of the upper jaw, see chapter 3), "D"-shaped "incisor"-like teeth at the front of the mouth, fused nasal bones along the top of the snout, extensive air spaces in the skull and lower jaws, a prominent "shelf" on the outer rear surface of the lower jaw, upper pelvic bones (ilia) that almost touch over the hip, vertebrae that bear a central vertical muscle attachment ridge, and a uniquely shaped thigh bone head (Holtz 2004; Brusatte et al. 2010b).

The first tyrannosauroids were small, relatively generalized carnivores that hunted small game while larger megalosauroids and allosauroids occupied large-bodied predatory niches. Some analyses have placed the Late Jurassic North American coelurosaurs *Coelurus fragilis* and *Tanycolagreus topwilsoni* (fig. 2.7A) as early members of the tyrannosauroid line (Senter 2007). This interpretation has not been widely upheld, but, even so, these small species probably had a body plan similar to the most archaic tyrant dinosaurs. It is more widely agreed that the earliest-diverging and oldest lineage of the tyrannosauroids were the Proceratosauridae, a group presently known from Europe and Asia (fig. 2.7B; Rauhut et al. 2010; Brusatte and Carr 2016). The majority of proceratosaurid species are newly discovered, such that their place in tyrant dinosaur evolution and uniqueness are relatively fresh information. They still retained a lot of early coelurosaur features, most notably their relatively long, three-fingered forelimbs and relatively weak bites (Johnson-Ransom et al. 2024), but proceratosaurids were one of the longest-lived tyrannosauroid radiations, persisting for around 50 million years (Brusatte and Carr 2016). Some, such as *Proceratosaurus bradleyi* (fig. 2.8A), *Guanlong wucaii* and *Yutyrannus huali* developed prominent midline snout crests, also making them the most ostentatious species

of the tyrannosaur line (Xu et al. 2006; Rauhut et al. 2010). Although generally small bodied, the 9–10 m long *Yutyrannus* and *Sinotyrannus kazuoensis* demonstrated that some proceratosaurids had graduated into larger predatory guilds by the Early Cretaceous (Ji et al. 2009; Xu et al. 2012). *Yutyrannus* and a fellow Cretaceous proceratosaurid, *Dilong paradoxus*, represent the only tyrannosauroids yet known to have definitely sported protofeathers (Xu et al. 2004, 2012).

THE MID-CRETACEOUS: TYRANNOSAUR WILDERNESS

Proceratosaurids appear to have gone extinct in the Early Cretaceous without any descendents, such that all their innovations in size and ornament had no bearing on further tyrant dinosaur evolution. It fell to their sister line of tyrannosauroids, which were at this point present in at least Europe and western North America, to carry the lineage forward. Our knowledge of early non-proceratosaurid tyrannosauroids is not extensive, but they are thought to be typified by small- to midsized (3–5 m long) species of the Jurassic and Early Cretaceous such as *Stokesosaurus clevelandi*, *Juratyrant langhami*, and *Eotyrannus lengi* (figs. 2.7C, 2.8B; Benson 2008; Brusatte and Benson 2013; Naish and Cau 2021). Tyrannosauroids of this grade were generally similar, if somewhat more robust and less ornamented, to proceratosaurids in appearance and ecology, with only the slender-snouted *Xiongguanlong baimoensis* hinting at anatomical diversification (D. Li et al. 2009). Compared to other exciting occurrences in theropod evolution at this time (including, for example, the proliferation of birds and other feathered dinosaurs, the development of fish-eating spinosaurids, the appearance of large carcharodontosaurids), Early Cretaceous tyrannosauroids might be regarded as relatively conservative. However, any generalizations about what tyrant dinosaurs were up to during this interval are tentative because a 20-million-year gap in the tyrannosauroid fossil record spans the middle duration of the Cretaceous (Brusatte and

A

B

C

FIGURE 2.7

Early tyrannosauroids. (A) The small (2 m long) Late Jurassic coelurosaur *Tanycolagreus topwilsoni*, a species that has been postulated as an early tyrannosauroid but may, instead, belong to another part of the theropod tree. The general form of *Tanycolagreus* was probably similar to the earliest members of the tyrannosaur lineage, in any case. (B) The 9 m long, crested proceratosaurid *Yutyrannus huali*, a large, protofeathered tyrannosauroid from colder regions of Cretaceous China; (C) *Eotyrannus lengi*, a small (4–5 m long) Early Cretaceous British tyrannosauroid.

Carr 2016). The recovery of fully formed, "advanced" tyrannosauroids on the other side of this intermission indicates that important developments in tyrant evolution must have taken place during this time: we simply lack the fossils to tell us what was happening.

Some authors posit that this gap in tyrant dinosaur evolution may be filled by an enigmatic theropod group: the Megaraptora (fig. 2.9A). The megaraptors were mid-to-large-sized predatory theropods that had a 60-million-year run lasting much of the Cretaceous, and their fossils occur across the Americas, Australia, and Asia. This apparent abundance contrasts with their relatively newly discovered status. The clade Megaraptora was only formally recognized in 2010 when Roger Benson and colleagues realized that several, hitherto difficult-to-place theropods such as *Aerosteon*

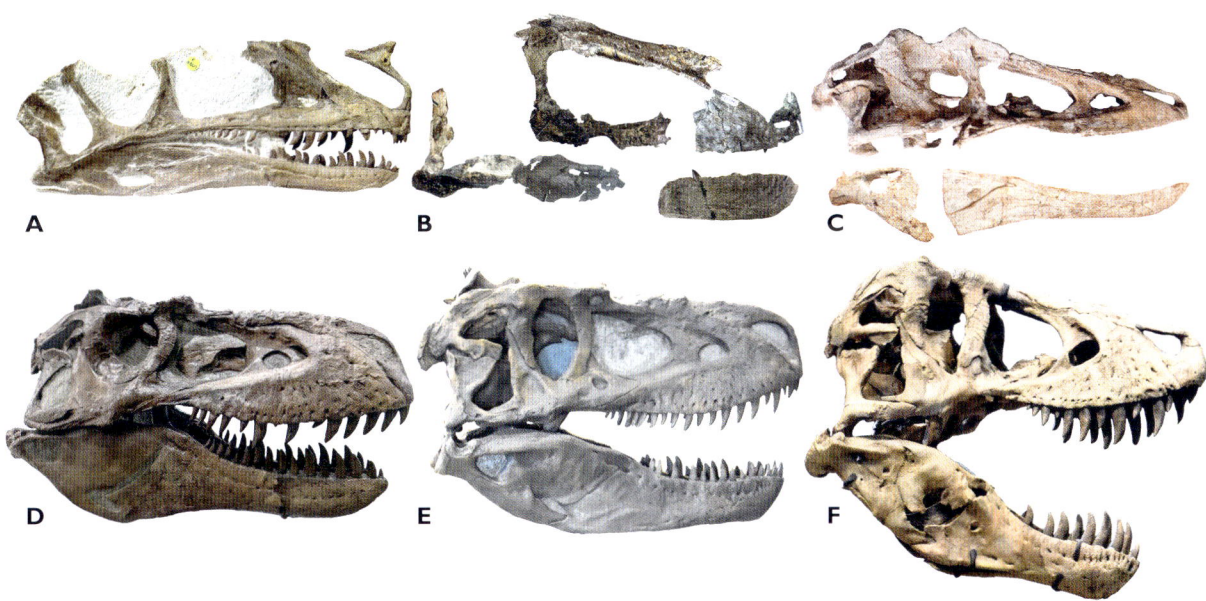

FIGURE 2.8

Skull material from select tyrannosauroids, representing the major grades of their evolutionary history. (A) The proceratosaurid *Proceratosaurus bradleyi*; (B) early tyrannosauroid *Eotyrannus lengi*; (C) alioramin *Qianzhousaurus sinensis*; (D) albertosaurine *Gorgosaurus libratus*; (E) tyrannosaurine *Daspletosaurus torosus*; (F) reconstructed holotype skull of *Tyrannosaurus rex*. Note the increasing skull depth in later tyrannosaurids, an adaptation to possessing longer, stronger tooth roots, increased skull rigidity, and larger jaw muscles. (B) Edited from Naish and Cau (2022). Images not to scale.

riocoloradensis, *Australovenator wintonensis*, *Fuku-iraptor kitadaniensis*, and *Megaraptor namunhuaiquii* belonged to a unique theropod lineage. The delayed identification of this group represents, in part, a dearth of complete skeletons. Even the best-known megaraptorans such as *Megaraptor*, *Australovenator*, and

Murusraptor barrosaensis are currently documented from very incomplete remains.

With scant fossils to work with, our understanding of megaraptoran anatomy, appearance, and ecology is relatively basic. Attaining up to 9 or 10 m body lengths, and with relatively long hindlimbs and shallow jaws

FIGURE 2.9

Possible and "intermediate" tyrannosauroids. (A) The megaraptoran *Australovenator wintonensis*. The affinities of megaraptorans are uncertain, but a number of researchers now hypothesize that they represent a lineage of unusual tyrannosauroids; (B) *Dryptosaurus aquilunguis*, a large-handed, late-surviving non-tyrannosaurid tyrannosauroid from the east coast of North America.

(Benson et al, 2010; Porfiri et al. 2014; White et al. 2015a; Aranciaga Rolando et al. 2022), their most characteristic features were long arms with enlarged, raptorial claws (White et al. 2015b; Novas et al, 2016). The relationships of megaraptorans to other theropods are controversial, with potential homes including Allosauroidea (Smith et al., 2008, Benson et al., 2010, Carrano et al., 2012; Zanno and Makovicky 2013; Coria and Currie 2016), the base of Coelurosauria (Apesteguía et al. 2016; Porfiri et al. 2018; Delcourt and Grillo 2018) or—of most interest to us here—as members or close relatives of the tyrannosaur line (Novas et al. 2016; Porfiri et al. 2014; Aranciaga Rolando et al. 2019, 2022; Naish and Cau 2022). If the latter is borne out, it would not only add a major, 60-million-year branch to tyrannosauroid phylogeny, but also place members of the tyrannosaur lineage in South America and Australia, continents hitherto considered beyond their reach. It would also demonstrate success with an unexpected body plan, the gracile skulls and long, powerful arms of megaraptorans being anatomically antithetical to the large heads and small arms that tyrant dinosaurs are famous for. As relatively large theropods, megaraptorans may also shed light on the evolution of larger size in later tyrannosauroids (Naish and Cau 2022). These implications must be regarded as provisional, however, as further discoveries and research are needed to elucidate exactly what sort of dinosaurs megaraptorans were, and where they fit in theropod evolution.

TYRANNOSAURIDAE: THE ROAD TO *T. REX*

The tyrannosauroid fossil record representing the final 25 million years of the Mesozoic is far superior to that of the mid-Cretaceous (Brusatte et al. 2010b; Brusatte and Carr 2016). Captured within this high-resolution window into tyrant dinosaur evolution were the last of the relatively archaic tyrannosauroids, *Suskityrannus hazelae* and *Moros intrepidus*, which occur in rocks of 100–90 million years old in western North America

(Nesbitt et al., 2019; Zanno et al., 2019). *Jinbeisaurus wangi*, meanwhile, lived roughly contemporaneously in China (Wu et al. 2020). The future of tyrannosauroid evolution belonged to other clades, however. Different models of tyrant dinosaur evolution cloud the exact trajectory of anatomical development in this part of the tree (e.g., Brusatte and Carr 2016; Voris et al. 2020; Brownstein 2021), but all predict that features characterizing the final and most famous members of the group, such as greater size and longer, running-adapted hindlimbs, started development during that 20-million-year gap in the tyrant dinosaur record.

The path to tyrannosaurs becoming large, arch predators from their smaller Early Cretaceous relatives was via "intermediate" forms like *Appalachiosaurus montgomeriensis*, *Bistahieversor sealeyi* and the dryptosaurids, including *Dryptosaurus aquilunguis* (fig. 2.9B; Carr et al. 2005; Carr and Williamson 2010; Brusatte et al. 2011; Brownstein 2021). These tyrants remain relatively poorly known, and their basic proportions are only broadly understood. Ranging from 6 to 9 m in length, they had seemingly begun the trend of sustained body size increase evident during the final stages of tyrannosauroid evolution. The skulls and jaws of *Appalachiosaurus* and dryptosaurids were still shallow, but some of the more iconic aspects of tyrannosaur cranial shape were becoming apparent, including their ornamented snouts and the prominent, horn-like bones around their eyes (Carr et al. 2005). *Bistahieversor* sported deepened jaws, indicating development of this feature also began among "intermediate" tyrants (Carr and Williamson 2010). The hindlimbs of dryptosaurids possessed compressed, narrow foot bones (Brownstein 2021), perhaps hinting at increased running capacity (chapters 3 and 4). It remains unclear when the long, three-fingered forelimbs of tyrannosauroids shrank to their shorter, two-fingered variant, but *Dryptosaurus* hints at a halfway condition where the arm bones reduced to the proportions of later tyrannosaurs, but the hand (possibly now two fingered?) remained large (Brusatte et al. 2011). This "intermediate" phase of tyrant dinosaur evolution remains one of the most interesting phases of their development, although further discoveries are needed to flesh out our knowledge of their anatomy and their roles in the emergence of their larger, later cousins.

The final major phase of tyrannosauroid evolution concerns the Tyrannosauridae (figs. 2.6, 2.10): the group of large, sometimes gigantic predators that, after tens of millions of years as small- and midsize carnivores, now filled top predatory roles in western North America and Asia. An excellent fossil record means that tyrannosaurid skeletal anatomy is understood in detail despite their geological longevity spanning a surprisingly short 15 million years (Brusatte and Carr 2016; Brownstein 2021). Phylogenetic predictions posit that tyrannosaurid origins occurred deep in the Cretaceous, perhaps over 100 million years ago (fig. 2.6; Loewen et al. 2013; Brusatte and Carr 2016; Brownstein 2021), but the early story of their evolution remains untold from fossils. Instead, our geological tyrannosaurid narrative starts with them already occupying vast regions across North America, from Alaska to Mexico, as well as in Asia, where their remains occur in Mongolia and China. By this point, tyrant dinosaurs were not only grand in stature (at least 6 m or more in length as adults, with most species reaching over 10 m), but they often represented the sole large predators in their respective environments (Holtz 2021). The circumstances behind their attainment of arch-predatory niches remains unknown: Did early tyrannosaurids drive their main competition, the carcharodontosaurids, to extinction? Or, alternatively, did they step into arch-predator niches after the carcharodontosaurid lineage ended through other means?

Along with grand body size, tyrannosaurids are defined by a number of features related to large-scale carnivory (Holtz 2004). Their jaws were deep and housed long-rooted teeth. These were no longer blade-like, as is the case for most predatory theropods, but thickened into lance- or spike-like shapes. Heightening and broadening the rear of the skull created enlarged spaces for jaw musculature, increasing their bite forces (Johnson-Ransom et al. 2024) to the extent that some

skull regions had to be reinforced (e.g., Henderson 2003; Rayfield 2004; Snively et al. 2006). As a consequence of remodeling the posterior skull, tyrannosaurids had elevated degrees of forward-facing vision (Stevens 2006). Neck musculature was also enhanced, necessitating upward expansions of the rearmost skull bones to accommodate larger muscles. Elsewhere on their bodies, all tyrannosaurids possessed short, two-fingered arms, although some were smaller than others: the Asian giant *Tarbosaurus bataar* currently holds the record for the smallest tyrant dinosaur arms relative to body size (Brusatte et al. 2011). In contrast, their legs were especially long and bore the proportions of runners, even in species that likely exceeded the biomechanical capacity for rapid locomotion (Paul 1988a; Holtz 1995; Hutchinson and Garcia 2002; Persons and Currie 2016; Dececchi et al. 2020; also see chapter 4). Their feet, in particular, hinted at enhanced speed and agility through development of the arctometatarsal condition: the state where the ankle-end of the middle metatarsal (the long bones making the shaft of the foot; see chapter 3) was pinched between its neighbors (Holtz 1995; Snively et al. 2004). Shortened bodies and tails (Newman 1970) added to this unique configuration. This body plan combined adaptations for agility and speed with the strongest bites of any dinosaur (e.g., Johnson-Ransom et al. 2024) and a propensity for giant size: a predatory dinosaur group like no other.

Currently, thirteen to fourteen genera of tyrannosaurids are recognized, divided into a few tribes. The Albertosaurinae comprises *Gorgosaurus libratus* (figs. 2.8D, 2.10B) and *Albertosaurus sarcophagus*: two relatively gracile, although still very large (8–9 m long) tyrants that are known primarily from Alberta, Canada, but that also occur in the northern United States (Russell 1970; Currie 2003a). The albertosaurines have one of the best fossil records of any theropod, rivaling that of *Tyrannosaurus* for specimen abundance and quality. They appear to have died out before the end of the Cretaceous, with *T. rex* filling the void left by their departure. Another group, the Alioramini, is

FIGURE 2.10

Examples of tyrannosaurids. (A) The Chinese *Qianzhousaurus sinensis*, representing the long-snouted alioramin branch of tyrannosaur evolution; (B) the albertosaurine *Gorgosaurus libratus*, a relatively gracile form of tyrannosaurid.

only known from Mongolia and China. Represented by two genera and three species, *Alioramus remotus*, *Alioramus altai* and *Qianzhousaurus sinensis*, the alioramins are perhaps the most unusual tyrannosaurids (fig. 2.10A; Brusatte et al. 2009, 2012; Lü et al. 2014; Foster et al. 2022). Smaller than their close relatives in only reaching 6–7 m in length, alioramins had especially low, long skulls that are, for their body size, 35 percent longer than expected. Their skulls were also unusually well decorated, with the ancestral tyrannosaur skull ornament exaggerated into large hornlets atop their snouts and around their eyes (fig. 2.8C). Their slender jaws indicate a departure from typical tyrannosaurid roles as arch-predators, perhaps better suiting mid-tier pursuers of small game. Larger prey was likely pursued by contemporary giant tyrannosaurids, such as *Tarbosaurus* and potentially *Zhuchengtyrannus magnus* (Lü et al. 2014). The co-occurrence of more than one tyrant species in a geological unit is unusual (Holtz 2021), further evidencing a deviant ecology for alioramins.

Where these strange tyrannosaurs fit within Tyrannosauridae is disputed (fig. 2.6B–C). Albertosaurines are generally regarded as the first branch to deviate from mainline tyrannosaurid evolution, but alioramins might represent a second early divergence of tyrannosaurids, forming a clade with the non-albertosaur tyrants, Tyrannosaurinae (fig. 2.6A and C; Lü et al. 2014; Brusatte and Carr 2016; Voris et al. 2020; Brownstein 2021). Alternatively, they might not be true tyrannosaurids at all, instead being close relatives (fig. 2.6B; Loewen et al. 2013; Dalman et al. 2024). These varied arrangements of tyrant clades affect hypotheses of tyrannosaur geographic origins. During the Late Cretaceous, North America was connected to Asia via a northern land bridge, forming a single landmass, Asiamerica, and, with a distribution across western North America and east Asia, it is clear that tyrannosaurid development involved dispersal across both continents. How often this occurred, however, is unclear. Some posit that tyrannosaurids arose in America and then moved across to Asia at least twice (e.g., Brusatte and Carr 2016; Voris et al.

2020; Brownstein 2021) if not on multiple occasions (Zheng et al. 2024), but others suggest that tyrannosaurid evolution began in Asia before shifting to North America, whereupon they developed distinct north-south faunas. Eventually, one branch then returned to Asia as the Cretaceous drew to a close (Loewen et al. 2013; Dalman et al. 2024).

This discussion has bearings on the origins of the *Tyrannosaurus* genus itself. The third main line of tyrannosaurid evolution consists primarily of North American species characterized by large size and particularly heavyset anatomy. It is from this line that *Tyrannosaurus* originated. The shape of this phase of tyrannosaurid evolution remains contested, with taxa from the southwestern United States, *Dynamoterror dynastes*, *Teratophoneus curriei*, *Lythronax argestes*, and the fragmentary Alaskan *Nanuqsaurus hoglundi* arranged differently in competing phylogenies (Carr et al. 2011; Fiorillo and Tykoski 2014; Brusatte and Carr 2016; McDonald et al. 2018; Voris et al. 2020; Brownstein 2021; Scherer and Voiculescu-Holvad 2023; Dalman et al. 2024). Some schemes consider the "intermediate" tyrannosauroid *Bistahieversor* a member of this lineage as well (Loewen et al. 2013), and others pull *Teratophoneus* outside of Tyrannosauridae (Naish and Cau 2022). A recently named Chinese species tentatively considered related to *Nanuqsaurus*, *Asiatyrannus xui*, looks set to complicate this arrangement further, not the least for implying that this phase of tyrannosaur evolution was not entirely confined to North America (Zheng et al. 2024). Some data also hints that *Asiatyrannus* may have been a "midsized" tyrannosaurid even as an adult, bucking the evolutionary convention of tyrannosaurids generally being large, arch carnivores. A mature specimen of this species is needed, however, to verify their actual adult proportions.

Dynamoterror, *Teratophoneus*, *Lythronax*, and *Nanuqsaurus* occupied large predator niches in western North America, from the northernmost regions down to southern states. From somewhere in this stock arose *Daspletosaurus* (fig. 2.8E), represented by three species, *D. torosus*, *D. wilsoni*, and *D. horneri* (Russell 1970;

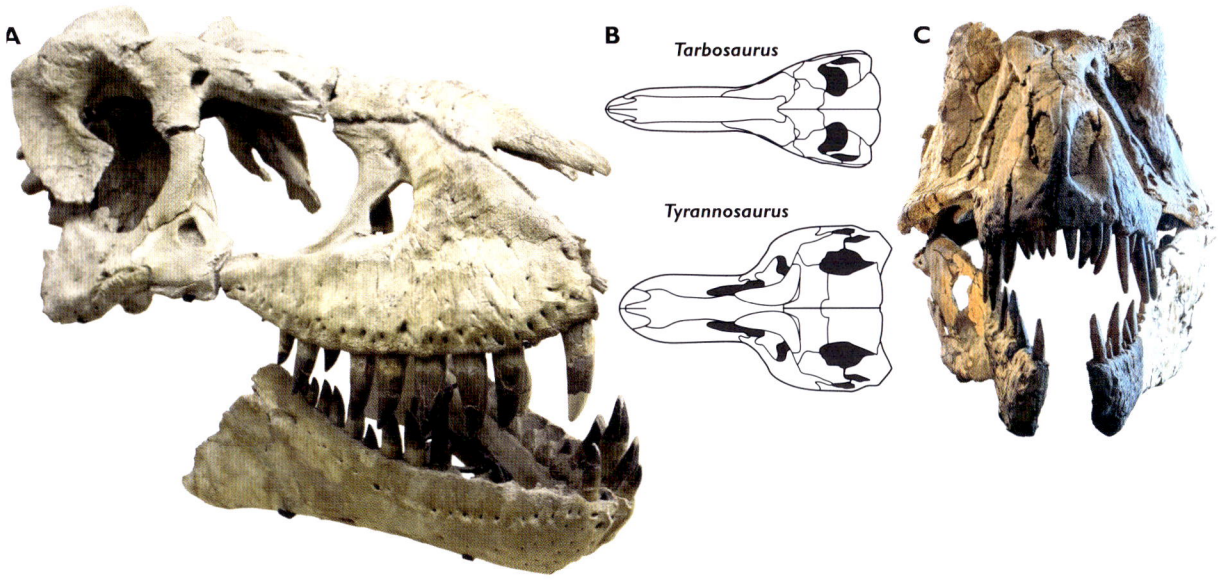

FIGURE 2.11

The closest known relative of *Tyrannosaurus*, the Mongolian tyrannosaurid *Tarbosaurus bataar*. (A) *Ta. bataar* holotype skull, demonstrating the strong similarity of this species to their American cousin; (B) comparison of *Tarbosaurus* and *Tyrannosaurus* skulls viewed from above; note the distinct bone arrangement as well as the difference in skull widths; (C) a large *Tyrannosaurus* skull housed in the Los Angeles Natural History Museum, demonstrating the extreme width of the posterior skull compared to the (also broadened) muzzle. (B) After Hurum and Sabath (2003).

Carr et al. 2017; Warshaw and Fowler 2022), as well as *Thanatotheristes degrootorum* (Voris et al. 2020). Together, these may form the clade Daspletosaurini. Their provenance in the northern United States, and the possibility that the three *Daspletosaurus* taxa represent a single evolving lineage (Warshaw and Fowler 2022, though also see Scherer and Voiculescu-Holvad 2023), hint at northern states having distinct tyrant faunas from the southern regions occupied by *Bistahieversor, Dynamoterror, Teratophoneus*, and *Lythronax*.

Daspletosaurus is considered a close relative of *T. rex* and may have been directly ancestral to the king tyrant lineage (fig. 2.8F). A twist in tyrannosaurid paleobiogeography complicates the origins of *Tyrannosaurus*, however. Despite being a North American taxon and the evidence of tyrannosaurids diversifying for millions of years in North America, the closest known relatives

of the *Tyrannosaurus* genus are, in fact, the Asian giants *Tarbosaurus bataar* (Maleev 1955a, 1955b; Rozhdestvensky 1965) and *Zhuchengtyrannus magnus* (Hone et al. 2011a). Together, these three species form the clade Tyrannosaurini (Scherer and Voiculescu-Holvad 2023; Dalman et al. 2024). *Zhuchengtyrannus* is a very poorly known species, but *Tarbosaurus* (figs. 2.11–12) is represented by a number of excellent specimens, from great adults to tiny juveniles (Tsuihiji et al. 2011). A juvenile specimen of a Mongolian tyrannosaurid initially interpreted as an adult from a dwarf species, *Raptorex kriegsteini* (Sereno et al. 2009), may represent an early growth stage of *T. bataar* (Fowler et al. 2011a), or else might be a juvenile of a species unknown from adult remains (Carr 2023). At estimated body lengths of 12 m, both *T. bataar* and *Z. magnus* were especially large and robust tyrannosaurids that rivaled *T. rex* in size.

FIGURE 2.12

Tarbosaurus bataar wanders home in the rain with a therizinosaur meal. Like *Tyrannosaurus*, *Tarbosaurus* had a wide posterior skull, although it was not as broad as that of *T. rex*.

Tarbosaurus also approached, if not quite matched, king tyrants in bite power (Johnson-Ransom et al. 2024). Although similar in build and form, *Tarbosaurus* can be easily distinguished from *Tyrannosaurus* by their narrower skulls, and also their smaller arms (Hurum and Sabath 2003; Carr et al. 2023).

With close relatives occurring in Asia, some question exists over where *T. rex* originated (figs. 2.6B–C). Some researchers (Loewen et al. 2013; Dalman et al. 2024) propose that *Tarbosaurus* and *Zhuchengtyrannus* form a clade within Tyrannosaurini, representing an independent branch of giant Asian tyrannosaurids that split from the *Tyrannosaurus* line within North America almost 80 million years ago. Other phylogenies, however, point to *Tarbosaurus* and *Tyrannosaurus* forming an exclusive group that only diverged c. 70 million years ago, with *Zuchengtyrannus* their next closest relative (e.g., Brusatte and Carr 2016; Voris et al. 2020; Brownstein 2021; Zheng et al. 2024). This implies that the lineage begatting *T. rex* lived in Asia, and that the distribution of *Tyrannosaurus* represents king tyrants, or their immediate ancestors, returning to regions once occupied by other American tyrannosaurids. Fossil data suggests that the *Tyrannosaurus* lineage existed in the southwestern United States about 70 million years ago (Dalman et al. 2024; though also see chapter 5), but it remains to be seen whether these fossils represent innovations of American tyrannosaurs or recent immigrants from Asia. An improved understanding of early tyrannosaurins, including better resolution of their geological ages, are necessary to resolve this biogeographical conundrum.

TYRANNOSAURUS REX, THE SPECIES

WHEREVER IT ORIGINATED, North America eventually became home to the animal that we call *Tyrannosaurus rex*, and it is consideration of this species that will wrap up our assessment of king tyrant taxonomy. Paleontologists have a reputation for being fussy and indecisive about the names of organisms. Sometimes, it seems that we just can't seem to agree on what certain taxa should be called. Should *Suchomimus* be considered a variant of *Baryonyx*? Is *Torosaurus* a form of *Triceratops*? Are *Stygimoloch* and *Dracorex* junior names for *Pachycephalosaurus*? Other times, we resurrect archaic names after decades of suppression, with a famous recent example concerning the iconic *Brontosaurus*. After almost a century of being regarded as an invalid, subsumed name for *Apatosaurus*, scientists decreed it appropriate and necessary to start using *Brontosaurus* again in 2015 (Tschopp et al. 2015). To those not directly involved in dinosaur research, this fussiness can seem trivial, even neurotic. Do dinosaur names really matter *that* much?

This obsession with names reflects an important discussion about how best to catalog the history of life. The terms we give to species and clades have bearing on their uniqueness, and thus their relationships to other organisms. With our knowledge of the fossil record constantly growing, new data sometimes demands that the labels applied to certain specimens and species need to change, too. Such considerations are the fields of taxonomy (the identification of biological species, genera and higher groups) and systematics (understanding how species relate to one another). They operate within a series of rules and guidelines established by governing bodies to ensure practicality, stability, and fairness among the millions of names applied to living and extinct species. Because Mesozoic dinosaurs were animals, their names and classifications follow guidelines outlined by the International Commission on Zoological Nomenclature, or ICZN.

Within this system, *Tyrannosaurus rex* has proved to be a robust and noncontroversial species name. The *T. rex* label pertains to Carnegie Museum specimen 9380 (figs. 1.8, 2.1), the partial skeleton described and named by Henry Fairfield Osborn in 1905. CM9380 is the *T. rex* holotype: a specimen of an organism to which a scientific name is anchored, and upon which the species is defined by a unique set of characteristics. Holotypes, sometimes called "type specimens" are critical to taxonomic practices as, once established, they act as permanent landmarks for specimen identification and categorization of their designated species. Accordingly, the fate of a species name is tied to the fate of its holotype. If a type specimen is found to be nondiagnostic (i.e., it is found to have no characters distinguishing it from other species), then the attached name becomes a *nomen dubium* (from Latin, "dubious name") and is no longer used by scientists. A name can also be invalidated if a holotype specimen is found to belong to another species. In this case, the older, more senior species name takes priority, and the younger, junior name is regarded as a synonym (this situation was, for a long time, considered the case for *Brontosaurus* until more detailed analyses showed that the *Brontosaurus* type material was anatomically distinct from *Apatosaurus*). Whatever the cause, these invalidated names tend to be written with quotation marks to stress their doubted nature (as with, for example, the defunct dinosaur genera "*Antrodemus*" [= *Allosaurus*] and "*Trachodon*" [= *Edmontosaurus*]).

The validity of holotypes constitutes a lot of discussion among paleontologists because experts frequently disagree over which are truly taxonomically distinct, which are over-interpreted variants of other taxa, and which are too poorly preserved to bear diagnosable features. Happily for *Tyrannosaurus* aficionados, CM9380 is a universally undoubted, reliable type specimen. It not only comprises multiple bones characterizing several parts of the *T. rex* skeleton, but represents an adult animal (Carr 2020), this being helpful because mature skeletons tend to bear more distinguished, diagnostic anatomy than juveniles. The properties that make *T. rex* unique have thus been known since the early 1900s, allowing us to diagnose the species through a unique combination of characters found on the holotype and, eventually, other specimens referred to the same species. This is not to imply that the defining characters of *T. rex* were an unchanging list of features written on stone tablets by Osborn when he first described the species in 1905. On the contrary, the diagnosis of *Tyrannosaurus*

rex has been sharpened as we've learned more about theropod diversity, making our list of features that distinguish *T. rex* more exacting and detailed. Osborn's initial 1905 attempt to characterize *T. rex* was short and inaccurate, reflecting misunderstandings about the nature of tyrannosaurs as well as, of course, his insistence on rushing out a description before the holotype was fully prepared (chapter 1). He merely defined *T. rex* as:

> Carnivorous Dinosaurs attaining very large size. Humerus believed to be of large size and elongate (Brown). No evidence of bony dermal plates (Brown). (Osborn 1905, p. 262)

Within a year, and now with fully prepared bones to work with, Osborn (1906) was able to refine this into something more precise and useful. Although his diagnosis was still hampered by a lack of complete specimens, he was able to list defining particulars from the skull, dentition, forelimb, belly ribs, pelvic girdle, and hindlimb bones, as well as the number and form of the vertebrae. These features allowed Osborn and subsequent workers to establish *T. rex* as an unquestionably valid animal, and outlined criteria by which other *T. rex* specimens could be distinguished from other species (e.g., Bakker et al. 1988; Carpenter 1990; Molnar 1991; Carr and Williamson 2000, 2004; Hurum and Sabath 2003; Paul et al. 2022). The diagnosing features of *T. rex* have continued to evolve until, at the time of writing, their most recent iteration prepared by a team of tyrannosaurid experts led by Thomas Carr (2022). Today, we distinguish king tyrants from their close relatives by at least ten features, mostly pertaining to minutiae of skull anatomy, but also their tooth counts and the proportions of their arms and legs.

Thanks to this robust holotype and description during a pioneering era of dinosaur science, the *T. rex* name has never been in significant danger of invalidation or overwriting by a senior species. The most serious risks of the latter concern the two genera established for *Tyrannosaurus* material around the turn of the twentieth century: Edward Cope's 1892 "*Manospondylus gigas*" and Osborn's second (1905) name for *Tyrannosaurus* material, "*Dynamosaurus imperiosus*." ICZN rules mean that the oldest name given to a species should take priority over any newer ones, and this would seem to give "*Manospondylus*" the advantage. However, whereas the *T. rex* holotype is very diagnostic, Cope's "*Manospondylus*" is not. Representing just two partial vertebrae, one of which is lost (fig. 1.5B), our remaining "*Manospondylus*"material is so badly weathered that we cannot even tell if it represents a part of the torso or the neck. Osborn first cast "*M. gigas*" as a *nomen dubium* in 1917 and any threat it presented to *T. rex* seemed over, at least until the year 2000. At this point, press reports suggested that the original "*Manospondylus*" quarry had been relocated, along with, potentially, more of the original "*M. gigas*" specimen (Brochu 2003). Would this have allowed "*M. gigas*" to rise from the grave and overthrow *T. rex*? As outlined by Brochu (2003), the answer to this is a straight "no." Not only would it be difficult to prove that the surviving "*M. gigas*" vertebra belonged to a newly unearthed specimen, but naming conventions dictate that scientific names can be abandoned if they have not been used in any meaningful manner for fifty years or more. Such a name is considered a *nomen oblitum*, Latin for "forgotten name." This situation clearly applies to "*M. gigas*."

"*Dynamosaurus imperiosus*" also poses no real nomenclatural threat to *T. rex*, on grounds of nomenclatural priority. Although Osborn named both *T. rex* and "*D. imperiosus*" in the same 1905 paper, *T. rex* was named on page 262, and "*D. imperiosus*" on page 263. That single page gives *T. rex* priority and means "*D. imperiosus*" has to be regarded as a synonym of *T. rex*, not the other way around. Furthermore, "*Dynamosaurus*" joins "*M. gigas*" in being considered a *nomen oblitum*. Neither of these historic names has a chance, therefore, of overthrowing the *Tyrannosaurus rex* label.

Stating that the *T. rex* name has been unchallenged by nomenclatural acts for the last century does not mean there has been no activity around king tyrant taxonomy, however. For the most part, *T. rex* has acted like a sponge, absorbing specimens initially identified as different tyrannosaurid taxa into an ever-expanding inventory of king tyrant remains. As our collections of *Tyrannosaurus* material have swollen, however, the tremendous variation in their remains has become apparent and some have doubted that all the fossils referred to *T. rex* truly represent a single species. This has led to several attempts to carve *T. rex* into different taxa, including the naming of potentially distinct tyrannosaurid species that lived alongside king tyrants. Simultaneously, some authors have treated the *Tyrannosaurus* genus as being expansive enough to encompass some Asian tyrannosaurs, chiefly, the Mongolian species *Tarbosaurus bataar*. Whether through new species or alternative means of generic classification, perhaps the name *Tyrannosaurus*, so long intimately associated with a single species, is due for expansion? This question is particularly pertinent as I finish editing this book in mid–2024. Following several years of relative stability, multiple studies have attempted, to greater and lesser success, to break *Tyrannosaurus rex* into more discrete taxonomic units since 2022 (Paul et al. 2022; Longrich and Saitta 2024; Dalman et al. 2024). At the time of writing, *T. rex* taxonomy is in an uncharacteristic state of flux, the outcomes of which will become apparent in years to come.

TYRANNOSAURUS BATAAR AND OTHER ASIAN "TYRANNOSAURUS"

Although *Tyrannosaurus* fossils are highly characteristic among those of other large theropods, they are not total anatomical outliers. Late Cretaceous Asia also housed gigantic, robust-skulled tyrannosaurids, some of which have been considered close relatives or even congeneric with *Tyrannosaurus* (fig. 2.6; Hurum and Sabath 2003). Among those to receive formal names are the Mongolian *Tyrannosaurus bataar* (Maleev 1955a) and the Chinese *Ty. lanpingensis* (Ye 1975), *Ty. luanchuanensis* (Dong 1979), *Ty. turpanensis* (Zhai et al. 1978), *Ty. zhuchengensis* (Hu et al. 2001) and *Zhuchengtyrannus magnus*

(Hone et al. 2011a). Most Chinese fossils labeled as "*Tyrannosaurus*" are fragmentary, nondiagnostic material such as teeth and portions of foot skeletons, and they are regarded as *nomina dubia* today (Holtz 2004; Hone et al. 2011a). They may represent pieces of more recently diagnosed taxa: "*Ty. zhuchengensis*" may belong to *Zhuchengtyrannus magnus* (Hone et al. 2011a), for example, and "*Ty. turpanensis*" is potentially a synonym of *Tarbosaurus bataar* (Shuonan et al. 1985; Holtz 2004).

The name *Tyrannosaurus bataar* has been more persistent in tyrannosaur literature. Named as part of a suite of new tyrannosaur taxa by E. A. Maleev in the mid-1950s (Maleev 1955a, b), *Tyrannosaurus bataar* was the original name for Mongolia's giant tyrannosaur, *Tarbosaurus*. The label *Tarbosaurus* was given to *Ty. bataar* in 1965 by A. K. Rozhdestvensky, who, simultaneously, interpreted all of Maleev's tyrannosaur "species" as growth stages of one, large-bodied taxon. Rozhdestvensky disagreed with Maleev's suggestion that this Mongolian tyrannosaurid was similar enough to *T. rex* to be placed in the same genus and instead used one of Maleev's other generic names, *Tarbosaurus*, to rehome the *bataar* lineage.

A somewhat complex taxonomic history followed, the technical details of which are unnecessary to relay here (Hurum and Sabath 2003 provide a review of the full history). We can instead summarize that most researchers have followed Rozhdestvensky's separation of *Tyrannosaurus* and *Tarbosaurus*, but not all, with some questioning whether *Tarbosaurus* should be kept as a distinct genus on grounds that it compares closely to *T. rex* in detailed anatomy (e.g., Paul 1988a; Carpenter 1992; Holtz 1994, 1995, 2001; Carr 1999, 2020; Carr and Williamson 2004). This view has some important implications for how we regard *Tyrannosaurus* evolution. If *Tyrannosaurus* and *Tarbosaurus* were congeneric, *Tyrannosaurus* would become an Asiamerican genus, not one restricted to North America. The geological range of *Tyrannosaurs* would also deepen thanks to *Tarbosaurus* possibly being slightly geologically older than *T. rex* (although also see below and chapter 5 for discussions of older *Tyrannosaurus* fossils in North

America). In some phylogenies at least, *Zhuchengtyrannus magnus* would be captured into the *Tyrannosaurus* genus, too, such that there would be at least three (or four; see below) valid *Tyrannosaurus* species across two continents. The paleoecological breadth of *Tyrannosaurus* would also expand, the dinosaur communities and environments of latest Cretaceous Asia being distinguished from those of North America in a number of ways. In short, what seems like trivial taxonomic reconfiguration actually has a lot of implications for what the *Tyrannosaurus* genus is and was!

Because the question of *Tarbosaurus* vs. *Tyrannosaurus bataar* concerns genus-grade classification, however, there is no right or wrong answer to this conundrum, nor a taxonomic "truth" to uncover. Genera are holdovers from Linnaean forms of classification, and, unlike considerations of species, their names have no bearing on the branching structure of evolutionary trees. Nor are there firm rules about their application, such as how many species a genus should contain, or how much variation is permitted between species before a new genus is warranted. To that end, personal preference plays a role in genus formulation and there is little consistency in their content. Some genera are enormous, with dozens of species, while others contain just one. There is not even a consensus about the use of genera among researchers working on the same types of organisms. Gregory S. Paul (1988a, 2016), for instance, has argued that *Tyrannosaurus* should not only house *bataar*, but also species contained within the genera *Nanuqsaurus*, *Teratophoneus*, and *Daspletosaurus*. This scheme has not been adopted by other researchers, but Paul is not objectively "wrong": he simply has a differently calibrated "genericometer" than other paleontologists.

There is thus nothing "incorrect" about replacing *Tarbosaurus* with *Tyrannosaurus*, but we might also consider this issue from another angle: that good taxonomic practice emphasizes stability in our classifications of organisms. While it is accepted that the names we apply to animal species are an artificial means of organizing life, we cannot take a nihilistic, "anything goes" approach to classification. Changes to taxonomic labels can be disruptive and confusing to researchers, especially if those names already have long, established, and widely understood meanings. Classification guidelines, therefore, suggest that specific and generic names should only be altered when there is good reason to do so, such as when resolving nomenclatural confusion or introducing new understanding to the evolution of a given clade. There are no such concerns to resolve with *Tarbosaurus* and *Tyrannosaurus*, however. Evolutionary studies repeatedly find them to be close relatives (e.g., Holtz 2004; Brusatte and Carr 2016; Voris et al. 2020; Brownstein 2021), but keeping the two as separate genera creates no complications for tyrannosaurid systematics. Neither is it peculiar for dinosaur genera to contain one species, as has traditionally been the case for both *Tarbosaurus* and *Tyrannosaurus*.

Such concerns, therefore, boil down to whether the benefits of expanding *Tyrannosaurus* to accommodate *bataar* outweigh any negatives. On one hand, such an act would leave no doubt about the close evolutionary affinities of *rex* and *bataar*. On the other, it clashes with the long-held understanding of *Tyrannosaurus* as an exclusively American genus, and means that more than fifty years of literature on *Tyrannosaurus* and *Tarbosaurus* requires a collective asterisk: "the classifications in these texts no longer apply." Perhaps there are more downsides to synonymizing *Tarbosaurus* and *Tyrannosaurus* than there are benefits, and, given that our application of the *Tyrannosaurus* label is ultimately arbitrary, it seems more sensible to retain their traditional uses. Such views may explain why the name *Tarbosaurus* remains widely used among contemporary tyrannosaurid researchers, while "*Tyrannosaurus bataar*" has only a few advocates.

"CRYPTIC" NORTH AMERICAN *TYRANNOSAURUS* SPECIES

Less subjective taxonomic matters concern proposals that unusual *T. rex* fossils from North America might represent "cryptic" or hitherto unnoticed species (fig. 2.13). Such ideas have a long history, with perhaps the first occurring in an unpublished 1972 thesis by then-student Douglas Lawson, who would later attain fame for finding the giant azhdarchid pterosaur *Quetzalcoatlus northropi* (Lawson 1975). Lawson's thesis named "*Tyrannosaurus vannus*" for a small jaw bone from the Texan Tornillo Group that seemed distinct from other *T. rex* specimens. Because this name was never published in a peer-reviewed journal, it failed to meet criteria for establishment of a new species and never became "official" in the eyes of zoological nomenclature. In any case, Lawson revised his interpretation soon after, referring the same bone to *T. rex* itself a few years later (Lawson 1976). Different opinions have been expressed on this specimen ever since. Some have agreed with Lawson that the specimen is *T. rex* (Brochu 2003; Carr and Williamson 2014; Carr 2020) or at least a very close relative (Brochu 2003; Wick 2014). In his 1990 review of *T. rex* variation, Kenneth Carpenter opined that this specimen might represent a different southern *Tyrannosaurus* taxon, while Thomas Holtz (2021) lists the specimen as an indeterminate *Tyrannosaurus* species. Sampson and Loewen (2005) were more skeptical, questioning whether the specimen can be confidently identified as *Tyrannosaurus* at all. Over fifty years on, this maxilla remains an intriguing specimen because it probably stems from rocks older than most other *T. rex* material (Fowler 2017; also see chapter 5) and, as we shall see below, it is among these older sediments that the most promising evidence for novel *Tyrannosaurus* species is found.

More buzz circulated around *Tyrannosaurus rex* representing multiple species in the 1980s. Horner and Lessem (1993) and P. Larson (2008b) give Robert Bakker credit for noting features in *Tyrannosaurus* skeletons that might delineate new species during this decade and suggest that it was only a lack of specimens to verify these observations that prohibited the idea from progressing further. It was perhaps these conversations that prompted Gregory S. Paul (1988a) to muse on multiple *Tyrannosaurus* species in his 1988 book *Predatory Dinosaurs of the World*. Despite acknowledging variation in *T. rex* dentition and limb robustness, Paul

Proposals in *T. rex* taxonomy c. 1970-late 1990s, with juveniles often recognised as new taxa (various authors)

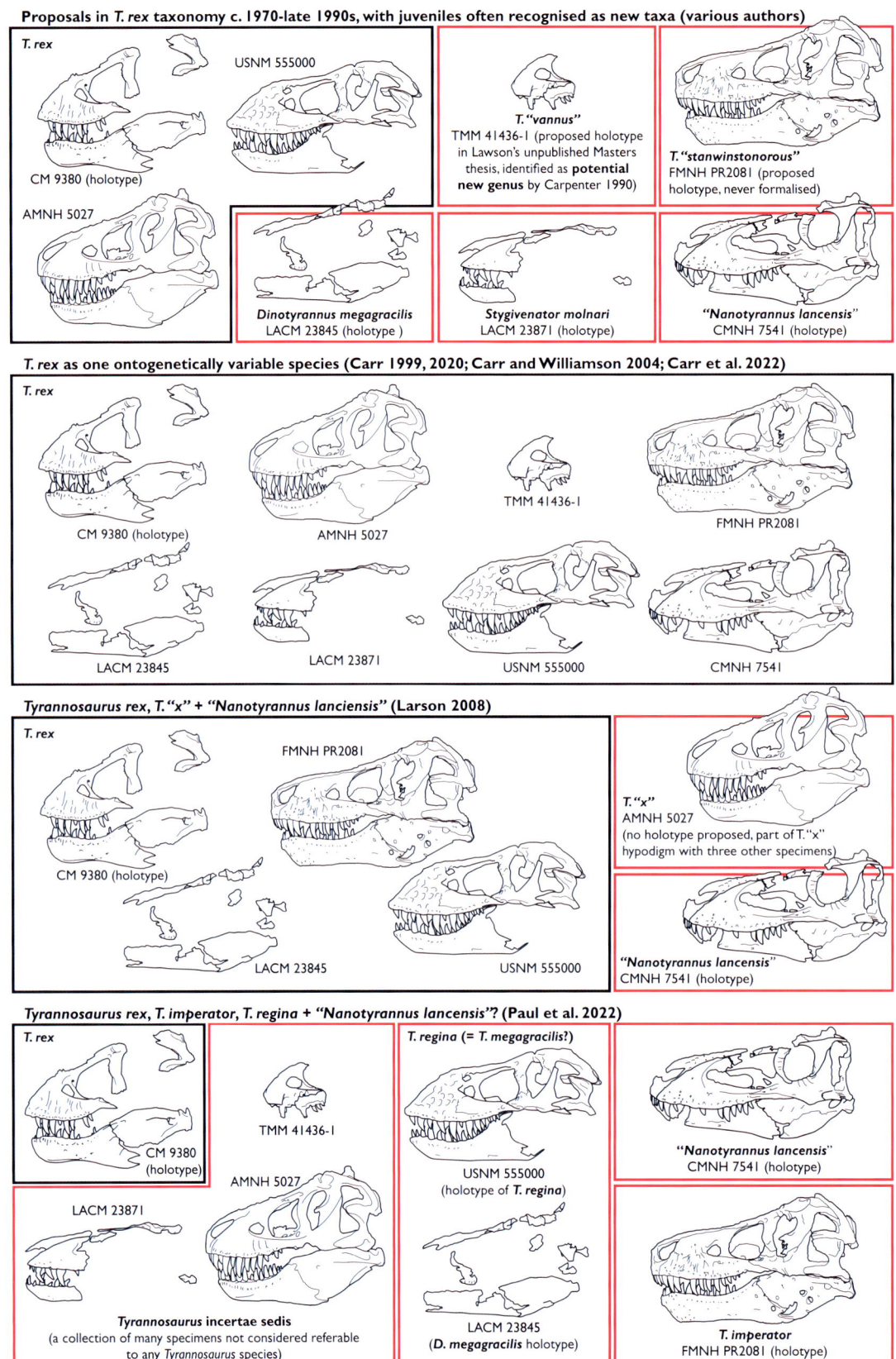

T. rex

USNM 555000

CM 9380 (holotype)

AMNH 5027

T. "vannus"
TMM 41436-1 (proposed holotype in Lawson's unpublished Masters thesis, identified as **potential new genus** by Carpenter 1990)

T. "stanwinstonorous"
FMNH PR2081 (proposed holotype, never formalised)

Dinotyrannus megagracilis
LACM 23845 (holotype)

Stygivenator molnari
LACM 23871 (holotype)

"Nanotyrannus lancensis"
CMNH 7541 (holotype)

***T. rex* as one ontogenetically variable species (Carr 1999, 2020; Carr and Williamson 2004; Carr et al. 2022)**

T. rex

CM 9380 (holotype)

AMNH 5027

TMM 41436-1

FMNH PR2081

LACM 23845

LACM 23871

USNM 555000

CMNH 7541

***Tyrannosaurus rex*, *T. "x"* + *"Nanotyrannus lanciensis"* (Larson 2008)**

T. rex

FMNH PR2081

CM 9380 (holotype)

T. "x"
AMNH 5027
(no holotype proposed, part of T."x" hypodigm with three other specimens)

LACM 23845

USNM 555000

"Nanotyrannus lancensis"
CMNH 7541 (holotype)

***Tyrannosaurus rex*, *T. imperator*, *T. regina* + *"Nanotyrannus lancensis"*? (Paul et al. 2022)**

T. rex

CM 9380 (holotype)

TMM 41436-1

AMNH 5027

LACM 23871

Tyrannosaurus incertae sedis
(a collection of many specimens not considered referable to any *Tyrannosaurus* species)

T. regina (= *T. megagracilis*?)

USNM 555000
(holotype of ***T. regina***)

LACM 23845
(*D. megagracilis* holotype)

"Nanotyrannus lancensis"
CMNH 7541 (holotype)

T. imperator
FMNH PR2081 (holotype)

FIGURE 2.13

Graphic summary of select *Tyrannosaurus* specimen classifications, exemplified by cranial material. This image does not represent the entirety of *T. rex* specimens known at any one time, focusing instead on holotypes or examples that have been regarded as taxonomically distinct from *T. rex*, even if they remain unnamed. Illustrations not to scale.

concluded that *Tyrannosaurus* specimens known in the late 1980s were not diverse enough to represent more than one species. Other researchers were not so sure: Molnar (1991) responded to Paul's conclusion by questioning whether we had enough *T. rex* specimens to deduce their real taxonomic diversity, and the question would remain open, Molnar argued, until more fossils gave a superior view of their variation. Some felt that *Tyrannosaurus* inventories were already sufficient to detect multiple species, however: Donald Glut (1997) reported that researcher Stephen Pickering had privately issued manuscripts in 1996 proposing that the "Sue" *T. rex* specimen, among others, represented "*Tyrannosaurus stanwinstonorous*," a name honoring the special effects pioneer responsible for the animatronic dinosaurs in *Jurassic Park* and its sequels. Pickering argued that multiple features of the skull and a 6–7 percent larger body size than either *Tarbosaurus* or other *T. rex* distinguished his new species. As with Lawson's *T.* "*vannus*," lack of formal publication prevented this name from entering scientific consideration.

The result was that, for all the buzz about multiple *Tyrannosaurus* species during the late twentieth century, no specimens were ever removed from *T. rex* catalogs. Simultaneously, other North American tyrannosaur fossils that we'd eventually recognize as *Tyrannosaurus* were receiving names. Because of their smaller stature and slender build, some of these were initially placed within the albertosaurine genera *Albertosaurus* or *Gorgosaurus* (Gilmore 1946a; Molnar 1980), or even referred to Dromaeosauridae, inferring a link to *Velociraptor* (Molnar 1978). Eventually, each were recognized as distinct tyrannosaur genera: the small-bodied *Stygivenator molnari*, known only from a skull (Paul 1988a; Olshevsky 1995), the larger *Dinotyrannus megagracilis*, represented by a partial skeleton (Paul 1988a; Olshevsky 1995), and, most famous of all, *Nanotyrannus lancensis*, another skull taxon (fig. 2.14; Gilmore 1946a; Bakker et al. 1988). The possibility that the *S. molnari* material represented a growth stage of *T. rex* was mentioned but rejected by Molnar (1978), it being considered that the immature growth stage of the specimen was not conclusive. The validity of *N. lancensis*, in contrast, hinged on it being presented as a subadult or adult that reached maturity at a smaller size and lighter frame than *T. rex* (Paul 1988a; Bakker et al. 1988). These attitudes perhaps reflect, in part, an age in vertebrate paleontology dominated by typological thinking, when our capacity to identify younger and older individuals of the same species was less

FIGURE 2.14

The holotype skull of "*Nanotyrannus lancensis*." Considered a juvenile *Tyrannosaurus* by most, a body of researchers maintain that this skull typifies a pygmy tyrant dinosaur that lived alongside *T. rex*.

developed than it is today. Specimens were accordingly sometimes considered new species without factoring the (sometimes substantial) influences of growth and individual variation on anatomy.

A more holistic philosophy toward classification of fossil species has, in the last few decades, become more dominant, and allows for consideration that young juveniles of extinct species might look very different from old adults. This approach that has characterized most recent taxonomic work on *Tyrannosaurus* (e.g., Rozhdestvensky 1965; Carpenter 1990, 1992; Carr 1999, 2020; Carr and Williamson 2004), and, collectively, these works have constructed a well-supported growth series that charts *T. rex* maturing from relatively small juveniles to supersized, robust adults. This growth series regards "*S. molnari*" and "*D. megagracilis*" as juvenile and subadult specimens, respectively, and, while a minority favor *Nanotyrannus* as a distinct taxon (see below), it also plots within this system as a juvenile *T. rex* (Carr 1999, 2020; Woodward et al. 2020). The result is an interpretation of *Tyrannosaurus* fossil inventories as evidencing a remarkable amount of morphological variation in their lifetimes, rather than a late Cretaceous North America filled with numerous tyrannosaur taxa. Not everyone is convinced of this view, however: some maintain that our *T. rex* archives contain multiple *Tyrannosaurus* species, and maybe even multiple genera.

THE *NANOTYRANNUS* CONTROVERSY

Although two of the three juvenile *Tyrannosaurus* specimens named during the late twentieth century have been absorbed into *T. rex* without much further remark, one taxon has not been so readily subsumed: *Nanotyrannus lancensis*. This proposed diminutive tyrannosaur species has attained fame among dinosaur enthusiasts thanks to several documentaries, TV series, and online communities trumpeting its legitimacy. Away from the cameras and internet forums, however, the *Nanotyrannus* hypothesis is not widely accepted

by theropod researchers and far less academic certainty surrounds it than some popular sources imply.

The story of *Nanotyrannus* is, nevertheless, an interesting one that has helped shed light on other aspects of *T. rex* biology. The type specimen of *N. lancensis* is a nearly complete, 57 cm long skull and lower jaw that was collected from the Hell Creek Formation in 1942. It was posthumously described and named by Charles W. Gilmore in 1946 as *Gorgosaurus lancensis* (fig. 2.14; Gilmore 1946a). With a low, long, and relatively narrow snout, and somewhat compressed, blade-like teeth, it's easy to understand why Gilmore thought this was a species distinct from the more massively-built *T. rex*. A further trait sealed this interpretation: the fusion of the skull bones. Animal skulls are made up of multiple bones which typically fuse together during growth (see chapter 3), such that the amalgamated cranial bones of the *G. lancensis* holotype were thought to indicate some degree of skeletal maturity. Russell (1970) agreed with Gilmore's identification of the skull as a mature individual but classified it as *Albertosaurus* instead, a product of the disputed convention of regarding *Gorgosaurus* as synonymous with the older name *Albertosaurus*.

The *lancensis* skull was reinterpreted again as a whole new genus, *Nanotyrannus*, by Robert Bakker and colleagues in 1988. They contextualized many of the same features identified by Gilmore against an improved understanding of theropod anatomy to conclude that *Nanotyrannus* was not just another Late Cretaceous albertosaur, but an entirely new and exciting taxon. Features said to separate *Nanotyrannus* from other tyrants were many, including the configuration of the rear of the skull, the shape and number of the teeth, and the slender snout being married to an expanded posterior skull region. Those fused skull bones indicated that growth was concluded, or was nearly so, making *Nanotyrannus* a small-bodied, pygmy tyrannosaur species. Moreover, the presence of compressed, relatively blade-like teeth—a feature not found among adult tyrannosaurids, but seen elsewhere in Theropoda—suggested that *Nanotyrannus* was a late-surviving archaic tyrannosaurid (Bakker et al. 1988).

As intriguing as these ideas were, the proposals of Bakker et al. (1988) were never without controversy. Notions that the *lancensis* skull was merely a growth stage of *T. rex* had been suggested as early as 1965 by A. K. Rozhdestvensky after his analysis of growth in *Tarbosaurus*. Kenneth Carpenter (1992) provided a confirmatory but more detailed assessment, noting that *lancensis* skull fusion had been overstated and that the shape of various skull bones and openings fit models of theropod maturation founded on other species. Although drawing no firm conclusion on the specimen, Carpenter (1992) asked why, all details considered, should *lancensis* not be regarded as a juvenile *Tyrannosaurus*? This question was answered in thorough detail by Thomas Carr in 1999, who performed an in-depth analysis of the *lancensis* skull in context with growth sequences derived from other tyrannosaurs. Carr (1999) found that none of the coalesced bones noted by Gilmore or Bakker et al. were actually knitted together, and also demonstrated the presence of fibrous surface bone textures across the skull: both features characteristic of juvenile reptiles. This, and a suite of anatomical observations linking the *lancensis* skull condition with other immature tyrant dinosaur specimens, was damning evidence against *Nanotyrannus* being a "pygmy" adult.

Carr's assessment remains the most robust justification yet produced for interpreting the specimen as a young *T. rex* and has proven persuasive to most dinosaur workers. *Nanotyrannus* has never wholly fallen from use among students of tyrannosaurids, however. It remains an accepted, or at least a potentially valid taxon among several researchers, and several projects have specifically aimed to investigate its legitimacy. The discovery of new small tyrant specimens—juvenile *T. rex* to some, *Nanotyrannus* to others—has only fanned the flames further. Currie (2003a, b) agreed that *Nanotyrannus* might be juvenile *T. rex*, but also questioned whether differences in tooth count might preclude full synonymization of these taxa. Although preferring to keep the question of *Nanotyrannus* open rather than making a firm call, Witmer and Ridgely's

(2009, 2010) examination of tyrannosaurid braincases and cranial pneumatic structures stressed how aberrant the *lancensis* skull is compared to those of *T. rex*. The shape of the brain, inner ear, and sinuses bear some similarities, but not as much as might be expected, given the general conservatism of these regions during growth in most reptiles. Schmerge and Rothschild (2016a, b) argued that a groove along the lateral surface of the lower jaw distinguished *Nanotyrannus* from *T. rex*, although other scholars have disagreed (Brusatte et al. 2016; Carr 2020). More dissent stemmed from Woodward and colleagues (2020), who looked at growth lines in referred *Nanotyrannus* limb bones to find that they represented immature animals, not pygmy adults.

Some researchers have especially championed *Nanotyrannus* in these long-running discussions. Among them is Peter Larson, who has argued in several papers (P. Larson 2008a, b, 2013) for its validity. This body of work relies less on the *Nanotyrannus* holotype and more on a referred *Nanotyrannus* specimen nicknamed "Jane," a nearly complete, 6.4 m long skeleton recovered in 2002 by the Burpee Museum of Rockford, Illinois (fig. 2.15; P. Larson 2008a, b, 2013). Like the *lancensis* holotype, "Jane" is a juvenile, estimated at perhaps twelve or thirteen years old from bone growth lines (P. Larson 2013; Woodward et al. 2020). A lean, lanky-looking animal with long, slender legs, Larson (2013) argued that this age was too old for the specimen to have transformed into a gigantic, gnarly adult *T. rex*. He further identified a suite of anatomies across the skull, teeth, shoulder, and pelvis that distinguished *Nanotyrannus* specimens from other tyrants. The conclusions drawn in these works were twofold: first, that *Nanotyrannus* was a distinct taxon characterized by dozens of features, and second, that no true juvenile *T. rex* had yet been discovered.

Nicholas Longrich and Evan T. Saitta (2024) have also argued that *Nanotyrannus* is a valid genus. Like Larson, Longrich and Saitta argue that *Nanotyrannus* is too different from adult *T. rex* to grow into them. They posit that *Nanotyrannus* is not only anatomically distinguished

FIGURE 2.15

The thirteen-year-old *Tyrannosaurus* specimen known as "Jane." The identity of this specimen has proven controversial, with some regarding it as an example of "*Nanotyrannus*" and others interpreting it as an adolescent *T. rex* about to embark on a dramatic growth spurt.

from adult *T. rex* by more than 150 features, but also point to particulars of skull anatomy that distinguish them from isolated bones considered by these authors to represent genuine *Tyrannosaurus* infants. Arm size is cited as particularly significant with adult *T. rex*, according to Longrich and Saitta, having smaller forelimb claws than *Nanotyrannus* specimens (fig. 2.16A). Arguments about maturity have also been revived, with details of bone texture and skull fusion reportedly pointing to *Nanotyrannus* specimens approaching maximum size.

Moreover, decreases in the diameters of growth rings within *Nanotyrannus* limb bones (known as lines of arrested growth, LAGs, see chapter 4), suggest that *Nanotyrannus* specimens were approaching maturity when they died (fig. 2.16C). Statistical analyses that distinguish *Nanotyrannus* from other *T. rex* and a phylogenetic analysis that recovers *Nanotyrannus* as a more archaic form of tyrannosaur make Longrich and Saitta's efforts the most detailed arguments for the validity of *Nanotyrannus* to date.

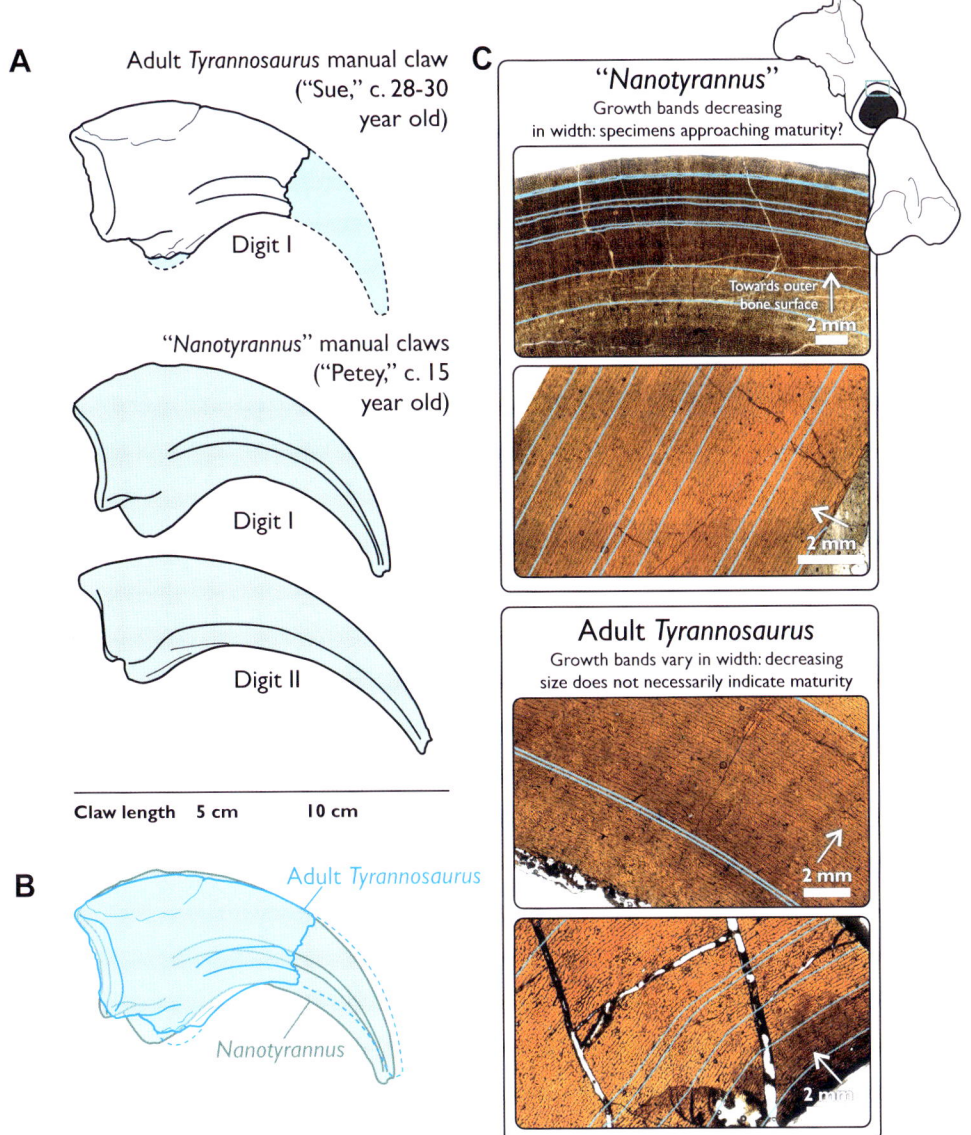

A Adult *Tyrannosaurus* manual claw
("Sue," c. 28-30 year old)

Digit I

"Nanotyrannus" manual claws
("Petey," c. 15 year old)

Digit I

Digit II

Claw length   5 cm   10 cm

B Adult *Tyrannosaurus*

Nanotyrannus

C *"Nanotyrannus"*
Growth bands decreasing
in width: specimens approaching maturity?

Towards outer
bone surface

2 mm

2 mm

Adult *Tyrannosaurus*
Growth bands vary in width: decreasing
size does not necessarily indicate maturity

2 mm

2 mm

FIGURE 2.16

Complications with the evidence for *"Nanotyrannus."* (A–B) Claw sizes in *"Nanotyrannus"* specimens and adult *T. rex*. Some authors report that the former are larger than the latter, but this is probably incorrect: the partial thumb claw of the large adult "Sue" is as large as the equivalent in *"Nanotyrannus"* (overlaid in [B]); (C) *"Nanotyrannus"* and adult *Tyrannosaurus* bones in cross section with growth bands, marked by LAGs (lines of arrested growth) highlighted in blue. *"Nanotyrannus"* specimens show decreasing growth band size toward their bone margins, said to indicate approaching maturity at small body sizes. Adult *T. rex* specimens, however, show that band thicknesses are inconsistent throughout life and are unreliable indicators of maturity. (A) Redrawn from Longrich and Saitta (2024); and Brochu (2003); (B) photos from Woodward et al. (2020).

WHAT, IN ACTUALITY, IS A *T. REX*?

What, considering all this academic back-and-forth, might we conclude about the legitimacy of *Nanotyrannus*? The idea of a second, small-bodied tyrant living alongside *Tyrannosaurus* is not unreasonable, and, undoubtedly, considering *Nanotyrannus* as a growth stage of *T. rex* implies that king tyrants changed almost beyond recognition during their lifespans. This might, intuitively, seem strange or suspicious, so is it simply more plausible to classify the *lancensis* skull and associated specimens as different species? This perspective overlooks that all the key arguments substantiating *Nanotyrannus* are equivocal, questionable, or simply flawed. That matters like skull textures and skull fusion are assigned contrasting significance for specimen maturity by different researchers implies either some degree or subjectivity in their interpretation, a need for greater investigation and documentation of these features, or both. Other matters are more demonstrably problematic. Holly Woodward and colleagues (2020), for instance, have demonstrated that *Tyrannosaurus* bone growth rings are inconsistently sized even into adulthood and that the decreasingly sized growth bands in *Nanotyrannus* bones are not reliable indicators of maturation (fig. 2.16C). Elsewhere, Longrich and Saitta's (2024) claim that the forelimb claws in adult *T. rex* were smaller than those of *Nanotyrannus* is simply erroneous: the hand claws of the fifteen-to-sixteen-year old purported *Nanotyrannus* specimen "Petey" are no larger, or are even slightly smaller, than the thumb claw known from the large adult "Sue" (fig. 2.16A–B). We might also note that statistical efforts to distinguish pygmy tyrant specimens from those of *Tyrannosaurus* find that large *Nanotyrannus* fossils stubbornly plot with *T. rex* adults even in datasets optimized to highlight their distinctions (Longrich and Saitta 2024).

We can, in addition, view this from the pro-juvenile *T. rex* angle. Drastic changes through growth seem to be the norm for tyrannosaurid species, not the exception (Rozhdestvensky 1965; Russell 1970; Currie 2003b; Woodward et al. 2020; Holtz 2021; Voris et al. 2021), just as they were for other dinosaurs (e.g., Varricchio 2011; Codron et al. 2013; Schroeder et al. 2021).

The seemingly aberrant anatomies of "*Nanotyrannus*" should thus not be considered unusual: to the contrary, all expectations are that juvenile king tyrants should look different from adults in most aspects of shape and proportion. Moreover, and perhaps most critically, *Nanotyrannus* proponents have yet to explain why, after a century of intense collecting and the recovery of hundreds of *T. rex* specimens, we've never found a single "*Nanotyrannus*" adult and no (P. Larson 2013) or virtually no (Longrich and Saitta 2024) genuine *Tyrannosaurus* juveniles. Why is one taxon, "*Nanotyrannus*," only known as immature specimens up to 6–7 m long, while *T. rex* is essentially unrepresented by specimens younger than subadults or adults of the same dimensions? This debate would be settled conclusively if a juvenile tyrannosaur that better matched *T. rex* or an unambiguously mature "*Nanotyrannus*" were discovered, but such specimens have hitherto remained elusive.

For now, the wider tyrannosaur community remains skeptical that "*Nanotyrannus*" is a valid taxon, despite recent efforts to revive the proposal (Elbein 2024a; Pare 2024). But with one of their particularly long, slender feet permanently in the paleontological rumor mill, perpetual hearsay exists of specimens that will finally prove the existence of "*Nanotyrannus*." It is thus unlikely that we've heard the last word on these pygmy tyrants. Indeed, at the time of writing, the study of Longrich and Saitta (2024) is relatively fresh and a formal response has not been published, but this will surely change before long. An important technicality is worth mentioning about any future "*Nanotyrannus*" developments before we move on, however. Even if the *concept* of a pygmy tyrant may continue, the *Nanotyrannus* label might not. As noted above, scientific names are attached to holotype specimens and if the consensus view holds that the *lancensis* holotype skull is truly a juvenile *Tyrannosaurus*, then the "*Nanotyrannus*" label will forever be sunk into *T. rex* and cannot be used for other specimens. The only way to overturn this would be to demonstrate that the *lancensis* skull belongs to a distinct, non-*Tyrannosaurus* species, giving any new "pygmy" specimens or future "*Nanotyrannus*" studies a job beyond simply authenticating the concept of a small-bodied tyrant: they also have to rescue the *lancensis* holotype from the bowels of *T. rex* synonymy. If they can't achieve this second goal, any valid small tyrant species would need a new taxonomic designation, and *Nanotyrannus*, in name at least, might finally be put to rest.

TYRANNOSAURUS "X"

Controversies about *T. rex* representing more than one species have not only concerned juvenile specimens. The rapid attainment of tyrannosaur remains from the 1980s onward led to a modern dataset comprising dozens of partial skeletons, mostly from adults. This is a large enough sample that trends or patterns have a chance of being teased out from statistical noise, potentially giving perspective on whether our *T. rex* inventories represent one species or several. In 2008(b), Peter Larson addressed this quandary in a discussion

of variation within *Tyrannosaurus*, noting that some adult *Tyrannosaurus* skeletons can be much stockier and more heavily built than others (fig. 2.17), while also documenting potentially consistent variation in their cranial sinuses, tooth counts, and the number of small "incisiform" teeth at the front of their lower jaw, where some specimens have one, others two (fig. 2.18B). Might, Larson proposed, these features represent sexual dimorphism (on which, see chapter 6) or taxonomic differences between two *Tyrannosaurus* species? The latter view was tentatively preferred, but Larson erred on the side of caution by not establishing a new taxon. Instead, he gave this potential second *Tyrannosaurus* species a nickname, "*Tyrannosaurus x*" (fig. 2.13).

This "*T. x*" proposal has not been developed further because subsequent statistical studies show that Larson's (2008b) purportedly consistent differences between *Tyrannosaurus* specimens are not, in actuality, consistent at all. Assessment of *T. rex* dental characteristics do not find predictable size variation in the second

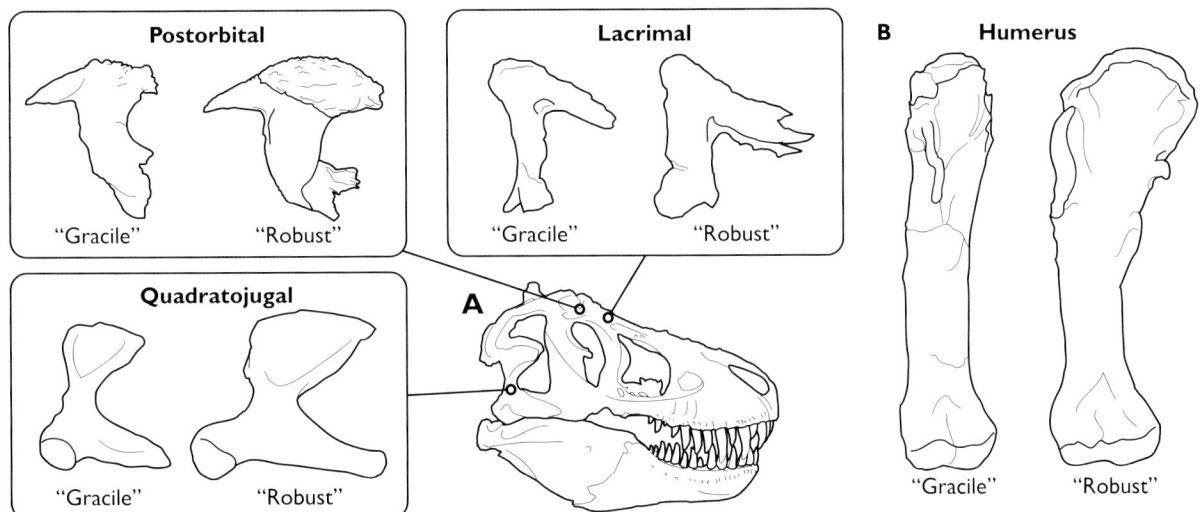

FIGURE 2.17

Bones suggested to exemplify "robust" and "gracile" *T. rex* morphologies from the skull (A) and forelimb (B). The concept of distinct robust and gracile *Tyrannosaurus* body plans remains in play among a minority of *Tyrannosaurus* researchers, but in-depth studies suggest that they likely represent different growth stages or individual variation rather than discrete body types. Drawings of individual bones adapted from Larson (1994).

WHAT, IN ACTUALITY, IS A *T. REX*?

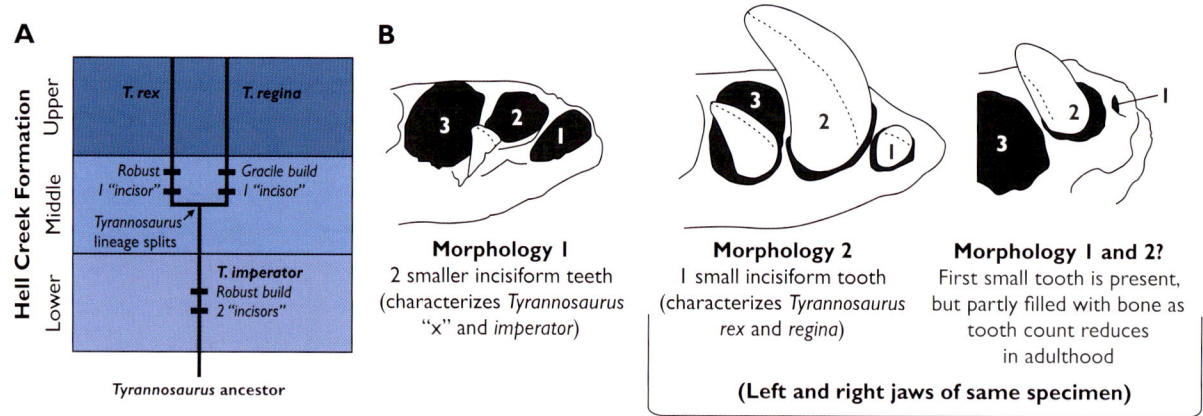

FIGURE 2.18

The *Tyrannosaurus* classification proposed by Paul et al. (2022). (A) The evolution and trait acquisition of *Tyrannosaurus*, from the geologically older *T. "imperator"* into two younger species, *T. rex* and *T. "regina"*; (B) proposed species distinctions in the lower jaw dentition of *Tyrannosaurus*, as argued by Paul et al. (2022) and Larson (2008b). Carr et al. (2022) suggest this distinction is flawed, noting the existence of an intermediate dental configuration in some specimens. (B) Based on Paul et al. (2022) and Carr et al. (2022).

lower jaw tooth, for instance (Smith 2005; Carr 2020; Carr et al. 2022). Instead, a transition from two to one smaller teeth at the tip of the lower jaw seems to occur through maturity, where the frontmost tooth socket was shortened, eventually leaving a remnant tooth "divot" filled with bone (fig. 2.18B; Carr et al. 2022). Other features of the skull said by Larson to split out *"T. x"* do not identify a distinct group of *T. rex* fossils in specimen-level assessments of *T. rex* relationships (Carr 2020).

The notion of "robust" and "gracile" *T. rex* morphs has also been criticized. This concept originated well before Larson's work on *"T. x,"* first being proposed and discussed as soon as king tyrant inventories grew beyond a handful of specimens (e.g., Paul 1988a; Carpenter 1990; Molnar 1991; P. Larson 1994; 2008b; Brochu 2003). These body morphs have been interpreted as having varying significance, such as taxonomic differences (P. Larson 2008b; Paul et al. 2022) or sexual dimorphism (P. Larson 1994, 2008b), but careful analyses of *Tyrannosaurus* bone metrics fail to find any reliable clustering of robust and gracile body types

(Brochu 2003; Horner and Padian 2004; Smith 2005; Mallon 2017; Carr 2020; Carr et al. 2022). Rather, the proportions of king tyrant limb bones, vertebrae, and skulls vary between similarly sized individuals, creating a spectrum of body types rather than discrete groups. Indeed, even dedicated efforts to quantify the schism between heavyweight and slender specimens (P. Larson 2008b; Paul et al. 2022) show little agreement on which specimens are "robust" and which are "gracile." So blurred are *Tyrannosaurus* proportions that even relatively complete specimens cannot be referred to either body type in some schemes (e.g., Paul et al. 2022). The identification of heavyset and slender forms, it seems, is essentially arbitrary, and lacks any compelling statistical backing.

Even if *T. rex* does not group into neat "robust" and "gracile" types, however, is it possible that we are overlooking some other significance of proportional variation within *T. rex* datasets? To address this, Thomas Carr and colleagues (2022) compared the variation of king tyrant bones to those of living species. They found that *Tyrannosaurus* was no more deviant in proportion than extant animals, and that they only seem unusually variable because we have a greater number of *T. rex* skeletons than other large theropods. The simplest explanation seems to be that robustness is chiefly a product of maturation and individual variation, and another consequence of the radical changes occurring between *T. rex* juvenility and adulthood (chapter 4; Erickson et al. 2004; Hutchinson et al. 2011; Carr 2020). That the degree of king tyrant skeletal robustness tells us more about the age and skeletal maturity of *T. rex* than taxonomic distinction accounts for some of the skepticism that "*T. x*" has encountered, although this not stopped others from continuing to claim that "robust" and "gracile" builds should be considered in attempts to revise *Tyrannosaurus* taxonomy.

TYRANNOSAURUS IMPERATOR, REX AND REGINA

Following decades of authors treading softly around the idea of more *Tyrannosaurus* species, not formally erecting new names until robust evidence for a second taxon comes to light, a team of authors led by Gregory S. Paul made a case for not just one, but *two* cryptic *Tyrannosaurus* species existing within existing *T. rex* datasets in 2022 (figs. 2.13, 2.18). Paul et al. used characters familiar to us from Larson (2008b; skeletal robustness, and the number of smaller incisiform teeth on the lower jaw) and approximated geological ages of specimens to identify an evolving *Tyrannosaurus* lineage. They proposed that an older, robust form of *Tyrannosaurus*, named *T. imperator*, evolved into two different, younger taxa, the robust "true" *T. rex* and the more gracile *T. regina*. Along with robustness, the front teeth of their lower jaws also varied, *imperator* sporting two slender incisiform lower teeth and *rex* and *regina* having one. *T. rex* (sensu Paul et al. 2022) and *T. regina* coexisted in western North America, making them a rare instance of two large tyrannosaurs inhabiting the same environment (chapters 5 and 7; Holtz 2021).

Carving *T. rex* into three species is the boldest claim yet in our efforts to interpret the variation of king tyrant fossils. A direct response to Paul et al. was penned by Thomas Carr and colleagues (Carr et al. 2022), within which several major flaws were identified in the *imperator-rex-regina* scheme. Not only are the characters pertaining to robustness and lower jaw tooth configuration of questionable taxonomic utility and statistically doubtful, having already been examined by prior studies (see above), but the species diagnoses for *T. imperator*, *rex*, and *regina* are problematically vague. From animal skeletons thought to have thousands of variable features (Carr 2020), Paul et al. only provided three features to distinguish their new taxa. Two pertained to the *general* robustness of the skeleton (e.g., whether it is a "robust" or "gracile" morph, an idea no better substantiated by Paul et al. than any previous researchers), and the *usual* number of incisiform teeth at the front of the lower jaw. That these features are only *sometimes* present makes them poor distinguishing features, and leaves the only consistent character relating to a ratio of thigh bone length and circumference. This is said to be above 2.4 in the robust *rex* and *imperator*, but less than 2.4 in gracile *regina*. No convincing clustering of specimens can be found based on this metric, however (Carr et al. 2022), and the problematic characterization of *imperator, rex*, and *regina* is demonstrated by over 50 percent of *Tyrannosaurus* specimens analyzed by Paul et al. (2022) being classed as *incertae sedis*: in other words, of "uncertain placement." Among these are excellent, relatively complete remains like the American Museum of Natural History specimen 5027 (fig. 1.18). It is not uncommon for animal species, living and extinct, to be distinguished by minutiae of anatomy that can leave specimens in a frustrating taxonomic limbo, but a classification system that cannot assign taxonomic identities to substantial dinosaur skeletons is clearly of questionable utility. Further issues with this study have been identified, such as a lack of precise dating to support the evolutionary proposal of a robust species dividing into two "gracile" and "robust" taxa, and that specimen-level examinations of *Tyrannosaurus* variation do not recover the fossils identified as distinct species by Paul et al. in discrete clusters (Carr 2020; Carr et al. 2022). It will suffice to say here that the existence of "*T. imperator*" and "*T. regina*" has not been viewed as convincing by tyrannosaur researchers, and they seem destined to join "*Dinotyrannus megagracilis*," "*Stygivenator molnari*," and "*Nanotyrannus lancensis*" on the *Tyrannosaurus rex* synonymy list.

TYRANNOSAURUS MCRAEENSIS

The most recent (at the time of writing) effort to erect a new *Tyrannosaurus* species is far more straightforward than some of the *T. rex* taxonomy studies that immediately preceded it. Rather than attempting to carve king tyrant inventories into several new taxa or resurrect controversial genera from the synonym waste bin, Sebastian G. Dalman and colleagues (2024) highlighted the distinctive properties of a single T. rex specimen from New Mexico. The bulk of this specimen, comprising a partial skull, lower jaw, and tail elements, has been known since 1983 and has mostly been regarded as *T. rex* (Lozinsky et al. 1984; Gillette et al. 1986; Wolberg et al. 1986; Carr and Williamson 2000), although some have long suspected that it represented a new genus (Lehman and Carpenter 1990). Dalman and colleagues agree with this latter assessment, considering the dentition, skull ornamentation, skull joint shape, and the upward curving, posteriorly slender lower jaw of taxonomic significance (fig. 2.19). The specimen's provenance from relatively old rocks has also drawn attention. While the exact age of the specimen is difficult to confirm (see chapter 5), Dalman et al. estimate it to be over 70 million years old, perhaps 2 or 3 million years older than verified *T. rex* remains. Collectively, these make a case that the specimen is a new species: *Tyrannosaurus mcraeensis*, named after its geological home, the McRae Group.

T. mcraeensis is potentially an important taxon, throwing useful light on the final stages of tyrannosaurid evolution. As potentially the oldest member of the *Tyrannosaurus* line, it stretches the longevity of the *Tyrannosaurus* genus to c. 4 million years and demonstrates that some aspects of the *T. rex* body plan were already developed well before the start of the *T. rex* fossil record. The McRae tyrant holotype is, alas, too fragmentary to reveal much about early *Tyrannosaurus* anatomy, except perhaps that the shallow lower jaw might infer a smaller region for jaw muscle attachment and, thus, a slightly lessened bite strength. More usefully, it provides data in constraining where and when the *Tyrannosaurus* genus evolved. Dalman et al. (2024) posit that *mcraeensis* vouches for an American origin for *Tyrannosaurus* in the biogeographic debate mentioned above, although they also note that this result depends on the analytical specifics used to establish relationships between tyrannosaurins. We are still, as noted earlier in this chapter, awaiting more late-stage tyrannosaurid specimens, particularly from Asia, to resolve where *T. rex* evolved.

Accepting these implications from *mcraeensis* are, of course, contingent on it withstanding the gauntlet of scientific scrutiny that befalls all newly described dinosaur species. As this chapter is finalized in summer 2024, insufficient time has passed to assess whether this will be so. Responses from tyrannosaurid experts in press reports promoting *T. mcraeensis* have ranged from relatively warm (e.g., Elbein 2024b) to skeptical (Dunham 2024). Historic considerations of the *T. mcraeensis* holotype might predict that disagreement awaits: we should not forget that researchers have already expressed opinions that it does, or does not, represent *T. rex* (Lehman and Carpenter 1990 vs. Carr and Williamson 2000). Critical focus is sure to focus, once again, on whether the *mcraeensis* fossils are genuine anatomical outliers from other *T. rex*, as well as the uncertainty of their geological age. As we will detail in chapter 5, estimating the antiquity of *Tyrannosaurus* fossils is difficult and Dalman et al. (2024) were forced to use imperfect methods to predict a 70-million-year age for *mcraeensis*. There may, in short, be plenty of discussion about *mcraeensis* on the horizon, and, given how rapidly *Tyrannosaurus* research moves, developments may have already taken place by the time you read this.

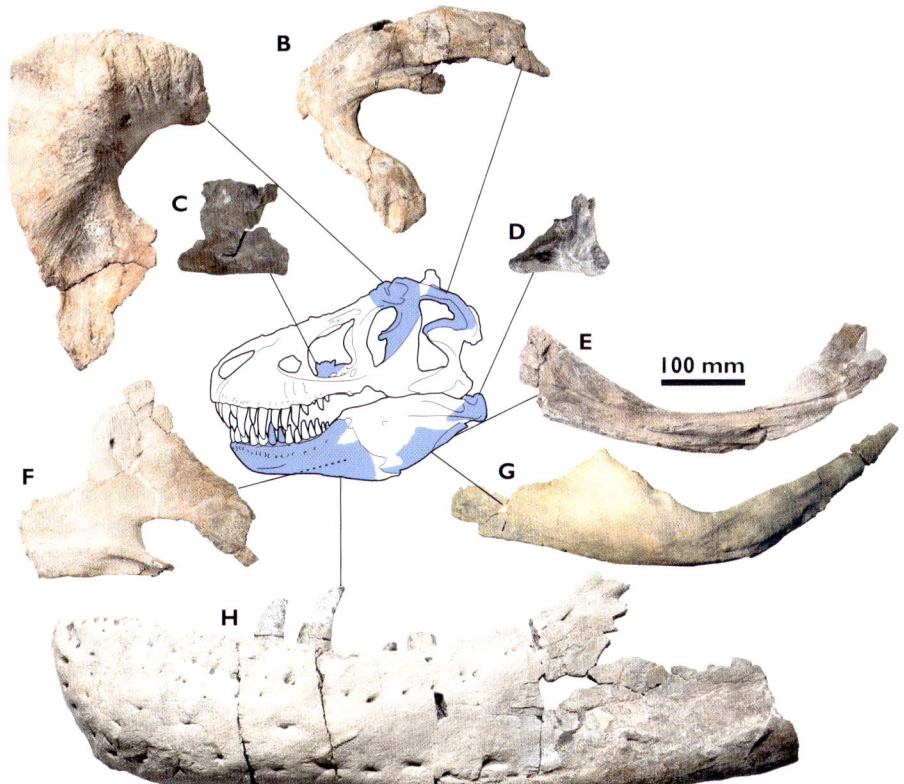

FIGURE 2.19

The fragmentary holotype skull of the McRae tyrant, *Tyrannosaurus mcraeensis*, mapped onto a schematic *T. rex* skull to show their location in life. (A) Postorbital; (B) squamosal; (C) palatine; (D) articular; (E) prearticular, (F) splenial; (G) angular; (H) dentary. Photographs from Dalman et al. (2024).

With controversy surrounding every modern taxonomic proposal applied to *T. rex*, will we ever see king tyrants satisfactorily segregated into two or more species? The idea of multiple *Tyrannosaurus* species living alongside one another is not, in itself, problematic or heretical. As we have seen, researchers have been toying with the idea for decades because *Tyrannosaurus* specimens have great anatomical, geographic, and, possibly, geological range, all of which could evidence more than one species. The challenge, however, is identifying truly distinguished specimens within *T. rex* inventories. That some degree of opinion and subjectivity have crept into efforts to break *T. rex* apart is demonstrated by both P. Larson (2008b) and Paul et al. (2022) citing nearly identical distinguishing characters for "*T. x*," *imperator*, and *regina*, but their respective studies drawing contradictory conclusions about which specimens belong to their proposed taxa. Throw in some uncertainty about

the exact age of specimens, and we have cause to wonder whether our datasets are currently good enough to tease apart *T. rex* remains, or whether we need a substantial increase in specimen numbers, ideally with good locality and geological range data, to see the metaphorical forest for the trees.

But we must also be wary of approaching this question with a predetermined answer and of thinking that, one day, new discoveries *will* validate proposals of multiple *Tyrannosaurus* species. As good scientists, we must also entertain the possibility that we already have the "right" answer, or at least the best and most defensible one: that *Tyrannosaurus rex* really was a single dinosaur species. Time will tell if our attempts to group king tyrant specimens into more granular categories says more about our innate tendency to look for patterns and clusters in large, noisy datasets than the biological reality of 66 million years ago.

INSIDE AND OUT

3

HOW *TYRANNOSAURUS REX* is viewed through the lens of biological classification tells us a lot about their evolution and variation but offers a limited perspective on what king tyrants were like as animals. Appreciation of this starts with an overview of their anatomy: their bones and muscles, their internal organs, and, finally, the skin that covered their bodies.

THE ESSENCE OF FORM: BONES AND MUSCLES

ENCOUNTERS WITH *T. rex* anatomy are arresting no matter how familiar you may be with them. It is difficult to accurately remember how truly *enormous* adult *T. rex* were, an experience heightened by visiting especially big individuals like "Sue" or "Trix." Their ludicrous proportions are also striking. The great pelvic girdle and robust legs are juxtaposed with slender ribs and waisted vertebral processes, while heavy-looking elements like vertebrae and skulls are actually, on closer inspection, more like frameworks of bony sheets. Their structural ethos is quite different from that of a giant mammal, like a whale or elephant, which have a solid, robust quality that almost reminds one of wood sculpture. Dinosaur skeletons are metalwork-like mosaics of bony columns, sheets, and struts, and *Tyrannosaurus* is a maximized expression of this evolutionary approach to constructing animal bodies.

For all their distinctiveness, the basics of *Tyrannosaurus* appearance are clouded in many minds by poor, overly monsterized or generic depictions in popular culture. To understand what *T. rex* really looked like,

and to appreciate how it functioned as a real animal, we have to cast these interpretations aside and focus our attention on raw king tyrant data: their fossils. We can learn more from these bones than the basic proportions of *Tyrannosaurus* bodies. Bones anchor muscles, and these frequently leave marks that allow anatomists to predict their extent and size. A well-preserved *T. rex* skeleton gives a fairly good idea of its entire musculoskeletal system, as well as other details, including unexpected ones, such as details of the respiratory system and skin (fig. 3.1). Osteology does not provide us with a complete picture of a living animal, but it goes a surprising distance toward this goal.

Dealing with fossil bones, however, rather than the skeleton of a freshly dissected animal, introduces challenges to this process. Fossils are not merely buried remains, but are once-organic structures that have been chemically and physically altered by the fossilization process. Entombment in sediment can skew, crush, distort, and color bones in distinct ways (fig. 3.2), and not all preservation is of the same quality. Some bones

---

FIGURE 3.1

There are two ways to experience the detailed anatomy of king tyrants. One is demonstrated by this unfortunate juvenile *Edmontosaurus annectens*; the other involves study of their fossil remains. Happily for us, it's the latter route that we'll be using here.

FIGURE 3.2

The crushed skull of "Sue." Deformation affects most *T. rex* skulls, contributing to the variation in shape along with biological factors. Some degree of reconstruction is thus required to "correct" most skulls for display or use as artistic reference.

are far more delicate or coarsely preserved than others, even if excavated and prepared with care. These factors, along with variation within the species, are why no two *T. rex* specimens look exactly alike.

Discussions of anatomy can be intimidating because they are associated with a lot of technical jargon. With perhaps three hundred bones in a *Tyrannosaurus* skeleton (N. Larson 2008), the list of names and features is extensive for their osteology alone, and that's before introducing musculature and other soft tissues! We will attempt to cover some of the more useful terms here, but, if this sounds concerning, it may help to know that the fundamental musculoskeletal structure of *T. rex* is the same as that of all tetrapods, including ourselves. Thus, a lot of osteological terminology familiar to us from everyday life (such as the humerus and femur, the upper bones of the arms and legs; the metatarsals, the long bones of the foot that soccer players often injure) also apply to dinosaurs (fig. 3.3). Moreover, we don't need to learn every anatomical component to deepen our understanding of fossil animals. Getting to grips with even just a handful of features will not only aid our

later discussions of *T. rex* natural history, but open up the wider world of knowledge for budding enthusiasts of vertebrate paleontology and zoology. Especially keen anatomists will find additional osteological and myological features labeled in this chapter's illustrations.

The osteology of *T. rex* is very well documented through a series of technical papers and monographs (e.g., Osborn 1906, 1912, 1917a; Molnar 1991; Carr 1999; P. Larson 2008a; Witmer and Ridgely 2010), of which special mention should be made of Christopher Brochu's 2003 bone-by-bone description of the "Sue" *Tyrannosaurus* specimen. Virtually every bone of the *Tyrannosaurus* body has been discovered, with only a complete hand, the bones supporting the tongue, and the tail tip remaining unreported in scientific literature. Details of predicted musculature, too, are well studied, with particular emphasis placed on their jaws, necks, arms, and legs (e.g., Hutchinson et al. 2005, 2011; Carrano and Hutchinson 2002; Carpenter and Smith 2001; Snively and Russell 2007a, b; Molnar 2008; Lipkin and Carpenter 2008; Persons and Currie 2011; Gignac and Erickson 2017). Much of the following overview stems

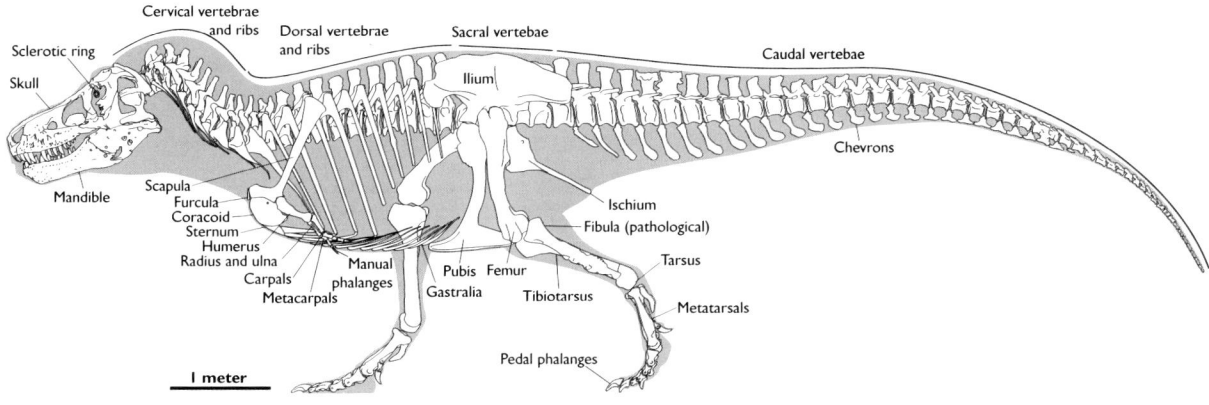

FIGURE 3.3

Labeled osteology of the "Sue" specimen of *Tyrannosaurus rex*. Gray bones indicate elements unknown in this specimen.

from these works. As we will see in chapter 4, the unusual amount of change experienced by *Tyrannosaurus* throughout growth means that single specimens represent snapshots of *T. rex* anatomy from specific life stages, so the proportions of juveniles were very different from those of mature adults (Carpenter 1990; Carr and Williamson 2004; P. Larson 2008b; Carr 2020; Carr et al. 2022). We will comment on some of this here, but further discussions of these growth stages can be found in the next chapter.

THE HEAD

The fossils of adult *Tyrannosaurus* skulls are remarkable objects (fig. 3.4). Many excellent, complete and uncrushed examples are known, and we have been able to document every bone of the skull and lower jaw in detail (Osborn 1912; Bakker et al. 1988; Carr 1999; Brochu 2003; P. Larson 2008a). As with human skulls, *T. rex* crania were made from a symmetrical series of bones that fused together, to a greater or lesser extent, throughout life. The skulls of some unfused specimens have been found in disarticulated states, allowing us to see their individual cranial bones from every angle.

Technologies such as CT scanning have permitted even deeper analysis, revealing internal bony structures such as sinuses and nerve pathways without having to cut into, and thus destroy, these amazing fossils (Brochu 2003; Witmer and Ridgely 2010; Kawabe and Hattori 2022; Bouabdellah et al. 2022).

Up to 1.4 m long (Brochu 2003), adult *Tyrannosaurus* skulls were enormous, with deep jaws and a robust constitution. They had a distinctive shape that differs markedly from the boxy, cuboid representations common to popular culture. Compared to other theropods, even their close relative *Tarbosaurus*, the skull of *T. rex* was very wide (fig. 2.11), with the rear half of the skull especially so. Large skulls approach one meter across, making the skull almost two-thirds wide as long and giving them the shape of an overweight "T" when viewed from above (fig. 3.4B). This configuration was present even in juvenile *Tyrannosaurus*, although their skulls were neither as broad nor robust as those of adults (fig. 2.14; Carr 1999, 2020; Witmer and Ridgely 2010). The top of the snout was narrower than the base, giving *T. rex* a characteristically "pinched" rostrum in front of a boxy braincase region.

Taking the major regions of the skull in turn, the snout was formed of three bones. Two bore teeth: the

INSIDE AND OUT

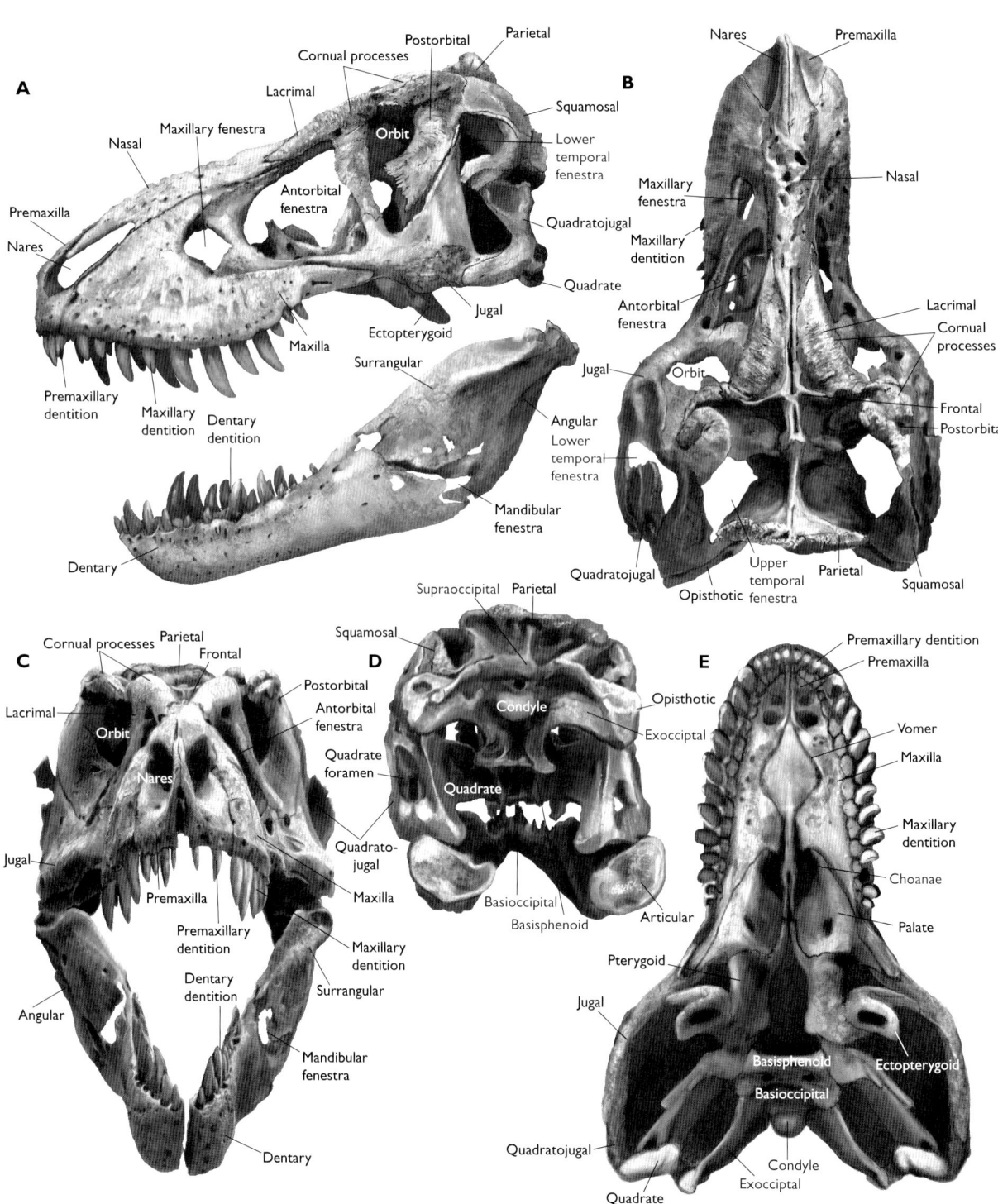

premaxilla, the tooth-bearing bone at the front of the skull; and the maxilla, one of the largest skull bones and the main tooth bearer of the upper jaw. This formed the side of the snout, and was capped by the final component of the rostrum: the paired nasals. All these bones bracketed the frontmost skull opening, the nares, which housed the nostrils. The snouts of all tyrannosauroids, except very young juveniles, terminated with a blunt, subvertical face rather than tapering to a point. The nasal bones were fused along their midlines, another tyrannosauroid trait. This, combined with the vaulted cross section of the nasals themselves, conferred strength to the upper jaw during biting (see chapter 4). A number of large bony openings, fenestrae, perforated the maxilla and are of disputed function. The largest of these is known as the antorbital fenestra, that feature common to (or at least ancestral for) all archosaurs (chapter 2). This opening has been linked with bulging, forward-reaching jaw muscles (e.g., Paul 1988a; Molnar 2008) but others interpret it as part of an extensive air-sac system (Witmer 1997; Witmer and Ridgely 2008; also see chapter 4). The latter interpretation, which is well grounded in studies of living archosaurs and the more likely of these two hypotheses, sees these fenestrae and the surrounding antorbital fossa, a shallow depression in the side of the maxilla, as occupied by pneumatic tissues: air sacs that lightened the head (on which, see below). The snout also had many smaller openings, known as foramina, providing passageways for blood vessels and nerves. The most conspicuous of these form a row just above the teeth.

Under the rostrum were the tooth sockets, with *T. rex* teeth largely constrained to the front half of the skull (fig. 3.4A, E). Each premaxilla had four teeth but the number in the maxilla varied with age. Young king tyrants had fifteen, rising to sixteen in slightly older juveniles, before falling to twelve or eleven teeth in adulthood (Carr 2020). This change in tooth count was associated with a

deepening of the maxilla to accommodate increasingly deep tooth roots (fig. 3.5A, D), such that juvenile snouts were much shallower than those of adults. The underside of the rostrum comprised a shallow cavity walled in by the premaxillae and maxillae. In adults, the roof of the mouth was solid rather than a series of bony struts, as it was in juvenile king tyrants (Carr 1999) and other theropods (fig. 3.4E, 3.5B–C). The underside of theropod skulls were complex and are of interest to specialists, but we won't dwell on their structure here.

Behind the antorbital region was a variably shaped opening: the orbit, or eye socket. In juvenile king tyrants and most tyrannosauroids, this was a round or oval opening (fig. 2.14) but, in mature *T. rex*, the rear margin was invaded by a sheet of bone that altered the orbit into a "B" or keyhole shape. The eyeball itself was positioned toward the top of this aperture, with the lower portion occupied at least partly by an air sac (Witmer and Ridgely 2008). The orbits faced forward on account of their position on the leading edge of the expanded region of the posterior skull, with implications for *T. rex* vision (see chapter 4). Although none have been reported from *T. rex* specimens yet, the eye would have been supported in life with a sclerotic ring: a ring of plate-like bones that surround the front of the eye in most vertebrates, including reptiles and birds.

Like the snout, the orbital region was made of a number of bones. The frontals were situated on the top of the skull and, like the nasals, they were fused along their midlines. They were situated between bones that bracket the front of the eye socket: the lacrimals. These strut-like elements separated the orbit from the antorbital fenestra and changed shape markedly through growth, bulging forward in adults. The top of the lacrimal had a low swelling, a cornual process, that formed the forward-most part of the ornament above *T. rex* eyes. Like the rest of the lacrimal, this process changed shape with age, being relatively small, narrow, and

INSIDE AND OUT

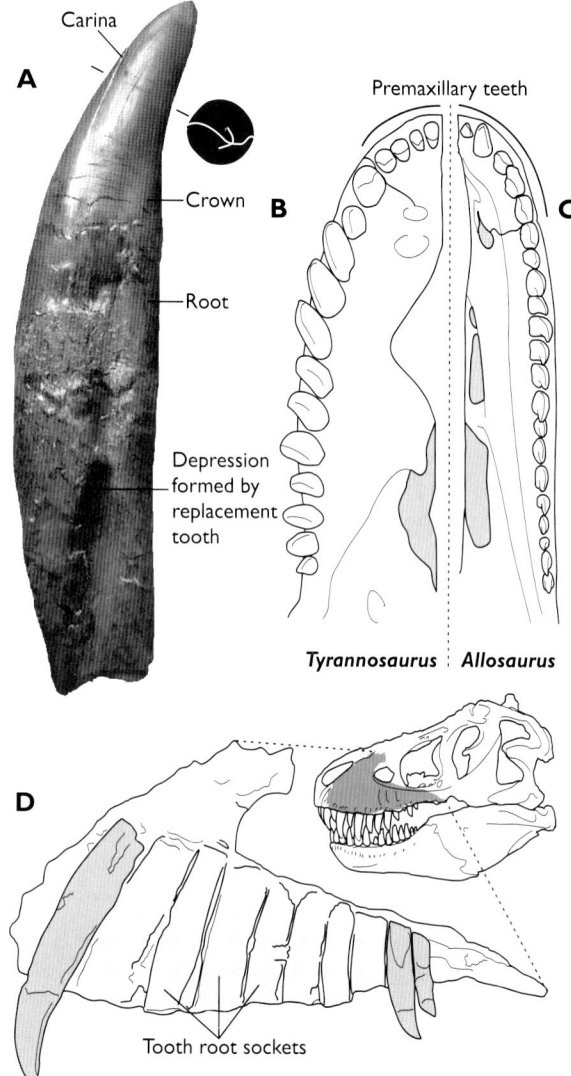

Carina

A

Crown

Root

Depression
formed by
replacement
tooth

Premaxillary teeth

B

C

Tyrannosaurus | *Allosaurus*

D

Tooth root sockets

FIGURE 3.5

The famously enormous teeth of *Tyrannosaurus rex*. (A) Large dentary tooth showing the enamel-covered crown and enormous tooth root; adjacent diagram shows the distribution of carinae on the front and rear crown surfaces; (B) comparison between the broad upper jaw of *Tyrannosaurus*, where the premaxillary teeth are aligned to form a broad scraping or chiseling surface, and (C) the more typically theropodan jaw of *Allosaurus*, where the premaxillary teeth form a cutting surface continuous with the maxillary dentition; (D) depth of the tooth sockets in *Tyrannosaurus* maxillae: similarly enormous dental alveoli occur in the premaxilla and dentary.

and somewhat outward during maturation, as well as the attachment of an adjacent osteoderm: a bone that grows within skin (Molnar 1991; Brochu 2003; P. Larson 2008a; Carr 2020). The formation of this dermal bone was one of the last major changes to happen in the growth of *T. rex* skulls, appearing as a thin sheet of bone in young adults (fig. 3.6; P. Larson 2008b) and developing into a rounded, overhanging hemisphere in more mature specimens. A second, smaller osteoderm grew between the two cornual processes to complete the ornament arrangement (fig. 3.6). This initially resembled a small horn but flattened as it fused into the ornament complex in older animals. Immature king tyrant specimens have yet to be reported with these osteoderms (it seems possible that they were present as small elements, but not preserved with their fossils), and they are sometimes either unfused or not preserved in fossils of adults (Carr 2020), accounting for some of the variation we see in different *T. rex* skull specimens.

Beneath the orbit was the jugal, a broad bone that formed the "cheek" of the *T. rex* skull. These were very large, complex elements, and their flared, sweeping shape was responsible for giving *T. rex* their laterally expanded rear skull region. In juveniles this curve was relatively gentle, but, in adults, it was more pronounced. How angular this region became is differently interpreted (e.g., Osborn 1912; Paul 1988a, 2016 vs. Molnar

crest-like in juveniles (Carr and Williamson 2004; Carr 2020) and becoming a swollen, low boss that overhung the side of the skull in adults. This feature contributed to mature *T. rex* having scowling "eyebrows," a look that was completed with the aid of another bone, the postorbital. This formed the rear margin of the orbit and was also capped with a cornual process. The postorbital also morphed throughout growth, the lower portion growing into the eye socket to give adult *T. rex* their distinctive "keyhole" or "B"-shaped orbit. The postorbital cornual process was more prominent than that of the lacrimal, a consequence of it expanding upward

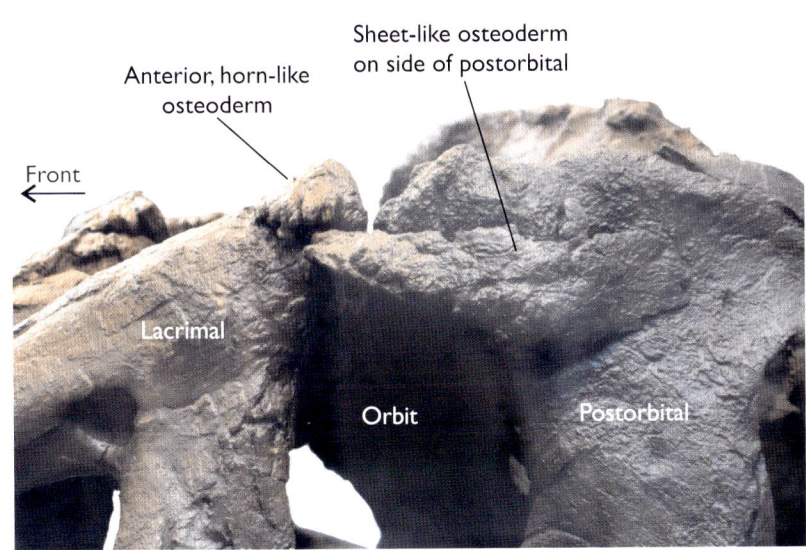

Anterior, horn-like osteoderm

Sheet-like osteoderm on side of postorbital

Front

Lacrimal

Orbit

Postorbital

FIGURE 3.6

Unfused osteoderms forming the cornual processes in a young adult *T. rex* ("Stan"). The horn-like structure between the lacrimal and postorbital cornual structures flattened and expanded with maturity, while the sheet-like osteoderm became more bulbous and fused to the side and top of the postorbital.

1991; Brochu 2003), and it seems possible that both gentler and more pronounced angles could be correct, given that the *T. rex* skull widened with age. The underside of the skull from the jugal region rearward was relatively open and complex, with deep cavities and thin scaffolds of bone contrasting against the flat bony sheets spanning the underside of the snout. Two very prominent elements extended downward from the skull between the jugals: the ectopterygoids. These were wrapped in jaw muscles that extended through the large openings in this area (see chapter 4). Much of this musculature originated at the back of the skull (fig. 3.7).

Two further sets of openings were found at the rear of the skull: the temporal fenestrae. One set, the upper, occured on the top of the skull, while the lower set occupied the side. Most reptiles have these paired temporal openings, and they differ from mammals, which have just one, but in both groups they serve the same function: housing the adductor muscles that pull our jaws shut. The temporal fenestrae were unusually big and cavernous in *T. rex*, as was everything related to jaw musculature, and the lower set changed shape with maturation. As with the orbit, juveniles had broadly open lower temporal fenestrae, but an invading growth of bone made them "B"-shaped in adults. The jugal,

postorbital, and frontals bracketed the forward region of the temporal openings, with the latter midline element shaped into a deep crest where it separated the two sets of bulging jaw muscles. Behind the frontal, and forming the topmost-rear portion of the skull, was the parietal. These bones, also fused along their midline, defined the back of the upper temporal fenestrae and formed a steep wall that projected well above the surrounding skull bones. This structure formed the upper rear surface of the skull and has deep, rearward-facing cavities for the insertion of powerful neck muscles and ligaments (figs. 3.4D, 3.7). Deeper within the temporal openings, the frontals and parietals comprised the top of the braincase, which was squeezed between the jaw muscles in life. Laterally, the upper rear of the skull was bracketed by the squamosal, which extended downward to contact its lower counterpart, the quadratojugal. This somewhat obscured the quadrate, which formed the majority of the jaw joint, with a minor contribution from its quadratojugal neighbor.

The back of the *Tyrannosaurus* skull was a broad, complex surface that connected the enormous, stout neck with the head. This region was well fused, at least in adults, such that determining where one bone stops and another begins can be challenging. The mid-region was marked by two structures: the foramen magnum,

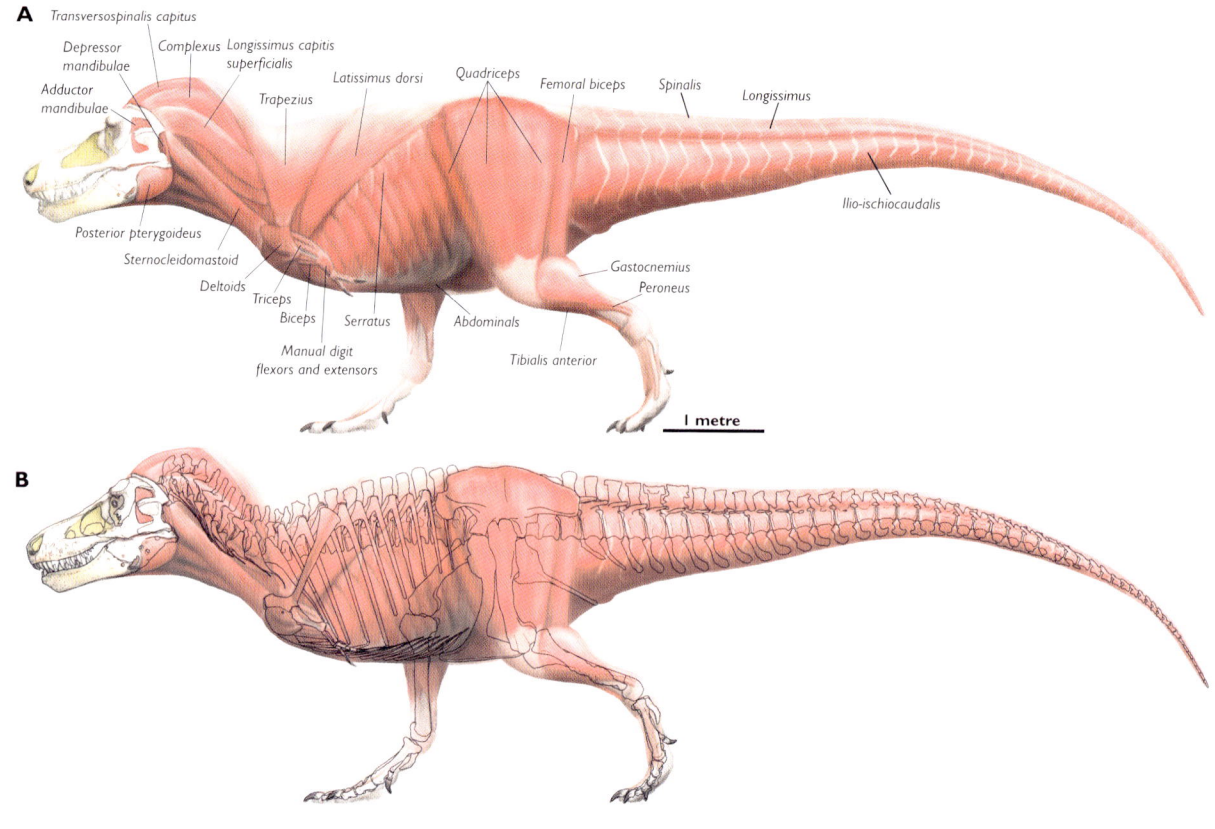

FIGURE 3.7

(A) Myological study of *Tyrannosaurus rex* based on "Sue," with major muscles labeled; (B) with embedded skeletal reconstruction to show relationship between osteology and muscle structure.

the opening through which the spinal cord connected with the brain, and the occipital condyle, the ball-like projection that articulated with the first neck vertebra. Great bony wings, the exoccipitals, projected sideways from these to receive powerful muscles that moved the head sideways, while the supraoccipital, a relatively narrow, swollen bone, occupied the region immediately above the foramen magnum. The occipital condyle represented the upper part of the basioccipital, a bone that flared outward as it extended downward. The basioccipital bore deep recesses from the insertion of specific neck muscles that were unusually large in *T. rex*, even compared to other tyrannosaurids

(Bakker et al. 1988). This element, in turn, connected with the basisphenoid, a complex bone characterized by concave, perforated surfaces. It projected forward to meet the complex of bones within and under the *T. rex* skull, including those that formed the lower portion of the braincase. We will not detail these here (interested readers should check out Osborn 1912, Molnar 1991, and Brochu 2003 for more details), but it would be remiss not to mention the stapes, a tiny bony rod of bone that, even in the biggest *T. rex*, was just 2 mm in diameter. Forming part of the ear, these elements are very rare in dinosaur skulls because their preservation potential is low, and their size makes them easily overlooked by researchers. It was only a comment by an observant preparator that informed Christopher Brochu, conducting his otherwise exhaustive scanning and descriptive work on the "Sue" *T. rex*, about missing the tiny *Tyrannosaurus* ear bone (Brochu 2003).

Like the skull, the lower jaw, or mandible, comprised multiple bones (fig. 3.4). Although of a similar length to the skull, the lower jaw was narrower and it nestled within the teeth of the upper jaw when the mouth was closed, such that there was no significant contact between the teeth in life (Gignac and Erickson 2017). As with the upper jaw, *T. rex* mandibles changed markedly through growth, deepening and becoming more robust with age (Carr 1999, 2020; Currie 2003a). A tall rear region, comprising the surangular (above) and angular bones (below) was surprisingly thin, and *T. rex* mandibles are often damaged in this area because of this delicacy. In life, however, the back of the lower jaw would have been thick with muscle (fig. 3.7). Adductor musculature descended from the temporal region to insert on the upper and inner surface of the lower jaw, while another set of muscles, the pterygoideus, ran from the antorbital region to attach on the lower and, possibly, outer surfaces (Snively and Russell 2007b; Molnar 2008; Bates and Falkingham 2012, 2018; Gignac and Erickson 2017). This formed a swelling around the back of the jaw that likely obscured the bony contour of the mandible in life. The depressor mandibulae, or jaw-opening muscle, was much smaller and ran from the posterior tip of the squamosal to a large, dish-shaped attachment scar on the back of the lower jaw. The largest mandibular element was the dentary, which, as its name implies, housed the mandibular teeth. It occupied the front two-thirds of the mandible length and a roughened area at the tip records the cartilaginous connection between each jaw. As with the maxilla, the number of dentary teeth changed with age: juveniles had sixteen to seventeen, decreasing to thirteen to fourteen in subadults and, finally, to thirteen in fully mature specimens (Carr 2020).

Among the most famous, but also most unusual features of *Tyrannosaurus* are their teeth (fig. 3.5). Ranging from a total of seventy-two to seventy-four teeth as juveniles to fifty-eight to sixty teeth as adults (Carr 2020), king tyrant teeth had long roots that were socketed deeply into their jaw bones, with the crown being the emergent, visible component. As noted in the detailed overview of *T. rex* dentition by Joshua B. Smith (2005), their teeth share some basic features with those of other theropods, such as having somewhat recurved crowns that, in cross section, were longer than wide. They also possessed carinae, serrated ridges that assisted with cutting flesh (fig. 3.5A). These features characterize "ziphodont" dentitions and they often see theropod teeth, including those of *Tyrannosaurus*, compared to steak knives (e.g., Horner 1994). The dentition of *Tyrannosaurus* only partly meets this analogy, however. Their teeth were highly specialized compared to those of other theropods, and none were especially knife-like, being far more robust and generally having straighter, swollen crowns. A "classic" *T. rex* tooth crown was less like a knife than it was a lance, or the huge, spike-like chisels we use to demolish concrete. Rather than ziphodont, this morphology is termed "incrassate."

Tyrannosaurus dentitions weren't merely brutish spikes, however (Smith 2005). They were heterodont, in that they had several tooth types adapted to specific functions, instead of "homodont," where only one tooth type exists. Among theropods, even other tyrant dinosaurs, *T. rex* teeth were especially differentiated, although this comes with a caveat: no dinosaur, even *Tyrannosaurus*, exhibited the degree of heterodonty seen in the especially complex, multi-cusped dentition of mammals, who remain the champions of tooth differentiation among vertebrates. Nevertheless, the different regions of king tyrant dentition were shaped to maximize effectiveness at slicing, scraping, and crushing animal body tissues (fig. 3.5B; Smith 2005). The premaxillary crowns had a "D"-shaped cross section where the flat surface faced rearward and the corners bore serrated carinae. Contrasting with most theropods, where the premaxillary teeth extended some way down the length of the jaw, *T. rex* premaxillary teeth were arranged in a gentle arc, creating a cookie cutter–like edge at the front of the mouth (fig. 3.5B–C). This condition recalls that of mammalian incisors and the premaxillary teeth of tyrannosaurids are appropriately described as "incisiform." They probably served a similar function to our mammalian incisors, too: cutting into and stripping soft foodstuffs, as is useful when pulling flesh from bone (see chapter 6). The lower jaw lacked truly incisiform dentition, but the first dentary tooth had some incisor-like qualities in adults (Carr 2020).

The first maxillary and secondary dentary teeth were transitional forms between these incisor-like morphologies and the incrassate, spike-like teeth that made up the bulk of *Tyrannosaurus* dentition. These teeth were at their largest in the front half of the jaw, and, including their roots, they grew to almost 30 cm long, of which nearly 12 cm was crown (fig. 3.5A, D; Smith 2005). Their carinae were generally positioned along the front and rear crown surfaces, although the rearward crest sometimes curved a little around the outer tooth surface. As adults, the large maxillary and dentary crowns were almost circular in cross section. All tyrannosaurids possessed this feature to some extent, but the teeth of *T. rex* embodied the incrassate condition more than any other member of the group. The small teeth at the rear of the toothrow were somewhat more typically theropodan and ziphodont, and may have primarily cut, rather than punctured, animal flesh. This function may have characterized the teeth of juveniles as well, as it was only in later life that *Tyrannosaurus* fully developed their long-rooted, swollen-crowned dentition, a change that correlated with increased bite force (Carr 2020; Johnson-Ransom et al. 2024).

Like all dinosaurs, *T. rex* continually grew new teeth, replacing old, worn dentition with fresh, sharp, new ones. Their dental replacement rate can be calculated using growth lines (so-called "von Ebner lines") created by incremental daily increases in tooth size (Erickson 1996). This value differs substantially between theropod groups, with some churning through teeth every 100 days or fewer (D'Emic et al. 2019) but others taking a year or more (Erickson 1996). *Tyrannosaurus* tooth replacement rates changed throughout their lives, being

quicker in juveniles than adults, but it was consistently slower than that of other theropods at every growth stage (Erickson 1996). Average tooth growth rates in juveniles and subadults took nine months to a year, and this slowed to over two years in large adults: 777 days. This has been interpreted as the large size, strength, and robustness of *T. rex* teeth taking longer to develop; a "quality over quantity" approach to tooth replacement demanded by high bite forces and specialized tyrannosaur feeding habits (chapter 6).

VERTEBRAL COLUMN AND TORSO

To look at a *Tyrannosaurus* is to first notice the great head, the small arms, and the powerful legs, but these units were held together by a tremendous vertebral column that should not be overlooked (fig. 3.3). The spine of *T. rex* was no mere bony thread from which more interesting anatomy hung, but an organ as specialized and unusual as the rest of the animal. Like all limbed vertebrates, it was divided into sections: the cervical vertebrae of the neck, of which there were ten; a series of thirteen dorsal vertebrae that formed the trunk; five sacral vertebrae over the hips, and at least thirty-seven caudals in the tail. The extent of the *T. rex* tail represents one of the few remaining osteological mysteries about king tyrants, with estimated caudal counts ranging from thirty-seven to fifty-three (Osborn 1917a; Newman 1970). The true number was likely forty to forty-five (Brochu 2003).

For anatomy fans, the fossilized cervical vertebrae of adult *Tyrannosaurus* are complex and satisfying structures to observe (fig. 3.8). Already somewhat stout in juveniles, the neck vertebrae of mature individuals were extremely short and robust, such that the neck was as abbreviated as it was bulky. Cavities within the vertebrae, connected to the pneumatic system (see below), ensured it was deceptively lightweight. Viewed from the front, the cervicals progressed from relatively narrow structures to broader configurations, with ever-larger intervertebral articular surfaces (zyga-pophyses) and increasingly large rib attachments extending outward and somewhat downward. Regardless of their position in the series, all possessed deep scars and roughened areas, reflecting the attachment of tremendous interspinous ligaments and muscle tissues, a feature seen throughout the entire king tyrant spine (fig. 3.9). These ligaments were likely capable of static support: that is, of holding the neck in place without muscular effort, much as elastic ligaments passively hold horse necks upright (Snively and Russell 2007b). The distinctive "S"-shape common to all theropod necks was especially pronounced in *T. rex* because the vertebrae were not symmetrical front-to-back, with vertically displaced articular surfaces forcing neighboring bones up or down. Many were also wedge-shaped, with narrow bases and expanded upper regions. So exaggerated were these features that one wonders if the neck could entirely straighten in life and, indeed, tyrannosaur neck mobility seems more limited than that of other theropods (Snively and Russell 2007a, b; Samman 2013). Long, double-headed cervical ribs articulated with the main vertebral body, the tails of their elongate, triangular shapes sloping toward the torso.

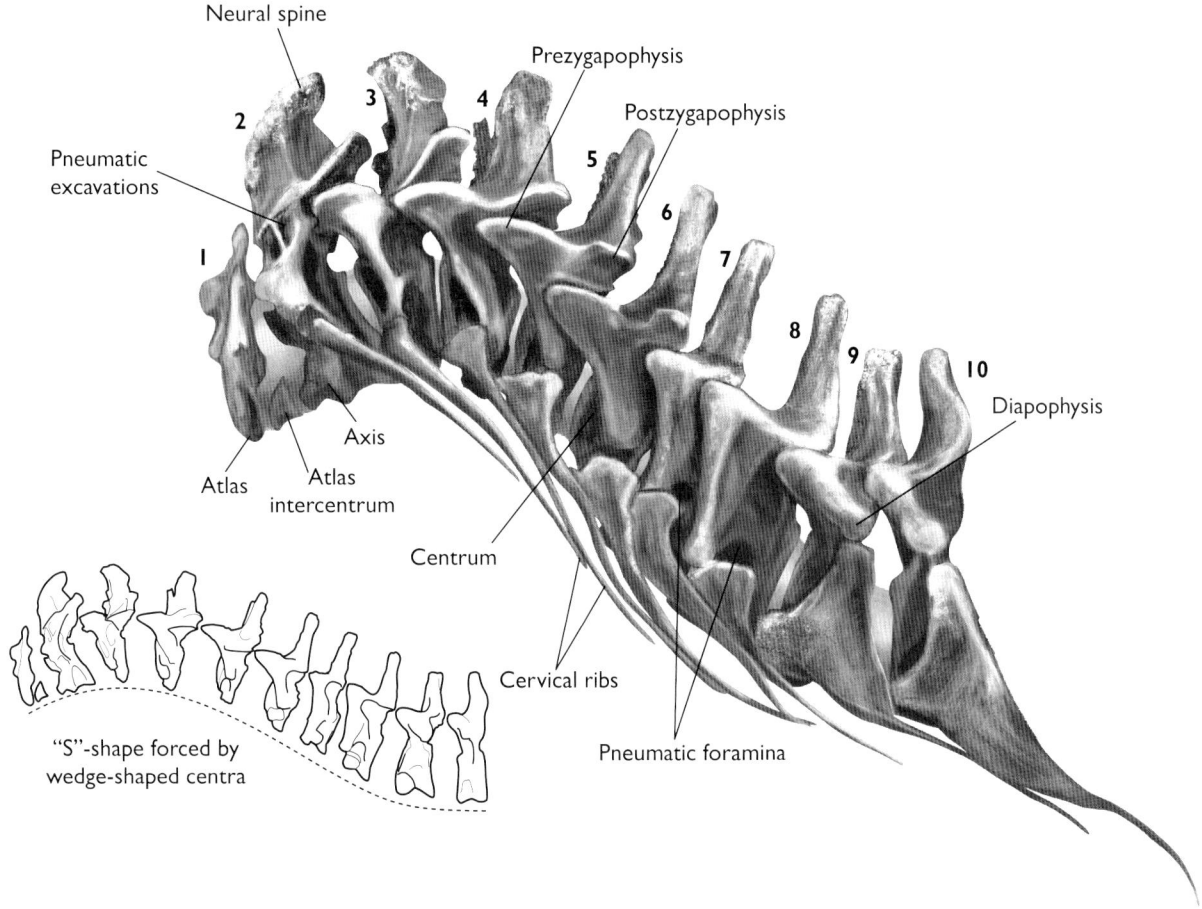

FIGURE 3.8

The cervical vertebrae and ribs of an adult king tyrant. Note the incredibly pleasing "S"-shape forced into the neck by the vertebral dimensions (inset), accommodated for by the orientation of the cervical ribs.

FIGURE 3.9

Roughened areas on the front and aft surfaces of *Tyrannosaurus* dorsal vertebrae, recording the presence of substantial interspinous ligaments. Similar surfaces occur on most large *T. rex* spinal elements.

An enormous amount of neck musculature is evidenced around the *Tyrannosaurus* neck by the robust and sculpted nature of the cervical series as well as the broad and expanded muscle attachment surfaces on the back of the skull (figs. 3.4D, 3.7–8; Snively and Russell 2007a, b). Studies of theropod cervical musculature indicate that their neck myology was more like that of crocodylians than birds and, even among tyrannosaurs, *T. rex* had large neck muscle volumes. They seem to have been especially well equipped for powerful upward extensions and rapid lateral movements of the head, both of which may have been important in predation and feeding (Snively and Russell 2007a, b). Restorations of *T. rex* should thus apply a particularly thick layer of musculature to the upper region of the neck, which may have bulged even above the elevated posterior surface of the parietal bones (see, for example, Figure 3.1). Conversely, muscles associated with pulling

the head downward were less developed than in allosauroids (Snively and Russell 2007a).

As with many reptiles, the division between the king tyrant cervical and dorsal vertebrae was blurred by the rear-most cervicals having many aspects of the forward-most dorsals (Osborn 1917a). More uniform in shape than the cervicals but still possessing the same lightening internal cavities and indications of powerful interspinous ligaments (fig. 3.9), each dorsal anchored a pair of twin-headed ribs that defined the extent of the torso (fig. 3.10). Most theropods were surprisingly narrow-bodied (e.g., Paul 2016), and artists sometimes carry this convention over to their *T. rex* depictions. In fact, like all tyrannosaurids, the dorsal ribs of *T. rex* bowed outward markedly, such that the chest of *Tyrannosaurus* was broad and barrel-like, the widest anatomical region of the entire body (fig. 3.10A). The rotund chest was augmented by the presence of wide, splaying belly ribs, or gastralia,

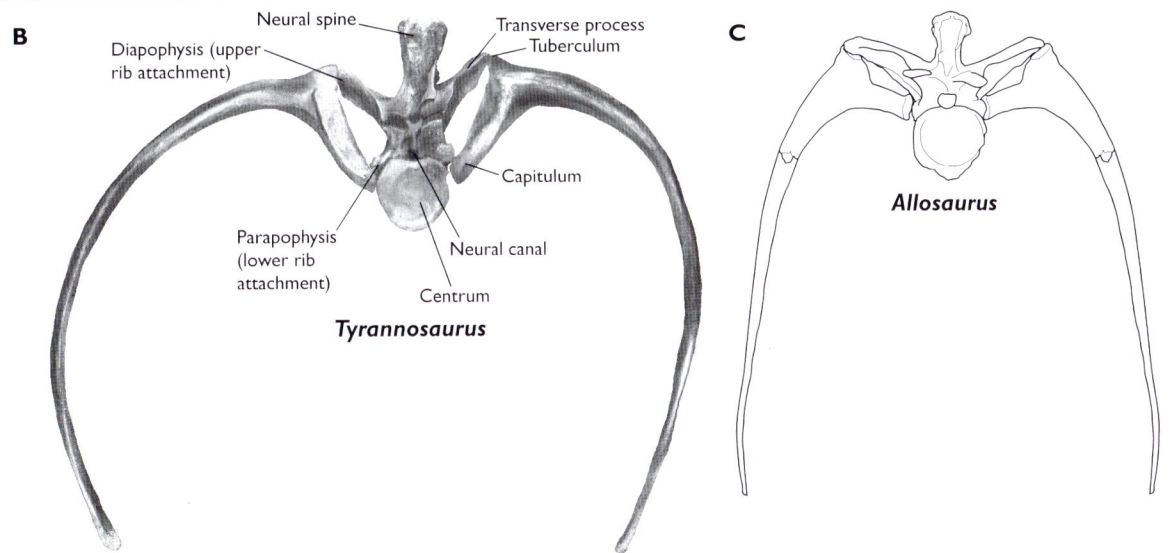

FIGURE 3.10

Unlike most theropods, *T. rex* had an enormous ribcage. (A) Mounted specimen of "Sue" showing the barrel-like ribcage of adult *Tyrannosaurus*; (B) the fourth dorsal ribs and vertebra of "Sue" in articulation, demonstrating how the strongly bowed ribs and elevated articulation contribute to the torso's girth (deformation of the bones has not been corrected); (C) reconstruction of the flat-sided ribs of *Allosaurus* (fourth dorsal vertebra), showing the more typical theropod condition. (B) and (C) are not to scale. (B) After Brochu (2003); (C) after Gilmore (1920).

that spanned the space between the shoulder girdle and hips. The exact number of gastralia in *T. rex* is unknown, but around nineteen are predicted (Brochu 2003).

Behind the torso, and between the enormous hip bones of *Tyrannosaurus*, lay five vertebrae fused into a single unit, the sacrum. This fusion reflected the pivotal mechanical role that these vertebrae played in holding *T. rex* together. Like the towers on a suspension bridge, they anchored the hypaxial muscles and ligaments that stopped the body and tail of *T. rex* from drooping downward, and they also fused with the pelvis, thus ensuring that propulsive motions from the legs were directly transmitted to the body rather than being wasted on bending the spine. As with the dorsals and cervicals, the sacrum was penetrated by hollowing pneumatic tissues.

The tail represented approximately half the body length of *Tyrannosaurus* and was a deceptively complex, multifunctional organ. One role was counterbalancing the torso and head, which may explain the absence of lightening pneumatic cavities beyond the hips (Benson et al. 2012). In being relatively dense and heavy, the tail could perform the same counterweighting service at a reduced length than if it had continued the lightened vertebral morphology seen elsewhere in the spinal column. The caudal series also played an important role in locomotion. As is common to most long-tailed reptiles, the tail base of *T. rex* was bulked out by enormous tissues responsible for retracting the hindlimbs: the caudofemoralis muscles (fig. 3.11; Persons and Currie 2011). These extended for about two-thirds of the tail

length and occupied the cavity between the vertebral transverse processes and the base of the chevrons, narrow bars of bone that articulated underneath each vertebra (figs. 3.3, 3.11B). Most theropods enlarged their accommodation space for the caudofemoralis musculature by elevating their transverse processes relative to the dinosaur norm—a reflection, no doubt, of the importance of rapid locomotion in predatory behavior—but the caudofemoralis volume in *T. rex* was especially huge thanks to especially long chevrons (Persons and Currie 2011).

Toward the end of the *Tyrannosaurus* tail, the transverse processes of the caudal vertebrae diminished entirely (their final reduction is known as the "transition point" on a reptilian tail) and the remaining bones assumed a different form: low, waisted tubes with relatively long, overlapping articulator prongs (zygapophyses). These structures are common to all vertebrae and help guide intervertebral movement, sliding past one another as the spine flexes and moves. At the end of the *Tyrannosaurus* tail, however, the zygapophyses were so long as to limit flexibility. The first assessments of *T. rex* tails assumed that their tips were long and flexible (Osborn 1917a), but with fewer vertebrae forming the end

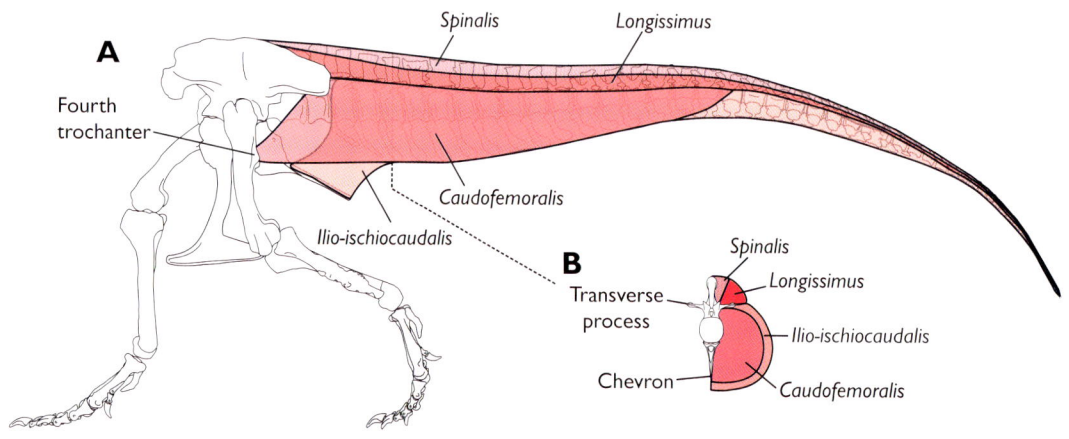

FIGURE 3.11

Deep muscles of the *T. rex* tail, showing the enlarged caudofemoralis musculature: an important hindlimb retractor. (A) Hindlimbs and caudal region in lateral view; (B) cross section through anterior tail. (B) Adapted from Persons and Currie (2011).

of the tail than initially expected, and the limitations on mobility incurred by the end-caudal zygapophyses; neither of these predictions has been borne out.

PECTORAL GIRDLE AND FORELIMB

Making *Tyrannosaurus* just as famous as their size and ferocious jaws are their arms: those proportionately tiny, two-fingered appendages that hung from their slender shoulders, dwarfed by the neighboring chest, head, and legs (fig. 3.12). In being roughly the same length as the forelimbs of an adult human, much effort has been spent mocking and lampooning *T. rex* arms over the last century. Even Osborn and Brown, when discussing the *T. rex* holotype for the first time in 1905, could not believe that its solitary humerus, the upper arm bone, belonged to such a large animal (Osborn 1905). Singling out *T. rex* for having small arms has always been somewhat unfair, however. The diminutive forelimbs of other large predatory theropods were appreciated years before king tyrants were unearthed (e.g., Cope 1866, 1868; Marsh 1884) and proportionally small arms convergently developed in several groups of large predatory dinosaurs, including carcharodontosaurids and, particularly, the ceratosaurians, where the forelimbs were even more reduced than in *Tyrannosaurus*. Moreover, as noted in the previous chapter, *Tyrannosaurus* does not even have the smallest arms of a tyrannosaurid, this record actually being held by *Tarbosaurus* (Currie 2003b; Brusatte et al. 2011). Viewed objectively, *Tyrannosaurus* is just another small-armed member of an often small-armed clade. Perhaps the fame of *Tyrannosaurus* focuses more attention on their small forelimbs than is due? Even so, scientists are not immune to their intrigue either, devoting much time to discussing tyrannosaur forelimb function and evolution—as we will see in chapter 4.

For all the thousands of *Tyrannosaurus* bones in museums, some elements of the *Tyrannosaurus* forelimb still remain unreported. Their long arm bones are documented from a number of specimens (Carpenter and Smith 2001; Lipkin and Carpenter 2008) but a complete hand and wrist has remained elusive until recently (fig. 3.12C). A newly discovered juvenile *T. rex* skeleton, part of the famous "dueling dinosaurs" specimen, looks set to plug this osteological gap as it preserves, for the first time, an entire forelimb skeleton. At the time of writing, this specimen is still under study and specifics of its arms, as well as its other anatomy, remain under wraps.

Other parts of the king tyrant forelimb apparatus have also historically been difficult to track down. It was only in recent decades that the *Tyrannosaurus* furcula, or wishbone, was discovered and the osteology of the shoulder girdle became entirely known (Carpenter and Smith 2001; Brochu 2003; Lipkin et al. 2007; Lipkin and Carpenter 2008). The furcula was a "V"-shaped bone that, in life, lay at the front of the shoulders between the two largest components of the pectoral girdle, the scapulocoracoids. These bones were composed of two elements that fused together in late life (Brochu 2003). The largest was the scapula, or shoulder blade, which were narrow, strap-like bones that curved around the front of the chest, spanning from the underside of the body toward the upper region of the torso. The lower portion of the scapula expanded into a broad, flat plate that articulated with the coracoid, a similarly broad, roughly semicircular bone that formed the lower element of the scapulocoracoid. The glenoid, or shoulder joint, was found on the rear of the scapulocoracoid at the junction of these two bones. It faced primarily rearward and, as we will see in chapter 4, did not permit an enormous range of motion for the forelimb. The furcula and the lower ends of the scapulocoracoids were bound tightly together in life, forming a narrow and low-slung set of shoulders (fig. 3.12A; Carpenter and Smith 2001; Lipkin et al. 2007).

Collectively, the length of the scapulocoracoid dwarfed the arm. This was an unusual condition because theropod forelimbs are usually longer than their shoulder elements and, even in other tyrannosaurids, their arm lengths usually at least matched the shoulder

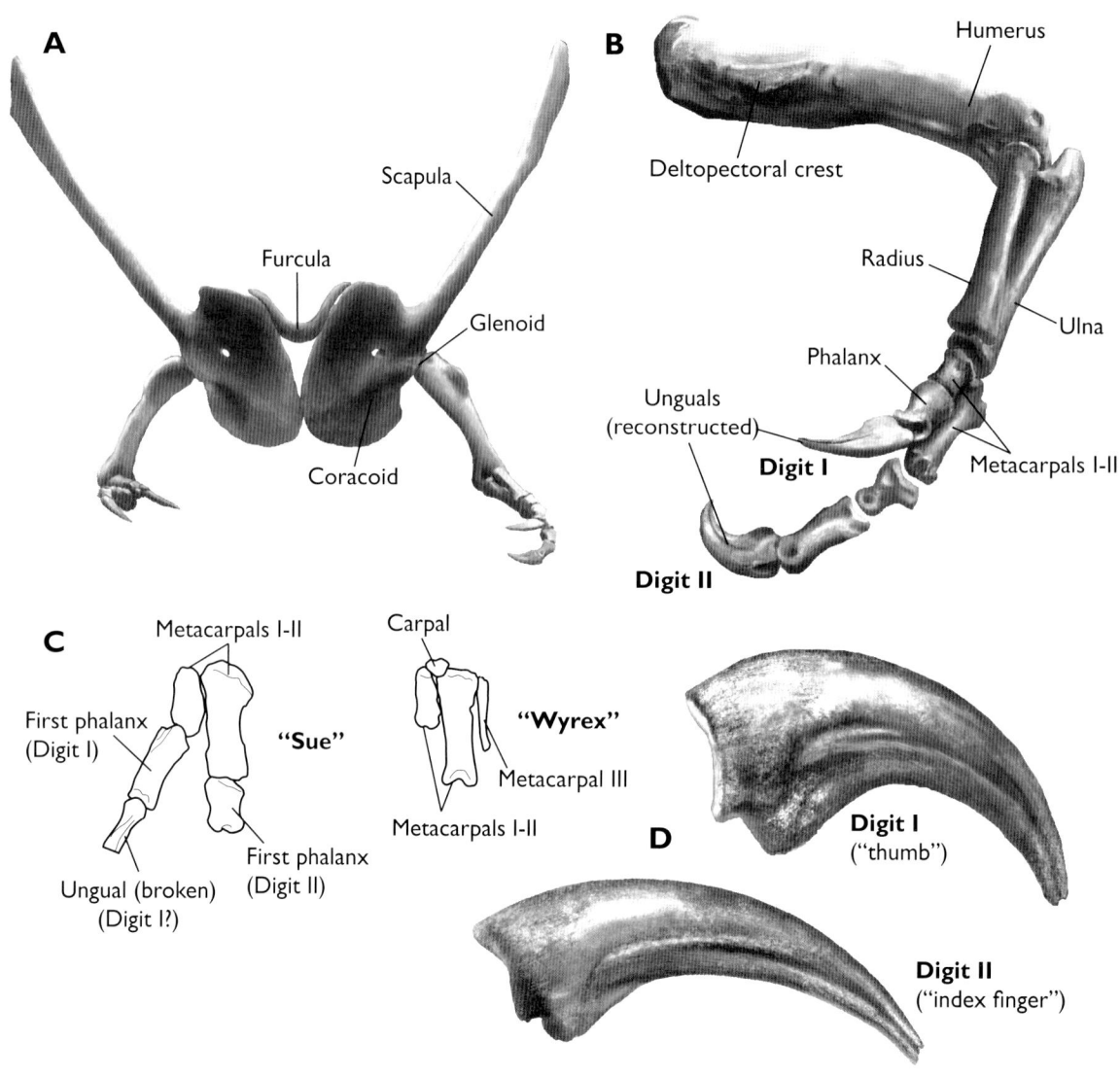

FIGURE 3.12

Pectoral girdle and forelimb osteology of *Tyrannosaurus*. (A) Articulated pectoral girdle and forelimbs;
(B) forelimb with reconstructed hand; (C) line drawings showing known extent of *Tyrannosaurus rex* hand
anatomy; (D) *Tyrannosaurus* forelimb claws, showing the distinct shape of the first and second digit unguals.
(C) After Brochu (2003) and Lipkin and Carpenter (2008); (D) after Longrich and Saitta (2024).

girdle dimensions. The retention of a large pectoral girdle in *Tyrannosaurus* probably reflected, to some extent, the role of the shoulders in anchoring neck musculature, such as the trapezius and levator scapulae, which lifted the neck and head (fig. 3.7; Carpenter and Smith 2001). *T. rex* arms had not been developmentally forgotten, however, the great shoulder girdle also inferring the existence of powerful muscles attaching to the forelimb (Carpenter and Smith 2001; Lipkin and Carpenter 2008). The broad, dished coracoids are particularly noteworthy, once anchoring a substantial set of deltoids that extended the arm forward and outward from the body. Large forelimb muscles are also reflected by the arm itself, which, despite appearing tiny compared to the body, was robust and stoutly constructed (fig. 3.12B). The uppermost element, the humerus, was the longest bone in the limb. Like all dinosaur humeri, the shoulderward region of the humerus bore a prominent, triangular-shaped structure known as the deltopectoral crest. This prominence was an important site of muscle attachment that anchored, among others, both the deltoids and the pectorals. The latter were the major muscles of the chest that connected the arm to the sternum, a skeletal element nestled between the coracoids that, in most dinosaurs, was cartilaginous.

The forearm was comprised of two short bony bars, the radius and ulna, which were controlled at the elbow by large biceps and triceps muscles. They led to a simple wrist formed of perhaps four or five carpals, or wrist bones, this number being determined from other tyrannosaurids as *T. rex* carpals are yet to be documented in their entirety (Brochu 2003; Lipkin and Carpenter 2008). In part because of their wrist construction, theropod forearms could not rotate relative to the elbow, such that the palms of king tyrant hands always faced inward. The hand, or manus (fig. 3.12C), could neither elevate or depress to form sharp angles to the wrist, instead projecting more or less directly from the shaft of the forearm (Carpenter and Smith 2001; Lipkin and Carpenter 2008).

The manus itself was formed of metacarpals, the long bones of the palm, and the phalanges, the individual bones of the digits. Although *T. rex* only possessed two fingers, they had three metacarpals, the third being a small splint of bone that nestled alongside the two larger, finger-supporting neighbors (Lipkin and Carpenter 2008). The digits represented the first and second fingers, or thumb and index finger (something to bear in mind next time you mime "*T. rex* hands": our default configuration uses our index and middle fingers and distresses nearby experts on theropod digit homology). The fingers of *T. rex* remain undocumented in entirety, but unpublished specimens and other tyrant dinosaurs show that the first digit was composed of two phalanges, the second of which formed the claw, or ungual. The second digit had three phalanges, including the claw. The shape of king tyrant claws has been only partly evidenced until recently, with the first digit unguals forming a deep, hooked structure (Molnar 1978; Brochu 2003; Longrich and Saitta 2024). Newly described specimens show that the second claw was somewhat longer and straighter (fig. 3.12D; Longrich and Saitta 2024), an intriguing finding that will further fuel interest in *T. rex* forelimb function. In life, the claws were covered and extended by cornified sheaths: the same material constituting our fingernails. Their tips may have been sharper than the claws of the feet on account of them not being routinely worn against the ground.

PELVIC GIRDLE AND HINDLIMB

While the arms of *Tyrannosaurus* were perplexingly small, the legs and pelvic girdle were huge. Viewed from the front of a mounted skeleton, the enormous head and barrel-chest eclipse the hips and legs, which are, despite their size, relatively narrow compared to the rest of the body (fig. 3.13). And yet, the entire hindlimb region is marked by such large surfaces for muscle attachment that, looking at the pelvis of *T. rex*, one is struck by the large quantity of "negative anatomical space": enormous, empty voids either side of the hip and hindlimb bones that were once occupied by huge muscles and powerful ligaments (fig. 3.7).

FIGURE 3.13

Behind their broad skulls and barrel-like chests, *T. rex* were surprisingly slim, especially around their pelvic regions. The width of their pelvises has been debated because some especially narrow specimens may have been affected by crushing, but even unquestionably undistorted pelvises are not wide structures.

The pelvises of adult *T. rex* are some of the most arresting parts of their skeletons (fig. 3.14). Comprising three bones in a triradiate arrangement, the ilium (top), pubis (lower front) and ischium (lower back), the pelvis of a large *Tyrannosaurus* measures over 1.4 m long and almost 1.9 m deep. The area of the ilium is 88 percent larger than that of *Giganotosaurus* (Snively et al. 2019), which had a smaller pelvic structure despite being of similar stature to *Tyrannosaurus* (fig. 3.14C). Indeed, to find larger pelvises, we need to examine the largest sauropod dinosaurs (fig. 3.14B), and even these are in the same height and length ballpark as *T. rex* (if, admittedly,

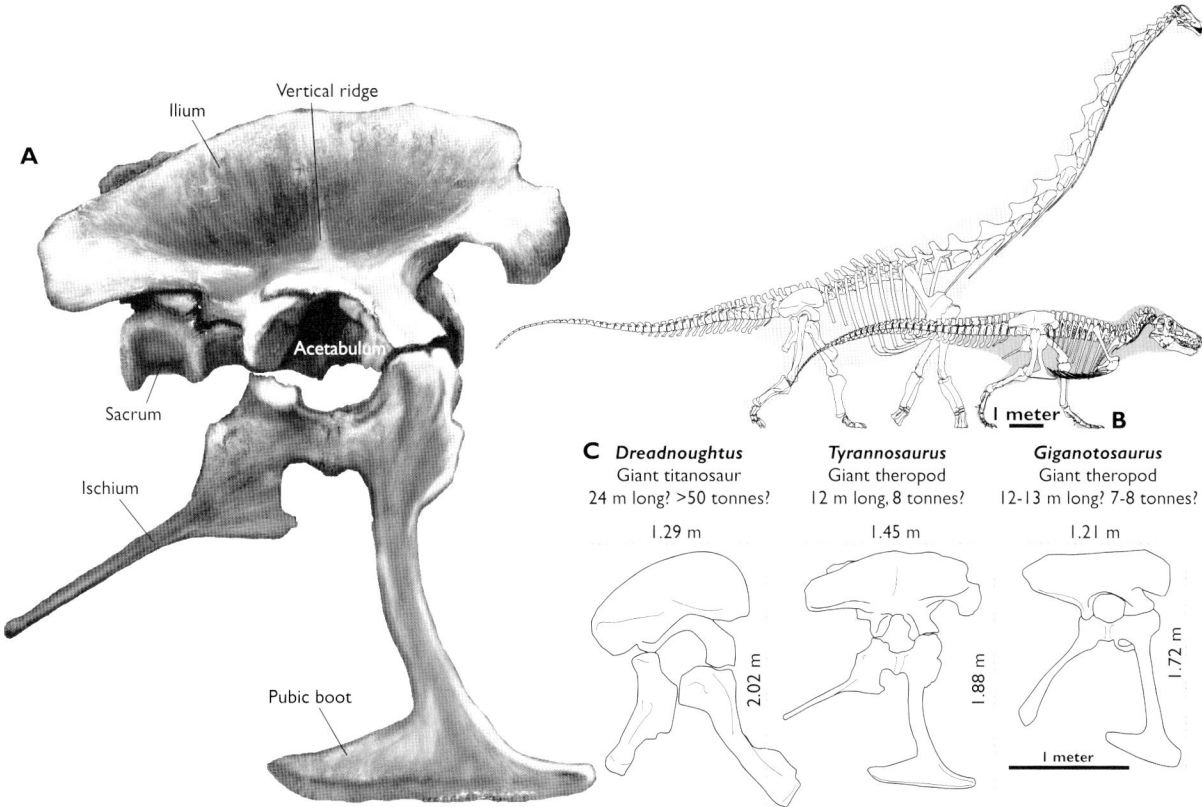

FIGURE 3.14

The enormous pelvis of *Tyrannosaurus*. (A) Adult pelvis in lateral view; (B) size comparison between a giant titanosaur (*Dreadnoughtus schrani*) and *T. rex*: note their similar lateral pelvic areas despite their contrasting overall size; (C) the *T. rex* pelvis compared with examples from other giant dinosaurs.

also more robust and of greater width, which contributed to their pelvic muscle volume). The three bones of the pelvis joined one another around the circular hip joint, or acetabulum, and may have fused together in very old individuals (Brochu 2003). The center of the acetabulum was, as in all dinosaurs, not sealed with bone, although cartilage likely spanned the gap in life to form a complete socket.

The ilium was a long, broad plate that overgrew the sacral vertebrae, reaching over the spine to almost touch the opposing element (fig. 3.13). This configuration recalls the pelvises of ostriches, living theropods with large, powerful legs. Deep depressions and a pronounced vertical crest in the ilium mark the presence of large upper leg muscles, with the bulk of this space occupied by the quadriceps: the tissues that made up the majority of the bulging *T. rex* thigh. These wrapped around the front of the knee to lift the hindlimb and extend the lower leg, while the femoral biceps anchored at the rear of the ilium and inserted at the top of the shin, pulling the knee closed when contracted. Housing muscles on the inside of the thigh were the pubes and ischia, long bones that contributed to the bottom of the acetabulum and formed the lower region of the hip. The ischia, which also anchored tail base musculature (fig. 3.7, 3.11), remained unfused through life as two closely aligned but distinct rods, while the two pubic prongs fused to form the pubic apron. This structure marked the rear of the torso cavity housing the guts and other internal organs. Below the apron lay a deep, narrow expansion known as the pubic boot, which king tyrants probably rested on when sitting or lying down. The overlying skin may consequently have been thick and calloused (Paul 1988a).

The largest bone in the adult *Tyrannosaurus* leg was the thigh bone, or femur (fig. 3.15A–B). This long, slightly bowed bone connected to the hip via a prominent femoral head, a hemispherical structure that extended from the top of the bone perpendicular to the main shaft. Typical of all dinosaurs, this condition enforced an upright hindlimb pose and distinguishes dinosaur femora from those of modern reptiles, in which

the femoral head extends at a lower angle and generally creates a sprawling stance. The mostly smooth femoral shaft had locally prominent crests and rugosities that record the attachment of powerful muscles. Of these, the fourth trochanter was one of the most noticeable. Located about one-third of the way down the femur, it projected rearward from the shaft and received the caudofemoralis muscle in life (fig. 3.11), pulling the leg backward when these voluminous, tail-based tissues contracted. At maximum retraction, the femur could move to a vertical position (relative to the pelvis), but it could swing more freely forward of this point. Like most non-avian theropods, king tyrants held their femora in a sloping posture rather than, as in avians, horizontally, birds having a strange configuration where leg control operates more from the knee than the hip (Farlow et al. 2000). The joints at the end of the *Tyrannosaurus* femur seem to have precluded total straightening of the limb, such that their legs were probably always flexed to some extent (Paul 1988a). This finding has been borne out by biomechanical analyses of hindlimb function as well, although these point to giant theropods having somewhat straighter limbs than their smaller cousins as a means to more efficiently use hindlimb muscles in supporting their weight (e.g., Hutchinson et al. 2005, 2007).

Beneath the flexed knee were the two bones of the lower leg: the stouter tibia and the relatively gracile fibula (fig. 3.15A–B). Attached to one another with bony buttresses and strong ligaments, non-avian dinosaurs were transitional between earlier reptiles and birds in their reduction of the fibula, this reflecting a lessened use of their inner and outer toes in weight bearing (Hutchinson and Allen 2009). The tibia, therefore, was the main weight bearer of the lower hindlimb and also the principal bone operated by the thigh muscles. A large lever at the top of the bone, the cnemial crest, received the great quadriceps and swung the shin forward when contracted. The lower hindlimb bore powerful muscles, too, these operating the ankle, foot, and toes. A bulbous gastrocnemius was the largest of these and would have given *Tyrannosaurus* limbs a drumstick-like

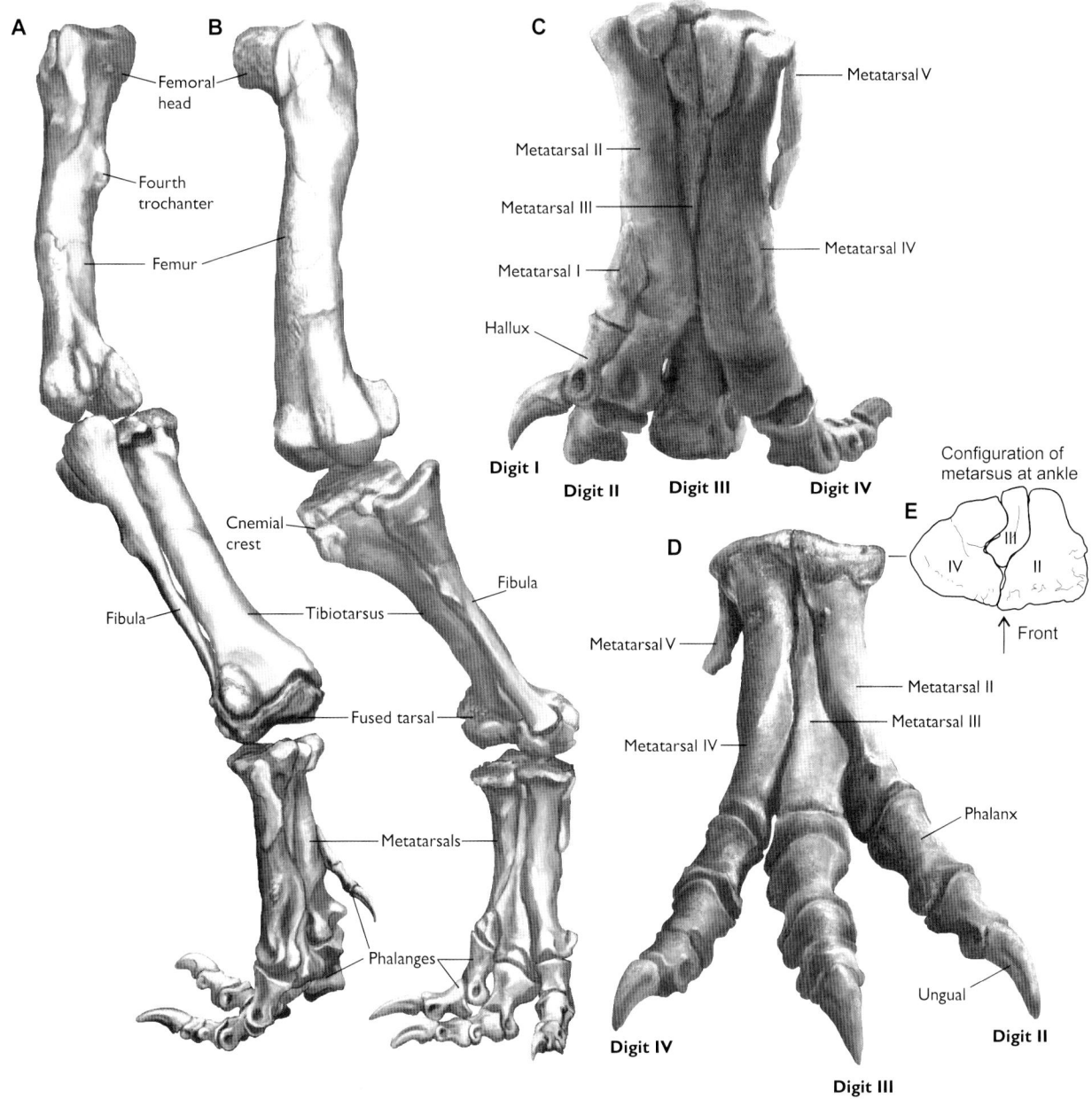

FIGURE 3.15

King tyrant hindlimb bones. (A–B) Entire adult hindlimb anatomy in rear-lateral (A) and front-lateral (B) views; (C) adult pes in aft and (D) fore view; (E) view of ankle from above. Among the distinguishing features of these powerful feet is the pinched upper region of the middle, or third, metatarsal: the arctometatarsal configuration. (E) After Brochu (2003).

appearance. The lower region of the tibia was fixed into, but maybe not fused with (Brochu 2003), two ankle bones. Along with other ankle elements, these are known as the tarsus. Like other dinosaur feet, the *T. rex* foot could only be swung fore and aft, without the rotation possible in modern reptile ankles.

Bearing the tonnage of *Tyrannosaurus* on the ground were surprisingly slender, gracile feet, or pedes (fig. 3.15). Much of their length stemmed from the metatarsals, the long bones forming the shaft of the foot that connected the pedal digits with the ankle. *Tyrannosaurus* was a typical theropod for being digitigrade, supporting their weight on their toes rather than their ankles (as in, plantigrade, like the "rauisuchians" discussed in chapter 2). They stood on the middle toes of their feet, representing the second, third, and fourth digits, with the outer and inner elements (equivalent to our big and pinkie toes) comparatively reduced. The small inner toe, digit I, or the hallux, was supported by a wedge-like metatarsal I, attached to the lower, inner-rear surface of metatarsal II. This toe, bearing just two phalanges, terminated in a curved claw that faced somewhat inwards, presumably giving king tyrants some gripping capability when they clenched their toes together. The fifth metatarsal, meanwhile, was curved and splint-like, and located high on the outer side of the foot. It bore no digit and was likely involved in supporting ligaments communicating between the foot and lower leg.

Metatarsals II–IV thus made up the main body of *T. rex* foot anatomy. The gracility of the pes was generated by a condition unusual among large theropods wherein the middle element, metatarsal III, was compressed into a narrow blade that only just contributed to the ankle joint (fig. 3.15C–E). This is the arctometatarsal configuration, a pedal anatomy common to all tyrannosaurids and convergently developed by several other theropod clades, including ornithomimids and troodontids. As we will see in the next chapter, arctometatarsal feet are instrumental to many locomotory hypotheses concerning *Tyrannosaurus* (Holtz 1995;

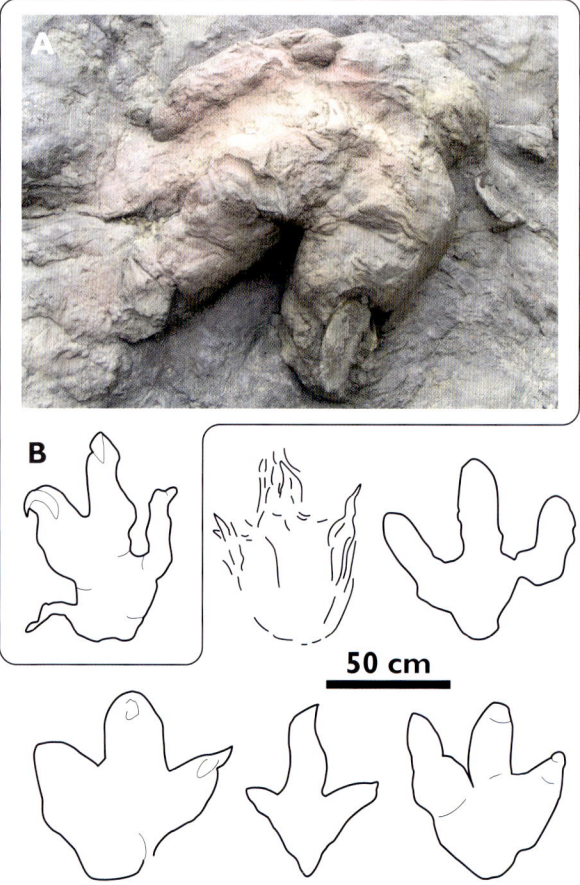

FIGURE 3.16

King tyrant tracks. (A) *Tyrannosauripus pillmorei*, a large theropod footprint likely created by *Tyrannosaurus*. Although not preserving fine details of the foot, the bulk of the print implies substantial padding on the king tyrant pes, as expected for a giant terrestrial animal; (B) outline drawings of footprints referred to *Tyrannosaurus*, of which only *T. pillmorei* (top left, grouped with photograph) is unquestionably theropodan; the rest probably pertain to large hadrosaurs. Outline drawings after Lockley et al. (2011).

Snively and Russell 2002; Snively et al. 2004). The narrowing of the foot had no effect on the development of the weight-bearing pedal digits, however, which were large and bore stout articulations on each phalange. Slightly curved unguals capped each of the toes, which had progressively larger phalangeal counts: three in digit II, four in digit III, and five in digit IV. As with the claws of the hand, the toe claws would have been covered and extended with hard, cornified sheaths in life, but they were likely blunted on account of wear from ground contact. *Tyrannosaurus* footprints (traces given the name *Tyrannosauripus pillmorei*; Lockley et al. 1994) suggest that expansive soft-tissues adhered to the underside of their feet as well,

presumably reflecting substantial padding to cushion the hindlimbs (fig. 3.16). This is consistent with other large theropod footprints, which, when well-preserved, show generous amounts of soft-tissue both below the foot and around the toes (Witton 2018). Indeed, king tyrant feet were sufficiently padded that their prints lost the more angular, gracile quality typical of smaller theropod footprints, and are hard to differentiate from those of the large, bipedal and herbivorous hadrosaurs that they lived alongside (Lockley et al. 2011; see chapter 5 for more on hadrosaurs). For all their ferocious reputation, it seems that *Tyrannosaurus* were not beyond developing some natural relief from sore feet.

A (PARTIAL) INTERNAL TOUR

WHEREAS THE SKELETONS and muscles of extinct animals can be reconstructed or inferred with some reliability, the same is not true for their internal anatomy. Internal organs are almost entirely devoid of hard tissues and they decay rapidly after death, meaning they rarely enter the fossil record. Our knowledge of *Tyrannosaurus* innards is accordingly sparse, but this is not to imply that we have no knowledge of these structures whatsoever. Details of *T. rex* lungs and air sacs can be deduced from their surprisingly intricate interaction with the skeleton, and their fossilized waste clarifies some properties of their digestive system. Amazingly, unmineralized remnants of *Tyrannosaurus* vascular systems and cells have also been discovered, with great implications for future research. The size and shape of *T. rex* brains are also somewhat known from internal details of their skulls, and we will discuss these as part of a wider conversation about *T. rex* cognition and sensory adaptations in the next chapter. Though limited in scope, these collective insights into *T. rex* innards not only add to our picture of their anatomy, but reveal important aspects of king tyrant behavior and physiology.

LUNGS AND AIR SACS: THE SECRET TO ENGINEERING A GIANT

Of all their internal organs, the dinosaur respiratory system is perhaps the best understood thanks to their surprisingly close association with skeletal anatomy. Unlike mammal lungs and airways, which are simple, bellow-like structures that occupy our chests and throats, dinosaur lungs were connected to a series of air sacs that interacted with their bones, even invading them to make them mostly hollow. Birds have this same system, and comparisons of their skeletons to those of extinct reptiles suggests that this anatomy originated early on the pterosaur and dinosaur evolutionary line (e.g., Wedel 2003, 2009; O'Connor 2004, 2006; Claessens et al. 2009). Our understanding of this complex network of airways, lungs, air sacs, and their outgrowths, collectively known as the pulmonary system, is sufficient that we can identify air sac–influenced features of dinosaur fossils in even fragmentary remains. Such findings are integral to considerations of many paleobiological matters, from body masses and thermoreg-

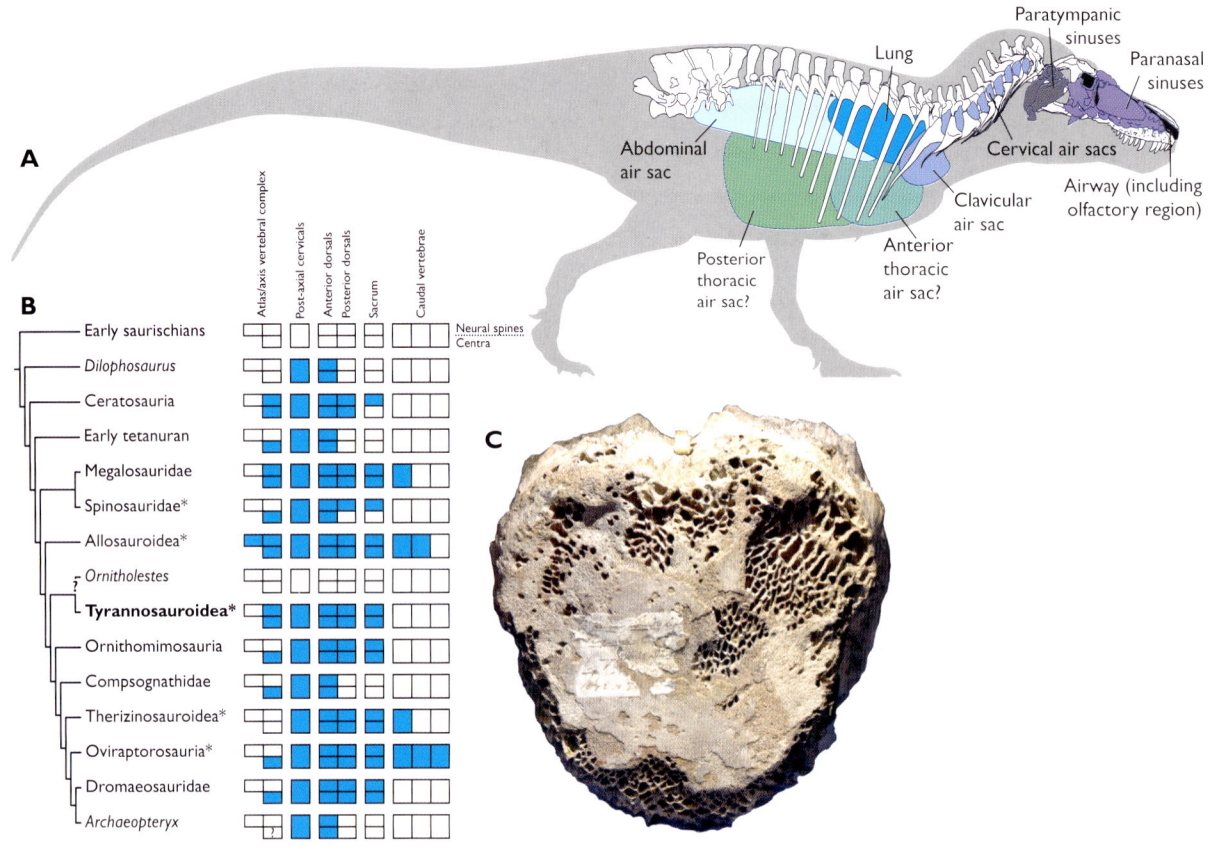

FIGURE 3.17

Skeletal pneumaticity in *Tyrannosaurus*. (A) Reconstruction of the *T. rex* pulmonary system, showing pneumatized bones and their associated air sacs; (B) distribution of pneumatized vertebral elements across Theropoda. Blue squares represent pneumatized elements; taxa marked with asterisks (*) evolved giant size; (C) the "honeycomb" internal texture of *Tyrannosaurus* vertebrae demonstrated by the "*Manospondylus*" holotype (see Chapter 1), likely recording the presence of internal pneumatic tissues.

ulatory abilities to energetics, dinosaur voices, and behavioral ecology. Thus, appreciating *Tyrannosaurus* or any other dinosaur to the fullest extent requires assessment of their respiratory system (fig. 3.17).

The secret to understanding dinosaurian pulmonary organs concerns bony structures, particularly on their vertebrae, where air sacs and associated structures impressed themselves against bone surfaces, sometimes even penetrating their outer walls. These hollows, holes, and internal cavities are evidence of

pneumatic structures, or body tissues filled with air. When pneumatic tissues invade bones, they are, intuitively enough, known as skeletal pneumaticity. Broken pneumatized bones might be largely hollow or, more often, develop a "honeycomb" like network of bony walls and chambers (fig. 3.17C). But we don't need to rely on broken fossils to see evidence of skeletal pneumatization: perforations in bone walls, known as pneumatic foramina, are openings that record the point at which an outgrowth of the pulmonary system, a pneumatic diverticula, invaded the bone (see examples in Figure 3.8).

Many animals, including us, have pneumatic bones. In mammals, these are mostly limited to the sinuses around our noses. Dinosaurs possessed air-filled sinuses too, but they were more extensive, extending throughout their nasal regions and also around their ears (fig. 3.17A). Most sauropod and theropod skeletons have evidence of postcranial skeletal pneumaticity (that is, pneumaticity outside of the skull) but ornithischians, for enigmatic reasons, do not seem to have developed postcranial pneumaticity at all. Conversely, the most extensively pneumatized animals, the birds and pterosaurs, show that just about all parts of the skeleton can be invaded by diverticula under the right evolutionary scenarios (O'Connor 2004, 2006; Claessens et al. 2009). In living species, the development of skeletal pneumaticity follows a pattern where certain body regions are reliably pneumatized by the same parts of the pulmonary system. Cranial pneumaticity, for instance, always develops via diverticula originating from nasal airways. Where present, postcranial pneumaticity is exclusively associated with the development of air sacs within the body cavity. Pneumatized cervical vertebrae are invaded by diverticula from a pair of cervical air sacs, while the dorsal and sacral vertebrae are perforated by outgrowths of the abdominal air sacs. Thus, when we find these vertebrae pneumatized in dinosaurs, we can assume that they had these air sacs as well.

Examination of the *Tyrannosaurus* skeleton shows all these hallmarks, indicating that they had an extensive air sac system throughout the head and body (fig. 3.17A–B; Brochu 2003; Benson et al. 2012; Gold et al. 2013; Witmer and Ridgely 2008). In addition to pneumatizing much of their skull and mandible, diverticula invaded every cervical vertebra, except the first, as well as the entire dorsal series and three sacrals (Brochu 2003). This allows us to make some predictions about king tyrant lung structure and breathing efficiency. In birds, the presence of air sacs in the fore and aft region of the torso allows them to move air around their bodies in a

manner that delivers oxygenated air to a pair of small, solid lungs during both inhalation and exhalation. This efficient system contrasts with mammalian lungs, where oxygenated air is only received during inhalation. Given the shared pneumatic skeletal features of many Mesozoic dinosaurs and birds, non-avian dinosaurs may have had similarly effective respiratory systems. It remains to be determined whether their breathing apparatus was identical, as birds have another set of air sacs lower in their torsos (the clavicular, anterior thoracic, and posterior thoracic air sacs), which also assist in breathing. The clavicular air sac pneumatizes the shoulder girdle and forelimb and apparently did so in a few dinosaur species (Wedel 2009), so one may have been present in *T. rex*, although such a structure is not yet directly evidenced by tyrannosaurid fossils. Thoracic air sacs do not interact with bones at all, however, so it is uncertain when they evolved. Even so, the evidence for a partly or wholly bird-like lung in dinosaurs is strong, and this has clear implications for predicted dinosaur activity levels (see chapter 4).

The functions of the pulmonary system extend beyond respiration. Skeletal pneumaticity was also an integral factor in how dinosaurs and pterosaurs managed to become so large and develop some of their most characteristic and exaggerated anatomy (e.g., Wedel 2003; O'Connor 2004, 2006; Claessens et al. 2009). This reflects, in part, nuances of how skeletal pneumaticity develops. The diverticulae that invade bone do not, strictly speaking, remove bony tissues from the body. Instead, they redistribute them, expanding skeletal elements almost like air inflating a balloon (Witmer 1997). Intuitively, our expectation might be that such expansions make bones more delicate, and this is true against buckling forces, but an expanded, hollow bone can be as strong, or stronger, against bending than a solid bone of the same mass (Currey 2006). This allows for the development of very large, but also proportionately lightweight, skeletal structures, and lineages with extensive postcranial pneumaticity have exploited this in a variety of ways. Birds and pterosaurs maximized their wing proportions; the long-necked sauropods grew substantial but lightweight necks, and several ptero-saur and dinosaur lineages exploited economical bone structures to attain enormous size.

The advantages that non-avian theropods gained from skeletal pneumaticity have not been investigated in as much detail as those of pterosaurs, sauropods, and birds, where the development of air-filled bones was undeniably integral to attaining certain body plans and ecologies (e.g., Wedel 2003, 2009; O'Connor 2006; Claessens et al. 2009). The evolutionary picture of theropod pneumaticity is also somewhat messy, such that it's difficult to point to one factor shaping the de-velopment of this feature across the clade (fig. 3.17B; Benson et al. 2012). It is surely significant, however, that very large predatory theropods consistently developed relatively extensive skeletal pneumaticity (Benson et al. 2012), and provisional studies have elucidated how this may have benefited giant dinosaurian carnivores.

One likely advantage was economization of mass, a useful adaptation for any large animal. Every gram of an organism represents an energy investment: energy to grow the tissue, energy to nourish it, energy to re-pair it following damage, and energy to carry and move it around. Thus, giant animals are often adapted to be economical with their bulk and reduce it wherever pos-sible. *T. rex* seems to have been no exception. Wedel (2004) calculated that king tyrant skeletons were 10 percent lighter because of skeletal pneumaticity and that the associated air sacs may have accounted for a further 10 percent of overall mass. Witmer and Ridgely (2008) similarly found that almost 8 percent of the *T. rex* skull and cranial soft-tissues was airspace. With all this mass reduction located in the torso and skull, not only was the entire animal somewhat lighter, but the di-mensions and mass necessary for the counterbalancing tail were also minimized. Both contribute to *T. rex* hav-ing a shorter body and tail that, in turn, made for a more agile predator (Henderson and Snively 2004; Hutchin-son et al. 2007; Snively et al. 2019).

Tyrannosaurus probably also mechanically benefit-ed from "inflating" some bones to greater dimensions. The pneumatically enlarged cervical vertebrae and expanded posterior skull face, for instance, potential-ly provided greater leverage for the already powerful neck muscles (Snively and Russell 2007a, b). Per vol-ume, some 17 percent of an adult *T. rex*'s cranial bones were filled with air (Witmer and Ridgely 2008), so some aspects of their jaw length and gape probably owes something to pneumatic expansion in the same way that pneumatization facilitated the long necks of sauropods or the vast wings of birds and pterosaurs. These air-filled skull regions had other mechanical benefits, too, lowering rotational inertia during king tyrant head movements by 20 percent compared to a cranium entirely filled with solid tissue (Snively and Russell 2007b). Such a difference has implications for the precision of tyrannosaur head movements when, for instance, tracking and striking at prey. Oth-er adaptive advantages are sure to become obvious as theropod pneumaticity is investigated further. The *Ty-rannosaurus* pulmonary system may not get the same attention as their sharp teeth and claws, but it had an equally important role in making *Tyrannosaurus* effec-tive, gigantic predators.

THE GUT OF *T. REX* ... REVEALED BY POOP

The pulmonary system is unusual among internal or-gans for impressing itself onto bones; generally speak-ing, other internal tissues leave far fewer osteological features for us to find in fossils. This includes our guts, which, as a series of warm, soft tubes and chambers full of microbes, are among the first organs to decay once we die. They are accordingly very rare fossils, although half-digested food is not uncommon in well-preserved specimens, including some tyrannosaurids (Varric-chio 2001; Therrien et al. 2023). Some insight into the king tyrant digestive system is instead afforded by coprolites, fossilized fecal matter, that can be reliably referred to *Tyrannosaurus* (fig. 3.18; Chin et al. 1998). *T. rex* coprolites are exciting fossils because, while fossil feces are abundant in the geological record, it is

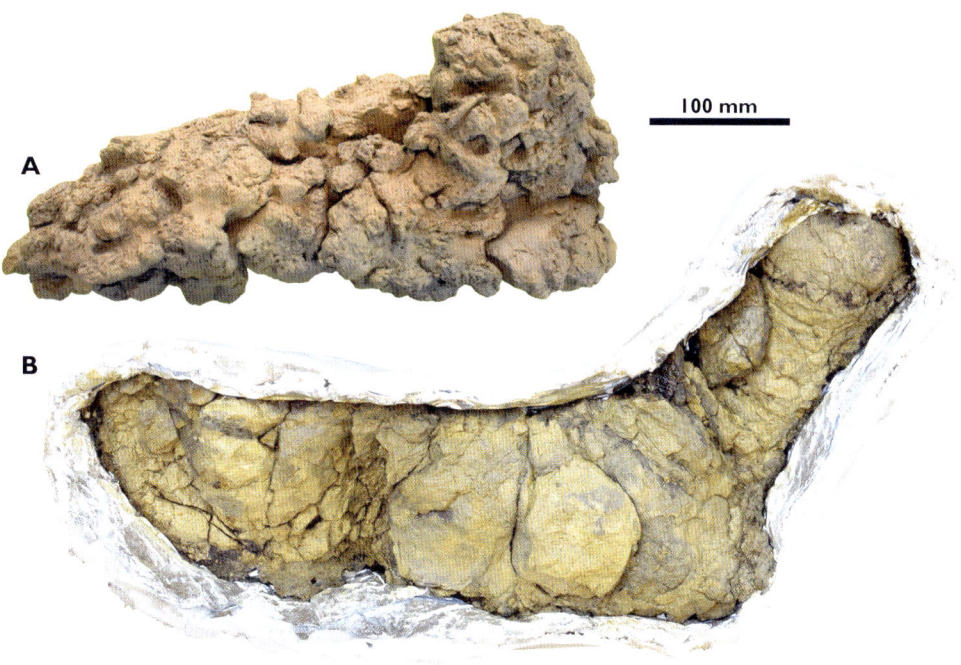

FIGURE 3.18

Tyrannosaurus coprolites. (A) A 44 cm long, 2.4 L example filled with partially digested dinosaur bone; (B) the even larger specimen known as "Barnum," a *T. rex* coprolite considered to be the largest carnivore scat known from the fossil record. "Barnum" is named, as you may have guessed, after Barnum Brown.

mostly near impossible to identify which species created them. *Tyrannosaurus* coprolites, however, can be identified by their size, their evidence of carnivorous diets, and their geological context, the specific circumstances of which make *T. rex* the only candidate. At least two king tyrant coprolites are known: a 44 cm long, 2.4 L specimen described by Karen Chin and colleagues during the 1990s (Chin et al. 1998) and another, even larger 67.5 cm example known as "Barnum," recognized as the "largest coprolite from a carnivore" by the Guinness World Records in 2020 (Guinness World Records 2020). Both examples may have been larger when fresh, as feces tend to dry out, compact, and fracture readily during the process of fossilization and excavation (Chin et al. 1998).

Each of these droppings have similar, elongate shapes and widths: just 15–16 cm across, despite their lengths. This implies that *Tyrannosaurus* cloacas, the openings under their tails that passed waste and eggs, were small compared to overall body size (this, in turn, has implications for egg size; see chapter 6). Some 30 to 50 percent of these coprolite masses comprise bone that passed through the *T. rex* gut, the microstructure of which is consistent with that of a juvenile or subadult ornithischian dinosaur in one example (Chin et al. 1998). These bone fragments ranged from no larger than a sand grain to chunks 34 mm across. Although evidence of acidic and enzymatic digestion is seen from some rounded and degraded bones, the *T. rex* gut evidently did not totally dissolve bone. This contrasts with the digestive process of their living cousins, the crocodylians, where only tough, proteinaceous material like hair and feathers cannot be digested entirely. The recovery of so much bone in *T. rex* scat hints at rapid passage of food through the gut (Chin et al. 1998), and numerous, sharply fractured elements imply that these bones were pulverized by teeth rather than broken down during digestion: *T. rex*, it seems, did not have an avian-style gizzard (a muscular organ housing "stomach stones," or gastroliths). Further evidence of this stems from the coprolite composition, which is primarily a blend of digested bone and phosphate (the latter being typical for carnivore droppings), without any indication of exotic pebbles or stones that could represent excreted gastroliths.

THE MICROSCOPIC, NON-MINERALIZED SOFT-TISSUES OF *T. REX*

Details about the guts and pulmonary system of *T. rex* are inferable through indirect evidence of their existence, but some of the most remarkable internal tissues of king tyrants have survived for millions of years to be directly observable. These are also among the smallest in their bodies and, despite their size, they have caused a storm of controversy, their mode of preservation challenging some fundamental assumptions

about fossilization processes. These microscopic remains were first found during the 1990s when paleontologist Mary Schweitzer identified unusual tissues within *Tyrannosaurus* bones. In 2005, she led a team reporting that they represented original, still soft, still spongy, translucent blood vessels and bone matrix from a 66-million-year-old *T. rex* fossil (Schweitzer et al. 2005a). Because soft tissues, like bones and shells, normally fossilize through mineralization, the proposal that pliable, virtually unmodified *Tyrannosaurus* tissues, even microscopic ones from inside a bone, might have survived for tens of millions of years was immediately controversial. A flurry of academic papers—far too many to review here—have been published in the last twenty years debating this matter. Some researchers remain skeptical, presenting data suggesting that the specimens used by Schweitzer and colleagues were contaminated by modern microbes or laboratory tissue samples (e.g., Kaye et al. 2008; Bern et al. 2009; Peterson et al. 2010; Buckley et al. 2017). But further work on *Tyrannosaurus* and other vertebrate fossils, as well as studies by other teams, continues to find evidence for these proteinaceous elements being genuine and recoverable from a number of bones, including non-*Tyrannosaurus* species (Schweitzer et al. 2007,

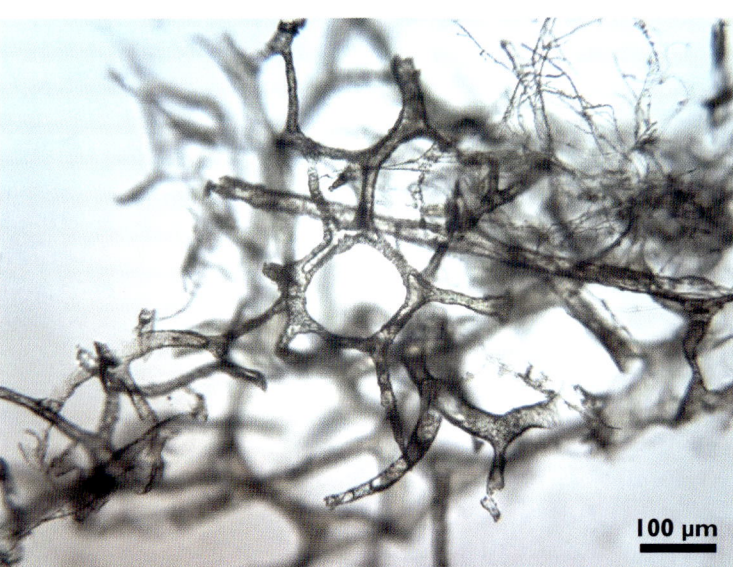

100 μm

FIGURE 3.19

Microscopic view of the hollow, pliable, vascular structures found within *T. rex* limbs. The discovery of these tissues has proved highly controversial as they challenge long-held assumptions and data regarding fossilization. Continued analysis of these and similar tissues suggests that soft, pliable tissues can survive for millions of years in the right circumstances. Image from Boatman et al. (2019).

2009; Lindgren et al. 2011; Boatman et al. 2019; Ullmann et al. 2021; Anné et al. 2023a), even if such findings are seemingly highly susceptible to contamination (Bern et al. 2009).

With so much investigation necessary to simply verify the identity of these structures, relatively little work has explored the implications of their discovery. Some focus has been applied to establishing their mechanism for preservation, with one conclusion being that freshly buried bones must be totally sealed from their surrounding environments for their internal soft tissues to escape mineralizing processes and lithification (e.g.,

Boatman et al. 2019; Ullman et al. 2021). Needless to say, if they can be verified beyond all doubt, the future possibilities of these findings are enormous and go far beyond merely providing more insights into *T. rex* biology. Among the reported tissues are bone cells with nuclei, ancient proteins, blood vessels (fig. 3.19), structures akin to red blood cells, and a reproductive-specific tissue type known as medullary bone (see chapter 6). As methods to identify and isolate these tissues from contaminants become increasingly robust, they will represent a whole new frontier in the study of fossil vertebrates.

CLOTHING KING TYRANTS: THE LIFE APPEARANCE OF *T. REX*

OUR TOUR OF *Tyrannosaurus* bodies has mostly focused so far on anatomy that was, unless something went badly wrong, located internally. We also, however, have some insights into what *Tyrannosaurus* looked like in life, from details of their skin to specifics of their facial appearance. Even aspects that we have no direct evidence of, like their colors, may not be entirely unpredictable. Addressing these topics relies on tissues that rarely fossilize in any dinosaurs, and, at the time of writing, most aspects of *T. rex* life appearance remain unknown. As we have already seen from other anatomical components, however, there are many subjects that we can make inferences about if we look at relevant data from *T. rex*, their fossil relatives, and the right modern animals.

SKIN: SCALES AND ... PROTOFEATHERS?

Even just a few decades ago, any query over *Tyrannosaurus* skin would be simply answered with the word "scales," as this was the only skin type known for all but the birdiest dinosaurs. Today, however, feathers, or their protofeather precursors, are known from a variety of distinctly non-birdlike dinosaur species, including

early tyrannosauroids (see Campione et al. 2020 for review). With 1.5 tonne tyrannosauroids like *Yutyrannus* sporting extensive feathering (fig. 2.7B, Xu et al. 2012), should we also imagine *Tyrannosaurus* covered with a fibrous integument?

The most direct means to address this matter are the fossil impressions of *Tyrannosaurus* skin described from the "Wyrex" specimen by Phillip Bell and colleagues in 2017 (fig. 3.20). *T. rex* skin data is not well represented and even the examples reported by Bell et al. are very small, some no larger than a postage stamp. While lacking surface area individually, however, their recovery from the neck, tail, and the upper pelvic region gives a patchwork picture of *T. rex* skin across the neck and torso, and each skin patch suggests the same integument type: tiny scales. Like joining the dots in a puzzle book, we might, from these, infer that *Tyrannosaurus* had scales covering much, if not all, of their bodies. Like most dinosaur scales, *T. rex* scales were tessellating, slightly raised ("tuberculate") nonoverlapping polygons, but they were also far smaller than most popular dinosaur depictions would suggest: just a millimeter or less across. Diminutive scales were common across Dinosauria to the extent that many species, viewed from afar, would probably have looked smooth-skinned or leathery rather than conventionally "scaly," in the manner of certain lizards and crocodylians (Witton 2018). Other tyrannosaurid scale samples record the existence of "feature scales" within this group (fig. 3.20B), which are larger (centimeter-wide) scales situated within the wider expanse of smaller "basement scales" (Bell et al. 2017). It's plausible that *T. rex* possessed these as well, because feature scales occur widely across Mesozoic dinosaurs (Campione et al. 2020).

It is not unreasonable to wonder if these small, widely spaced samples represent the complete story of *Tyrannosaurus* skin, however. Other dinosaurs, both living and extinct, show that skin types can vary considerably across their bodies. Consider, for instance, how all birds blend feathery bodies with scaly feet, and how some species, like ostriches, also have patches of skin devoid of either scales or feathers. The Cretaceous ornithischian dinosaur *Kulindadromeus* shows an even greater mosaic of skin types, bearing numerous types of protofeather across their faces, necks, and bodies, scales of varying kinds on their arms and legs, as well as heavy, almost crocodylian-grade scales on their tails (Godefroit et al. 2014). Wholly scaled and more "conventional" bird-like feather distributions are also known from dinosaur fossils, so *Kulindadromeus* and their mosaic-skinned kin were not the exclusive rule for dinosaurs, but they demonstrate the need for comprehensive skin data before declaring a species entirely scaly or otherwise. Alas, the sediments preserving *Tyrannosaurus* are not the fine-grained muds that lend themselves especially well to the preservation of delicate skin structures, hampering potential for learning about their integument to the same level as some other dinosaurs. This said, the recent recovery of protofeathers from rocks that are not typically regarded as conducive to such fossilization (van der Reest et al. 2016) suggests that we have a chance of recovering skin fibers in *T. rex* fossils, if they existed.

In the meantime, we are left to ponder the possibility of *T. rex* bearing protofeathers in the regions between those scale impressions. Tyrannosauroids could at least grow monofilamentous—single-fibered, or hair-like—protofeathers (Xu et al. 2004, 2012) and there is no reason to think that this ability was denied to *Tyrannosaurus*.

Dense coverings of these, even over just part of the body, might be questioned on grounds of thermoregulation, however: Would a gigantic, multi-tonne animal overheat if it were covered in hair-like structures or feathers? Key to this concern are animal thermal tolerances, which we can discuss through the concept of thermal neutrality (Kingma et al. 2014): the range of environmental temperatures at which animals expend little or no energy to keep warm or cool off (fig. 3.21). An animal's thermal neutral zone is chiefly determined by mass (Porter and Kearney 2009) with larger animals having a wider "comfort zone" than smaller ones, and also having a greater capacity to warm themselves by increasing their metabolic rate when external temperatures drop below their thermal neutral threshold. Conversely, larger animals also have a narrower tolerance between their neutral zone and hyperthermia, where they are lethally warm. Body mass is the main driver of changes to thermal neutrality, but it is also affected by the ratio of skin surface area to body volume and the presence of insulating tissues, such as feathers, fur, or fat (Porter and Kearney 2009). The impact of insulation becomes more pronounced with size. Adding a thin layer of fur or feathers to a large animal lowers its thermal neutrality far more than the same thickness of insulation would for a smaller one (Porter and Kearney 2009).

Some of our best data on animal thermal tolerances comes from livestock, as farmers need to know which temperatures their animals are comfortable living in (National Research Council 1981). Establishing the neutral zones for some of these species gives us a sense of how quickly animal temperature tolerances change relative to body size. Small animals are relatively sensitive to temperature changes with a narrow band of thermal neutrality. Commercial hens of c. 1.5–2 kg, for instance, are thermally neutral between 18 and 23°C. Emus, which average 30–40 kg, are more cold tolerant, being thermally neutral down to 10°C (Maloney 2008). Animals massing hundreds of kilograms, such as horses and dairy cattle, are unphased by temperatures between 5 and 25°C, and can even tolerate freezing conditions if their hair is left to grow (Morgan 1998).

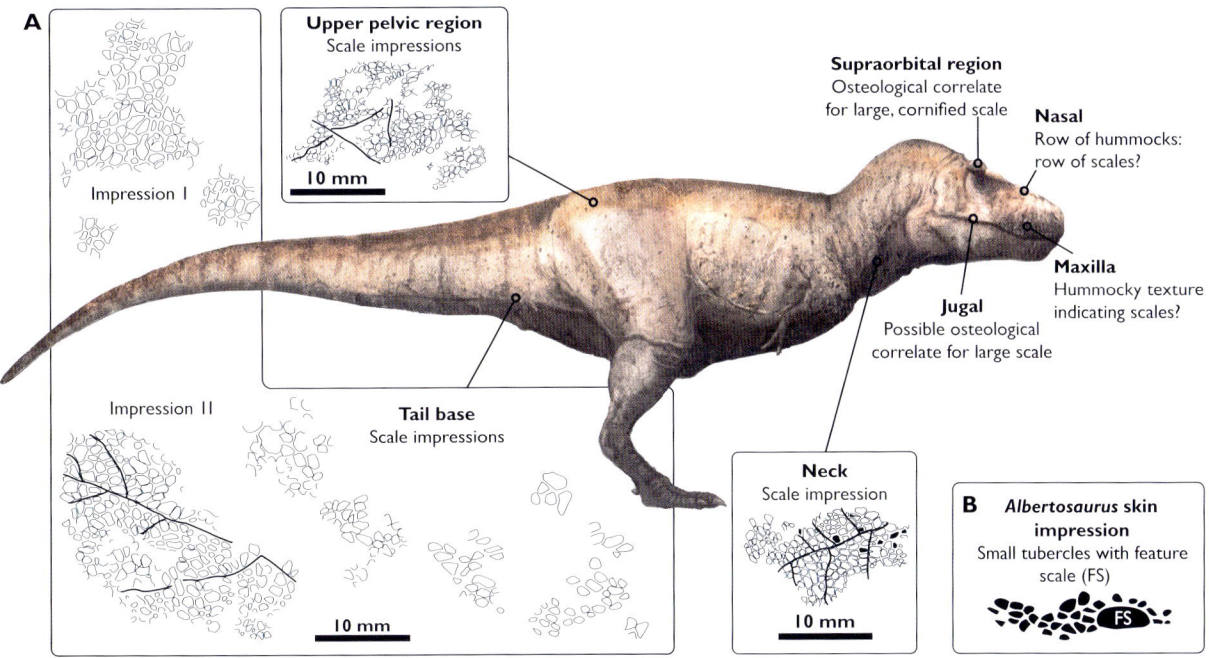

FIGURE 3.20

Tyrannosaurid integument. (A) *T. rex* skin data, comprising skin impressions from *Tyrannosaurus* specimens that record tiny, nonoverlapping scales across much of the body, as well as epidermal correlate data from the skull (see Figure 3.23); (B) *Albertosaurus* skin impression showing the presence of larger "feature scales" within Tyrannosauridae, a common scale morphology found widely across Dinosauria. Skin details redrawn from Bell et al. (2017).

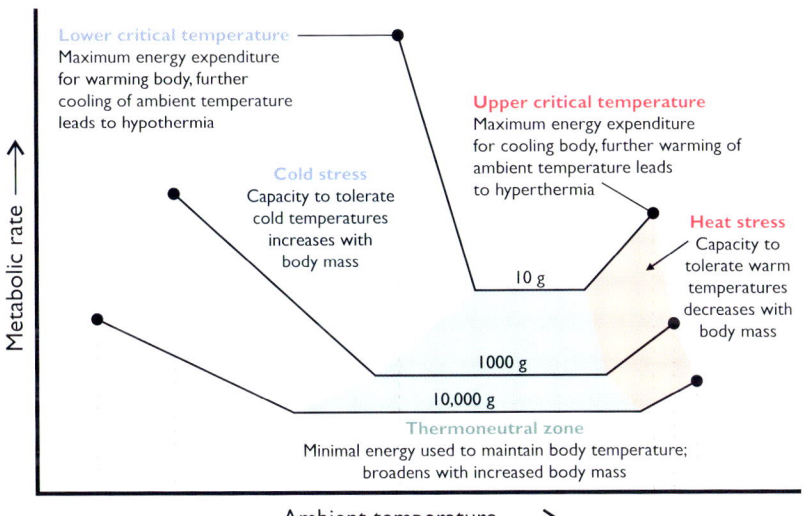

FIGURE 3.21

The principles of thermoneutrality and "thermal neutral zones." Larger animals have broader thermal tolerances than smaller ones, and a greater energetic capacity to warm themselves when chilled. They also, however, have less resilience to overheating. Such observations have implications for our consideration of extinct animal appearance, especially regarding the presence of insulating fibrous integuments.

Elephants, conversely, have real problems staying thermally neutral even in cool conditions. Studies on their heat dissipation show that they generate heat twice as fast as it can be dispelled in ambient temperatures of just 8°C, and their heat production exceeds dissipation rates more than fivefold if external temperatures are above 30°C (Rowe et al. 2013). In such conditions, it is predicted that just four hours of exercising could be fatal. Elephants, it seems, are sufficiently large to generate their own microclimate, a narrow envelope of warm air around their bodies which slows the flow of heat through their skin (Rowe et al. 2013). Elephants thus have a suite of adaptations, including their famously large ears and wrinkled skin, to help cool down, and they also employ certain behaviors in high temperatures to negate the impact of extreme heat. It may seem that animals at such risk of overheating would have little need for any body hair, but, surprisingly, sparse body fibers channel heat up from the body, helping them to shed heat rather than trap it in (C. Myhrvold et al. 2012). Despite all these adaptations, elephants spend much of their time with core body temperatures well above the mammalian average, approaching conditions that would be dangerous for most species. Their comparability to other giant animals is limited by the fact that elephants are unusually poorly adapted for staying cool, especially for creatures that inhabit warm climates: they cannot sweat or pant, have dark, heat-absorbing skin, and a low surface area to volume ratio. But other multi-tonne animals with more conventional heat dissipating skills, like rhinos, also routinely employ cooling behaviors (Rowe et al. 2013).

These trends in thermoneutrality paint a clear picture of the relationship between increasing mass and a decreasing ability to shed heat. Models of dinosaur heat dissipation point to the same conclusions (e.g., Henderson 2013; Hartman et al. 2022), even when these models assume sub-mammalian metabolic rates (Rowe et al. 2013). This relationship may factor into the lack of evidence for fibrous coats on dinosaurs larger than *Yutyrannus*, a species that, in addition to being a dinosaur of a mere 1–1.5 tonnes, also inhabited a mountainous

habitat with cold, frozen winters (Zhang et al, 2021). Perhaps dinosaurs larger than this, and especially those that dwelt in warmer climates, simply had lessened needs for fibrous insulation? A possible caveat to this suggestion is that some skin fibers, vaned feathers, can *shade* skin as well as prevent heat loss, allowing birds that live in extremely hot settings to forage when the local mammals are forced into shelter from sunlight (Dawson and Maloney 2004). Such a system could be beneficial to a giant tyrannosaur, but it's unlikely that *Tyrannosaurus* bore the sort of complex feathers that can shade skin. Rather, the monofilamentous protofeathers predicted for *Tyrannosaurus*-grade theropods may have operated more like mammal fur, which is not known to produce the same shading effect.

None of these points inform us directly about whether *Tyrannosaurus* had protofeathers in body regions currently unrepresented by fossil skin, of course. They can, however, guide our predictions of their life appearance. Given that adult *T. rex* were comparable in mass, if not heavier, than any living terrestrial animal (chapter 4), lived in warm climates (chapter 5), and may have been active for long durations when foraging (Dececchi et al. 2020), our expectations should be for minimized insulation, perhaps limited to a sparse covering of erect, hair-like structures to instigate a cooling benefit (C. Myhrvold et al. 2012). This may not have been the case for juvenile *Tyrannosaurus* that had yet to reach the intense thermoregulatory demands of multi-tonne adulthood, however. Modern animals weighing hundreds of kilos or less often have fibrous coverings, even in warm climates, so, in addition to a model where juveniles had similar, mostly scaly integuments to their parents, we can also entertain ideas of juvenile *T. rex* having large patches or even near full-body protofeather coverings. Indeed, we know that *Ornithomimus*, an ornithomimosaur genus that lived alongside *Tyrannosaurus* (chapter 5) and was of similar size and running capacity to *T. rex* juveniles (see chapter 4), bore a blanket of fibers on their backs, arms, and the tops of their tails (van der Reest et al. 2016). Below this, in a condition that recalls ostriches, the skin on their legs and bellies was naked, even lacking scales. This created an arrangement where fibers could cover exposed skin when *Ornithomimus* reclined or sat, but would expose the naked legs and belly when the animal was active, permitting open heat exchange (van der Reest et al. 2016). Such a configuration seems possible for juvenile *Tyrannosaurus*, although future discoveries are needed to show how accurate any of these predictions are.

THE FACE OF *T. REX*: PORTRAIT OF A LARGE PREDATORY DINOSAUR

Few prehistoric faces have been drawn by artists more than that of *Tyrannosaurus*. There is a tendency among artists and filmmakers, even those attempting to follow tyrannosaur science, to make king tyrant faces as awesome and monstrous as possible, outfitting them with horns, permanently visible teeth, and large, all-seeing eyes. Because features like eyeballs, nostrils, lips, ear openings, and details of cranial ornaments are lost to time, it might seem that such depictions are permitted by scientific uncertainty and that, short of discovering a mummy-like *T. rex* head with extensive skin preservation, we're unlikely to know what *Tyrannosaurus* faces were really like. But analysis of extinct animal skulls, correlated against those of living species, allow us to constrain the basic appearance of extinct facial features of all kinds, from eye sizes to nostril position to facial skin types (Witton 2018). This approach allows us to reconstruct several aspects of *Tyrannosaurus* faces with results that often contradict many of their most famous depictions.

Helping to establish the general look of *T. rex* are the surface textures of their skulls (fig. 3.22). Skin grows directly over bone in several areas on animal faces and, in some cases, this leaves distinctive textures on bone surfaces. Comparing these features, known as epidermal correlates, with those of living species gives an insight into the sort of skin coverings present in fossil animals even when skin remains are not present (Hieronymus 2009; Hieronymus et al. 2009; Hieronymus and

Witmer 2010; Carr et al. 2017). Working with epidermal correlates can be tricky as they are only visible on very well-preserved bone, tend to be less prominent or absent in younger individuals, and they are prone to distortion by injuries. However, species with tough or armored skin types tend to have more obvious epidermal signatures, and Carr et al. (2017) noted that the tyrannosaur *Daspletosaurus* possessed robust skin of this nature. Ongoing work has drawn similar conclusions for *T. rex* (Witton and Hone 2018).

The skin over *Tyrannosaurus* faces seems to have interacted a lot with their underlying bone, suggesting a fairly snug association (fig. 3.23). *T. rex* heads probably weren't, therefore, especially spongy or fleshy. The roughness of their epidermal correlates implies that their facial skin was thick and reinforced with extra keratin, a process known as "cornification." Keratin is the protein that makes skin tough and waterproof, and adding or reducing this protein alters skin properties. Scales, for example, can be made more resistant by enhancing their keratin quotient, and especially dense keratin layers create horn coverings, beaks, fingernails, and claws sheaths. Happily, epidermal correlates seem to distinguish between these skin types, and they suggest that *T. rex* faces possessed sheets of cornified tissues in some areas, and tough scales in others.

Scales are primarily evidenced on the snout (fig. 3.23). The maxillae of adults are covered with a rough, slightly lumpy (or "hummocky") texture, sometimes with honeycomb-like arrangements of depressions above the larger teeth. These correspond well to the textures found under lizard and turtle scales (Hieronymus et al. 2009). A single row of low hummocks on the nasal bones and top of the premaxillae suggests a row of larger scales: we can infer, from living reptiles, that each of these bumps was shaped by a single scale (Hieronymus et al. 2009). The development of irregular bony burrs in the same regions, similar to those seen under bovid horns, implies that, in older individuals, some of these scales became cornified. A further scale correlate is seen on the jugal, although this is sometimes hard to distinguish in big adults. Even in very mature

FIGURE 3.22

The excellently preserved skull of "Trix," a skeletally mature, perhaps fairly old *T. rex* with well-developed skull surface textures. In concert with exceptional bone preservation and expert preparation, these textures might be informative of different skin types covering the *T. rex* face (see Figure 3.23).

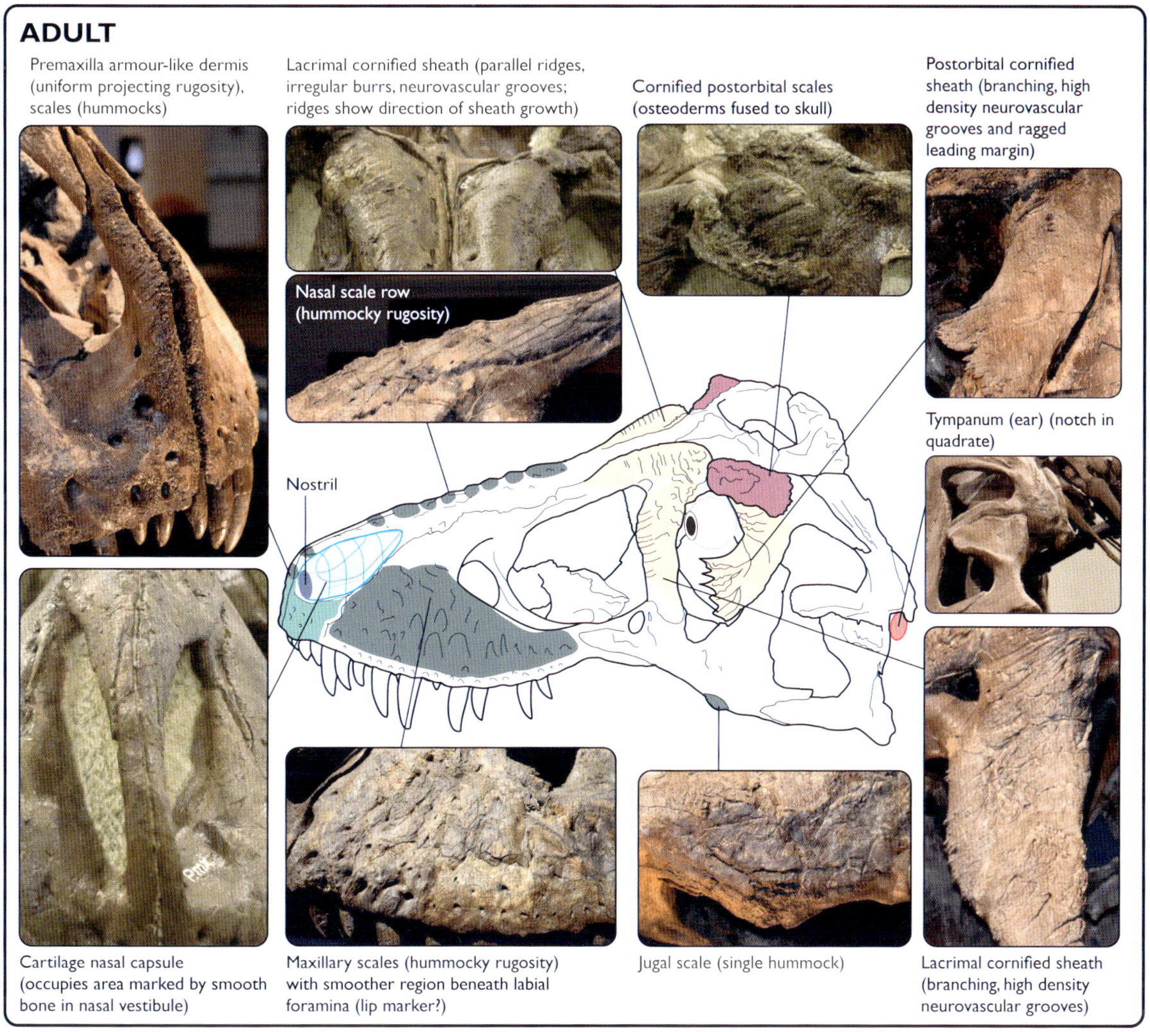

ADULT

Premaxilla armour-like dermis (uniform projecting rugosity), scales (hummocks)

Lacrimal cornified sheath (parallel ridges, irregular burrs, neurovascular grooves; ridges show direction of sheath growth)

Nasal scale row (hummocky rugosity)

Cornified postorbital scales (osteoderms fused to skull)

Postorbital cornified sheath (branching, high density neurovascular grooves and ragged leading margin)

Tympanum (ear) (notch in quadrate)

Nostril

Cartilage nasal capsule (occupies area marked by smooth bone in nasal vestibule)

Maxillary scales (hummocky rugosity) with smoother region beneath labial foramina (lip marker?)

Jugal scale (single hummock)

Lacrimal cornified sheath (branching, high density neurovascular grooves)

JUVENILE

Cartilage nasal capsule (occupies area marked by smooth bone in nasal vestibule)

Nasal scale row (hummocky rugosity)

Lacrimal cornified sheaths (neurovascular grooves)

Postorbital cornified sheath (branching, high density neurovascular grooves and irregular leading margin), osteoderms not yet fused

Nostril

Maxillary scales (hummocky rugosity) with smoother region beneath labial foramina (lip marker?)

Jugal scale (single hummock)

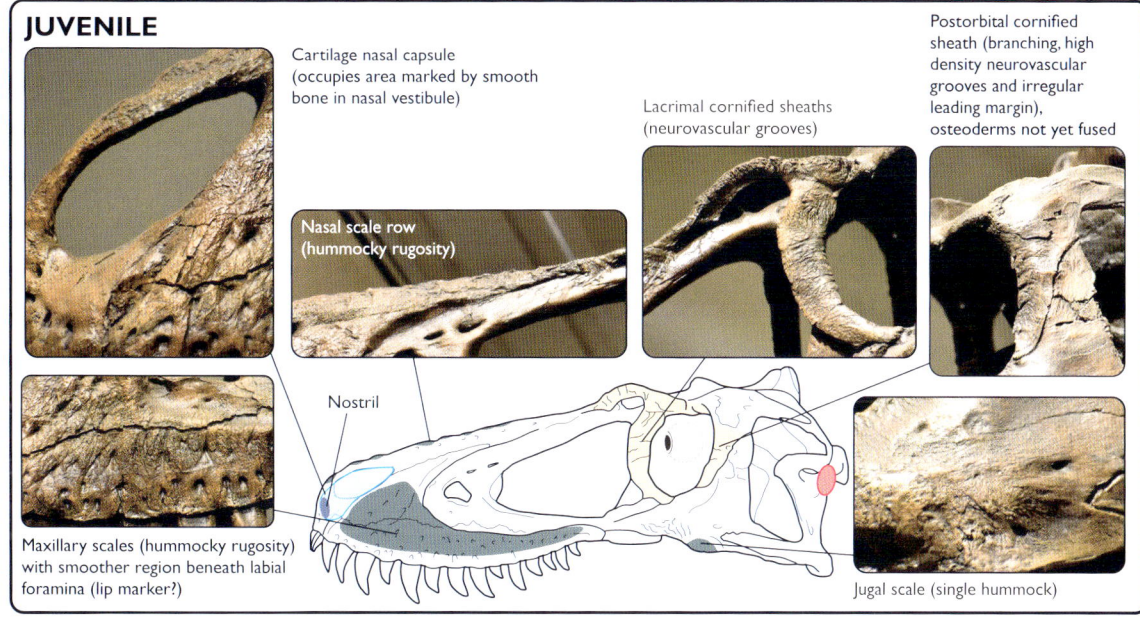

specimens, these jugal scale and nasal scale rows are relatively flat, contrasting markedly with the horn-like structures seen in some tyrannosaurids, such as the alioramins (fig. 2.8C; Lü et al. 2014).

T. rex skulls hint at reduced ornament in other regions as well. In adults, the cornual processes above the eyes are not, as they are in albertosaurines and alioramins, horn-shaped (figs. 2.8, 3.4): they are blunt, flattened, and boss-like. The lacrimal cornual process has an upper surface marked by parallel-sided ridges and grooves, recalling the bones under bovid horns where bony lineation aligns with horn growth trajectories. We might infer, therefore, that T. rex had thick cornified sheaths comparable to those covering cow horns in this region, although the underlying bone does not suggest it attained a pointed shape. Further cornified sheath correlates occur on the side of the lacrimal and postorbital, which are covered in impressions of branching blood vessels, much like those found under bird beaks and bovid horns. The irregular margin of the lower postorbital bone, the same region that divides the orbit into a "B" shape, is probably a reflection of interaction with skin as well. This bone has the same sharp, ragged termination that occurs at the bony horn tips of especially large bovid horns, such as those of bighorn sheep. This jagged postorbital margin only develops in old adults, but even young juveniles possess evidence of cornified tissues around their eyes (fig. 3.23). Such features may have been typical of tyrannosaurids in general, as the skulls of albertosaurines and *Daspletosaurus* also show evidence of toughened tissues in front of and behind the orbit (Carr et al. 2017).

The osteoderms that contribute to the postorbital cornual processes of T. rex are also informative of skin types. In modern reptiles, osteoderms are almost always covered by a single, well-cornified scale, and their shape is largely dictated by the underlying bone. The smaller, frontmost cranial osteoderm in adult *Tyrannosaurus*

becomes a low hummock in adults, while the larger osteoderm that forms the bulk of the postorbital ornament is an oval-shaped, bulbous hemisphere. Together, these imply that T. rex postorbitals were adorned with rounded bosses, not horns. As with other large scale correlates on *Tyrannosaurus* skulls, irregular surface textures hint at these osteoderm scales cornifying over time, and they joined the lacrimal cornual process and cornified orbital tissues to create a ring of toughened skin around the eye (fig. 3.24). The absence of pointy horns on *Tyrannosaurus* contrasts with the cranial condition of closely related tyrannosaurids and implies, perhaps, different functionality: we'll return to this topic in chapter 6.

Within the shielded orbital area of the *Tyrannosaurus* head were their eyeballs, structures that almost never fossilize. Eye structure reveals a wealth of information about the behavior and ecology of extinct animals, however, so predicting the properties of *Tyrannosaurus* eyes has proven of interest to several researchers (Horner and Lessem 1993; Horner 1994; Stevens 2006; Holtz 2008; Carpenter 2013). Efforts to ascertain T. rex eye size have formed two sets of opinions, with some predicting that their eyes were "small and beady," perhaps to the extent of limiting vision (Horner and Lessem 1993), and others arguing that T. rex eyes were absolutely large and would only appear small if measured in proportion to their enormous skulls (Stevens 2006; Holtz 2008; Carpenter 2013).

Two methods have been used to estimate the size of T. rex eyes (fig. 3.25). The first is by simply comparing the diameter of the orbit to the space occupied by eyeballs in living animals (Stevens 2006). Eyeball-to-socket proportions vary between species because some eye sockets merely house eyes and small muscles to move them around, whereas others might require space and musculature to withdraw the eye into the skull. Animals without retractable eyes, like a horse, might have

Previous page **FIGURE 3.23**

Epidermal correlates on the skulls of adult and juvenile T. rex specimens. These rugose bone surfaces characterize different skin types interacting with bone, allowing for prediction of distinct facial skin regions, and thus some details of life appearance.

FIGURE 3.24

Reconstructed face of an old adult *Tyrannosaurus*; note the ring of cornified tissues, blunt bosses above the eye and low scales on the snout. In terms of skull structure, *T. rex* were not especially "showy" dinosaurs.

eyeballs that occupy about 80 percent of the orbit diameter, while the eyeballs of a species with retractable eyes, like crocodylians, are around 65 percent of the same measurement. *Tyrannosaurus* probably didn't have retractable eyes, but, even so, these ratios allow us to make upper and lower estimates of *T. rex* eyeball size from the diameters of their orbits. With eye sockets measuring about 140 mm across in adult specimens, we can predict eyeball diameters of 91–110 mm using the above proportions (Stevens 2006).

A second approach is to use the eye skeleton: the sclerotic ring. Fish, most reptiles, birds, and some amphibians have eyes that are supported by these rings of bone or cartilage plates that lie around the pupil and iris. Delicate as they are, these structures can fossilize in well-preserved dinosaur skulls. Sclerotic rings tell us a lot about eyes, including their size, shape, and

how much of the eye was visible on the living animal. The outer margin of the ring gives a minimum diameter of the eyeball, and the internal margin corresponds to the size of the cornea, the transparent front region of the eye that covers the iris and pupil. The eye thus effectively peers through this bony ring, while other tissues—eyelids, eye-moving muscles, and so on—surround it, meaning that the internal sclerotic ring diameter roughly corresponds to the externally visible portion of the eyeball. Thus, whenever we find a sclerotic ring in a fossil skull, we have a good idea how visible the eyes of that animal were.

Sclerotic rings are most commonly preserved in smaller dinosaurs, but a number occur in large dinosaur specimens, too. As noted above, *T. rex* ocular skeletons have yet to be discovered, but they are known for the tyrannosaurid *Gorgosaurus*. Using these, Carpenter (2013)

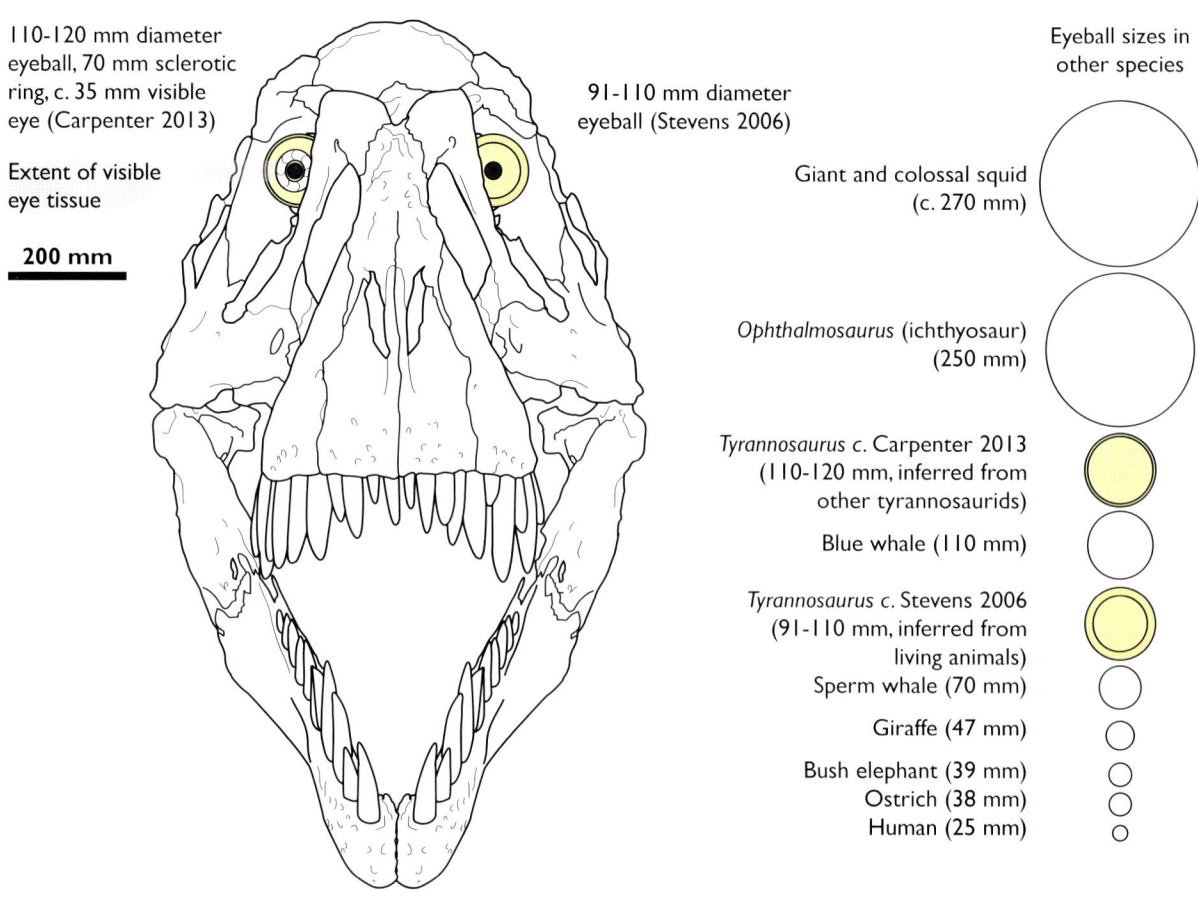

110-120 mm diameter eyeball, 70 mm sclerotic ring, c. 35 mm visible eye (Carpenter 2013)

Extent of visible eye tissue

200 mm

91-110 mm diameter eyeball (Stevens 2006)

Eyeball sizes in other species

Giant and colossal squid (c. 270 mm)

Ophthalmosaurus (ichthyosaur) (250 mm)

Tyrannosaurus c. Carpenter 2013 (110-120 mm, inferred from other tyrannosaurids)

Blue whale (110 mm)

Tyrannosaurus c. Stevens 2006 (91-110 mm, inferred from living animals)

Sperm whale (70 mm)

Giraffe (47 mm)

Bush elephant (39 mm)
Ostrich (38 mm)
Human (25 mm)

FIGURE 3.25

The predicted eye size of *Tyrannosaurus*. Different methods point to *T. rex* having absolutely enormous eyes, comparable in size to those of whales, although the head is so large as to make them appear small.

was able to estimate the size of the same element in *Tyrannosaurus* and predict the size of the king tyrant eyeball: 110–120 mm across, a similar value to Stevens's (2006) estimate. Both studies point, therefore, to *Tyrannosaurus* having eyes larger than any living land animal (fig. 3.25)—so much for "small and beady." This does not equate to gigantic, highly visible eyes in king tyrants, however. Carpenter's (2013) study noted that the projected *T. rex* sclerotic ring was much smaller than the eyeball: 70 mm across the external diameter and a mere 35 mm across the internal opening. This means that *T. rex*, with their 1.4 m long heads and substantial eyeballs,

were peeping through eyelids only a little wider than an inch across. If their pupils were like those of living reptiles, they would have occupied about 90 percent of that opening, limiting the (potentially) more colorful iris to a mere slender ring around the margin of an otherwise dark eye (Carpenter 2013). In this model, we have to wonder how obvious the eyes of *Tyrannosaurus* might have been in life, especially given that the orbits were overshadowed by projecting lacrimal and postorbital ornamentation. Perhaps, in contrast to film and TV *Tyrannosaurus*, with their large, expressive eyes, real *T. rex* eyes would have looked small, dark, and shaded.

A

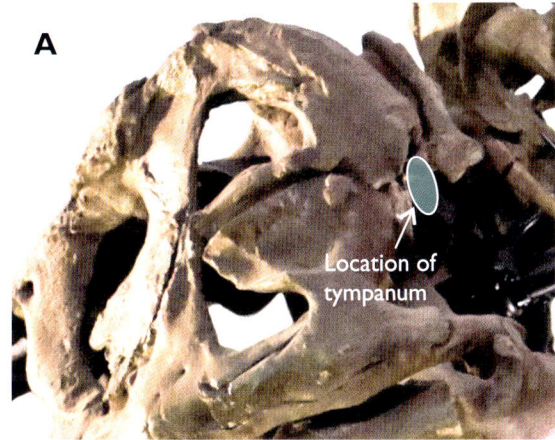

Location of tympanum

B

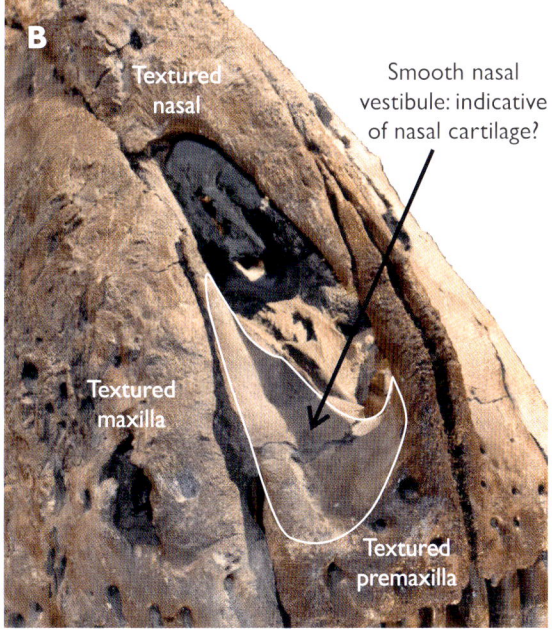

Textured nasal

Smooth nasal vestibule: indicative of nasal cartilage?

Textured maxilla

Textured premaxilla

FIGURE 3.26

T. rex ears and noses. (A) Lateral view of posterior skull seen from somewhat below, showing the position of the outer ear (tympanum); (B) detail of the nares showing the distinction between the highly textured, perhaps skin-influenced bones surrounding the nares, and the smoother internal surface, possibly indicating the presence of a cartilaginous nasal capsule in life.

Elsewhere on the cranium, we can also deduce something about *T. rex* nostrils (fig. 3.26B). Artists have depicted king tyrants with a few different nose configurations, from small, simple openings to a bird-like, see-through narial "window" through the snout, as well as with sealable nostrils, like those of crocodylians. Dinosaur skulls do not directly tell us much about their nostril types, but we can attempt to narrow these possibilities. Studies of reptile nostril positions show that they almost always occur at the very front of the nasal opening, which places those of *T. rex* right at the

tip of their snouts (fig. 3.23; Witmer 2001). We cannot easily determine whether the nostrils were sealable or not, but most reptile and all bird nostrils are simple, immobile openings. Moreover, closable nostrils tend to develop among swimming or burrowing animals to prevent invasion by water or debris, or else in species that live in dry climates that want to carefully regulate the moisture level of their nasal cavities. Given that *T. rex* was not specialized for any of those lifestyles, perhaps their nostrils were simply permanently open, as they are in most reptiles.

We can also probably rule out a see-through, window-like nasal configurations. This condition, known as "pervious nostrils," occurs in birds (most famously, the turkey vulture) that lack a cartilaginous septum dividing their nasal chamber. To see this septum in the first instance, however, requires minimized nasal tissues, and we can probably rule this out for *Tyrannosaurus*. Like all dinosaur nares, the more obvious nasal opening into the *T. rex* skull is actually the *internal* margin of the narial apparatus, and the depression into the surrounding skull bones represents the full extent of the narial tissues. This dished region, the nasal vestibule, is entirely smooth, contrasting with the roughly textured bones of the surrounding maxillae and premaxillae (fig. 3.26B). As explored above, these textures were probably derived from skin interacting with bone, implying that the smooth nasal vestibule represents a region where tyrant skin didn't directly contact the skull. Perhaps, instead, some sort of cartilage capsule occupied this space, filling the nasals either side of *Tyrannosaurus* snouts to give them a more rounded, fuller rostral pro-file. Such cartilage structures are found in living reptiles, filling the expansive narial openings of monitor lizards and iguanas, for instance, adding volume to snouts that would be shrunken and hollow-looking without them (Witton 2018).

The position, if not the appearance, of *T. rex* ears is also known. Reptiles house their ears in notches on their quadrate bones, immediately in front of the depressor mandibulae muscles that open their jaws. Molnar (2008) located this recess under the overhanging squamosal region found at the back of *Tyrannosaurus* skulls, placing the ear roughly level with the nostril in a horizontally held head (figs. 3.23, 3.26A). Alas, what the ear looked like externally isn't clear. Modern reptiles give us a few options: we may have been able to see the ear membrane (the tympanum) directly, or this may have been hidden within an opening into the side of the head. The tongue is another structure we only have a loose handle on. The tongue-supporting bones, or hyoids, of toothed carnivorous theropods seem to indicate that a fleshy tongue akin to that of crocodylians was anchored to the base of the mouth by strong connective tissues (Z. Li et al. 2018). This is our best guess for *Tyrannosaurus*, it also being unlikely on anatomical and phylogenetic grounds that dinosaurs had forked tongues like lizards (Witton 2018).

One of the most contested aspects of reconstructing theropod faces concerns the tissues covering their mouths (fig. 3.27). This is a topic that particularly applies to *Tyrannosaurus* because, largely thanks to the *Jurassic Park* movie, they are frequently depicted without extra-oral tissues: that is, structures like lips, cheeks or other soft parts that shield teeth from view. A number of researchers have attempted to elucidate the presence or absence of such tissues in dinosaurs (e.g., Galton 1973; Bakker 1986; Paul 1988a, 2016, 2019; Ford 1997; Knoll 2008; Morhardt 2009; Keillor 2013; Witton 2018; Nabavizadeh 2020; Cullen et al. 2023) and the "lips debate" has been thrashed out ad infinitum by paleoartists and scientists online, with Hartman (2019) being a notable contribution among the many web articles dedicated to this topic. Among specialists who have contributed to this debate, explicit support for the lipless model, despite its mainstream popularity, is a minority view (Ford 1997), and most authors have concluded that dinosaurs, including *T. rex* (Keillor 2013; Paul 2019; Cullen et al. 2023), covered their teeth with lips.

The most compelling reason to think Mesozoic dinosaurs had lipless mouths and exposed teeth concerns their relationship with modern crocodylians and birds. If dinosaurian cousins, the crocodylians, lack lips, and modern dinosaurs lack them too, might lips have been lost deep in archosaur history (fig. 3.27B)? Though seemingly the most parsimonious interpretation of archosaur facial evolution, crocodylians and birds have unusual, highly specialized jaw tissues, whereas Mesozoic dinosaur jaw bones, in properties considered relevant to extra-oral tissues, are more lizard-like (Soares 2002; Knoll 2008; Keillor 2013; Witton 2018, Cullen et al. 2023). For instance, both birds and crocodylians have large, sometimes enormous, numbers of foramina covering their jaw bones, but dinosaur jaws tend to

FIGURE 3.27

Lips on *Tyrannosaurus*? (A) Competing reconstructions of *T. rex* as a lipless and lipped reptile; (B) different assumptions of the evolution of reptilian lips and their implications for dinosaurs; (C) the identical relationship between tooth size vs. jaw length in lizards and theropod dinosaurs, showing that species like *T. rex* do not have, relative to jaw size, unusually large teeth; (D) schematic dental cross-sections of alligator and the tyrannosaur *Daspletosaurus*, where the outer surface of the crocodylian tooth has been weakened and worn by perpetual exposure to air, unlike the tyrannosaur; (E) distribution of neurovascular foramina across the jaws of a lipless American *alligator* and a lipped Komodo dragon; (F) distribution of neurovascular foramina in *T. rex*: note the greater similarity to lizards than crocodylians. (C–D) Redrawn from Cullen et al. (2023).

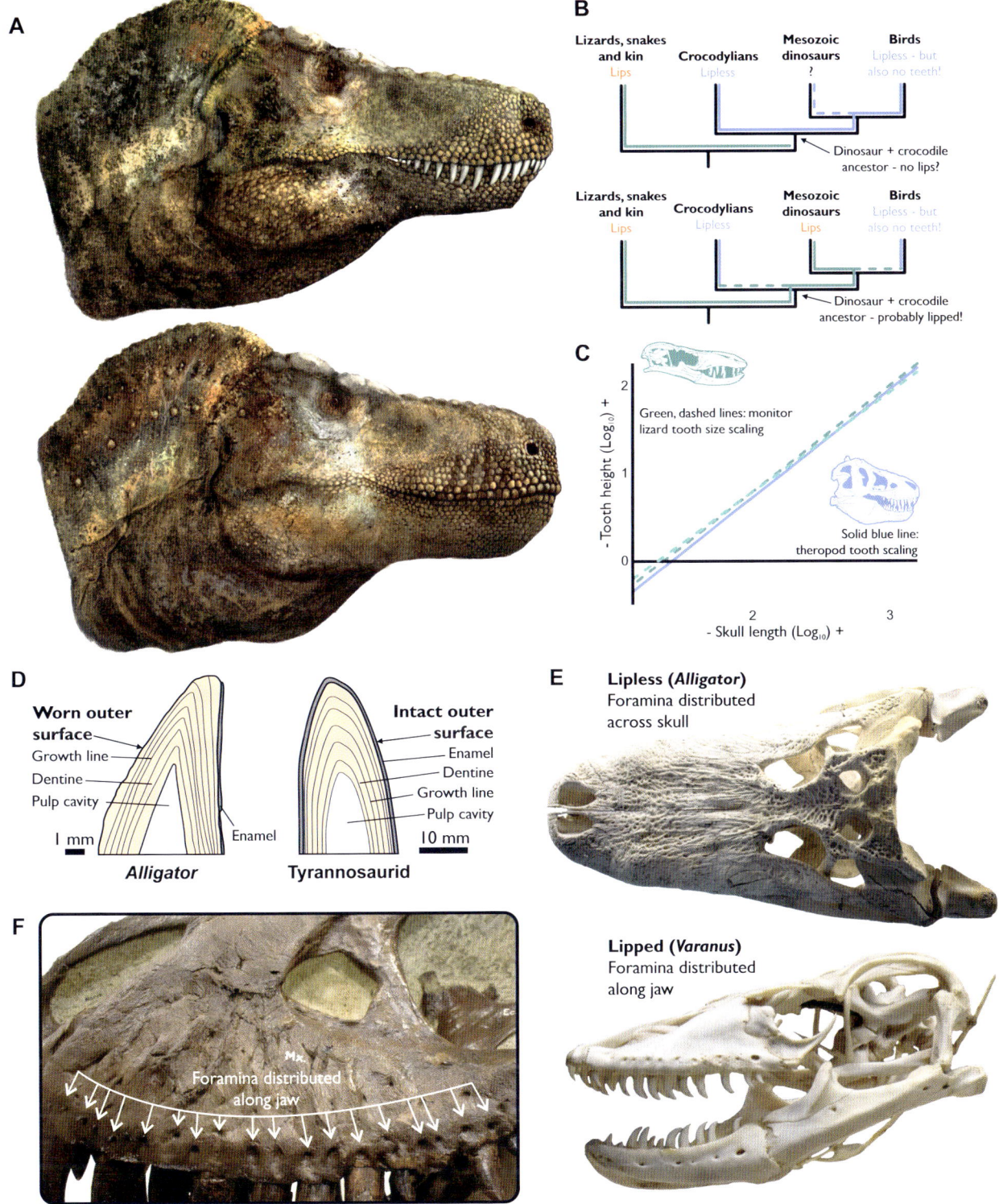

A

B

Lizards, snakes and kin
Lips

Crocodylians
Lipless

Mesozoic dinosaurs
?

Birds
Lipless - but also no teeth!

Dinosaur + crocodile ancestor - no lips?

Lizards, snakes and kin
Lips

Crocodylians
Lipless

Mesozoic dinosaurs
Lips

Birds
Lipless - but also no teeth!

Dinosaur + crocodile ancestor - probably lipped!

C

- Tooth height (Log₁₀) +

Green, dashed lines: monitor lizard tooth size scaling

Solid blue line: theropod tooth scaling

- Skull length (Log₁₀) +

D

Worn outer surface
Growth line
Dentine
Pulp cavity
Enamel
1 mm
Alligator

Intact outer surface
Enamel
Dentine
Growth line
Pulp cavity
10 mm
Tyrannosaurid

E Lipless (*Alligator*)
Foramina distributed across skull

Lipped (*Varanus*)
Foramina distributed along jaw

F

Mx.
Foramina distributed along jaw

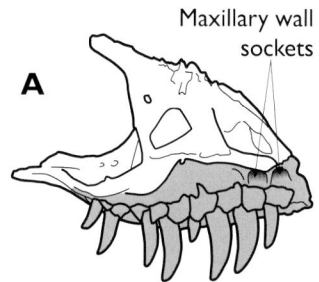

Maxillary wall sockets

A

B

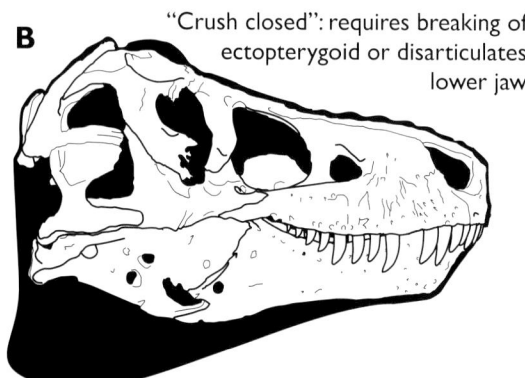

"Crush closed": requires breaking of ectopterygoid or disarticulates lower jaw

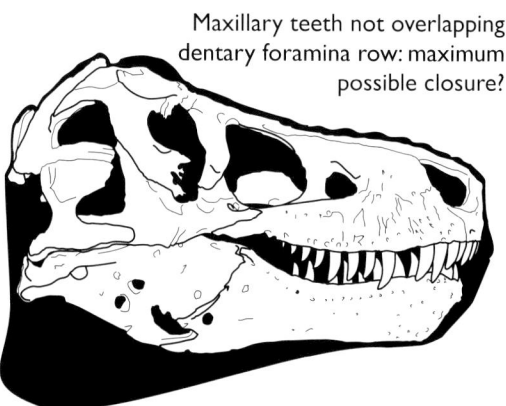

Maxillary teeth not overlapping dentary foramina row: maximum possible closure?

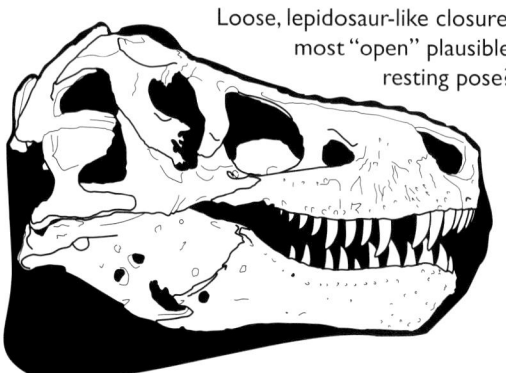

Loose, lepidosaur-like closure: most "open" plausible resting pose?

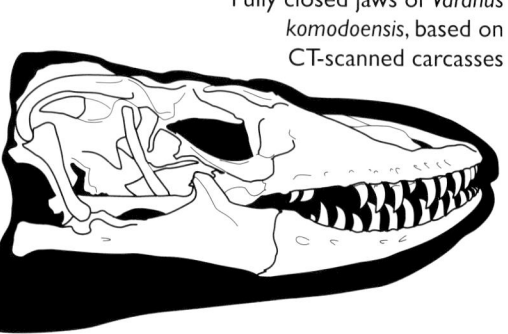

Fully closed jaws of *Varanus komodoensis*, based on CT-scanned carcasses

FIGURE 3.28

Theropod jaw closure and its relevance to the presence of lips. (A) Maxillary wall sockets on the inner surface of the maxilla, postulated by some as representing recesses for the lower dentition when theropod mouths were closed; (B) different postures of the lower jaw, including the anatomically impossible "crush closed" posture required to push the dentary teeth into the maxillary wall sockets. Lizards, represented here by a Komodo dragon, hold their jaws agape even when their mouths are shut, and this may have been true of theropods, too.

have fewer foramina distributed in a row adjacent to their teeth (known as "labial," or lip, foramina; Figure 3.27E–F; Bakker 1986; Keillor 2013; Cullen et al. 2023). *T. rex* jaws have more foramina than average for theropods, but they are still well short of the numbers seen in crocodylians, and they still possess an obvious labial foramina row (Morhardt 2009; Cullen et al. 2023). Furthermore, close observation of *Tyrannosaurus* jaws shows that the rough textures likely indicative of epidermal contact terminate at this row, suggesting separation of skin and bone at this juncture by other tissues: lips are the most likely cause.

The closure of the theropod jaws is also significant (fig. 3.28). Theropod mandibles slid within their upper jaws when closed, matching the condition of lizards rather than meshing their jaws and teeth together, like crocodylians. This leaves a lot of gaps around even tightly shut theropod jaws, raising the question of how they sealed their oral cavity to prevent moisture loss. Bony recesses, maxillary wall sockets, high within *Tyrannosaurus* mouths have been proposed as pockets for the lower teeth to nestle into when the jaws are tightly closed (fig. 3.28A; Molnar 1991; Currie 2003a), but raising the lower jaw to reach these forces the mandible into the ectopterygoid, either crushing the bones under the eye or dislocating the jaw joint (Keillor 2013; Cullen et al. 2023). CT scans of lizards, however, show that they only loosely close their jaws even when their mouths are fully shut, and those complications with tightly closing theropod jaws mean this was almost certainly the configuration they used, too. A loosely closed mouth can only be sealed by soft tissue, again implying the presence of lips. It may seem unreasonable that teeth with 12 cm long crowns could be hidden by extra-oral tissues, but theropod teeth, even those of *T. rex*, are smaller, proportionate to jaw length, than those of the fully lipped monitor lizards (fig. 3.27C; Cullen et al. 2023). *T. rex* would not have needed oversize, comically inflated lips to seal their mouths.

That tyrannosaurs kept their teeth within a well-hydrated environment is evidenced by their well-maintained tooth enamel (fig. 3.27D). Saliva-hydrated enamel is more resistant than that found on exposed, dry teeth, which tends to become brittle and wear away. Perpetually emergent crocodylian teeth develop a characteristic wear facet on their outer surface and are frequently broken, forcing a rapid tooth replacement rate (around forty-five to fifty times in a lifetime) to maintain a full set of teeth (Grigg and Kirchner 2015). Examination of a fully emerged, 510-day-old tyrannosaurid tooth shows no such damage, however, implying that its owner probably kept it within a lipped, hydrated oral environment (Cullen et al. 2023). The capacity to maintain good dental condition for sustained periods surely also factors into the tyrannosaurid trait of having very low tooth replacement cycles; another way in which they contrast with lipless crocodylians.

Collectively, these observations of jaw structure, tooth size, tooth wear, and jaw closure mean that theropod jaws don't make much anatomical or functional sense without lips of some kind, and *T. rex* was almost certainly a lipped animal. Reptile lips are extensions of the scaly skin covering their faces, and, with only a handful of exceptions, they are not muscular, so our lipped king tyrants are not expected to have snarled, dog-style, at one another (sorry, Hollywood). How obscured the teeth would be when the mouth is open is a related, and unexpectedly interesting question. Lizards tend to have fleshier gums than mammals, and some (e.g., monitors and snakes) have such deep lips and gums that even large-toothed species obscure their dentition entirely. Paul (2019) speculates that tyrannosaurs were similar, and that we would see virtually no teeth even in gaping *T. rex*. Cullen et al. (2023) argued for slightly more dental visibility in open mouths, noting that most lizard gums only cover the basal 25 percent of tooth crowns, leaving the upper region visible (this interpretation is followed in Figure 3.1). Even this more conservative prediction is a radical departure from the traditional visions of *T. rex*, however and, combined with its blunted ornaments and armored facial features, sets the scientific vision of *Tyrannosaurus* well apart from the monsters often seen on film and TV.

WHAT COLOR WERE KING TYRANTS?

Any discussion of dinosaur appearance would not be complete without the ageless question of their colors and patternings. For decades, this query has effectively been non-answerable, even for well-studied dinosaurs like *T. rex*. Typically, even top paleontologists have only managed a shrug of their shoulders and some arm waving about camouflage colors helping to avoid predation or sneak up on prey, and the use of display patterns for social signaling. Today, however, advances in analyses of fossilized soft tissues have allowed us to reconstruct some aspect of color for several dinosaur species as well as those of some other fossil animals, including pterosaurs and marine reptiles (see Vinther 2015 for a review). This new frontier in dinosaur science has helped to flesh out not only the life appearance of dinosaurs, but also their ecologies, as elements of habitat preference, daily activity pattern, level of predation concern and so on are reflected in details of skin colors and patterns. Deducing dinosaur color requires exceptionally high-quality preservation of their skin so that their pigment cells ("melanosomes") can be identified. Unfortunately, this excludes the vast majority of dinosaur species from such analyses as skin fossils are not only rare, but they are mostly preserved as sediment impressions of skin texture, not actual fossil skin itself. This is the case for our few tyrannosaurid skin impressions, including those of *Tyrannosaurus* (Bell et al. 2017). At this time, none of our new science on color restoration can be applied to tyrant dinosaurs.

This is not to say, however, that there is nothing intelligent to say about *T. rex* color. Innumerable studies have been conducted on the function of living animal color, and we should not entertain an "anything goes" approach to coloring tyrannosaur reconstructions. We can, instead, be guided by the adaptive constraints on color in modern, large terrestrial predators to make informed predictions about the appearance of *Tyrannosaurus*. Such an exercise is limited in many ways: no living creature is ecologically or phylogenetically close to *Tyrannosaurus*, our modern environments are not identical to those of the latest Cretaceous, and the controlling factors of animal coloration are highly complex. We are still unable to explain the patterns and colors of many living species, especially for predators (e.g., Caro 2013; Pembury Smith and Ruxton 2020). To that end, our efforts here can only paint *T. rex* with the broadest brush and we might, at best, find the margins of probability rather than pin down specifics of color and patterning.

In the most general sense, animal coloration is a conflict between natural selection, which promotes color configurations that help animals remain undetected, avoid temperature stress, and generally survive from day to day, and sexual selection, which promotes the adoption of bold, broadcasting colors and patterns that attract mates and deter rivals (Darwin 1871). Across vertebrates, we see that the former dominates color expression in our largest living predators, which are consistently colored for concealment: that is, they have camouflaging colors and patterns that hide their presence from their prey (fig. 3.29). This applies to carnivorous mammals, which are a relatively drab group overall and often lack good color vision (Caro 2013), as well as clades that have the adaptive capacity to produce and see the most brilliant and striking color schemes in nature, such as lizards, snakes, birds, crocodylians, and

FIGURE 3.29

A selection of the largest predatory animals alive today and their color schemes: all are adapted for concealment using different camouflage schemes. (A) Komodo dragon (background matching); (B) lion (background matching); (C) tiger (disruptive coloration); (D) polar bear (background matching); (E) saltwater crocodile (background matching); (F) golden eagle (background matching); (G) great white shark (countershading).

fish. Research on the impact of body size on predatory ecologies may explain this observation. From crustaceans and insects to fish and mammals, bigger predators are less able to hide from their prey, this even being the case in species with extremely adept camouflage capability, like chameleons (Cuadrado et al. 2001; Pembury Smith and Ruxton 2020). Prey species also respond more vigorously to predator size, reacting sooner, fleeing farther, or responding more aggressively (Stankowich and Blumstein 2005). These phenomena are so universal across living animals that they were surely also true in deep time. On account of their unprecedented size, big theropods probably faced even greater concealment challenges than those experienced by animals today and they may have exploited every advantage to remain inconspicuous. Might, therefore, *Tyrannosaurus* coloration have prioritized concealment rather than advertising their sociosexual status with vivid colors or conspicuous patterning?

This proposal is difficult to test in lieu of *T. rex* skin pigment data, but it may be supported by the limited extent of *Tyrannosaurus* cranial ornamentation (see above). Indeed, giant theropods were generally less ornamented than small- or midsized species, and, because faces are important sites for signaling structures and colors in modern species (e.g., Darwin 1871; Caro 2005), we might wonder if this indicates a lessened emphasis on visual display in giant dinosaur predators. An obvious outlier to this observation is the strange *Spinosaurus* (fig. 2.4), but perhaps this is the exception that proves the rule, the fishing habits of *Spinosaurus* clearly segregating it ecologically from the terrestrial predator-prey dynamics of tyrannosaurids and other giant theropodan carnivores. In any case, a caveat here is that elaborate osteological features are only suggestive of sexually selected coloration, not confirmatory: there are many examples of horns or casques that are drably colored, just as there are numerous, osteologically unassuming animals with brilliant patterns and hues.

If we continue on this line of assuming a predominantly camouflaged *T. rex* hide, what sort of patterns and colors might we expect? In modern animals, camouflage strategies are adapted to specific habitats and predation styles as no system is universally effective across all environments and prey types (Pembury Smith and Ruxton 2020). There are thus many variants on the commonest camouflage configurations, including background and pattern matching, where animals have skin tones that approximate their surroundings (fig. 3.29A–B, D–F); disruptive coloration, where high-contrast colors, often perpendicular to the body margins, break up body outlines and disguise distinctive features such as eyes (fig. 3.29C); or countershading, where a dark upper region and lighter underside disrupts the distinctive formation of shadows (fig. 3.2G). Factored against these are functional properties of skin pigmentation. Darker pigments, for instance, can protect skin from harmful UV rays and may have antibacterial properties, but, conversely, they also absorb more solar heat and increase an animal's thermal load (Walsberg 1983). Juggling all these adaptive influences, and many others, is why the study of color in living animals can be so complex and why we are also denied simple, robust color predictive methodologies for extinct species.

Nevertheless, if we cross some general predictions of *T. rex* physiology and ecology (chapters 4–6) with the basics of effective camouflage systems in living animals, we may attempt to narrow the scope of uncertainty for king tyrant color. Adult *Tyrannosaurus* were probably inhabitants of relatively open settings on account of their size and prey preferences; if so, lower contrast, more uniform coloring may have helped them blend into these relatively clutter-free environments. This strategy is employed by our largest open-country apex predators today, such as lions, polar bears, wolves, and Komodo dragons. Sharp countershading, with a rapid transition between the darker top and paler lower region, may have been useful in such settings, too (a look modeled by *Giganotosaurus* in Figure 2.5). Although uniform colors do not generally work well at concealing animals in wooded environments (Pembury Smith and Ruxton 2020), they may not have precluded hunting in denser habitats: flat-gray elephants are reportedly remarkably difficult to find once they enter forests or woodlands (Caro 2013). Of course, the broad geographic and habitat range of *T. rex* (chapter 5) allows us to consider whether they had color morphs adapted to different environments, such as plainer, duller variants in open terrain and higher-contrast patterning among those living in more mountainous, highland regions.

Muted colors in adult tyrants also match the observation that, in scaly animals, color vividness tends to reduce with size and age (Olsson et al. 2013). Indeed, living animals give us reason to imagine that juvenile king tyrants may have had entirely different colors and patterns from adults, their size and ecology demanding a unique color response. As potential prey items themselves, juvenile *T. rex* may have relied on patterns that reduced predation risk. Modern juveniles of many species are covered in stripes, spots, bars, and other features that provide pattern matching to surrounding environments as well as, in some taxa, a means to create a "motion dazzle" effect that confuses predators about their speed and direction, or else draws focus to less critical anatomies, like tail tips (Murali & Kodandaramaiah 2016). Whatever they looked like, we should not imagine that *Tyrannosaurus* only had two color schemes: "juvenile" and "adult." It took decades for king tyrants to reach full size and the route to adulthood was likely via several different ecological niches (e.g., Holtz 2021), each of which may have had different adaptive pressures on coloration. It's these changes with growth, and other aspects of physiology and raw functional anatomy, that we will consider in the next chapter.

BREATHING LIFE INTO BONES

4

THE REPUTATION OF *Tyrannosaurus* as a "model species" is not only owed to their relatively abundant specimens and well-documented anatomy, but from the application of these to the study of physiology, growth, sensory capabilities, locomotion, and other biological properties. The findings of this research transforms the base anatomical data of king tyrants into hypotheses about their condition as living animals, research that has also helped to shine light on the paleobiology of dinosaurs in general.

RUNNING HOT, LIVING LARGE

HOT-BLOODED TYRANTS

Transitioning from understanding *Tyrannosaurus* as a study in grand-scale anatomy to appreciating them as living animals (fig. 4.1) requires appreciation of their physiology, the fundamental biochemical processes that operate living beings. Physiology encompasses the chemical and physical workings of cells, the conditions that are maintained within an organism, and metabolic energy demands. The latter has dominated discussions of dinosaur physiology since the late 1960s and has traditionally boiled down to debates over whether they were "cold-blooded," or ectothermic, using their environment to control their body temperature, or "warm-blooded," or endothermic, maintaining a steady,

relatively high internal body temperature by rapidly processing energy derived from food. Ectothermic species have relatively slow metabolisms that can neither rapidly assimilate food energy nor quickly process metabolic waste produced during muscle action, and they are thus limited to low activity levels with occasional bursts of high energy. Endothermic physiologies are the functional inverse of this condition. Elevated metabolic rates, or tachymetabolism, allow for sustained periods of high activity via quick assimilation of food energy and efficient mopping up of physiological waste, but this comes at the cost of much higher fuel (food) requirements.

Traditionally, this physiological distinction has seen vertebrates divided into "higher" and "lower" classes,

FIGURE 4.1

Four chapters in, *Tyrannosaurus* finally roars! ... but not in the way we usually imagine. This *T. rex* is using a closed-mouth vocalization, where expanded throat tissues and a vast nasal cavity act as resonation chambers. Such vocalizations are common among large reptiles and birds and may have been used by dinosaurs as well. Such considerations of king tyrant functional anatomy are how researchers transform fossil bones into concepts of living, breathing animals.

with endothermic birds and mammals viewed as "higher" beings than the lowly ectotherms: amphibians, reptiles, and fish. The graduation of dinosaurs from being "lower" ectotherms to "higher" endotherms, or near-endotherms, played some role in making them more interesting to researchers during the late twentieth century, as if tachymetabolic animals are more "worthy" of attention than their metabolically slower, cold-blooded relatives. In truth, there are no properties of animal physiology that make them "higher" and lower" species. Endothermy and ectothermy each have advantages and drawbacks, and the metabolic line between them is blurred. Some ectothermic lineages include species with elevated, endotherm-like physiologies, just as some endotherms are capable of letting their body temperatures vary in response to environmental conditions.

The last fifty years have seen a shift of the principal question asked about dinosaur physiology, from *"Did dinosaurs have elevated metabolisms?"* toward *"How elevated were dinosaur metabolisms?"* This reflects an overwhelming body of evidence demonstrating that non-avian dinosaurs, and maybe their early archosaur ancestors, were not ectotherms. The cumulation of this data is not, however, a *Tyrannosaurus* story, instead relying on a multitude of observations pertaining to many aspects of dinosaur biology. This includes their endotherm-like upright limb postures, large brains, specifics of bone microstructure, rapid locomotion evidenced from dinosaur trackways, accelerated growth rates, the existence of insulating protofeathers and feathers in a number of lineages, the capacity of some herbivores to chew and thus rapidly process food, bird-like and highly efficient lung systems, the existence of high-latitude, cold-adapted dinosaur species, and the incontrovertible origin of birds from theropods (see Benton 2000, 2021; Naish 2021; Currie 2023 for reviews and discussion). The thrust of this evidence points to most or all dinosaurs being tachymetabolic endotherms of some kind, even as research continues into the specifics and nuances of this topic.

Many features indicative of endothermy are determinable in *Tyrannosaurus*, including upright limbs,

evidence of rapid and efficient locomotion, rapid growth, and a large brain (all discussed in more detail below). More direct evidence that *T. rex* bodies were maintained at a constant temperature is found in their bone chemistry (Barrick and Showers 1994). Vertebrate bone contains phosphate, a molecular construct of phosphorus and oxygen (PO_4). The type of oxygen isotope (variants of oxygen atoms with different atomic masses), incorporated into phosphate during bone growth depends, in part, on body temperature, such that consistent oxygen isotope composition across an animal skeleton indicates a consistent, stable temperature in the tissues of that individual. Investigations into the oxygen isotopes of an adult *T. rex* skeleton by Reese Barrick and William Showers (1994) concluded that king tyrant body temperatures varied by no more than 4°C, with the tail tip and extremities of the limbs showing greater temperature range than the core of the body. This is consistent with living endothermic mammals and birds, which generally maintain stable body temperatures that do not rise or drop by more than 2°C.

This, at face value, makes a good case for endothermy in *T. rex*. However, as noted in our discussion of *Tyrannosaurus* skin in chapter 3, *T. rex*–sized animals can produce a lot of heat on account of their immense body size, so this alone might be enough to maintain a sustained, elevated body temperature without being tachymetabolic. This state is known as "mass homeothermy," a mechanism that allows ectothermic species like leatherback turtles to maintain warm bodies. Mass homeothermy remains a possible physiology for large dinosaurs, but Barrick and Showers (1994) suspect that this was unlikely for *T. rex*, given how little temperature variation is evidenced from king tyrant oxygen isotopes. Even for large animals massing more than five tonnes, Late Cretaceous climates might be expected to create c. 20°C variance in core temperatures over the course of a year, something not evidenced in *Tyrannosaurus* bone geochemistry. Moreover, mass homeotherms rely on their extremities to lose heat in hot conditions, which leads to localized fluctuations in oxygen isotopes in some bones, a state also not recorded in *T. rex* fossils. Bone chemistry, therefore, points to *T. rex* being a tachymetabolic endotherm.

Such a physiology in *T. rex* has many implications for their biology, including the regulation of body temperature. Our discussion of skin insulation in chapter 3 outlined the issues that giant endotherms face with losing body heat; principally, that they exist closer to a state of dangerous hyperthermia than smaller animals, especially in warm climates. This may have factored into the seeming reduction of king tyrant insulation relative to that of earlier tyrannosauroids, causing a secondary reversion to a scalier body (Bell et al. 2017), but *Tyrannosaurus* may have had other tricks to avoid overheating. Movement of air through their pulmonary system, for instance, likely cooled the body through evaporation, moisture within the respiratory system helping to wick away heat in much the same way that sweat cools our skin. Increased vascular flows to areas adjacent to air sacs, such as the blood vessels covering the bones next to the antorbital fenestra and orbit (fig. 3.2), may have increased heat dissipation through this mechanism (Porter and Witmer 2020). This system is employed by living birds and is essential to thermoregulation in species that dwell in hot climates, such as ostriches (Schmidt-Nielsen et al. 1969). Panting and gular fluttering (rapid expansions and contractions of the throat) enhances evaporative cooling within the pulmonary system and the complex respiratory pathways of birds can be manipulated to enable panting without interfering with lung airflow, bypassing health risks imposed by overactive breathing (Schmidt-Nielsen et al. 1969). A king tyrant on a hot day may have panted vigorously, like a warm bird, to stay cool (fig. 4.2).

Elsewhere on the body, a recently discovered structure of the archosaur head may have also aided in shedding heat. Crocodylians possess shallow depressions at the front of the upper temporal fenestrae, termed frontoparietal fossae, that house highly vascular and somewhat fatty organs that cap the muscles occupying the underlying temporal space. Provisional investigations of these structures suggest that their chief function may be the radiation of heat (Holliday et al. 2020). Examination of dinosaur skulls shows that they, too, have frontoparietal fossae and, in *Tyrannosaurus* at least, they appear to have expanded with greater size and maturity, increasing their potential thermoregulatory effect for larger individuals (fig. 4.3; Holliday et al. 2020). The anatomy associated with frontoparietal fossae is newly described and its function is still being discussed (Carr [2020] disagrees with suggestions that they housed non–jaw muscle related tissues in *Tyrannosaurus*, for instance), but they may have more to tell us about dinosaur thermoregulation in future.

A further inference for an endothermic *Tyrannosaurus* concerns appetite. Tachymetabolic endotherms consume about ten times as much food as ectotherms of similar mass, so we might imagine king tyrants having the constant, voracious appetite of a mammal or bird rather than the monthly or annual intakes of a large modern predatory reptile. *T. rex* were probably never too far from the urge to find their next meal, and they must have needed a large pool of potential prey items to sustain them. Perhaps, like mammalian and avian predators, they existed in relatively low population densities to avoid depleting prey sources. The exact predator-to-prey ratios for *T. rex* remains something we can only speculate on (Farlow 1993), but such considerations typically involve modeling caloric intake as one of several factors influencing population density. One extremely serious study of *T. rex* dietary requirements (Brett-Surman and Farlow 1997) calculated how many lawyers it would take to maintain a healthy *Tyrannosaurus* over the course of a year, providing us with an especially relatable idea of *T. rex* dietary intake. It posited that an endothermic 4,500 kg king tyrant would consume 292 attorneys of 68 kg mass per annum, equating to almost one a day. A mass of 4.5 tonnes is on the light side for a mature *T. rex* however (see below), so this value would almost certainly surpass one lawyer per day in larger, fully grown animals. Assuming that lawyers represent the same energy value as humans without legal training, Brett-Surman and Farlow may

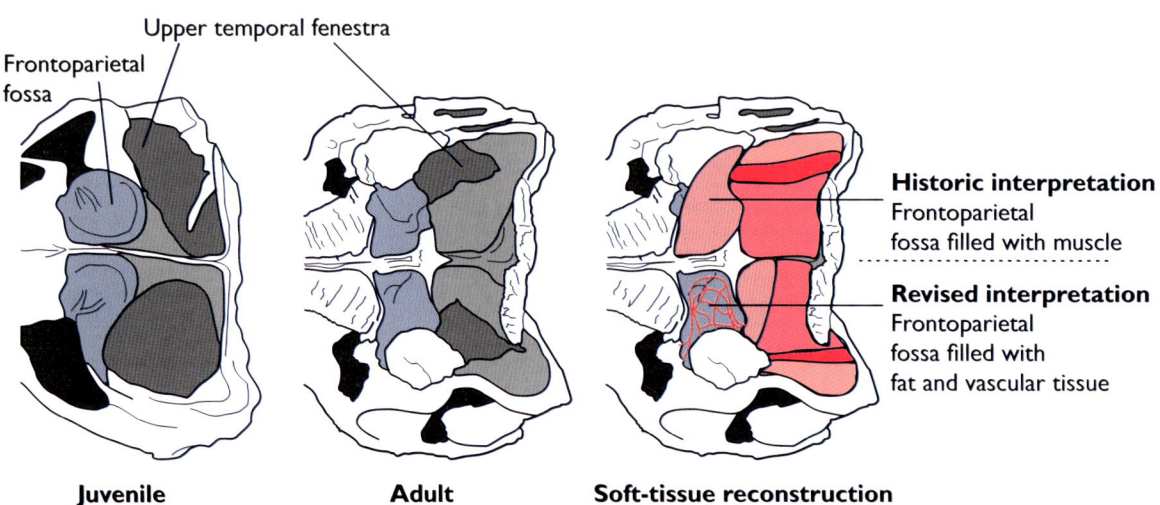

FIGURE 4.2

A warm day in the Cretaceous sees *Tyrannosaurs* employing cooling behaviors: panting, fluttering its throat, covering itself in mud, and reclining in water. Like living tachymetabolic giants, *T. rex* surely faced more challenges with staying cool than warming up, but their (seeming) reduction of insulating fibers and extensive air sac system may have helped to prevent overheating.

Upper temporal fenestra

Frontoparietal fossa

Historic interpretation
Frontoparietal fossa filled with muscle

Revised interpretation
Frontoparietal fossa filled with fat and vascular tissue

Juvenile   **Adult**   **Soft-tissue reconstruction**

FIGURE 4.3

Frontoparietal fossae, newly identified structures on archosaur skulls possibly linked with thermoregulation. These are particularly large in adult *Tyrannosaurus*, their extent potentially reflecting the challenges of heat dissipation at increased body size. Images after Holliday et al. (2020).

have cracked why movie *T. rex* are so single-minded in their pursuit of human prey: animals of our size would never be quite large enough to keep king tyrant bellies full for long.

GETTING BIG: THE REMARKABLE GROWTH STRATEGY OF *TYRANNOSAURUS REX*

Staying well fed would be especially essential for *Tyrannosaurus* as the majority of their existence was not spent maintaining a steady body weight, but adding to it. *T. rex* are famed for their remarkable size and power, but they deserve to be recognized for another component of their biology: the radical changes in body form that occurred from hatching to adulthood. Studies into organismal growth, or ontogeny, can be challenging for paleontologists because recognizing young and old animals of the same species is not always straightforward. As we saw in chapter 2, juvenile *T. rex* were so unlike adults that their fossils have confused and sparked controversy among paleontologists, frequently being identified as different theropod taxa rather than growth stages of one species. Today, most tyrannosaur researchers are satisfied that these fossils exhibit osteological features indicative of immaturity, and, in concert with consideration of their anatomical characteristics and geological provenance, they are best interpreted as young king tyrants. Equipped with these specimens, researchers have learned how *Tyrannosaurus* morphed from small, gracile juveniles into large, robust adults (fig. 4.4; Carr 1999, 2020; Carr and Williamson 2004; Horner and Padian 2004; Erickson et al. 2004; N. Myhrvold 2013; Woodward et al. 2020; Cullen et al. 2020; D'Emic et al. 2023). It seems that the *T. rex* growth strategy was an extreme example of allometric growth, whereby body parts developed at different rates, meaning that older individuals do not have the same proportions as younger members of the same species (contrasting, therefore, with isometric growth, where all body parts grow at the same rate).

A reasonably complete ontogenetic series has been reconstructed for *T. rex* that incorporates fragmentarily known, relatively small juveniles with skulls of around 40 cm long to gigantic adults with heads 1.4 m long (fig. 4.4, 4.5D). To understand how rapidly these young king tyrants became old, gerontic adults, an accurate assessment of *Tyrannosaurus* specimen maturity is required. Opinions on dinosaur growth rates have varied with changing opinions on their physiology. If typically "reptilian" metabolisms are assumed, *Tyrannosaurus* would have needed centuries to attain their adult size (Lee and Werning 2008). Such estimates are unrealistic in light of what we now understand about dinosaur physiology, however (Lee and Werning 2008; Erickson 2014), and precise means of estimating dinosaur specimen ages have started to constrain their lifespans and growth rates. Under microscopy, a sectioned bone looks much like a cut tree trunk with alternating bands of light and dark material. The dense, dark rings are lines of arrested growth, or LAGs: dense bone deposited during seasonally induced periods of lessened size increase (fig. 2.16C). These occur in animals of all kinds, from "cold-blooded" lizards and crocodylians to "warm-blooded" mammals. LAGs thus provide an annual record of growth and, in theory, we need only to count them to estimate the age of a fossil animal at their time of death. If these data can be mapped against measures of animal size, they provide insights into how rapidly fossil species grew and when certain anatomical features developed. With the right fossils, we can even predict the age when reproductive behavior began.

As usual, harsh reality pours cold water on elegant theory, and estimating dinosaur ages via LAGs is somewhat more complex than this. Fossilization can remove LAGs, while weight-bearing and non-weight-bearing bones sometimes have different age signals (Cullen et al. 2020). The most dependable bones for deducing specimen ages have also proven difficult to ascertain for dinosaurs, but, in large theropods, it seems that hip and leg elements provide the most reliable and complete LAG data (Erickson et al. 2001, 2004;

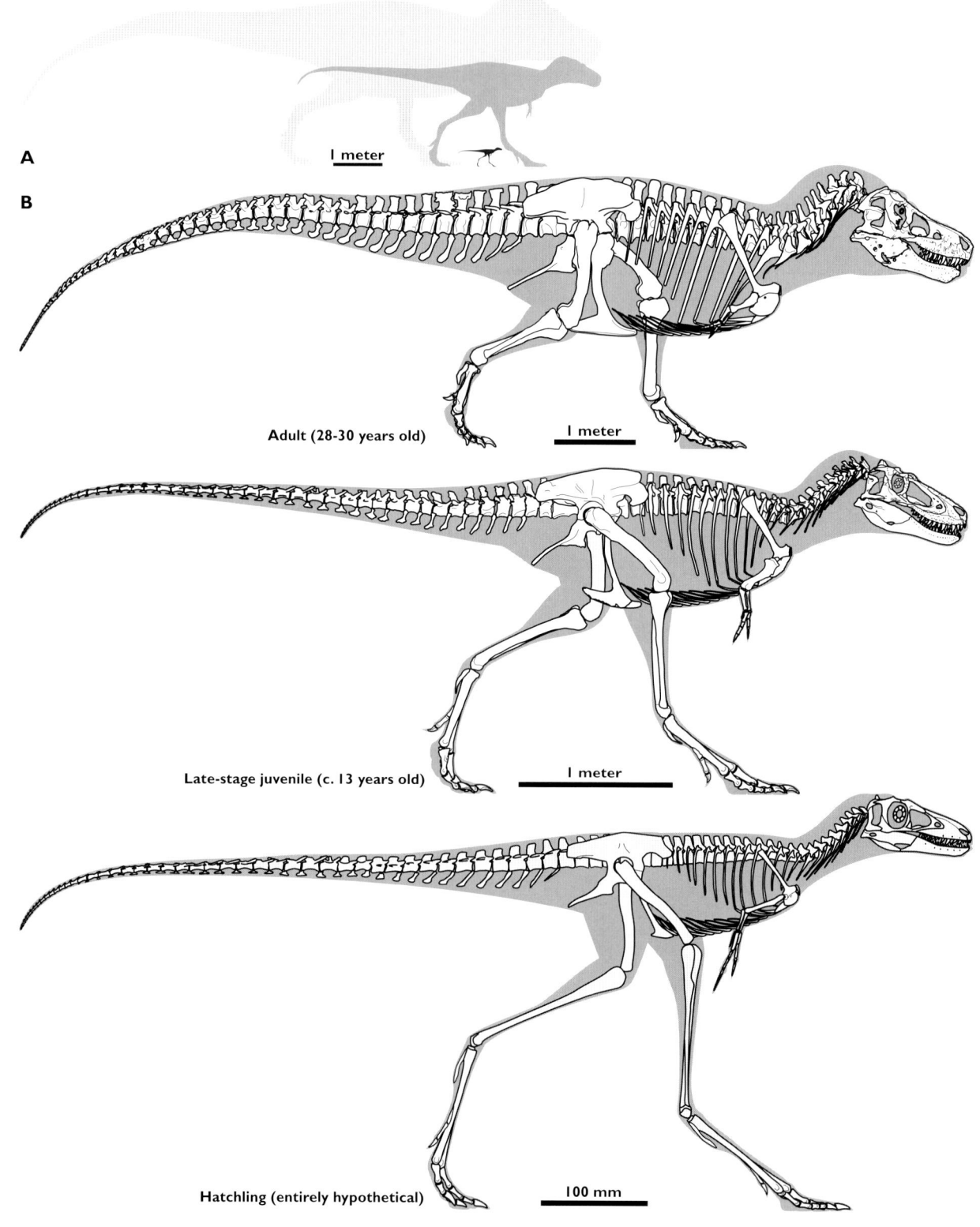

A

B

1 meter

1 meter

Adult (28-30 years old)

1 meter

Late-stage juvenile (c. 13 years old)

1 meter

Hatchling (entirely hypothetical)

100 mm

FIGURE 4.4

The dramatic shifts in body plan exhibited by growing *Tyrannosaurus*. (A) Sizes of a hypothetical hatchling, late-stage juvenile ("Jane") and large adult ("Sue") compared; (B) skeletal reconstructions of the same growth stages restored to the same body length.

Horner and Padian 2004; Cullen et al. 2020). Working through these issues, LAG studies have concluded that dinosaurs had diverse growth strategies (fig. 4.5C; N. Myhrvold 2016; Cullen et al. 2020; D'Emic et al. 2023), but that, in general, they grew quickly, more like birds and mammals than lizards, crocodylians, and other reptiles. Although perhaps not matching the fastest mammalian and avian growers, Mesozoic dinosaurs attained full size in a few years or decades rather than, as with our largest modern reptiles, many tens of years.

Tyrannosaurus were among the first dinosaurs to have their growth speeds estimated (fig. 4.5A; Horner and Padian 2004; Erickson et al. 2004) and continued study has made their growth regime especially well understood (Erickson 2014; Carr 2020; Cullen et al. 2020; D'Emic et al. 2023). Different approaches to calculating *Tyrannosaurus* growth rates have cast slightly different shades on their development rates, but, broadly, it seems that a period of accelerated maturation during their second decade was responsible for transforming relatively small juveniles into gigantic, robust adults. Bursts of rapid growth were common to many dinosaur clades, and, among large theropods, tyrannosaurids and some allosauroids exploited this tactic to achieve great size as soon as possible (Erickson et al. 2004; Cullen et al. 2020; D'Emic et al. 2023). Not all giant dinosaur predators relied on growth spurts to attain large size, however (fig. 45C), some instead simply growing at a steady rate over many decades (Cullen et al. 2020; D'Emic et al. 2023). At least some carcharodontosaurids were still growing at ages well beyond those of the oldest known *T. rex* specimens, perhaps attaining full size at thirty-five to forty-nine years, and then continuing to live for several years after (fig. 4.5B). *T. rex*, in contrast, seem to have largely stopped growing by their midtwenties and, to date, no

specimen older than about thirty years old has been identified (Erickson et al. 2004; Cullen et al. 2020). This has interesting biological implications for these different giant carnivores, but our knowledge of dinosaur ontogeny is still in its infancy and further studies are needed to ascertain the significance of these varied growth regimes.

Despite following the general dinosaurian trend of growing rapidly to full size, *T. rex* maturation was, in other respects, far from typical (Erickson et al. 2004; Erickson 2014; D'Emic et al. 2023). Most tyrannosaurids began their growth spurts before they reached ten years of age, and, at their peak, they gained less than 250 kg of body mass each annum (fig. 4.5A). *T. rex*, however, seems to have had an especially long juvenile phase, slowly reaching masses of 600–1,300 kg by their thirteenth year (Erickson et al. 2004; Woodward et al. 2020; Carr 2020). Thereafter, they stampeded toward giant size and adulthood, accumulating around 700–890 kg each year: 1.5 to 4 times faster than the likes of *Daspleteosaurus* or *Albertosaurus* (Erickson et al. 2004; D'Emic et al. 2023). Such a growth rate is not entirely unprecedented among other large theropods (the carcharodontosaurid *Mapusaurus* may have even grown slightly faster; D'Emic et al. 2023) but *T. rex* was among the fastest growing of all theropods, and was certainly the fastest-growing giant predatory species (fig. 4.5C). By their late teens, *T. rex* already surpassed most or all other tyrannosaur species in body mass (Carr 2020) and only then, as they approached or entered their twenties, did growth finally slow (Erickson et al. 2006; Cullen et al. 2020). Development finally ceased in their early to midtwenties when *Tyrannosaurus* growth bands became mere microns across, representing the end of measurable growth. These final, dense series of LAGs are known as the external fundamental

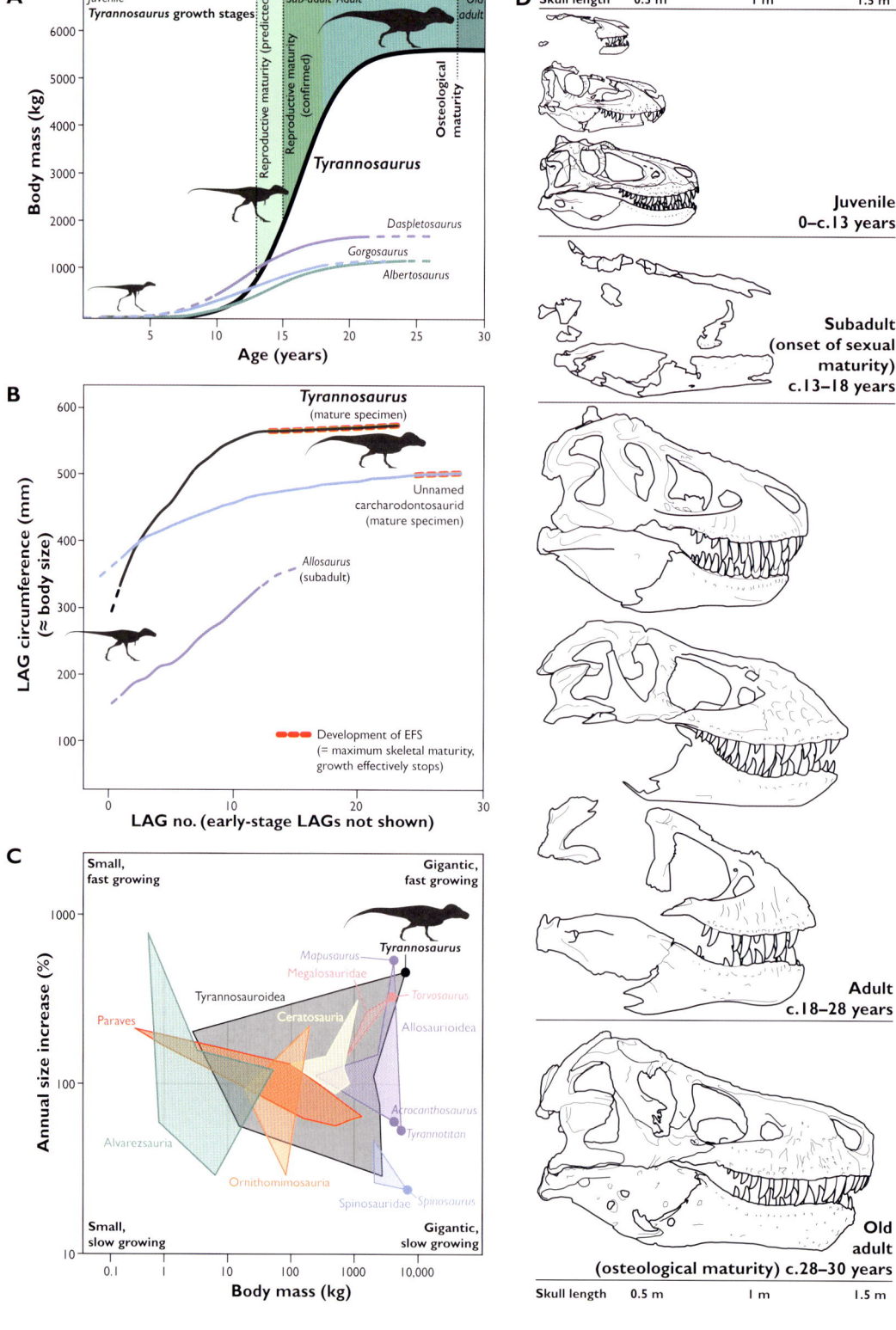

FIGURE 4.5

Growth strategies and cranial changes in *T. rex* ontogeny. (A) Growth rates of tyrannosaurids mapped against life stages and reproductive onset in *Tyrannosaurus*; (B) comparison of LAG count (≈ age) against LAG size among several large theropods, showing different growth strategies; (C) the variable rates of annual size increases of theropod dinosaurs, showing the multitude of growth strategies within the clade; (D) changes in *T. rex* skull shape with maturation. (A) Adapted from Erickson (2014); (B) after Cullen et al. (2020); (C) after D'Emic et al. (2023); subadult skull remains of (D) after Carr (2020).

system (EFS), a characteristic of animals that exhibit finite growth (as opposed to infinite or continuous growth, which is common in many living reptiles).

Understanding how rapidly *Tyrannosaurus* grew makes the pronounced changes exhibited during their development even more remarkable. Virtually every bone in the *T. rex* body underwent a dramatic shift in relative size and proportion during ontogeny, such that juveniles had a body plan almost unrecognizable from that of mature adults (figs. 4.4, 4.6). These differences dictate that characterizing the growth forms of *T. rex* is best done in stages, from young juveniles to old adults.

FIGURE 4.6

Small juvenile, subadult, and young adult *Tyrannosaurus* skeletons on display at the Natural History Museum, Los Angeles. The juvenile skeleton (lower left) is hypothetical except for parts of the skull (see Figure 4.7), while the subadult (lower right) is reconstructed from a partial skeleton (the holotype of "*Dinotyrannus megagracilis*").

Juvenility: 0–12 years. Only a few fossils of juvenile *T. rex* have been discovered, making the first phase of king tyrant lives the most mysterious. Our handful of specimens from this interval consists primarily of cranial material, among which is the broken skull and lower jaw of an estimated 40 kg two-year-old once known as "the Jordan theropod" or "*Stygivenator molnari*" (figs. 4.5D, 4.7; Molnar 1978; Carr and Williamson 2004; Carr 2020). The slightly larger, more mature "*Nanotyrannus*" holotype (fig. 2.14) may represent a four-year-old (Carr 2020). Fossils of younger animals remain elusive, as they do for virtually all tyrannosaurids, with the embryonic tyrant teeth and bones recently described by Funston and colleagues (2021) marking our first discoveries of such animals.

From these fragmentary remains we can deduce that juvenile king tyrant skulls were much shallower than those of their parents, although they already possessed broadened muzzles and expanded temporal regions relative to other theropods, as well as "D"-shaped premaxillary teeth (Molnar 1978; Carr 1999, 2020; Carr and Williamson 2004). Their maxillary and dentary teeth were also robust, although they were not yet expanded into the incrassate dentition of adult *T. rex*. This, combined with their smaller temporal volumes for jaw muscles, and more gracile, mechanically weaker skulls, suggests that juveniles had yet to develop powerful, bone-crushing bites (Carr 2020; Rowe and Snively 2021; Johnson-Ransom et al. 2024). Juveniles possessed three or four more additional teeth in their maxillae and dentaries (maximum total counts sixteen and seventeen, respectively), a trait that some have used to argue for these small skulls belonging to distinct taxa (chapter 2). Tooth loss through growth is evidenced in several living crocodylian and lizard species, however (Brown et al. 2015), so such changes in reptile growth are not unprecedented. Conversely, juvenile *Tarbosaurus* show that their tooth counts remained consistent throughout life (Tsuihji et al. 2011), one of several differences between juvenile *T. rex* and their Asian relatives. Other distinctions include the (estimated) 40 cm long skull of a juvenile *Tyrannosaurus* appearing

somewhat longer than those of the similarly sized (c. 30 cm) *Tarbosaurus* and *Raptorex* skulls of comparable growth stages (fig. 4.7B), and these Asian taxa also possess short, adult-grade arms in early juvenility (Sereno et al. 2009; Tsuiihji et al. 2011). As we'll explore more below, *T. rex* juveniles had proportionally longer arms than adults (fig. 4.4; Williams et al. 2010).

Frustratingly little is known about the body plans of the youngest king tyrants. With no fossils to work from, several authors have attempted to reconstruct the appearance of baby tyrannosaurs by extrapolating the skeletal scaling trends of larger specimens to diminutive sizes (Russell 1970; Currie 2003b; Funston et al. 2021). Although differing in some details, their results agree in all basic aspects: baby tyrants were more akin to stilt-legged monitor lizards than the podgy, plush-toy-like animals seen in some media. These models of long-legged, shallow-jawed, and small-bodied individuals contrast dramatically with the proportions of their parents, especially in the length of their shins and feet, and also by their shallower pelves and longer arms. A young juvenile *T. rex* may have been tiny compared to an adult at just one meter long, but they had—relatively speaking—the longest legs of any growth stage (fig. 4.4). Tyrannosaurid leg length changed dramatically with maturation so constraining the leg length of hatchlings has proven difficult, our scaling models proving especially sensitive to different predictive parameters. The most exaggerated estimated proportions stem from a study by tyrannosaur expert Phillip Currie (2003b) where, owing to our lack of juvenile skeletons to guide extrapolations from a dataset of large animals, baby tyrannosaurs are predicted to look like extras from *The Nightmare Before Christmas* (fig. 4.8). These predictions were understandably dismissed as unbelievable by Currie (2003b) as quickly as they were introduced, who cautioned that "there are limitations to what can be done in extrapolating this data" (Currie 2003b, p. 663). This remains the case today, and even with embryonic tyrannosaur material to work from (Funston et al. 2021), the error bars on predictions of juvenile tyrant dinosaur proportions are wide. Additional, newly

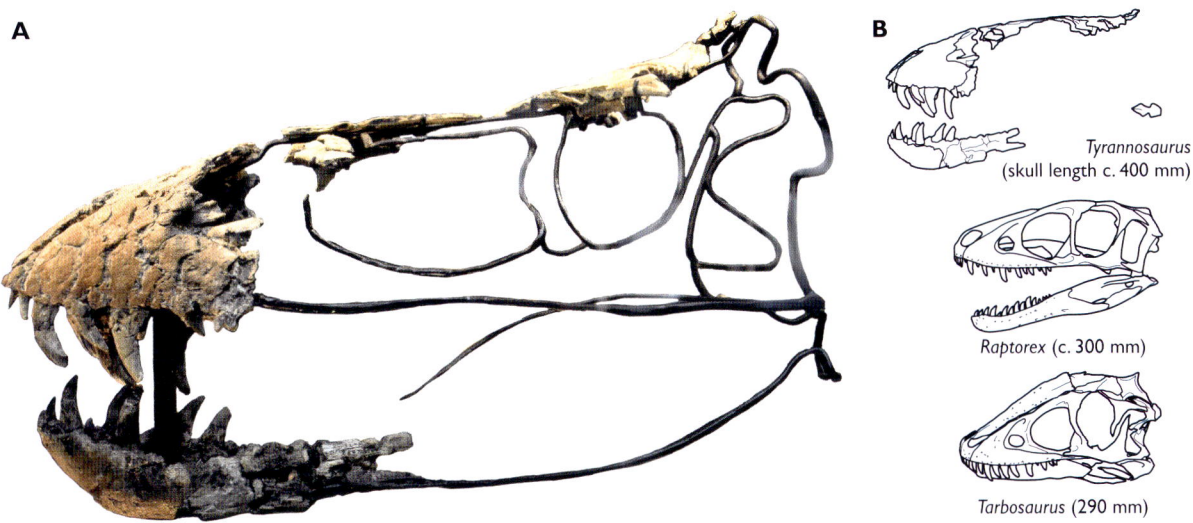

A

B

Tyrannosaurus
(skull length c. 400 mm)

Raptorex (c. 300 mm)

Tarbosaurus (290 mm)

FIGURE 4.7

(A) Cranial remains of one of the smallest *T. rex* fossils, the skull once known as the "Jordan theropod," and the holotype of "*Stygivenator molnari*." This specimen is now regarded as a two-year-old *T. rex*; (B) juvenile tyrannosaurid skulls, *Tyrannosaurus*, *Tarbosaurus*, and *Raptorex*, compared. *Raptorex* skull after Sereno et al. (2009); *Tarbosaurus* after Tsuihiji et al. (2011).

discovered *T. rex* juveniles await description at the time of writing: research on these specimens may shed light on the body plans of tiny king tyrants.

Late juvenility into subadulthood: 13–18 years. The middle point of *T. rex* lives appears to have represented a concentrated transition from lanky juveniles into gigantic, multi-tonne goliaths (Erickson 2006). This growth stage also has a sparse fossil record, but the late-stage juvenile *Tyrannosaurus* "Jane" has been transformative in our understanding of not only *T. rex* adolescence, but also *T. rex* ontogeny in general (figs. 2.15 and 4.4; Erickson 2004; Woodward et al. 2020; Carr 2020). Represented by a partial skeleton, including an excellently preserved skull, this thirteen-year-old *T. rex* shows that juveniles retained their leggy, shallow-jawed configuration into moderate body size: "Jane" is estimated to have been about 6 m long and between 600 and 1,300 kg (Hutchinson et al. 2011). Their legs were ideally suited to running, bearing the

long shins and metatarsals of rapid locomotors. This specimen also shows that king tyrant arms were proportionally longer during earlier growth stages, with claws already as large as those of adults (fig. 2.16) and a wider glenoid joint permitting greater shoulder mobility (P. Larson 2008b; Williams et al. 2010). This finding potentially has significance for the long-running discussion about arm function in *T. rex* (below).

Had "Jane" lived longer, they would have embarked on the most dramatic period of growth and bodily transformation in the entirety of *T. rex* ontogeny (Carr 2020). An especially intense series of changes saw large "Jane"-grade juveniles attain forms that are more readily recognizable as "classic" *Tyrannosaurus*. This started at around fourteen or fifteen years of age and is typified by specimens such as the *T. rex* once named "*Dinotyrannus megagracilis*" (figs. 4.5D, 4.6; Molnar 1980). King tyrants of this maturity started to develop the deeper skulls and fatter teeth of adults, with stronger bite forces to match (Carr 2020). Their body proportions

FIGURE 4.8

The challenges of predicting hatchling tyrannosaur proportions: restoration of a young tyrannosaurid juvenile based on Currie (2003b). Estimating hatchling measurements from adult specimens is very difficult and can lead to amusingly exaggerated forms such as this. Note, however, that the legs in this reconstruction are only slightly more ridiculous than tyrannosaur hatchlings may have possessed (Figure 4.4).

shifted too, transitioning from long-legged and (relatively) long-armed animals to individuals with longer bodies, shorter forelimbs, and stouter legs (Hutchinson et al. 2011; Carr 2020). This coincided, presumably, with a biomechanical shift from having legs adapted for fast locomotion (Erickson et al. 2004) to limbs better suited to supporting large body masses, which perhaps now approached two tonnes (Carr 2020).

These dramatic changes to body proportions and size coincided with an important life history threshold: sexual maturity (Schweitzer et al. 2005b, 2016; Woodward et al. 2020). Such determinations are possible because female dinosaurs appear to have developed

a distinctive bone tissue during reproductive phases known as "medullary bone" (fig. 4.9). Created within cavities of the limb bones, this unusual tissue type occurs exclusively today in female birds during egg-laying cycles, forming rapidly in response to increased estrogen levels (Schweitzer et al. 2016). As a highly vascular and ephemeral component of bone structure, medullary bone does not seem to enhance skeletal strength, instead serving as an easily accessed calcium reservoir for the creation of eggshells. Diagnostic properties from bone microstructure and skeletal location allow medullary bone to be identified in extinct birds and their relatives, including both non-avian theropods and ornithischian dinosaurs (Schweitzer et al. 2005b; Lee and Werning 2008; Hübner 2012). Determining the presence of medullary bone can be complicated by diseased skeletal tissues adopting superficially similar microstructures, but chemical and microstructural analyses can distinguish genuine reproductive tissues from aberrant bone growths in such cases (Schweitzer et al. 2016). The potential to recognize a female-specific reproductive tissue in Mesozoic dinosaurs is a powerful tool for understanding dinosaur biology, helping us to identify both the sex of individual specimens (see chapter 6) as well as, when factored into a growth series, the timing of sexual maturation.

Medullary bone has been found in at least two *T. rex* samples, the youngest of which was fifteen years old (Schweitzer et al. 2005b, 2016; Woodward et al. 2020). This implies that *Tyrannosaurus* was reproductively capable before the attainment of what we might traditionally consider their "adult" features, including gigantic size. This is unsurprising, given that reproductive maturity frequently develops well before the

attainment of full development in living vertebrates, including our own species. It is also unlikely that this fifteen-year-old individual represents the earliest possible onset of sexual maturity (Woodward et al. 2020), and *T. rex* may have been reproductively capable even earlier. Crocodylians, for instance, attain sexual maturity at around 50 percent of their adult proportions, which would equate to animals of around thirteen years old in *T. rex* (Woodward et al. 2020; Carr 2020).

Reaching sexual maturity between thirteen and fifteen years of age coincides with the onset of accelerated growth and skeletal modification, suggesting that these events may be linked. Noting this, Thomas Carr (2020) has suggested that king tyrants exhibited secondary metamorphosis: a period of concentrated, radical anatomical development associated with sexual maturation (as opposed to primary metamorphosis, which occurs before sexual maturity; see Rose and Reiss 1993). Researchers differ on exactly how metamorphosis should be defined, but *T. rex* meets most criteria that might be asked of a metamorphosing species. They exhibited a

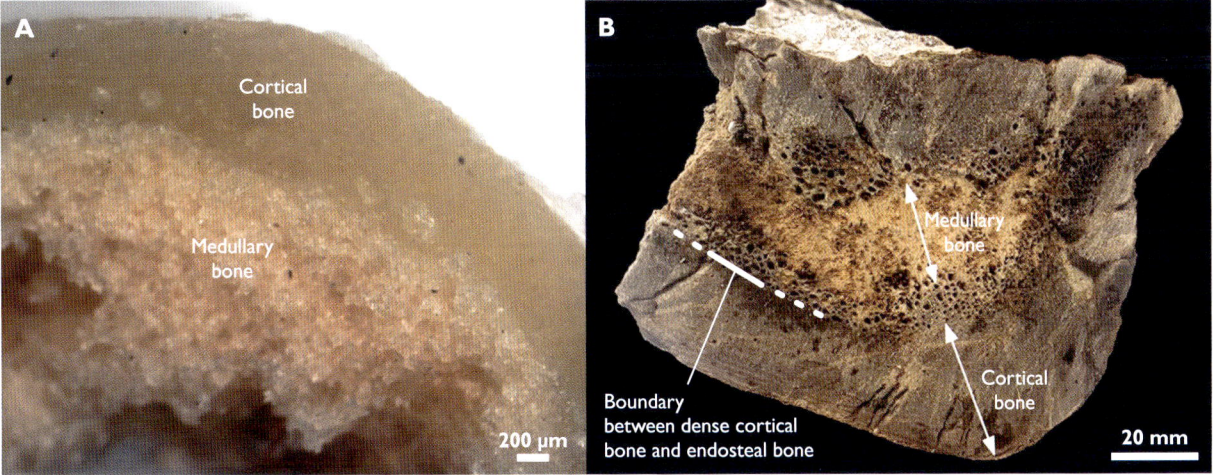

FIGURE 4.9

Reproductive bone tissue, medullary bone, in chickens *Gallus domesticus* (A) and *T. rex* (B). The identification of medullary bone in fossil reptiles is not without controversy, but, if correct, it captures reproductive periods in dinosaur life histories and allows specimens to be determined as female. Images from Schweitzer et al. (2016).

period of concentrated, whole-body, post-embryolog-ical change bracketed by periods of slower develop-ment, and probably also transitioned from one ecology to another as they matured (see chapter 6; Carr 2020). It is difficult not to equate such changes with the likes of salmonid fish, which radically alter every aspect of their bodies once they reach their reproductive age, in-cluding the shape of their skulls and physiology, while changing their ecology at the same time (in salmon, this metamorphosis is termed "smoltification").

Adults: *years 18 upward*. The concentrated ontogenet-ic alterations that defined the teenage years of *T. rex* slowed as they became young adults near their twenti-eth annum (Carr 2020). Now surpassing three tonnes, growth had not yet finished. Their skulls continued to deepen and their cranial features became more de-fined (fig. 4.5D), developing prominent chins and more conspicuous ornamentation (Carr 2020). Changes also continued in their bodies, including the attainment of their enormous pelvises, deep torsos, and hindlimbs further suited to weight bearing over running. The latter occurred via a reversal of the length of major leg bone elements where the femur became longer than the tibia, and the metatarsus became proportionally short. Young adult *Tyrannosaurus* include several famous specimens, such as the Nations/Wankel *T. rex* (fig. 1.2); the Amer-ican Museum of Natural History specimen 5027 (fig. 1.18); "Stan" (fig. 3.13); and "Thomas" (fig. 4.6).

Tyrannosaurus growth slowed considerably in their twenties and modifications across the skeleton became minor, mostly pertaining to greater demarcation of exist-ing features such as muscle scars and epidermally derived textures. Their bones finally developed external funda-mental systems during their mid-to-late twenties, mark-ing the cessation of growth. Only a few fully mature king tyrants have been discovered, including "Sue" (fig. 1.20); "Scotty"; and "Trix" (fig. 3.22). "Sue", at least, appears to have only lived for a few years after growth stopped, seemingly reaching 27–33 years before death (Erickson et al. 2004; Cullen et al. 2020). Complications with in-terpreting the LAG counts of other very large specimens

precludes knowledge of their age, however (e.g., Per-sons et al. 2020; Bjil 2016), and some disagreement ex-ists over whether we have discovered the upper limit of *T. rex* lifespans (Horner et al. 2011; Cullen et al. 2020). The development of medical conditions commensurate with advanced age in king tyrants of around thirty years old suggests that animals of this maturity were feeling their mileage, however (chapter 6). Even if their late twenties or early thirties were not the maximum lifespan of *T. rex*, those approaching their thirtieth birthdays were probably closer to the end than the beginning.

HOW LARGE WAS "LARGE"?

Considering how *T. rex* grew to their maximum sizes in-vites an obvious question: how large could king tyrants get? Statistics on dinosaur size are some of the most de-sired information among fans of the terrible lizards as their great dimensions are undeniably part of their ap-peal. Entire popular books are devoted to these figures (e.g., Molina-Pérez and Larramendi 2019), and many of us grow to command detailed mental lists of dinosaur proportions and measurements, instantly able to recall the biggest, the smallest, the heaviest. *Tyrannosaurus*, of course, dominates discussions of the largest thero-pod, even if we are considering some of the other con-tenders for this title (chapter 2).

Paleontologists share this keen interest in dinosaur size, but not (ostensibly) for bragging rights about who has discovered the largest species. Rather, size is es-sential to understanding much about animal biology. Without understanding their dimensions and masses, properties of growth, physiology, functional morphol-ogy (that is, the relationship between anatomical form and function; see below), and biomechanics cannot be investigated in detail. Body masses, in particular, are integral to many dinosaur studies. However, determin-ing the dimensions and masses of dinosaurs is a trickier business than is implied by news reports and popular works with pages of stats and figures, even for relatively well-studied species like king tyrants.

Total length is theoretically the most straightforward measure of overall size because it can be measured relatively objectively from fossil specimens. If we articulated the skulls and vertebral column of the largest known *T. rex* and measured them from end to end, we would recover a value of between 12.3 and 12.6 m (Brochu 2003; Persons et al. 2020), depending on how much space we left for cartilage between each bone (fig. 4.10). Other adult specimens generally fall a little below this, at lengths of 11–12 m long. These values are estimates, however, because we lack the tail tip from any *T. rex* (chapter 3). The error bars on the remaining tail dimensions are not enormous, but this fact still precludes precise knowledge of *T. rex* length. *Tyrannosaurus* is far from the only dinosaur with an incomplete tail skeleton, and, responding to this, tyrannosaur expert David Hone has wondered if we should abandon using snout-tail tip length metrics for comparing dinosaur size. Might a measure of skull + cervical, dorsal, and sacral vertebrae length (equivalent to the familiar "snout-vent" length used in zoology) be more reliable and useful, given that these elements are known in entirety for many dinosaur species (Hone 2012)? If we were to follow this concept, we would give a "snout-vent" length of c. 5.6 m to a big *T. rex*.

Reporting the height of *T. rex* is a subjective exercise because this dimension is determined by posing, and even subtle changes to the angles between bones can have dramatic effects. This is particularly so for dinosaurs because their hindlimbs were perpetually flexed in life, such that assumptions about the openness of their knees or ankles have a significant impact on height estimates. The poses shown in Figure 4.10 indicate *Tyrannosaurus* heights in different standing and sitting configurations. In a "neutral" standing pose, where the hindlimb was neither strongly flexed nor extended, the top of the pelvis in a large *T. rex* stood 3.3 m tall. When crouched down to their fullest extent, resting their pubic boot on the ground, the top of the pelvis in the same animal would still be taller than most people: about 1.9 m high. A lazy effort at standing upright, pitching the torso 35° upward without any considerable strain on the joints of the legs or tail, sees the top of the skull reaching almost 6 m, comparable to the height of the classic tail-dragging American Museum of Natural History mount (fig. 1.9B; Osborn 1917a).

Such length and height measurements are useful shorthands for us to mentally visualize dinosaur proportions, but linear metrics are not great scientific measures of body size—at least, not on their own. As three-dimensional beings, animal size is best expressed with a value that accounts for their entire form rather than a single, unidimensional measurement. Body mass is thus the preferred way to discuss and compare animal sizes among researchers because, as a measure of animal volume and density, it accounts for all three dimensions of organismal shape and bulk. Furthermore, multiplying body mass by the force of gravity gives us a measure of weight, a critical value for biomechanical considerations that cannot be replaced with any simple measurement of length or height. Thus, if we want to appreciate how large *T. rex* was in a meaningful sense, we need to predict their body masses.

Tyrannosaurus are no strangers to such studies, having been subjected to a particularly large number of body-mass estimates (e.g., Colbert 1962; Alexander 1985; Anderson et al. 1985; Campbell and Marcus 1992; Farlow et al. 1995; Paul 1997; Henderson 1999; Seebacher 2001; Erickson et al. 2004; Horner and Padian 2004; Christiansen and Fariña 2004; Therrien and Henderson 2007; Bates et al. 2009; Hutchinson et al. 2011; Campione et al. 2014; Persons et al. 2020; Reolid et al. 2021). These studies have produced a range of mass values despite working on the same species and, often, even the same specimens (fig. 4.11). The body mass of the mature king tyrant "Sue," for instance, has been predicted at less than 4,000 kg (Campione et al. 2014, using methods outlined by Campbell and Marcus 1992) or almost 10,000 kg (Hutchinson et al. 2011). Such deviance is not unusual for mass estimates of fossil organisms and they pose clear challenges for studies needing reliable, well-constrained mass predictions. With such widely ranging values, can we identify which are most reasonable?

BREATHING LIFE INTO BONES

3.3 m

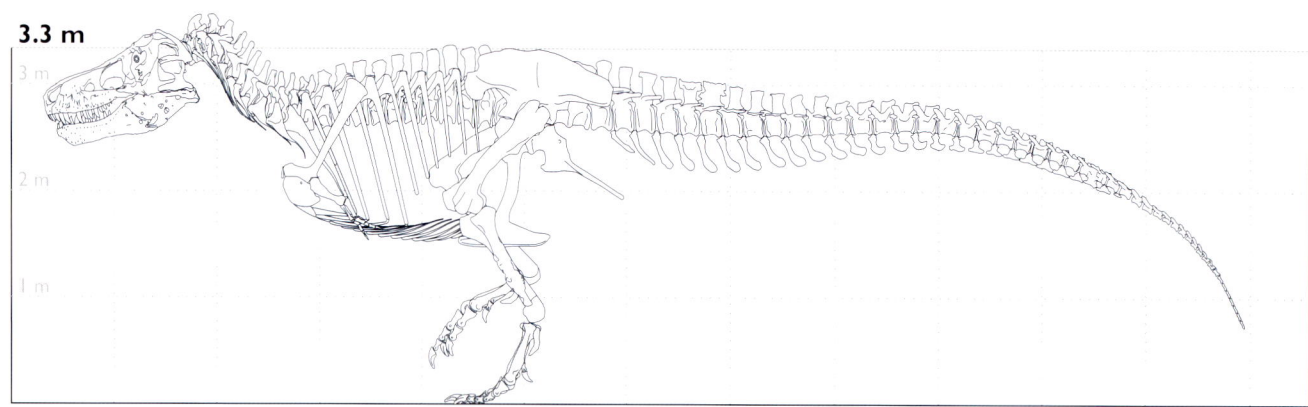

5.8 m

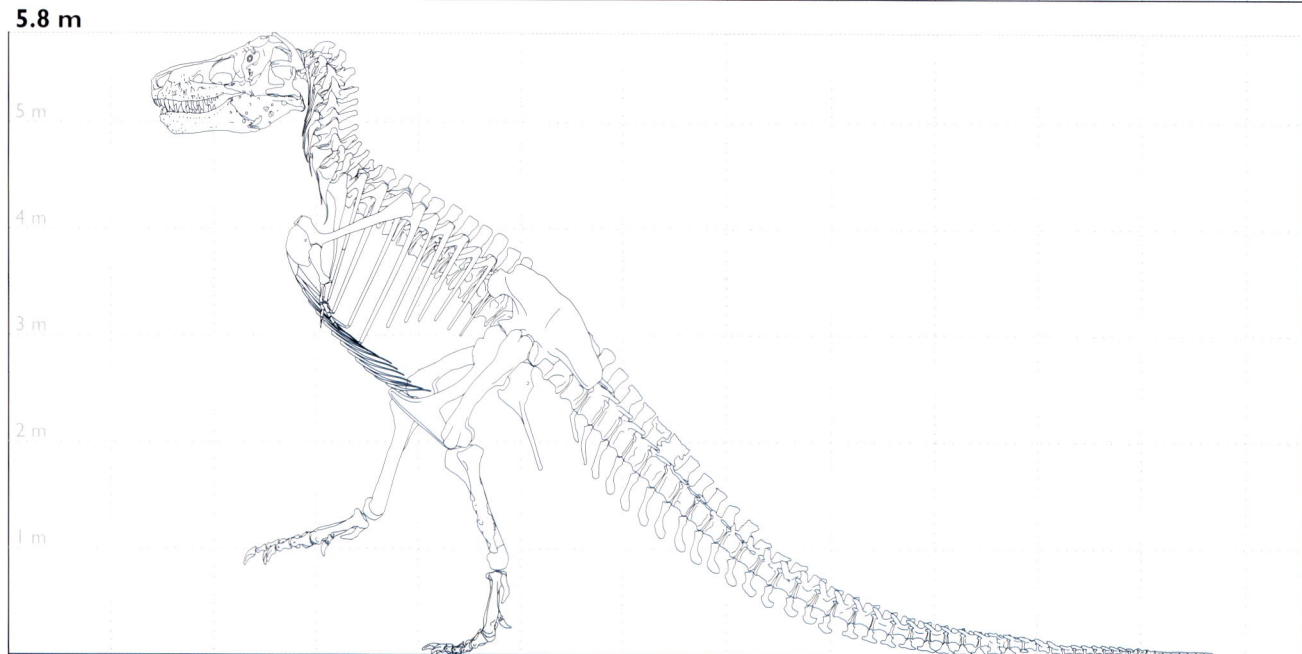

1.9 m

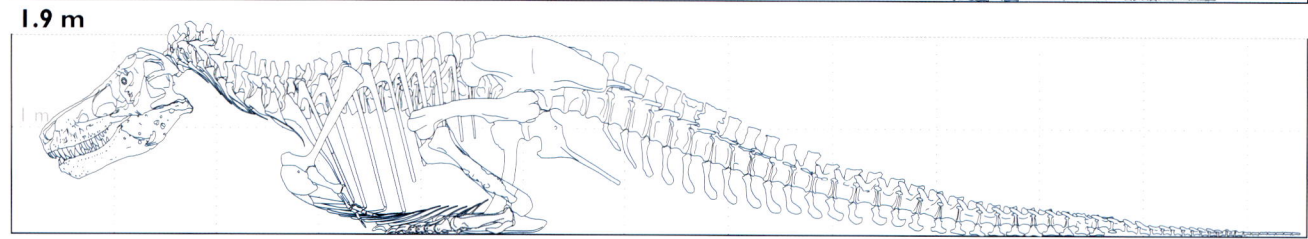

FIGURE 4.10

Some standing, rearing, and crouching dimensions of *Tyrannosaurus*.

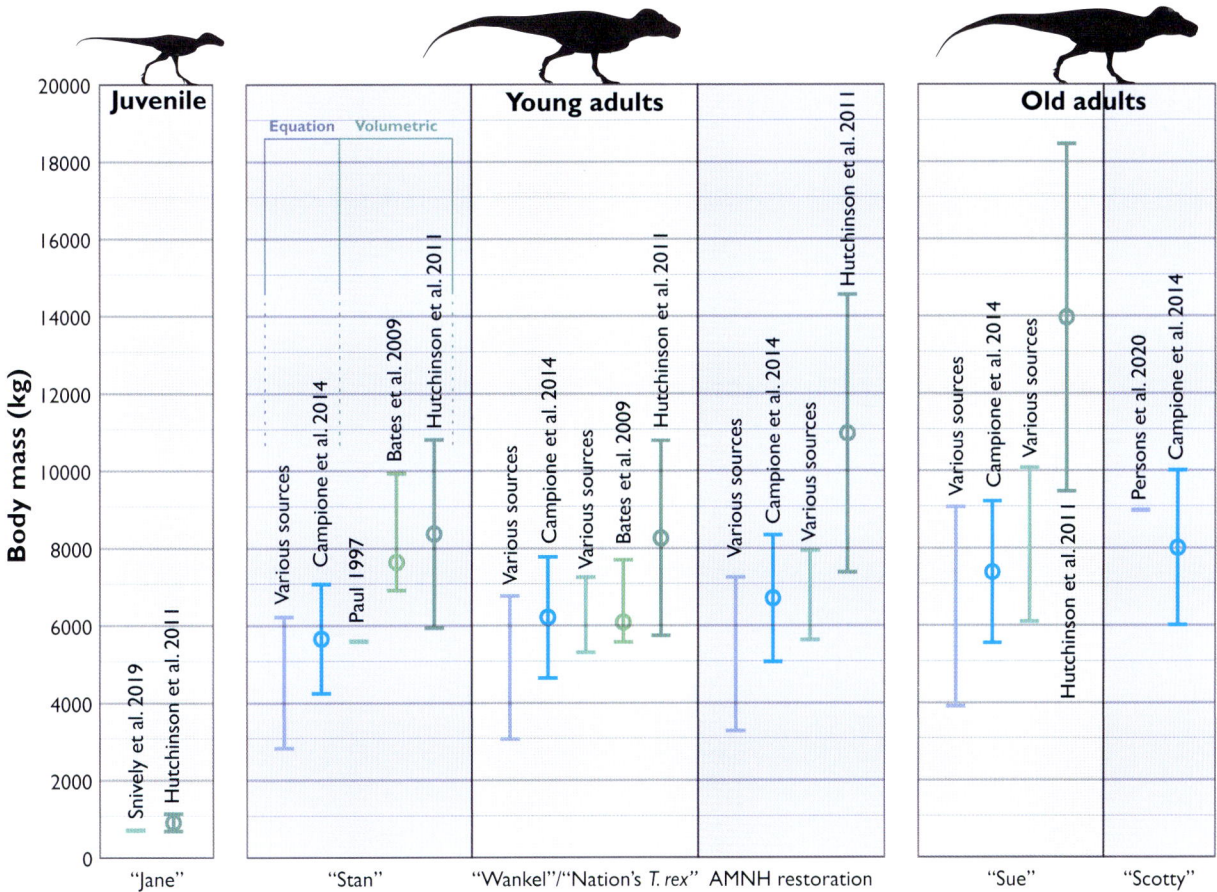

FIGURE 4.11

Summary of body mass predictions for select *T. rex* specimens, comparing volumetric (blue ranges) and equation-based (green ranges) outcomes. Note that, across the scatter of estimates, most point to a 6- to 8-tonne range for adult *T. rex*, with especially large animals tipping the scales closer to 9–10 tonnes. Adapted and expanded from Campione et al. (2014).

Our uncertainty about the masses of extinct animals reflects, in part, two different methodological approaches that have not always produced congruent results: scaling masses from living species, and calculating mass from estimates of body volume and density (Campione and Evans 2020). The former rely on compiling tables of body masses and measurements of weight-bearing bones, typically the circumference of the femur, from living animals. The theory behind such studies is a sound one: mass has a direct impact on the structure

of weight-bearing limb elements, and, if we can understand this relationship in modern animals, we can use limb bones to extrapolate the body masses for extinct taxa. Such approaches have great utility in that, once the groundwork with living animals is performed, mass calculations can be performed for any fossil specimen with the right body part (Campione and Evans 2020). Traditionally, however, equation-based estimates have been considered of doubtful reliability when applied to unfamiliar animals like *T. rex*. Can mass databases based on extant birds, which are far smaller, and stand in a perpetually crouched posture unlike that of any non-avian theropod, be used to predict king tyrant masses, for instance? Should datasets based on mammals, most of which are quadrupedal, be projected to reptilian bipeds? In general, *T. rex* body masses derived from such studies are lower than those calculated using other methods, with some predicting that large adult king tyrants massed a mere three to four tonnes; lighter, even, than modern elephants (Campione et al. 2014).

Mass scaling equations have been rehabilitated to some extent by Nicolás Campione and David C. Evans (2012; Campione et al. 2014). Across a series of studies, they have shown that the weight-bearing bones and body masses of extant animals can reliably correlate across different body plans if datasets are large enough and, furthermore, that comparing bipedal with quadrupedal species can work if appropriate data conversion calculations are used. Such adjustments predict that king tyrant adults ranged from 5.6 to 8 tonnes, with error bars around these estimates constraining the likely figures to between 4.2 to 10 tonnes (Campione et al. 2014). This study predicted that the two largest substantial *T. rex* specimens, "Sue" and "Scotty," had similar masses of 7.3 and 8 tonnes, respectively. Other equation-based estimates of these two specimens are of negligible difference, although "Scotty" is once again larger (8.4 vs. 8.8 tonnes; Persons et al. 2020). These seem to indicate that "Scotty" is the largest presently known *T. rex*, although the closeness of its mass estimate to those of "Sue" preclude placing too much emphasis on this ranking.

Our second approach has also been modernized in recent years. Volumetric mass estimates involve predicting the total volume of animal tissues and multiplying this by an appropriate body density. Because volumetric methods require engaging with the precise shapes of fossil skeletons, they have traditionally been viewed as more reliable than equation-based scaling, and, subsequently, they have often dominated dinosaur mass research. Although more laborious to perform than equation-based estimates, volumetric methods provide greater amounts of information for the effort, such as the volumes and masses of individual body segments or, in detailed modern studies, even the values of specific tissues. For all their utility, however, two questions complicate this conceptually simple approach: How bulky were dinosaurs, and how dense were their bodies? Extinct animal volumes have traditionally been calculated from reconstructions that seem intuitively "right" to the artists or scientists who create them, without much consideration for the margin of error. To account for this, some researchers now model a range of tissue depths to constrain likely masses within an envelope of possibility (fig. 4.12; Hutchinson et al. 2011). The second issue concerns density: How much air space existed within a dinosaur body? This can be derived from related modern animals, such as birds (e.g., Larramendi et al. 2021), which share pneumatic bodies with dinosaurs (chapter 3), but more sophisticated models circumvent this matter entirely by calculating separate volumes for different tissue types (e.g., bone, muscle, airspace, and so on), avoiding the need to rely on a modern analogue. Confidence that at least some volumetric predictions are plausible stems from ground-truthing methodological approaches on living animals of known masses (e.g., Henderson 1999; Hutchinson et al. 2011).

T. rex has been subjected to several generations of volumetric mass studies. Low-tech approaches include submerging models in water to measure their displacement, and thus volume (Colbert 1962; Alexander et al. 1985), or drawing their bodies in multiple views, dividing them into sections to figure out their individual volumes, and then adding them together to get a total (Paul 1997). More modern, high-tech versions involve digitizing mounted skeletons and modeling their tissues directly over their scanned osteology, improving the repeatability of such studies and removing subjective errors that might creep into translating three-dimensional bones into flesh models or 2D reconstructions (fig. 4.12; Hutchinson et al. 2011). Generally, volumetric methods have pointed to young adult *T. rex* massing between six and eight tonnes, with large specimens pushing beyond nine tonnes (fig. 4.11; Colbert 1962; Alexander 1985; Paul 1997; Henderson 1999; Seebacher 2001; Bates et al. 2009; Hutchinson et al. 2011).

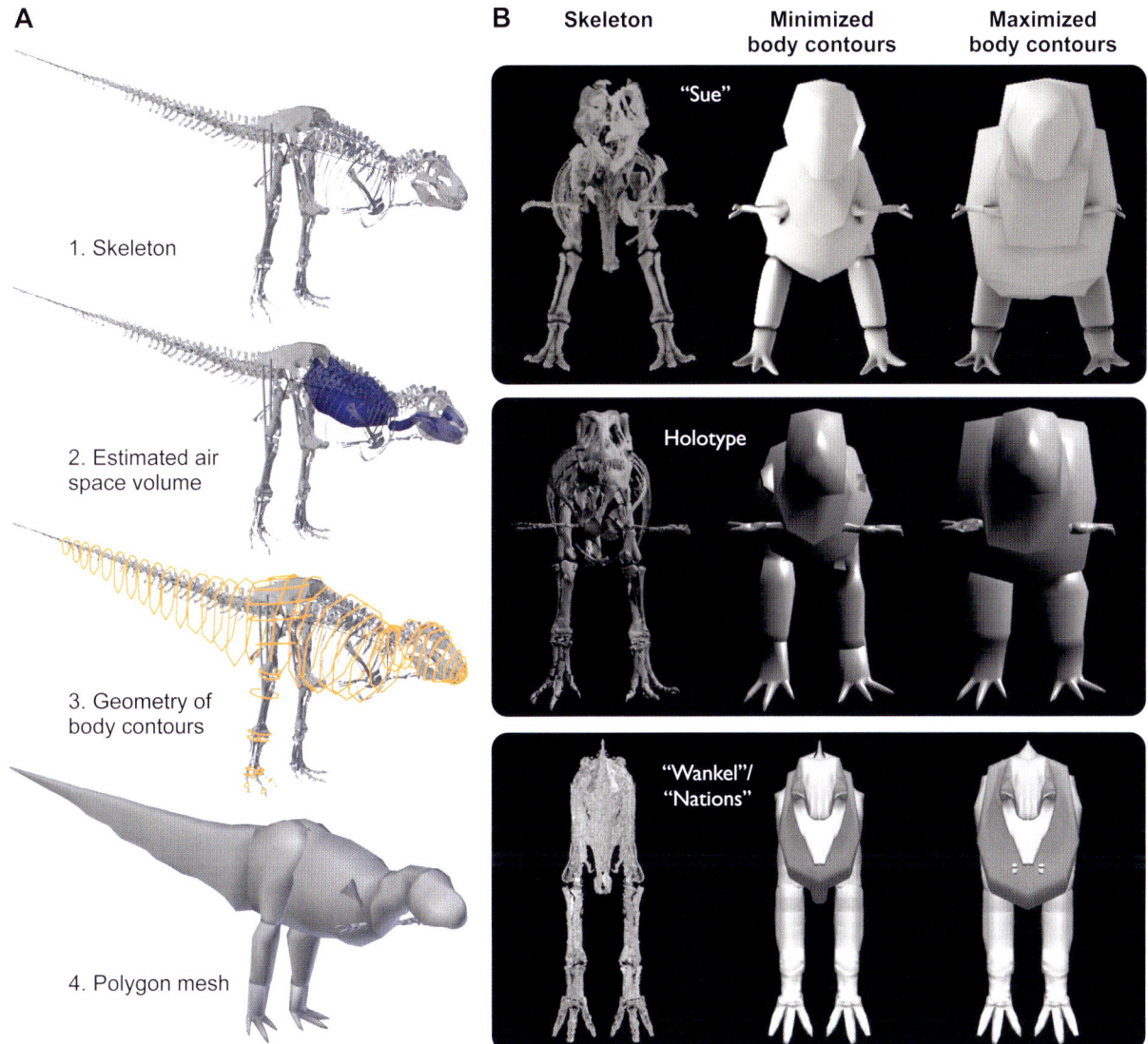

A

1. Skeleton

2. Estimated air space volume

3. Geometry of body contours

4. Polygon mesh

B

| Skeleton | Minimized body contours | Maximized body contours |

"Sue"

Holotype

"Wankel"/ "Nations"

FIGURE 4.12

The principles of volumetric body mass estimation, exemplified by digital methods used by Hutchinson et al. (2011). (A) Reconstruction of soft tissues over a scanned skeleton, including incorporation of torso air space; (B) variable body contour estimates to determine a range of volumes and masses for different specimens. Images from Hutchinson et al. (2011).

These volumetric values are similar to those produced by the adjusted limb bone–based mass estimates of Campione et al. (2014), an encouraging finding that might imply we are finally getting a grip on *Tyrannosaurus* mass. Even with the broad error bars of these mass estimates meaning king tyrants could have been a few tonnes lighter or heavier (fig. 4.11), they give researchers ballpark figures to use in functional and physiological studies: six to eight tonnes in young adult *T. rex*, and maybe more in very large, old individuals. They also give us a little more confidence in comparing *T. rex* masses to those of living animals, wherein we find that six to eight tonnes is comparable to large male African bush elephants (Macdonald 2009), with especially big king tyrants matching the ten-tonne size record for this species (Larramendi 2016).

EVEN BIGGER?

A six-, eight-, or even ten-tonne theropod is undeniably impressive, but is this as big as *Tyrannosaurus* could get? Even though the king tyrant fossil record is unusually good for a substantially sized dinosaur predator, our sampling remains too limited to assume that we've found their maximum adult size. Indeed, most dinosaur specimens were not skeletally mature at time of death, allowing for the possibility that some already gigantic species could have grown even bigger (Erickson 2014). Some *T. rex* fossils hinting at exceptionally large animals are known, but, alas, they are too fragmentary to reliably predict their body dimensions or masses. For example, Longrich et al. (2010) described a *Tyrannosaurus* toe bone that, at 13 cm long, is 17 percent larger than its equivalent in "Sue" (11.1 cm). This can be interpreted as representing an even larger *Tyrannosaurus* (e.g., Molina-Pérez and Larramendi 2019), and simple scaling suggests that this toe bone could represent a 14.5 m long king tyrant. However, extrapolating body sizes from such material is unreliable because some bones, especially minor ones from limb extremities, do not correlate tightly with body size: consider, by way of example, the range of hand and foot sizes among people you know of broadly similar height. Realistically, any specimen wanting to contend for the title of "largest *T. rex*" has to include elements that correlate more reliably with total size, such as major hindlimb bones, skull material, or stretches of vertebrae. A new specimen nicknamed "Bertha" may provide such material, but details of this specimen are still forthcoming at the time of writing.

The notion of *T. rex* that surpassed our biggest, more complete individuals like "Sue" or "Scotty" is not ridiculous. Among modern animal populations are individuals that far outsize the averages of their species. The typical shoulder height of male African bush elephants, for instance, is around 3.2 m, but the largest mounted skeleton of this species is over 4 m at the same dimension, and 4.4 m tall specimens—over a third larger than average—are recorded in scientific literature (Haynes 1991; Larramendi 2016). The largest captive saltwater crocodile measured almost 6.2 m long, 13 percent longer than the next largest, 5.5 m specimen (Grigg and Kirshner 2015). Our own species contains people who grow to over 2.2 m tall, 50 cm (28%) above the world average for men (171 cm) and over 60 cm (38%) taller than the global average for women (159 cm).

How large may a giant *T. rex* have been if it deviated in a similar way from current, presumably typically sized specimens? Jordan Mallon and David Hone (2024) have attempted to answer this question by comparing the size variation within alligator populations to the size distribution of *Tyrannosaurus* fossils. They found that absolutely enormous, fifteen-tonne, over 15 m long *T. rex* are *conservative* projections of maximum size. Scaled using the tyrannosaurid growth parameters outlined by Currie (2003b) and scaling equation of Campione et al. (2014), we can estimate that such tyrants would have stood almost 4 m tall at the hips, had especially large, deep lower jaws, enormous pelvises (figs. 1.1, 4.13), and limb bones with similar circumferences to those of sauropods. The ilium alone is predicted at 1 m tall! Alas, such individuals would be extremely rare and thus have a very low chance of fossilization, so we may never find record-breaking specimens of this kind.

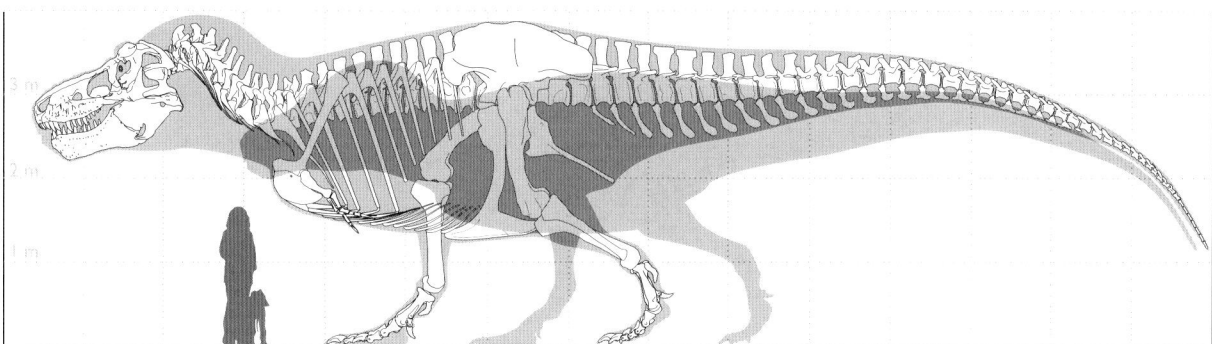

FIGURE 4.13

How big could *Tyrannosaurus* get? Skeletal reconstruction of a giant, 16–17 m long, 13-tonne *T. rex* based on the maximum size prediction of Mallon and Hone (2024). The skeleton has been modified beyond that of a large *T. rex* using scaling equations from Currie (2003b), resulting in an especially robust, thick-jawed, and large-hipped individual. The silhouette of one of the largest real *T. rex* specimens, "Sue," is superimposed for scale. This skeleton is the basis for the king tyrant illustrated in Figure 1.1.

It perhaps bears emphasis that size deviation of this kind would apply to *all* extinct animals: *T. rex* were not somehow "special" in having the potential to be much larger in life than their fossils suggest. That the maximum sizes attained by fossil species is difficult to gauge is worth considering in the never-ending discussion over which fossil animals were the biggest (chapter 2).

There is some scientific merit (and also plenty of fun) in comparing exceptionally big fossils to see which dinosaurs and other extinct organisms hold the title of longest, heaviest, or tallest, but without knowledge of their intraspecific size variation, we can rarely, if ever, say which species were truly largest of all.

THE SMARTS, THE SENSES, THE VOICE

OPERATING SEVERAL TONNES of *Tyrannosaurus* flesh and bone were brains that, compared to the sizes of their skulls, were relatively small (fig. 4.14). Located deep within the skull and squeezed between the enormous temporal regions, knowledge of these structures dates back to the earliest years of *T. rex* discoveries when fossils had to be cut open to reveal the brain chamber, or endocast (Osborn 1912). Broken skulls, through chance, revealed the same details in other king tyrants (Molnar 1978). Nowadays, CT scanning permits the same information without destructive investigations, and the digital endocasts of several *T. rex* and other tyrannosaurids have been studied in detail (Brochu 2000; Saveliev and Alifanov 2007; Witmer and Ridgely 2008, 2009, 2010; Hurlburt et al. 2013; Bever et al. 2011, 2013; Morhardt 2016; Kundrát et al. 2018; McKeown et al. 2020), giving them some of the most studied brains of any dinosaur group. Comparisons between these and extant animals, as well as with other dinosaurs, have been used to predict how *T. rex* perceived the world and how intelligent they were, as well as gain insight into their neurological adaptations for operating gigantic bodies (Kundrát et al. 2018).

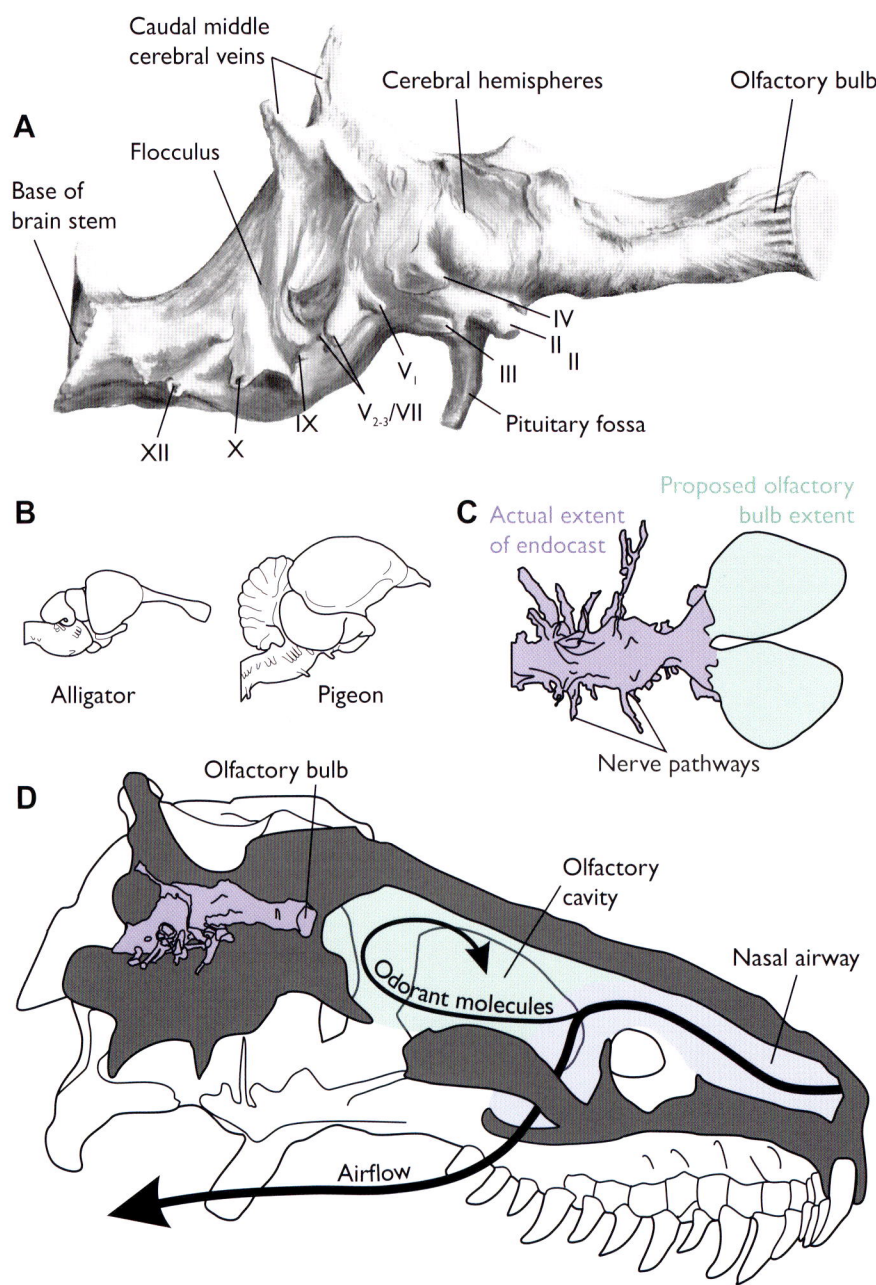

FIGURE 4.14

The brain of *T. rex*. (A) The king tyrant endocast; (B) line drawings of alligator and pigeon brains; note how the *Tyrannosaurus* endocast resembles the former more than the latter; (C) interpretation of the *T. rex* endocast where the olfactory chambers were conflated with the olfactory bulbs, leading to interpretations of an unprecedented sense of smell in king tyrants; (D) the actual relationship between the *Tyrannosaurus* endocast and the adjacent olfactory cavity. (A) After (and somewhat modified) from Osborn (1912); (B) after Witmer et al. (2003); (C) after Brochu (2000); (D) after Witmer and Ridgely (2009).

Dinosaur brains have a reputation for being small, and *T. rex* was no exception to this stereotype. The endocast of an adult king tyrant has a volume of almost 1.2 L, compared to over 500 L for the entire head (Witmer and Ridgely 2008). Early researchers marveled at these relatively diminutive brains and factored them into their visions of dinosaurs as evolutionary dimwits, blundering their way through the Mesozoic on instinct alone. Matthew (1915) viewed *Tyrannosaurus* as having no real consciousness or cognition, and wrote of how seemingly intelligent acts would only be performed through chance.

> We cannot doubt indeed that the Carnivorous Dinosaurs developed, along with their elaborately perfected mechanism for attack, an equally elaborate series of instincts guiding their action to effective purpose; and a complex series of automatic responses to the stimulus afforded by the sight and action of their prey might very well mimic intelligent pursuit and attack, always with certain limits set by the inflexible character of such automatic adjustments. (W. D. Matthew, 1915, p. 52)

The small brains of *T. rex* did not only inspire musings on their perceived low intelligence. To some, they represented the limits of evolution itself. Swinton (1934) regarded *Tyrannosaurus* as the largest possible bipedal animal that a reptile-grade brain could retain control of, their diminutive gray matter being unable to process the additional information from further bulk. For Osborn (1917b), the brain of *T. rex* formed the ultimate example of an evolutionary tussle between intelligence and brute strength: "The excessively small size of the brain ... indicates that in animals mechanical evolution is quite independent of the evolution of their intelligence; in fact, intelligence compensates for the absence of mechanical perfection" (pp. 214–215).

Views on dinosaur intelligence began to shift with seminal studies by Jerison (1973), who showed that, in fact, dinosaur brains were not small: they were actually in line with expectations of brain volume for large animals. Brain dimensions vary between animal lineages

but invariably become smaller fractions of overall body mass in larger-bodied species. Scaled up to many tonnes, an alligator would have a brain of similar size to *T. rex* (Hurlburt et al. 2013). These observations gave rise to the "encephalization quotient," a measure of brain or endocast volume against body mass, often shortened to "EQ" (Jerison 1973). Much emphasis has been placed on this as a measure of intellect in fossil species. Some, for instance, have noted that some EQ estimates for *T. rex* are comparable to those of famously intelligent chimpanzees (Brusatte 2018), but to imply that this equates to primate-like intelligence is an oversimplification (Hurlburt et al. 2013; McKeown et al. 2020). Our greatest risk is conflating brain size with endocast volumes, as these are not the same measurements: the relatively large brains of birds and mammals fill greater proportions of their endocast volumes than those of reptiles, such that endocast volume is a better proxy for brain size in birds and mammals than it is in reptiles. We should not, therefore, simply measure the volume of a dinosaur brain cavity and look for a matching EQ among modern species to start making deductions about intellect. Rather, we have to model several possible brain volumes, consider the influence of scaling on brain proportions, and make considered EQ comparisons among related modern taxa.

Applying this to *Tyrannosaurus* shows that they had some of the largest brains, proportionate to their body size, of any dinosaur, but also that they better fit the scaling patterns of reptilian EQ than the avian alternative (Hurlburt et al. 2013; Ksepka et al. 2020; Caspar et al. 2024). Indeed, *T. rex* brain sizes are below the EQ range of any living bird. This is in keeping with the position of *Tyrannosaurus* as an early coelurosaur because, while the evolution of an enlarged, avian-grade brain happened within Theropoda, most of this occurred within the maniraptoran and, later, avian clades (Kspeka et al. 2020; Caspar et al. 2024). *T. rex* thus belongs to a branch of theropod evolution that had yet to significantly move away from reptilian brain sizes or shapes. To that end, *Tyrannosaurus* endocasts are typical of non-maniraptoran theropods in resembling the long, tubular brains of crocodylians more than the rounded, bulbous braincases of birds (fig. 4.14B). This morphological difference reflects the avian development of the cerebral hemispheres beyond that of their reptilian ancestors, these brain regions being responsible for cognition and intellect. Measured against overall brain size, the cerebrum of *Tyrannosaurus* is slightly expanded compared to crocodylians and overlaps slightly with the lower end of avian cerebral proportions. The far smaller bodies of birds lessens the significance of this observation somewhat, however: measured against body size rather than brain volume, all living birds have much larger cerebral volumes than king tyrants (Hurlburt et al. 2013).

Do these observations tell us something about *T. rex* intellect? They may give some insight, but brain size is only somewhat indicative of cognitive ability. Despite having brains that weigh only a few grams, crows and parrots are incredible problem solvers, are tool users, can anticipate the behavior of other species, and generally possess many hallmarks of elevated intelligence (Olkowicz et al. 2016). Ants, with brain volumes of just microlitres, can craft and use tools, and demonstrate varying degrees of "advanced" cognition (Czaczkes 2022). The capacity for intelligent behavior thus seems less related to absolute brain size than the properties of the brains themselves, such as the size and density of information processing centers (Olkowicz et al. 2016). In recent years, it has been suggested that estimating this by way of neuron counts might give some insight into extinct dinosaur intelligence; a prediction of this value for *T. rex* has yielded startling hypotheses of crow or primate-like intellects that could permit the fashioning of tools and development of culture (Herculano-Houzel 2023).

As spectacular as this would be, such conclusions overlook the many complications of estimating neuron counts for fossil species (Reiner 2023; Caspar et al. 2024). First, neuron densities vary between animal groups, with reptiles bearing lower numbers than birds or mammals. Neuron counts for *T. rex*, if extrapolated from reptiles, are far lower than those that would be

predicted using mammals or birds, and do not point to primate levels of brain processing power (Caspar et al. 2024). Second, regardless of neuron density, the larger brains of bigger animals have greater neuron counts than smaller ones. Giraffe brains, for instance, have more neurons than those of tool-using birds and primates, and yet these smaller animals demonstrate far greater cognitive abilities (Caspar et al. 2024). Whatever link exists between neuron density, neuron counts, and cognition, it is complicated by body size (Caspar et al. 2024): needless to say, this factors into consideration of neuron estimates in prehistoric giants like king tyrants. Finally, any results about neuron density are difficult to interpret without detailed knowledge of brain anatomy and chemistry. Mammal brains, for instance, have some structural properties that allow for more rapid and efficient communication between neurons than those of other species, and we do not know whether dinosaurs, even big-brained ones like king tyrants, had similar adaptations (Reiner 2023). *Tyrannosaurus* intelligence thus remains not only an unknown quantity, but is perhaps *unknowable* without significant advancements in our understanding of dinosaur brains and the values that correlate with enhanced cognition in living species.

If brain size and structure offer only limited insights into *T. rex* cognition, are there other means to rationalize tyrannosaur smarts? Fossils may provide some answers, and some authors have already interpreted *T. rex* bites on less-appetizing parts of dinosaur carcasses as evidence of playful behavior, and thus elevated intelligence (Rothschild 2015; Snively and Samman 2015). These are interesting ideas, but are difficult to test scientifically (Caspar et al. 2024). Phylogeny offers another angle of investigation, although cognitive variation within clades only permits extremely broad-brush comparisons with extinct species. We can note, for instance, that the particularly dense neuron packing that seems to make crows and parrots so smart is restricted to the songbird and parrot bird clade (Psittacopasseres), a relatively recent branch of the bird evolutionary tree. More archaic bird lineages and their reptile cousins, in contrast, have neither the cognitive level nor neuron density of these especially clever birds (Olkowicz et al. 2016). While we cannot rule out the independent development of elevated cognition in tyrannosaurs or other dinosaurs, our phylogenetic baseline for *T. rex* intellect has to be somewhere between reptiles and birds like ostriches, ducks, and chickens, and the reptile-like structure and EQ of *T. rex* brains perhaps indicates something closer to the former. This needn't be seen as conforming to the stereotype of dinosaurs being slow or dimwitted, however. On the contrary, ongoing studies on living reptiles are revealing that their intelligence and behavioral complexity have long been underestimated (Lambert et al. 2019). They exhibit a number of emotional states, can form complex social groups, employ sophisticated means of intraspecific interaction and cooperation, demonstrate play behavior and parental care, construct nests and burrows, and generally exhibit as much behavioral complexity as any other animal group. We are still learning what it means to have a "reptile-grade" brain, and what this might imply for extinct species that married similar brain structures to warm bodies and sustained, elevated activity levels.

THE SENSITIVE SIDE OF KING TYRANTS

While intelligence remains a tricky matter to predict, *T. rex* neuroanatomy (that is, the configuration of their nervous system, including the brain) and other aspects of their skulls give more confident insights into their vision, sense of smell, and hearing (Brochu 2000; Witmer and Ridgely 2009; Morhardt 2016). There are limits to how precisely we can predict these attributes because some estimations of sensory capabilities rely on knowledge of brain structure and, as noted above, reptile brains only partially fill their endocasts. We thus do not know the precise shape of any dinosaur brain (this may change in the future: work by Ashely Morhardt [2016] discusses promising means to accurately predict dinosaur brain shape from endocast data) but, even so, com-

paring *T. rex* endocast morphology with those of other species, living and extinct, gives a rough idea of where king tyrants enhanced their brain shape to process sensory input.

Tyrannosaurus vision was probably very good. The size of their optic lobes (the part of the brain processing visual data) is ambiguous (Witmer and Ridgely 2009), but their semicircular canals, the brain structures associated with equilibrium, suggest a well-developed link to the muscles that controlled eye and head movement (Witmer and Ridgely 2009). This might indicate strong visual tracking capabilities, where moving objects could be reliably followed or stationary ones could be held in view as the rest of the tyrant body moved around. Strong neck musculature, with a capacity for quick, powerful and precise movements, are consistent with this inference (Snively and Russell 2007b).

T. rex also had very large, light-gathering eyes (chapter 3) that, unusually, faced forward. Most reptilian predators, including the majority of carnivorous dinosaurs, have laterally facing eye sockets and primarily monocular vision. In this configuration, lateral vision is emphasized and the visual fields of each eye only overlap marginally (Stevens 2006; Shimizu et al. 2009). Working out the visual fields of any animal, even living ones, is deceptively complex and requires knowledge of many features forever lost in extinct species (see Martin 2007 for discussion), so we will probably never know exactly how *T. rex* saw the world. Nevertheless, a rough insight was provided by Kent Stevens (2006) who, using fleshed-out models of theropod heads and measurements of pupil visibility, approximated the extent of their visual fields. The results possibly overestimate visual range as retina areas (the extent of light-detecting cells in our eyes) do not cover every angle that light might enter a pupil (Martin 2007), but, even with this caveat, *T. rex* probably had visual fields with substantial forward overlap. Stevens (2006) modeled a c. 40° arc of binocular vision for king tyrants, comparable to that of other tyrannosaurids and double that of a large allosauroid. Tyrannosaurid binocularity was not the most extensive of all Mesozoic theropods, however, with some predatory maniraptorans having overlapping visual fields of 60° (Stevens 2006). In all cases, overlaps were maximized by looking slightly above the snout. This may explain why *T. rex* semicircular canals, those equilibrium-related brain structures mentioned above, imply a somewhat downturned habitual head posture, allowing *T. rex* to consistently look forward with their broadest visual overlap (Stevens 2006; Witmer and Ridgely 2009).

Predatory birds and mammals benefit from binocular vision because it aids in judging distance and provides visual cues for grabbing or manipulating objects with jaws or appendages (e.g., Martin 2007, 2017; Cantlay et al. 2023). It is tempting to conclude that *T. rex* developed large overlapping visual fields for the same reasons and that they became more efficient predators for their efforts (Stevens 2006), although there are complications to this hypothesis: Do forward-facing eyes really equate to enhanced predatory prowess? Not only does the ubiquity of monocular vision in Mesozoic theropods suggest that binocular vision was not essential to their predatory acts, but widely overlapping visual fields are not common to avian predators in modern times. On the contrary, although many raptorial and diving birds have relatively developed binocular vision to help them chase prey, they retain somewhat laterally facing eyes that provide views above, below, and behind their heads (Martin 2007, 2017; Cantlay et al. 2023). Raptors with eyes that face entirely forward, like owls, are exceptional and their unusual visual fields may not be entirely adaptive. Instead, the development of elaborate, skull-altering ear anatomy may have forced owl eyes into a more anterior position, some visual awareness apparently being sacrificed for increased auditory capability (Martin 2007, 2017). We might wonder if the forward-facing eyes of tyrannosaurids were also a consequence of skull modification, their orbits rotating forward as their temporal regions and jaw muscles expanded outward. In this scenario, the perceived enhancement of tyrannosaurid vision was actually an exchange of sensory awareness for power; trading lateral and rearward vision to facilitate a stronger bite.

Perhaps, for all their popular reputation as dangerous, perpetually alert super-predators, *T. rex* were easier to sneak around than, say, *Allosaurus* or *Coelophysis*, where laterally facing eyes gave greater awareness of their surroundings. (Note to future time travelers: I accept no responsibility for your safety if you choose to field-test this hypothesis.)

This is not to imply, of course, that king tyrants were unable to utilize other senses to compensate for their forward-focused vision. Their sense of smell was probably well-developed, although it was perhaps not as spectacular as once thought. Analyses of *T. rex* endocasts once indicated that their olfactory lobes, structures at the front of the brain associated with processing smell stimuli, were enormous, even larger than the rest of the brain itself (fig. 4.14C; Brochu 2000). These melon-sized neurological expansions implied a phenomenal, perhaps unprecedented sense of smell in king tyrants, but subsequent analysis has shown that they actually represented parts of the nasal chamber adjacent to the brain, not elements of the brain itself (fig. 4.14D; Witmer and Ridgely 2009; Morhardt 2016). Even so, the olfactory region of king tyrant endocasts are large and their nasal cavities, a blind chamber in the snout linked to the nasal pathways, are cavernous; both imply a well-developed sense of smell. Evidence of enhanced olfactory capabilities is not present in early tyrannosauroids (Kundrát et al. 2018) but characterizes many later members of the group (Witmer and Ridgely 2009; McKeown et al. 2020), suggesting an increased importance over the course of their evolution. It remains to be seen how the *T. rex* sense of smell ranked more widely among dinosaurs, however: Carpenter (2013) noted that the olfactory capabilities of other dinosaurs, such as *Allosaurus* and *Edmontosaurus*, were also well developed. King tyrants may not have been unique among dinosaurs in their enhanced capacity to detect odors.

The nostrils may not have been the only sensitive part of king tyrant snouts. Examinations of tyrannosaurid jaw bones, including those of *Tyrannosaurus*, have revealed that a dense and complex network of nerves and blood vessels once ran throughout their maxillae, premaxillae, and dentaries (fig. 4.15; Carr et al. 2017; Kawabe and Hattori 2022; Bouabdellah et al. 2022). These were responsible for the tactile sensitivity of tyrant dinosaur jaws, and their distribution is typical of the nerve pathways found in other large theropods (e.g., Ibrahim et al., 2014; Barker et al. 2017). Similarities between theropod jaw neurovascularity and that of crocodylians has prompted repeated suggestions that predatory dinosaurs possessed hypersensitive facial skin like these reptiles (Brazaitis and Watanabe 2011; Ibrahim et al. 2014; Barker et al. 2017; Carr et al. 2017; Carr 2020; Kawabe and Hattori 2022). Crocodylian jaws and mouths (along with other body parts, depending on the species) possess hundreds or thousands of small, dome-shaped skin sensors that respond to temperature, acidity, and especially pressure, giving their armored faces even greater mechanical sensitivity than primate fingertips (e.g., Soares 2002; Leitch and Catania 2012). These structures, termed

"integumentary sense organs" or "dome pressure receptors," seem to have evolved after crocodylian-line archosaurs became semi-aquatic (Soares 2002), enabling them to detect pressure waves created by moving prey. Originating deep within crocodyliform evolution, the jaws of living crocodiles and alligators have become so sensitive that they can detect, catch, and manipulate prey without use of their vision.

It is possible that tyrannosaurs and other theropods possessed similarly amazing degrees of tactile sensitivity, and this may have played a role in catching aquatic prey in some specialist taxa (Ibrahim et al. 2014). It is hard to rationalize why terrestrial carnivores would possess this level of mechanoreception, however, as hypersensitive jaws are largely restricted to aquatic foragers or sediment probers in living species (Bouabdellah et al. 2022). Barker et al. (2017) suggest that non-aquatic theropods like the carcharodontosaurid *Neovenator* may have used sensitive snouts to check nest

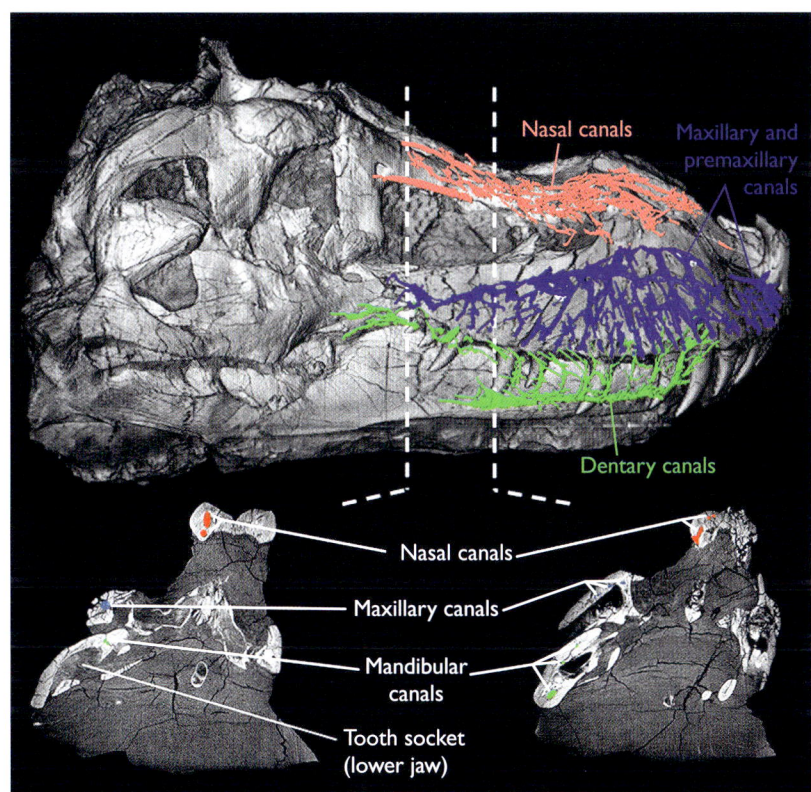

FIGURE 4.15

The distribution of neurovascular pathways in the jaws of the "Sue" *Tyrannosaurus*, in lateral (top) and cross-sectional (bottom) views. Imagery from Bouabdellah et al. (2022).

temperatures or avoid biting bone when defleshing carcasses, noting that carcharodontosaurid teeth show none of the heavy wear expected from routine tooth-on-bone contact. These are not entirely satisfactory explanations however, because neither of these behaviors particularly demands crocodylian-like degrees of facial sensitivity. Furthermore, the neurovascular structure of *T. rex*, who definitely did not care about biting bone when eating carcasses (chapter 6), appears to be just as complex as that of the bone-averse *Neovenator* (Bouabdellah et al. 2022).

It's in this spirit that some authors urge caution about the hypersensitive theropod face hypothesis. Bouabdellah et al. (2022) note that the neurovascular density of terrestrial theropods are less dense than those of their semiaquatic relatives, suggesting that there might be variation in facial sensitivity among predatory dinosaurs. *Tyrannosaurus* and *Neovenator* may have had "average" tactile sensitivity for the group (whatever that entailed) rather than the heightened mechanical perception of fish eaters like *Spinosaurus* (Bouabdellah et al. 2022). More fundamentally, caution is urged about overinterpreting theropod jaw neurovascular data before we have a more developed picture of their sensory capabilities. While we are increasingly able to image neurovascular complexes of theropod jaws with sophisticated scanning technology, it remains challenging to distinguish nerve pathways from those for blood vessels (Porter and Witmer 2020). Both systems operated in channels of similar sizes and had comparable branching patterns, and our usual aids for unraveling such anatomical tangles, modern species, are of little assistance. The neurovascular architecture of theropod jaws is unlike any living reptile or bird, a consequence of their deep-rooted teeth altering ancestral reptilian neurovascular pathways (Bouabdellah et al. 2022). These complications raise questions about the ratio of nerves to blood vessels in theropod jaws, and how sensitive their faces really were. Just as crucially, studies of theropod facial sensitivity have yet to compare neurovascular density against independent measures of this ability, such as the size of corresponding brain regions

or associated nervous pathways (Bouabdellah et al. 2022). The concept that *T. rex* and other theropods had unusually sensitive faces is certainly intriguing, but more work is needed to allay concerns that their internal jaw structure is being overinterpreted.

Happily, another *Tyrannosaurus* sense, their hearing, is somewhat more straightforward to investigate, and it seems that king tyrants had good auditory detection. The length of the inner ear, a trait roughly associated with auditory capability in living animals or, at least, the behavioral importance of hearing, is comparatively great in *Tyrannosaurus* among Mesozoic theropods (Witmer and Ridgely 2009). Relationships between animal size and hearing capacity, as well as the impact of extensive pneumatization around the ear (the paratympanic sinuses; see Figure 3.17), point to *Tyrannosaurus* having particular sensitivity to low-frequency sounds. These ear-adjacent hollow bones would augment detection of lower frequency noise as well as amplify these sounds over those of higher pitches (Witmer and Ridgely 2009). This complements the sharp vision and strong olfactory capacity evident for king tyrants, and suggests that their major environmental senses were finely attuned, as we might expect for a predatory species.

VOCALIZATIONS: COULD *TYRANNOSAURUS* ROAR?

Knowing that *Tyrannosaurus* had sensitive hearing raises questions about their voices: Did they make noise, and what may they have sounded like? Television shows and movies have answered these queries uncountable times with the same answer: mighty, eardrum-splitting, open-mouth roars. Indeed, *Tyrannosaurus* has been roaring on film since before the technology for conveying sound with motion pictures was developed. Whether vanquishing other dinosaurs or menacing the human heroes in Harry O. Hoyt's 1925 silent movie *The Lost World*, its stop-motion *T. rex* stops to roar and bellow at opponents with regularity. Following these muted screams, films and television shows have imagined a variety of *Tyrannosaurus* vocalizations, among them the instantly recognizable booming roar created for 1993's *Jurassic Park*.

Contrasting with a century of popular representation as a vociferous, bellowing character, science actually lacks any direct evidence into the sounds that *Tyrannosaurus* could make. Indeed, we generally lack this for all non-avian dinosaurs, owing to a near-total lack of fossilized throat tissues or indications about their vocal organs. This leaves us with a more fundamental question than whether or not *T. rex* could really generate an ear-splitting roar: Could non-avian dinosaurs vocalize *at all* (Senter 2008; Kingsley et al. 2018)? The idea that dinosaurs may have been non-vocal is not a product of overzealous scientific conservatism, but pertains to the surprisingly complex evolution of reptilian vocalization (fig. 4.16) and our extremely limited fossil data with bearing on this matter (Yoshida et al. 2023). Our insights on dinosaur voices are accordingly coarse and boil down to modeling the broad-strokes of their vocal organ evolution rather than predicting the specifics of dinosaur sounds.

One possibility we can probably rule out are avian-style vocalizations: *Tyrannosaurus* were probably not singing like songbirds. Birds have a unique vocalization structure known as the syrinx: a modification to the trachea and bronchial tubes (the passageways to the lungs) where portions of the airway walls are mobile. When air from the lungs rushes out, these membranous walls can be manipulated to vibrate, producing sounds. Syringes are rare fossils (Clarke et al. 2016) but, in modern birds at least, the syrinx is associated with two structures that are detectable in the geological record: a series of mineralized rings that reinforce the syrinx region of the airway, and the clavicular air sac, a pneumatic structure that leaves evidence on animal skeletons by invading forelimb and shoulder bones (chapter 3; Senter 2008; Clarke et al. 2016). We do not find evidence of the former outside of fossils from the modern bird group (Clarke et al. 2016), and the latter has only sporadic occurrences, mostly in Mesozoic birds, but also in a few exceptional non-avian dinosaurs (Wedel 2004; Senter 2008). This suggests that the syrinx evolved late among dinosaurs, perhaps even being an avian invention. Experimentation on living birds

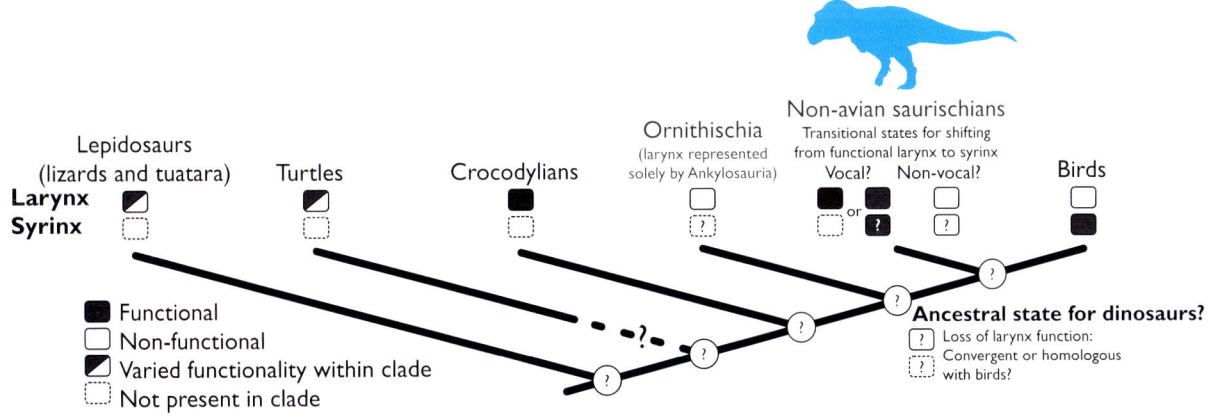

FIGURE 4.16

The distribution of vocal anatomy among reptiles, and what it means for dinosaurian acoustics. The varied state of vocal anatomy on the lizard and avian lines of reptile evolution complicates our ability to make confident predictions about the presence of a functioning larynx within Dinosauria. After, and somewhat modified from Kingsley et al. (2018).

shows that syringes can function without a clavicular air sac (Clarke et al. 2016), implying that a hypothetical "proto-syrinx" could have existed without leaving behind an osteological clue, but we can only speculate on when, why, and how such a structure evolved (fig. 4.16; Kingsley et al. 2018; though also read on). We can say with more certainty, however, that the phylogenetic position of *Tyrannosaurus* is far removed from anything we currently associate with syrinx development, and there is no reason to assume *T. rex* had this avian vocal organ.

Instead, *Tyrannosaurus* and other non-avian dinosaurs were likely equipped with a larynx: a cartilaginous structure located at the top of the throat that produces sound with vibrating vocal folds. A recently discovered larynx, the first for any dinosaur, in the ankylosaur *Pinacosaurus* goes some way to confirming this prediction (Yoshida et al. 2023). Larynges are, of course, our own sound-producing organs, and they are common to most reptiles and even birds, but they do not always function as vocalizers. Crocodylians, some turtles, and some lizards have functionally vocal larynges but birds, other lizards, as well as the lizard relative, the tuatara, do not (fig. 4.16; Russell and Bauer 2021). This confuses predictions of whether the reptile-bird line of evolution has always been able to vocalize. Did nonfunctioning larynges repeatedly evolve among reptiles (Kingsley et al. 2018), or, conversely, did a lineage of ancestrally nonvocal reptiles convergently gain voices (Russell and Bauer 2021)?

Even ignoring the complexities of lizards and turtle larynges to focus purely on archosaur vocalization still leaves ambiguity about larynx function. Non-avian dinosaurs are evolutionarily bracketed by "vocal larynx" crocodylians on one side, and "silent larynx" birds on the other. This leaves two possible models (Kingsley et al. 2018): (1) the ancestral archosaur was vocal, with crocodylians maintaining the "primitive" vocal larynx but birds augmenting or replacing the larynx with the syrinx for an unknown reason late in dinosaur evolution; or (2) archosaurs were ancestrally silent, with crocodylians and birds developing their distinct sound-producing organs in isolation. The latter implies that syringes evolved in response to birds having lost, or having

never had, a vocal organ (Kingsley et al. 2018). This scenario requires that some theropods, if not all dinosaurs, went through a nonvocal phase in their evolutionary history (Senter 2008). Interestingly, our sole Mesozoic dinosaur larynx, that of *Pinacosaurus*, is structurally more akin to the non-vocal avian larynx than the vocal alternative in crocodylians, and this has been used as evidence for pulling the evolution of syringes deep into Dinosauria (Yoshida et al. 2023). However, this explanation does not account for the absence of evidence for syringes discussed above, and an argument can be made that a non-vocal dinosaur larynx better supports a model where dinosaurs were voiceless rather than the early evolution of a syrinx. More data (ideally, more dinosaur larynges) are needed to clarify how this important specimen fits into wider models of dinosaur vocalization.

This caveat-filled discussion leaves us with two broad possibilities for *Tyrannosaurus* sound production. If *T. rex* lacked a functioning larynx, they would have been limited to creating nonvocal acoustics: that is, the noises animals create without using vocal anatomy. These include a wide array of sounds, such as hisses, snorts, and the percussive noises of slapping or rubbing of body parts against other objects. These noises are widely made by living animals, even those with functioning vocal organs, and were probably utilized by dinosaurs too, whatever other sounds they could produce. Indeed, hissing is such a common aggressive noise among reptiles and birds that we might speculate that angry, agitated *Tyrannosaurus* were more likely to hiss than roar. The large internal nasal cavities of *Tyrannosaurus* would have been excellent acoustic resonating chambers (Witmer and Ridgely 2009), and such hisses may have been very deep and growling, as they are in the resonator-equipped king cobra (Young 1991).

Conversely, if *T. rex* had a functioning larynx, a wider world of vocalizations opens up. Phylogenetic proximity suggests that reptilian vocalization, rather than avian or mammalian, should form our predictive foundation for dinosaur sounds. Reptiles have traditionally been regarded as relatively quiet and nonvocal, even in species with functioning vocal organs, but we are now learning much about their sophisticated vocal anatomy and communication strategies (Russell and Bauer 2021). In addition to emitting noises with open mouths, these include the capability for reptile-line animals to create deep, booming sounds by vocalizing into inflated throat cavities, a behavior precluded from mammals by our more rigid throat structure (Riede et al. 2015, 2016). These noises are referred to as "closed-mouth" vocalizations because, despite their unusually deep and far-reaching sounds, their creators mostly or fully seal their mouths to prevent significant air escape (fig. 4.17). Among birds and reptiles, these vocalizations are more commonly used to create deep, loud noises than open-mouth calls (Riede et al. 2016) and this may—in contrast to innumerable big screen *Tyrannosaurus* depictions—have been true for king tyrants as well. Perhaps a noisy *T. rex* was less likely to roar like a lion than to close its mouth, inflate its throat like an alligator or pigeon, and then emit a deep rumble or boom (fig. 4.1). Given that vocalization frequency scales negatively with body size (in other words, bigger animals produce deeper, lower-frequency sounds), any booming, resonant calls emitted by *Tyrannosaurus* were surely comparable in pitch, if not deeper, than the largest elephant bellows. A possible ability to create deep noises concurs with *Tyrannosaurus* ears being suited to hearing lower-frequency sound and would give *T. rex* the potential to communicate over long distances (Witmer and Ridgely 2009).

For a model of what *T. rex* acoustic communication could have been like, we might turn to crocodylians, which are relatively vocal living reptiles. This analogy is limited as no crocodylians are closely related to Mesozoic dinosaurs, but they at least demonstrate the range of acoustics that large predatory archosaurs are capable of. The alligatoroids, especially the American alligator, are especially talkative species and express a range of vocal and nonvocal sounds (Garrick et al. 1978). Several crocodylians produce deep, closed-mouth bellows by adopting a raised head and tail posture, taking deep breaths that inflate their chests, and then exhaling a protracted, rumbling call through a barely open mouth. Other loud vocalizations include various grunts and growls, and a variety of nonvocal noises are created by slapping their jaws on water, snapping their mouths closed, and hissing. If *Tyrannosaurus* were vociferous,

FIGURE 4.17

Closed-mouth vocalization in a familiar archosaur, the domestic pigeon. Note the lengthened neck and inflated neck of the individual on the left, which has created a resonating chamber in its throat to amplify its call. Such vocalization strategies are common to many bird and crocodylian species, and are the most commonly evolved means of producing loud, deep noises within living archosaurs.

they may have performed equivalent actions. We might also pay attention to *when* crocodylians vocalize because, in further contrast to on-screen *Tyrannosaurus*, their noises are not made randomly or constantly. Rather, they relate to specific behaviors and social functions and are often accompanied by ritualized poses (Garrick et al. 1978). Hisses are often used aggressively, such as to deter trespassers around crocodylian nests; bellowing mostly occurs during the spring to attract potential mates and intimidate rivals; and friendly-sounding "contact noises" emerge from young juveniles looking to remain in a group. We might predict similar, context-specific acoustic behavior for *Tyrannosaurus* and, indeed, for all dinosaurs.

COMETH THE MECHATYRANTS

ENERGETICS, GROWTH, AND senses are fine and fascinating paleobiological topics, but they do not bring animation to our extinct subjects. For this, we turn to functional morphology, the analysis of how anatomical form pertains to biomechanical function. It is here where we learn what ancient organisms were physically capable of: how they moved through the world, how fast they were, how hard they could bite and so on. *Tyrannosaurus* have become the poster children for such analyses, their well-preserved fossils and fascinating, unusual natural history proving ideal subjects for appreciating dinosaur functionality. We have touched on aspects of king tyrant function in our discussions already, but will focus here on how the major headlines of *T. rex* anatomical performance—locomotion, bite force, and arm use—have been investigated.

BODY OF A SPRINTER, MASS OF AN ELEPHANT

The enormous hips and long hindlimbs of *T. rex* have always been considered remarkable, portraying *Tyrannosaurus* as a powerful carnivore even before their skeletons were fully known (chapter 1). Just as long-lived are contrasting interpretations of how fast king tyrants could move. Osborn (1917a) established one side of this discussion in the early days of tyrannosaur research when, during an era in which dinosaurs were viewed as lumbering ectotherms, he saw king tyrants as capable of relatively high speeds. This view would be carried through the next century with some, especially researchers Gregory S. Paul and Robert T. Bakker, championing *T. rex* as 50–60 kph sprinters, making them theoretically capable of matching racehorses (Bakker 1986; Paul 1988a, b, 2008, 2016). This view has been hugely influential in dinosaur media, not least through a scene in 1993's *Jurassic Park* where an escaped king tyrant almost catches a speeding jeep.

Another view has also persisted since the early twentieth century. Although noting that *T. rex* hindlimbs were gracile and ostrich-like, William Diller Matthew (1915) regarded *Tyrannosaurus* as slow and ponderous, assuming that their great bulk would limit speed and agility. Later researchers have perhaps not viewed *T. rex* with quite the same functional pessimism as Matthew, but a notion that king tyrants were incapable of running faster than c. 30 kph has been maintained through a series of studies (e.g., Alexander 1976; Farlow et al. 1995; Christiansen 1998; Hutchinson and Garcia 2002; Hutchinson 2004; Sellers and Manning 2007; Gatesy et al. 2009; Sellers et al. 2017; Dececchi et al. 2020). Such discussions are complicated by conversations about what constitutes a running gait (Hutchinson et al. 2003): at sub 30 kph speeds, would *T. rex* even break a brisk walk? Running is traditionally defined as locomotion during which animals momentarily clear all their limbs from the ground, but some argue for a more expansive, biomechanics-based definition based on the degree of compression in limb joints and amount of forward inertia. Under this latter classification, elephants moving at their maximum recorded speed of 25 kph are running, their limb motion and momentum matching that of smaller species even if they keep at least one foot on the ground at all times (Hutchinson et al. 2003). Here, we will not worry too much about this distinction and focus instead on raw speed. Whatever the kinematics of their limbs, were *T. rex* high-velocity species like horses, or confined to slower speeds, like elephants?

To some extent, the sheer size of king tyrants gives pause to any thought of them moving especially rapidly. However, a suite of evidence suggests that the entire tyrannosaurid clade was characterized by hindlimb features associated with cursoriality: the capacity to run. The most obvious of these are their long, powerfully muscled legs (fig. 3.7). Tyrannosaurid shin and foot bones are much longer, relative to body size, than those of other large theropods, including early tyrannosauroids (Persons and Currie 2016) and their hindlimb muscle proportions are predicted to be greater than those of any living animals (Hutchinson et al. 2011). One of the largest and most important hindlimb retractor muscles, the caudofemoralis, was likely particularly big in *T. rex* thanks to their deepened tails (fig. 3.11; Hutchinson et al. 2011; Persons and Currie 2011, 2014; Snively et al. 2019).

Tyrannosaurids also possess arctometatarsal feet (see chapter 3), a condition that conferred several advantages for speed and agility in a number of unrelated theropod clades. These are related to reinforcing the structure of the foot, increasing the length of the leg, and, via a system of ligaments, storing and releasing energy during walking and running, making for more efficient and energetic locomotion (Holtz 1995; Snively and Russell 2002; Snively et al. 2004). Further agility was conferred by the tyrannosaurid body plan, which, with a short torso and powerful hindlimbs, lowered rotational inertia to the degree that they had a nimbleness comparable to theropods half their size (Henderson and Snively 2004; Hutchinson et al. 2007; Snively et al. 2019). Sprightliness was further augmented by the constantly flexed hindlimbs of giant theropods and their responsive elastic ligaments,

which recalled the legs of running birds and mammals rather than the subvertical, columnar limbs of heavyset species like elephants and sauropods (Paul 1988a, b; Hutchinson et al. 2005, 2007). Tyrannosaurid hindlimbs were as adapted for fast locomotion as might be expected in a large biped, and they must have been far more nimble and agile than any multi-tonne terrestrial animal alive today.

This said, while the tyrannosaurid propensity for speed and agility is undoubted among modern researchers, it remains to be demonstrated that these traits translated into 50 or 60 kph sprinting capabilities. Crucially, proponents of high-velocity king tyrants have yet to substantiate these speeds using biomechanical methods, instead relying on qualitative observations of cursorial features. These have merit, but all biomechanical models of *Tyrannosaurus* locomotion find that *T. rex* was restricted to, at best, "intermediate" speeds, these findings being substantiated by the same methods estimating, with reasonable accuracy, the measured speeds of living species (Christiansen 1998; Farlow et al. 1995; Hutchinson and Garcia 2002; Hutchinson 2004; Sellers and Manning 2007; Hutchinson et al. 2007; Gatesy et al. 2009; Sellers et al. 2017; Dececchi et al. 2020). One of the most vociferous proponents of sprinting *T. rex*, Gregory S. Paul, has dismissed these studies as mere simulations that cannot accurately model reality (Paul 1988b, 2016), but this is countered by the diversity of biomechanical approaches brought to this question as well as their ground-truthing on modern animals. The bulk of scientific discussion about *T. rex* speed has thus moved away from ideas of king tyrants charging around at breakneck velocities, and instead views them as more moderately paced animals. They were nimble for their size, certainly, but not anything to worry about if you were already in your *Jurassic Park* jeep.

This shift in focus reflects several major conclusions. The first is that, although adult *T. rex* limbs bear many cursorial features, their shin and foot segments are relatively short compared to other tyrannosaurids, and their leg skeletons are relatively robust, a reflection of their greater role in weight bearing (Farlow et al. 1995;

Christiansen 1998; Persons and Currie 2016; Sellers et al. 2017). Longer bones are more vulnerable to bending under heavy loads, and, in response to this, larger animals tend to shorten and broaden their limb skeletons to enhance their strength and minimize risks of failure. Even these augmented bones can be pushed beyond their limits, however. Considering such safety concerns in a 6,000 kg *T. rex*, Farlow et al. (1995) found that the risk of fatality from a fall was near certain in a 72 kph sprint, and, in any case, that their leg bones would approach or exceed their failure points from the stresses of generating such velocities.

Nor, for that matter, were king tyrant leg muscles as strong as those of smaller animals. This seems counterintuitive, given the capacious spaces for hindlimb muscles on *T. rex* leg and tail osteology (fig. 3.7, 3.11); surely their legs were *stronger* than those of smaller tyrant dinosaurs? While true in an absolute measure of muscle power output, *T. rex* legs could not avoid an inescapable biomechanical law: proportionate to body size, muscle performance in larger animals is lower than that of smaller ones. Muscle contraction speed and power scale at slower rates than muscle mass, such that greater muscle bulk leads to diminishing power returns. The enhancement of *T. rex* hindlimb muscle attachment sites in adult individuals is thus better viewed as an attempt to compensate against lessening muscle performance than the raw compounding of greater leg strength, and even this may have not been enough to maintain the same muscle power profile as juveniles. Relative to body mass, it is predicted that the leg muscle volumes of king tyrants *shrank* as they got bigger (Hutchinson et al. 2011). Muscle power, it seems, was generally limiting on *Tyrannosaurus* speed, with 7–10 percent of body mass devoted to hindlimb retractors in adults, but over 40 percent of body mass would be needed (for each leg!) to power a fast run (Hutchinson and Garcia 2002). Even 40 kph may have been a stretch given their available leg muscle power (Hutchinson and Garcia 2002).

Skeletal safety factors and muscle performance scaling, among other properties, explain why midsized terrestrial animals tend to be absolutely faster than either small or large species, regardless of clade or locomotor method (Hirt et al. 2017). "Intermediate"-sized animals have bodies with optimal configurations of muscle power, limb reach, and skeletal strength to attain high speeds, and it seems that this relationship was as true for dinosaurs as it is for living taxa (fig. 4.18; Dececchi et al. 2020). In theropods, it seems that maximum velocities (perhaps 60 kph) may have been reached by species massing around 100 kg, while animals exceeding 1,000 kg were limited to 40 kph or less (Dececchi et al. 2020). These data, and other methods at predicting the speed of adult *Tyrannosaurus*, point to maximum speeds of 25–29 kph, a stable pace where one foot could remain on the ground at all times (e.g., Hutchinson and Garcia 2002; Sellers et al. 2017; Dececchi et al. 2020). If this seems disappointingly slow and nonthreatening in comparison to tyrants running around with the velocity of racehorses, consider that most humans run at 15–16 kph: a striding *T. rex* would likely have easily caught any human that was not in training for an athletics competition.

Lower top speed would have some advantages, too, amplifying king tyrant agility through their bodies generating less inertia when cornering. Even moving at pace, therefore, king tyrants may have been able to position themselves in optimal positions for attacking prey. There is also evidence that adult *Tyrannosaurus* were particularly efficient walkers. Above a certain size threshold, the hindlimb features that made smaller tyrants excellent runners adopted a new role, optimizing striding motions over those of other big theropod predators (Dececchi et al. 2020). Further walking efficiency may have been facilitated from elastic energy stored in the interspinal ligaments of their tails, their pacing resonating with passive motions of the tail. This would provide extra spring in king tyrant steps, even at low speeds of 6 kph (van Bijlert et al. 2021). As we'll discuss further in chapter 6, trading speed for economy in adult *T. rex* has implications for how we view their lifestyles, implying that they may have been able to roam widely in search of prey. Where they may have struggled, it

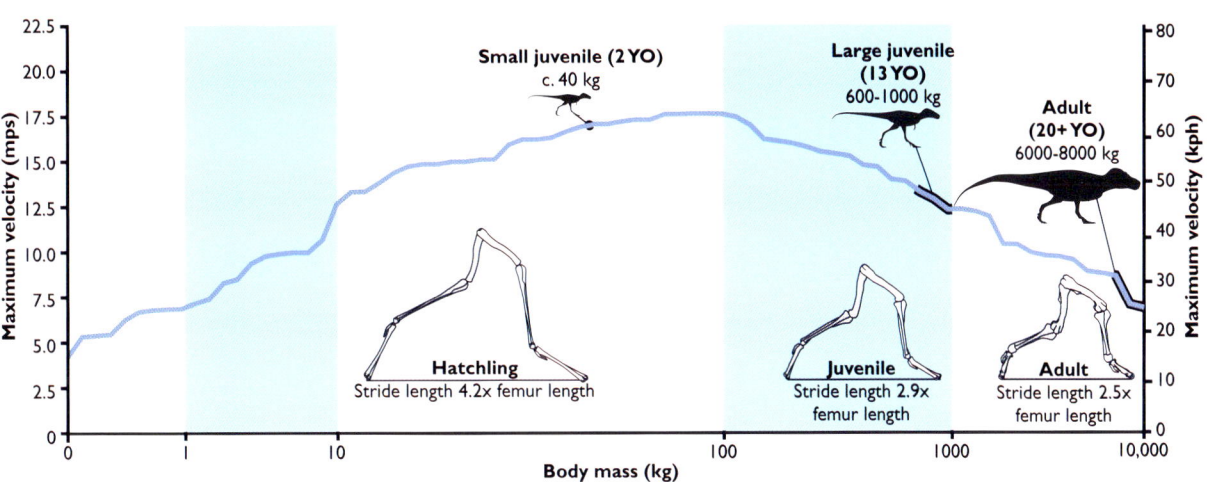

FIGURE 4.18

Maximum speed estimates for theropod dinosaurs compared to *T. rex* body mass estimates. As with living animals, peak speeds seem to have been reached by midsized theropods that had optimized stride lengths and muscle power outputs, factors that (among others) limited maximum velocity at more extreme body sizes. *Tyrannosaurus* limb graphics scaled to the same femur length. Graph based on Dececchi et al. (2020) and Hirt et al. (2017).

seems, was endurance at higher velocities: Persons and Currie (2014) suggest that the *T. rex* hindlimb configuration was better suited for short bursts of speed than sustained fast locomotion. Potential prey items not only needed to move quicker than *T. rex*, but perhaps outlast them in a chase.

A twist in our consideration of king tyrant locomotion, however, concerns those long-legged, lighter-bodied juveniles. Studies on the speeds of juvenile *Tyrannosaurus* are less common than those focused on adults, but provisional research points to them having been very fast (Henderson and Snively 2004; Hutchinson et al. 2011; Persons and Currie 2016; Snively et al. 2019; Dececchi et al. 2020). Much of the rationale for putting speed limitations on adult *Tyrannosaurus* outlined above would be reversed for younger individuals. Here, muscle power output would be relatively higher, their hindlimb retractors being larger proportionate to body mass, and the safety factors on their bones more conducive to high velocity travel. Their long legs, meanwhile,

belonged to small enough individuals that running fast, not mere power walking, better explains their cursorial quality. Juvenile *T. rex* hindlimbs are among the most cursorially adapted of all theropods, suggesting that they were some of the fastest land predators in the Mesozoic (fig. 4.19; Persons and Currie 2016). With their skeletons showing traits associated with rapid acceleration, young king tyrants may have been especially fast straight out of the gate (Dececchi et al. 2020). This, presumably, was useful in both pursuit of prey as well as avoiding predation themselves. Further work on the biomechanics of juvenile *Tyrannosaurus* promises to reveal more details of their capacity for fast running. Already, however, it is evident that *T. rex* embodied two known extremes of theropod locomotive functionality, being a gigantic, ultraefficient strider in adulthood while, in their youth, being one of the fastest of all predatory dinosaurs. This reminds us, again, of the remarkable transformation that these animals went through as they matured.

FIGURE 4.19

A juvenile *Tyrannosaurus* chases the ostrich dinosaur *Ornithomimus*. Young *T. rex* were among the fastest of all predatory dinosaurs, with hindlimbs strongly adapted for cursorial locomotion. As they aged, larger, more mature tyrants had increasingly little hope of catching ostrich dinosaurs in this manner, despite their absolutely greater size and bigger hindlimbs.

COULD *T. REX* SWIM?

Walking and running were clearly the de facto means of getting around for *Tyrannosaurus*, but their locomotion may not have been entirely restricted to land. The sediments housing king tyrant fossils were mainly deposited by large bodies of water, such as rivers, streams, and ponds, as well as estuaries and seaways (chapter 5). What happened when *T. rex* had cause to get their feet wet? For a time, it was considered that aquatic locomotion was a no-go for theropods. Herbivorous dinosaurs, it was assumed, could have avoided predators by launching themselves into nearby waterways, swimming away while their attackers stood, confounded, on dry land (e.g., Norman 1985; Milner and Lockley 2016).

Trackways created by swimming dinosaur carnivores pressing the tips of their toes into riverine and lake sediments show that these assumptions were incorrect, however; carnivorous dinosaurs undoubtedly entered and could move through water (Milner and Lockley 2016). Building on this footprint data, simulations of *Tyrannosaurus* buoyancy and floatation suggest that even this large theropod could float and swim without issue (fig. 4.20). As part of an investigation into the controversial swimming capabilities of *Spinosaurus* (chapter 2), Donald Henderson (2018) modeled the floating postures of a range of theropods, among them an adult *Tyrannosaurus*. Having successfully replicated the floating postures of alligators and penguins, (fig. 4.20A–B),

Henderson found that the low-density bodies of Mesozoic theropods were generally able to float with their noses and mouths above the waterline, allowing them to breathe, and that most were able to stay upright when buffeted by currents or crosswinds. The exception, contrary to predictions of a heightened swimming ability (e.g., Ibrahim et al. 2014, 2020b; see chapter 2), was *Spinosaurus*, the bone sail of which raised their center of buoyancy and made them prone to tipping (a finding replicated in later studies: Sereno et al. 2022).

FIGURE 4.20

Tyrannosaurus in water. (A–B) Three-dimensional models of floating extant archosaurs ([A] emperor penguin; [B] alligator) to ground-truth models of a floating *Tyrannosaurus* (C); (D) restoration of *T. rex* swimming, where thrust is provided by kicking legs. (A–C) Adapted from Henderson (2018).

A

B

C

50 cm

50 cm

2 m

D

Tyrannosaurus had no such issues, and, via kicking actions of their long, powerful legs and sculling motions of their tails, they could probably swim as well as most land-based vertebrates. There are no anatomical indications that aquatic habitats were a realm king tyrants spent much time in, however, and we might imagine that swimming was more a means to cross aquatic obstacles rather than a critical part of their ecology.

"EXCESSIVELY POWERFUL AND DESTRUCTIVE": THE BITE OF KING TYRANTS

Today, the incredibly strong bites of *Tyrannosaurus* are some of their most famous attributes. Their inflated, incrassate dentition, deep, broad snouts, and voluminous jaw muscle cavities were noted early on as signifiers of what Osborn called "excessively powerful and destructive functions" (Osborn 1912, p. 29). So conspicuous are these features that it might be imagined that no one could ever imagine *T. rex* jaws operating any other way. A few authors, however, have taken contrary opinions. British paleontologist Beverly Halstead crafted a biomechanically miserable vision of king tyrant jaws when he wrote that the "vicious-looking teeth were not as bad as they seemed: if [*Tyrannosaurus*] had tried to tackle living animals, the teeth would have snapped off in the struggle" (Halstead 1975, p. 78). Barsbold (1983) echoed these comments, regarding tyrannosaur skulls as mechanically weak and their teeth blunt, with little predatory potential.

Needless to say, such opinions have not withstood scrutiny. A number of studies have brought the bite strength of *Tyrannosaurus* into sharp focus in recent decades, making it one of the most explored aspects of their biology (Bakker et al. 1988; Farlow et al. 1991; Erickson and Olsen 1996; Erickson et al. 1996; Molnar 1998, 2013; Meers 2002; Henderson 2003; Rayfield 2004; Snively et al. 2006; Bates and Falkingham 2012; Gignac and Erickson 2017; Rowe and Snively 2021; Johnson-Ransom et al. 2024; Rowe and Rayfield 2024).

Every aspect of tyrannosaurid jaw and tooth structure points to incredible strength and resilience, including remarkably powerful, bone-piercing bites. *Tyrannosaurus* developed these traits even further, perhaps granting them the strongest bites of any land animal in history (fig. 4.21).

That a carnivorous dinosaur might be adapted for such powerful biting was, at one stage, little evidenced, but a series of studies over the last thirty years have provided a wealth of supporting data for this hypothesis. Contrasting with the typically blade-like, flesh-slicing teeth of other theropods, the stout, robust teeth of king tyrants were the first clue to their bone-breaking capability (Bakker et al. 1988; Farlow et al. 1991). A lack of tyrannosaur feeding traces on dinosaur bones initially seemed to argue against this interpretation, but, in the mid-1990s, Gregory M. Erickson and others provided irrefutable proof that *Tyrannosaurus* had sufficient dental resilience and bite power to penetrate even large bones (Erickson and Olsen 1996; Erickson et al. 1996). Entering the discussion was a large *Triceratops* pelvis covered in tooth marks, gouges, and even gnawed-off, chewed surfaces. A number of details, including the size and shape of the feeding traces, as well as their geological provenance, meant that they could only have been generated by a king tyrant (this gnawed *Triceratops* specimen is discussed further in chapter 6). Shortly after, another approach, structural engineering, was used by Molnar (1998) to demonstrate that *T. rex* skulls were remarkably resilient to powerful biting thanks to their wide snouts with subtriangular cross sections and reduced cranial openings.

These early studies set the tone for what would follow. Along with an accumulation of further *Tyrannosaurus* bite mark data (see chapter 6), increasingly sophisticated biomechanical analyses have pinpointed further features that made *T. rex* skulls so strong. In addition to their broad and robust construction, the fused, vaulted nasal bones provided particular resistance to bite stresses, absorbing peak strains under heavy loads (Rayfield 2004; Snively et al. 2006). This feature, as outlined in chapter 2, is an ancient characteristic of

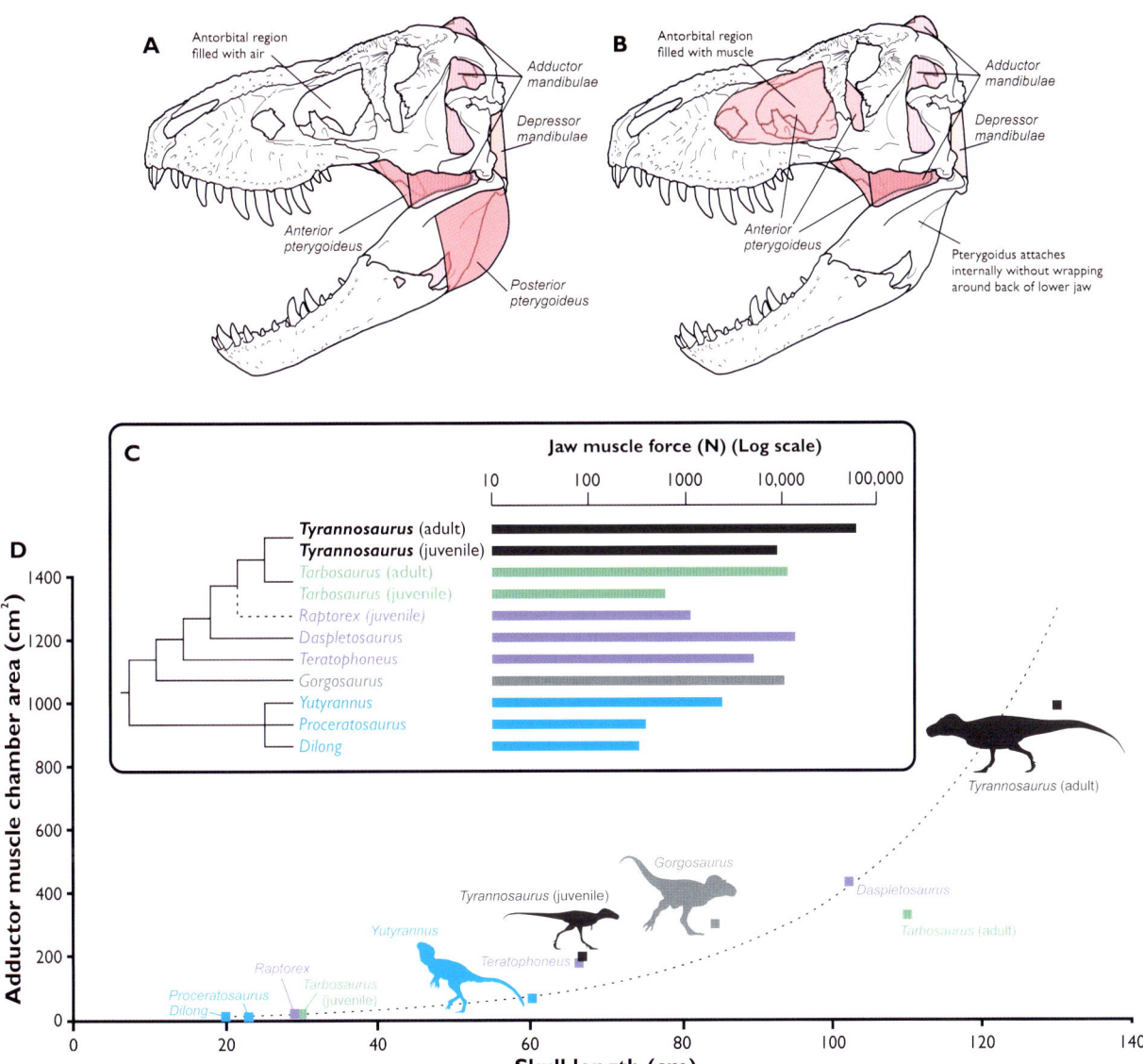

FIGURE 4.21

Competing jaw muscle reconstructions and bite force estimation in *Tyrannosaurus rex*. (A) Widely assumed jaw myology of *T. rex*, where the posterior pterygoideus muscle wraps around the rear of the lower jaw and the antorbital fenestra represents the development of pneumatic tissues; (B) alternative jaw myology based on Molnar (2008) where the mass of the pterygoideus is located within the antorbital cavity, without a component wrapping around the lower jaw; (C–D) estimated tyrannosauroid bite forces mapped to tyrant dinosaur evolution (C) and skull length (D). (C–D) Data from Johnson-Ransom et al. (2023).

tyrannosauroids and implies that strong snouts were important to tyrant dinosaurs from the earliest phases of their evolution, even though early species were not hard biters (Johnson-Ransom et al. 2024). Equally important to resisting bite stresses were cranial reinforcements against powerful pulling and twisting motions of the neck generated during feeding (Snively and Russell 2007a, b).

Parallel to these discussions of skull strength, however, have also been considerations of *Tyrannosaurus* skull flexibility. Some of the joints between *T. rex* jaw bones are fairly loose, suggesting a capacity for cranial kinesis: the ability of a skull to flex at certain points, like those of birds, snakes, and some lizards, during the apprehension or swallowing of food (Osborn 1912; Molnar 1991; Hurum and Currie 2000; Rayfield 2004, 2005a, b; Holliday and Witmer 2008; Larsson 2008; Cost et al. 2020). Against the evidence of their formidable bites, the potential of mobile, flexible *T. rex* skulls has been described as a biomechanical paradox (Cost et al. 2020) because jointed jawbones make skulls mechanically weaker. Stress distribution models of *T. rex* skulls confirm that cranial kinesis would lessen their jaw strength, although their effect was not entirely negative: some joints may have also absorbed shock (Rayfield 2004).

Potentially, however, king tyrant skulls are simply misleading us about their kinetic potential. Some living animals have skull joints that would be interpreted, from fossils, as indicative of cranial kinesis, but they act as solid, immobile units in life thanks to specifics of osteology and soft-tissue anatomy. The bulk of evidence suggests that this was likely the case for *Tyrannosaurus* (Osborn 1912; Hurum and Currie 2000; Cost et al. 2020). Analyzing the kinetics of a complete king tyrant skull suggests that their few loose joints were rendered immobile by the rigid, unbending struts of bone forming much of the cranium, the especially large and solid maxillae, and strongly sutured bone connections at kinetically significant points (Cost et al. 2020). Simply put, the few open, loose bone contacts in *T. rex* skulls were outnumbered by structures that inhibited movement,

suggesting that there is no paradox to resolve concerning jaw mobility and bite strength of king tyrants. Their skulls were functionally akinetic and immobile, even if some joints weren't strongly fused.

While the notion of an exceptionally powerful king tyrant bite is now universally accepted, constraining estimates of the forces exerted between their jaws has proved more problematic (Bates and Falkingham 2018). Bite forces are estimated in Newtons, N, with one Newton representing the gravitational force necessary to accelerate a one kilogram mass one meter in one second. This shakes out to 1,000 Newtons being roughly equivalent to 102 kilogram-force. The biomechanics of biting mean that the highest bite forces are generated at the back of the jaw, so it is between the rear teeth that *Tyrannosaurus* jaw strength is generally estimated.

The strongest measured bite from any living animal stems from the saltwater crocodile, *Crocodylus porosus*, which can generate over 16,000 N between their rear teeth (Erickson et al. 2012). *T. rex*, it seems, surpassed this, but by how much is difficult to say. Estimated bite strengths from different studies are wide ranging: 8,000–24,000 N (Gignac and Erickson 2017), 35,000–57,000 N (Bates and Falkingham 2012, 2018; Cost et al. 2020), 61,000–94,000 N (Rowe and Snively 2021), up to a staggering 183,000–235,000 N (Meers 2002). The latter, especially high estimates have been regarded as unlikely because they represent bite strength extrapolations from living animals rather than direct modeling of *T. rex* bite function (Meers 2002). As with the body mass estimations discussed earlier, scaling of this kind can produce questionable results unless all predictive parameters are carefully considered and corrected for. Bite force estimates of 200,000 N are probably excessive, even for *T. rex*.

More faith is placed in efforts involving precise modeling of *Tyrannosaurus* jaw muscle volumes and contraction trajectories. However, even these estimates differ by more than a factor of 10, prompting questions as to why we cannot find a common biomechanical ground for *T. rex* bites in the way researchers have for maximum speed or body mass (Bates and Falkingham 2018). Assumptions of jaw muscle configuration, it seems, are not to blame. To date, all studies use schemes where, in addition to the temporal adductor musculature (see chapter 3), the large posterior pterygoideus muscle wraps around the back of the lower jaw, and the antorbital region is filled with air (fig. 4.21A; e.g., Bates and Falkingham 2012, 2018; Gignac and Erickson 2017). An alternative reconstruction, where the pterygoideus muscle fills the antorbital cavity but does not extend to the outer surface of the lower jaw (fig. 4.21B; e.g., Molnar 2008) has been proposed but not biomechanically tested. Nor are these differences attributable to different *T. rex* specimens giving conflicting biomechanical signatures, as different research teams often base their models on the same skulls. This leaves assumptions about muscle strength and physiology as the likely cause of our poorly constrained bite strengths (Bates and Falkingham 2018; Cost et al. 2020), and continued research in this area is needed to refine our bite-force models. For now, a baseline figure for *T. rex* bite strength can be taken from physical simulations of a model king tyrant tooth being pushed into a large cow bone: this act requires at least 6,000 N, or 612 kilogram-force (Erickson et al. 1996). Whatever the exact bite force was, there is no doubt that *T. rex* had jaws far stronger than those of most carnivores, living or extinct.

As with everything pertaining to *Tyrannosaurus* anatomy and function, the bite forces and skull mechanics of juvenile king tyrants were different from those of adults (fig. 4.21C–D). With shallower skulls and lessened volumes for jaw musculature, juvenile *T. rex* had yet to develop the incredibly strong bites and reinforced jaws of adults (Bates and Falkingham 2012; Carr 2020; Rowe and Snively 2021; Johnson-Ransom et al. 2024). Bite strength seemed to increase markedly during the *Tyrannosaurus* growth spurt, with estimates for the smallest juveniles (fig. 4.7) rating at just over 1,000 N, large juveniles like "Jane" (fig. 2.15) attaining 14,000 N, and subadults producing up to 32,000 N (Rowe and Snively 2021). As with adult *T. rex*, the constraints on these bite forces are loose: other simulated bite

strengths for the "Jane" specimen are a mere 5,000 N (Peterson et al. 2021). Happily, physical evidence gives an unequivocal demonstration of juvenile bite strength even as we struggle to pin down their precise Newtonic capacity. Tooth marks created by juvenile *Tyrannosaurus* show that "Jane"-grade king tyrants had already developed bone-puncturing bites even before their skulls and teeth metamorphosed into more massive and robust structures (Peterson et al. 2021).

With both physical evidence and biomechanical simulations attesting to the power of *T. rex* jaws, we might wonder if they were the strongest biters of all time. The challenges of predicting bite forces, alas, preclude confident comparison with other species, but *T. rex* are surely contenders for this title. They vie for the top spot with a suite of aquatic animals, the giant-skulled pliosaurid plesiosaurs and giant extinct crocodylians, each predatory species that were also adapted for rending animals apart with powerful jaws (Foffa et al. 2014; Aureliano et al. 2015). Some studies predict that the giant megatooth sharks, *Otodus megalodon*, could bite even harder (e.g., Wroe et al. 2008), although these come with substantial caveats about predicting the bite forces of a species known only from vertebrae and teeth.

A final twist about this most famous aspect of *Tyrannosaurus* biology are recent studies that question whether *T. rex* bites were truly exceptional once we factor in body size. Bigger animals invariably have stronger bites than smaller ones because their jaw muscles are absolutely larger. This, it seems, may explain much about *T. rex* jaw strength. Analyses of body mass and bite forces across a wide range of vertebrates find that, for their size, *T. rex* bites were not actually remarkable (Sakamoto et al. 2019). Nor, for that matter, did any tyrannosaurid appear to enhance the evolution of their skulls to gain stronger bites than are expected for their size. What may have made tyrant dinosaurs unique are the anatomical innovations that exploited the heightened bite strength that their body mass already conferred, adapting their jaws and teeth to maximize application of powerful jaw muscles. Genuine enhancements of bite force beyond those predicted by body masses are more commonly seen among smaller animals, including maniraptoran dinosaurs, primates, and certain birds. Smaller-bodied species that need strong jaws cannot rely on giant bodies that come pre-equipped with powerful bites; instead, they need to bulk up their jaw muscles and skulls (Sakamato et al. 2019). Remarkably, some of the hardest biting animals, for their body masses, aren't giant predators or even carnivorous species; they are tiny, 30 gram finches. Proportionate to body size, the 70 N bite from a 30 g finch is 320 times stronger than the 57,000 N bite from a multi-tonne *T. rex*. Comparatively speaking, the most excessively powerful and destructive jaws in evolution do not belong to a giant reptile, but to a tiny, seed-eating bird.

TYRANNOSAURUS ARMS: WHAT THE HECK?

Our consideration of king tyrant functionality would not be complete without some examination of their most perplexing feature: their arms. The diminutive size of tyrannosaurid forelimbs has invited explanation since the early days of tyrant dinosaur research, even when their body plans were only vaguely known (Cope 1866, 1868). Famously, they are so short that they cannot touch one another, nor reach the jaws. Today, opinion is split on their function and even the value of giving them research attention; according to some, "the reduced size of [tyrannosaurid] forelimbs shows they were not important to their owners, so they should not be important to us" (Paul 1988a, p. 320). This judgment is evidently not a majority view as, over the last 150 years, researchers have published diverse ideas and hypotheses over tyrannosaur arm use, with a particular focus on the arms of *T. rex*. What their arms were capable of and why they were so small has formed the focus of several major studies, with some positing that they are even evidence of new mechanisms of biological evolution.

Broadly speaking, interpretations of tyrant dinosaur arm use fall into one of several camps. The first is that their arms, despite their size, still performed some sort

of mechanical function. Osborn (1906) was the first to make such a suggestion, inferring that powerful muscles once wrapped around the *T. rex* humerus and that their forelimbs may have served to clasp mates during copulation. Brown (1915) considered the arms rudimentary, but noted their potential utility in grasping and holding. Others have posited that they helped tyrannosaurids rise from a prone position (Newman 1970; Stevens et al. 2008) or, through studying the mobility and strength of the *Tyrannosaurus* forelimb, that they were involved in predation, helping to restrain struggling prey while the head delivered powerful bites (Carpenter and Smith 2001; Carpenter 2002; Holtz 2008; Lipkin and Carpenter 2008; Krauss and Robinson 2013). Rothschild (2013) presented possible evidence of aggressive arm use in the form of scars and other wounds on *Tyrannosaurus* skulls that could have been formed by piercing manual claws. Concerned about the limited reach of the arms for attacking purposes, however, Holtz (2008) emphasized a role in stabilizing carcasses or victims already being dispatched by the jaws, and speculated on another function: visual communication (Holtz 2007). This view somewhat echoes that of Mattison and Griffin (1989), who had similar concerns about forelimb reach and postulated that juvenile tyrannosaurs, as fast, active predators, may have used their arms more than slower, (in their view) carrion-foraging adults.

A second school of thought is that the arms were nonfunctional, atrophied structures that had no real role in tyrant lives. This is the oldest view of tyrannosaur arm function, with Edward Drinker Cope (1866) writing of "*Laelaps*" (= *Dryptosaurus*) that "The small size of the fore limbs must have rendered them far less efficient as weapons than the hind feet" and positing that these dinosaurs attacked prey and rivals by leaping into the air, brandishing taloned feet like a saurian cross between kangaroos and eagles (Cope 1866, 1868). This inspired one of the most famous pre-*Tyrannosaurus* restorations of any tyrannosauroid, Charles R. Knight's 1897 *Leaping Laelaps*, where leaping and rolling tyrant dinosaurs employ taloned feet against one another (fig. 4.22). Cope's views of hindlimb-dominant tyrannosaur predation have not endured, but the view that their arms were reduced to a point of uselessness, especially in predation, has been widely shared (e.g., Bakker 1986; Paul 1988a; Horner and Lessem 1993; Horner 1994; Lockley et al. 2008; Padian 2022). Taken to an extreme, the perceived redundancy of the arms has been touted as evidence for *Tyrannosaurus* being restricted to scavenging, it being reasoned that powerful forelimbs would be essential in holding live prey (Horner and Lessem 1993; Horner 1994). This suggestion, and the wider scavenging hypothesis, has been criticized at length (chapter 6).

A third view is that tyrannosaurid arm reduction has not occurred through functional redundancy, but because the evolution of smaller arms was promoted by specific behaviors or strange evolutionary mechanisms. Among these proponents is paleontologist Kevin Padian who, in 2022, concluded that ecological and behavioral causes were behind their reduction. He proposed that a need to keep tyrannosaur arms clear of bites from conspecifics during group feeding caused their shrinkage, an idea that, even at face value, is not without complications. As we'll discuss more in chapter 6, gregarious behavior in tyrannosaurs is far from well evidenced, and, moreover, *Tyrannosaurus* forelimb bones bear plenty of injuries: fractures, evidence of torn ligaments, wrenched muscles and even tooth marks from other *T. rex* (Rothschild and Molnar 2008; Lipkin and Carpenter 2008; Longrich et al. 2010; Carpenter 2013). Whatever benefits their peculiar arm evolution offered, it does not seem that reduced size kept *T. rex* forelimbs safe from injury.

Other efforts to understand the forces shaping *T. rex* arms propose that they were somehow liberated from the constraints of typical adaptive evolution. This was the view taken by Martin Lockley and colleagues (2008), who concluded that the small arms of *Tyrannosaurus* and other theropods developed through "morphodynamic compensation" and genetic "anteriorization." In essence, they posited that *T. rex*–grade theropods invested most of their genetic developmental resources on their heads, bypassing development of their arms. This sees reduced forelimbs evolving not through the conventional route of natural selection and adaptation, but rather as a genetic byproduct of growing large heads. Another genetics-based approach, so-called "evolutionary teratology," proposed that the short arms of *Tyrannosaurus* and other theropods were the result of "developmental anomalies ... over evolutionary times, which become integral parts of groups and taxa" (Guinard 2015, p. 1). Drawing on modern clinical studies, this idea posited that the proportions and reductions of tyrannosaurid arms were malformations that, by having no detrimental effect on their owners, became a part of the tyrannosaurid genome. There is, however, no evidence for any anomalous evolutionary processes influencing the development of *Tyrannosaurus* and the lofty concepts of Lockley et al. (2008) and Guinard (2015) have not been adopted by other tyrannosaur workers. Morphodynamic compensation has even been characterized as speculative and "useless" by some (e.g., Lipkin and Carpenter 2008; Farke 2009; Brusatte 2010), which seems a fair, if blunt, assessment. The fundamentals of evolution need not be rewritten because one group of dinosaurs has unexpectedly small forelimbs.

Thanks to a number of studies on the size and functional properties of tyrannosaur forelimbs, some of the ideas discussed above can be ruled out. The concept of *T. rex* arms being vestigial or unimportant does not, on balance, seem well supported. Although undoubtedly small for their body size (Middleton and Gatesy 2000) and sporting claws that are neither as large nor as curved as those of some other theropods (fig. 3.12), the general proportions of tyrannosaurid arms and their relative length remained unchanged throughout 15 million years of evolution (Currie 2003b; Lipkin and Carpenter 2008). As one of the last and most "advanced" of all the tyrannosaurids, king tyrants might be predicted to have particularly small, less developed and less mobile arms than their ancestors if they were truly vestigial, but this is not the case. Instead, early tyrannosaurids seemed to find a forelimb configuration that required no significant change throughout their known

FIGURE 4.22

One of the most famous paleoartworks in history, Charles R. Knight's 1897 *Leaping Laelaps*, is an early depiction of tyrant dinosaur forelimb redundancy. Overseen by Edward Drinker Cope, this artwork realizes Cope's interpretation of tyrannosaurs having a kangaroo-like capacity for leaping and eagle-like feet for striking other animals. This view was based, in part, on a mistaken referral of a large hand claw to the feet, making the feet seem more raptorial than they actually were, and the arms less useful.

CHAS. R. KNIGHT
97

evolutionary run. As noted earlier in this chapter, the most notable change in *T. rex* forelimb length occurred during growth, not evolution, as younger animals have proportionally much longer arms and hands than their parents (fig. 4.23C; Williams et al. 2010).

The range of motion of the forelimb joints also argues against vestigiality (fig. 4.23B; Carpenter and Smith 2001; Carpenter 2002; Lipkin and Carpenter 2008). King tyrant arms were less mobile than those of some other theropods, but not to the extent seen in the similarly short-limbed abelisaurs, where no significant motion was possible below the shoulder (Senter and Parrish 2006). Rather, *Tyrannosaurus* had forelimb arthrological ranges broadly similar to those of other large theropods, such as *Allosaurus* and *Acroncanthosaurus* (Carpenter 2002; Senter and Robins 2005). Briefly summarized, the upper arm of *T. rex* was restricted to movement behind the glenoid so the humerus could move up and down in a 40° arc. They could not, as is often shown, dangle straight down from the shoulders or reach forward. They could, however, be drawn unusually high up against the body, and the broad joint at the top of the humerus allowed for some rotation and sideways reach. More mobility may have been present in juveniles, as the shoulder joints of younger animals face sideways as well as rearward (P. Larson 2013). The elbow was not especially flexible, preventing the arm from straightening and only permitting about 60° of movement. The wrist was also relatively rigid and only allowed for some inward and outward motion, so the hand could not fold up against the forearm, bird-style. The fingers were more mobile, however. Unlike human fingers, they could hyperextend when the hand was opened (i.e., they arced upward), during which they also separated, giving *T. rex* a wide grasp between their two claws. When closed, the claw tips pointed back along the arm as well as inward, recalling the grip of a raptorial bird foot. Carpenter (2002) summarized this range of motion as that of a "clutcher," a theropod that lacked the strong grabbing ability of something like *Allosaurus*, but that retained sufficient mobility to grasp and hold objects to its chest.

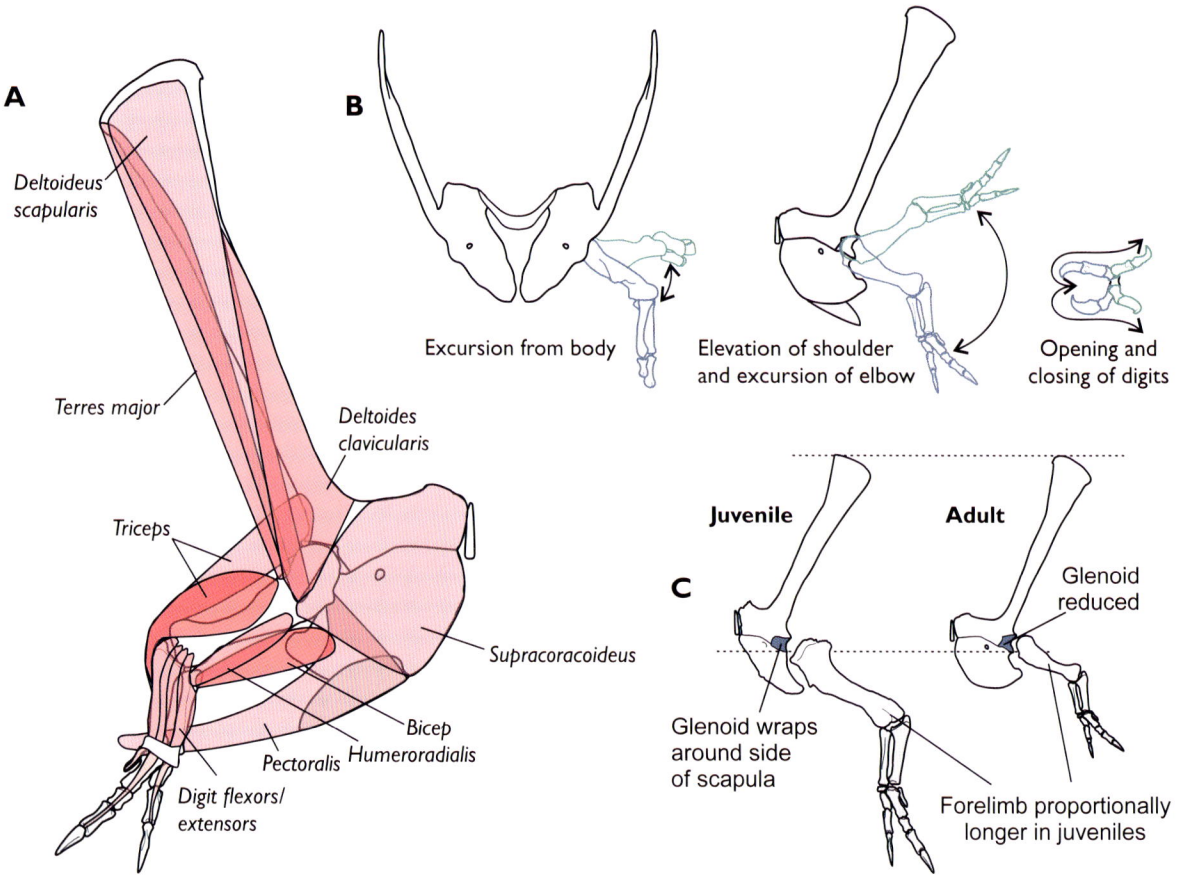

FIGURE 4.23

Forelimb function in *Tyrannosaurus*. (A) The surprisingly extensive musculature of *T. rex* arms; (B) forelimb range of motion in adult king tyrants; (C) variation in glenoid extent and relative forelimb length in juvenile and adult *T. rex*, scaled to scapula length. (A) Based on Lipkin and Carpenter (2008); (B) adapted from Carpenter and Smith (2001) and Carpenter (2002).

Studies on the muscles doing this grasping have also been performed (fig. 4.23A; Carpenter and Smith 2001; Lipkin and Carpenter 2008). Because the fore-arms were so short and the muscle attachment sites so relatively large, the effort exerted by their muscles was maximized. Thus, *T. rex* arms were much more pow-erful than those of a human, despite their similar size. Lipkin and Carpenter (2008) predicted that a single, 9.5 cm wide bicep of the "Sue" *T. rex* could lift almost 350 kg. This may seem unimpressive in comparison to their overall, multi-tonne body weights, but *T. rex* were probably not using their arms like weightlifters doing bicep curls. Rather, king tyrants looking to grasp or clutch objects probably used the full suite of their fore-limb, shoulder, and chest muscles, including the large, powerful deltoids and pectorals, to enhance their grip and holding capability. Working to an upper arm mus-cle reconstruction with a 25 cm diameter, Lipkin and Carpenter (2008) predicted that a single *T. rex* arm could exert 1,150 kg of force; a value that implies *T. rex* arms could be used for serious exertion. Such efforts, in turn, are reflected in the large number of injuries found across their forelimbs and pectoral girdles. These include stress fractures caused by repetitive actions, and shoulder muscles being torn from their insertions on the humerus (Lipkin and Carpenter 2008; Carpenter 2013; also see chapter 6).

This arthrological, ontogenetic, myological, and pathological data paints the most complete picture of *T. rex* arm function, and with such data we don't need to tie ourselves in knots interpreting their use. There are, after all, only so many applications for a set of small but strongly muscled and grasping-capable limbs. All things considered, the model of *Tyrannosaurus* arms as organs that clasped other animals seems best supported, perhaps with such behaviors being especially important in younger, smaller individuals with greater relative arm size. This greater reach and capacity to grasp prey may have compensated for weaker bites in these individuals, although we await detailed analyses of juvenile *Tyrannosaurus* arms to gain more perspective on this hypothesis. A role in assisting *T. rex* rising from the ground seems technically possible, but was perhaps limited by the reach of the forelimbs. Moreover, the strength of the hindlimb muscles, along with elastic energy stored in tendons when crouching, probably precluded any significant need for forelimb assistance during standing (Paul 1988a; Carpenter 2013).

If a clasping or clutching function seems likely, where may this have been useful? There is probably not one single answer to this. Limbs are multifunctional anatomies and can be deployed in a variety of contexts, even those they are not specifically adapted for: we might consider how turtles dig nests with their flippers, or how birds stabilize themselves with their wings when climbing or falling over. We will probably never know the full repertoire of behaviors that *T. rex* applied their arms to, but given the carnivorous ecology and aggressive interspecies interactions evidenced for *Tyrannosaurus* (see chapter 6), a base assumption might be that their arms were used for grappling other animals during close encounters, most likely prey during predatory acts (fig. 4.24) and conspecifics during fights. While tyrannosaur jaws and teeth were clearly their most important offensive equipment and the elements most likely to first contact another animal during an aggressive act, their arms may have helped to restrain their targets while they bit or ate other parts. This idea is somewhat analogous to the strategies employed by raptorial birds, which tear at their prey with their beaks while gripping them with their talons (Fowler et al. 2011b). Further analogy with raptor prey apprehension stems from the shape of *T. rex* hand claws (fig. 3.12; Longrich and Saitta 2024) where the thumb, like the inner major toe of a raptor foot, has a relatively deep, especially curved claw. Objections that the forelimbs were too short for such purposes might be rebuffed by not only their space for expanded forelimb musculature but also their strain-related injuries: these are hard to explain if *Tyrannosaurus* weren't doing *something* strenuous with their arms. We might also note that giant theropods with more obviously raptorial hands, like allosauroids and spinosaurids, were not gifted with substantial arm length or reach either. Grasping prey in front of their shoulders was also impossible in these dinosaurs and their clutching ability would, like *Tyrannosaurus*, have drawn objects close to their chests (Carpenter 2002; Senter and Robins 2005). If we are satisfied that these giant predators used their arms aggressively, we can surely entertain a similar role for the arms of *Tyrannosaurus*.

Overleaf: **FIGURE 4.24**

Did *Tyrannosaurus* forelimbs have a role in predation, such as capturing the hadrosaur *Edmontosaurus annectens*? Although limited in reach and unlikely to assist with apprehending prey before the jaws, the strength and clutching capability of *T. rex* arms may have helped to restrain prey while the mouth delivered more devastating injuries.

LANDS OF THE TYRANTS

5

TYRANNOSAURUS **ARE ICONS** of the American West, but exactly where, and for that matter when, did they live? Considering these questions brings us to discuss the nature of the *Tyrannosaurus* fossil record, the nature of the environments they dwelled in, and the organisms they lived alongside. Understanding the ecologies and communities that king tyrants occupied adds context to their physiology and functionality, and establishes details necessary to discuss their lives and behavior in the next chapter.

KING TYRANTS IN TIME AND SPACE

UNDERSTANDING WHERE AND when *Tyrannosaurus* lived (fig. 5.1) is a discussion requiring us to consider geology as much as paleontology, as evidence of their ancient habitats lies in the rocks that house their remains. This necessitates confronting what might be considered the less glamorous, "nuts and bolts" side of paleontological research. Dinosaurs, and *Tyrannosaurus* in particular, are often known to us through spectacular "trophy specimens": fantastic skeletons that become museum centerpieces and the subjects of detailed descriptive and biomechanical studies. But such specimens are single data points in efforts to appreciate the geographic and geological range of an extinct organism, and it is only through careful, systematic fieldwork, and a capacity to identify scrappy, broken shards of ancient organisms, that the geological and geographic extents of fossil species are determined.

THE *T. REX* FOSSIL RECORD

We mentioned in chapter 1 that the fossil record of *Tyrannosaurus* is excellent compared to other dinosaurs, and especially so compared to other larger theropods. Indeed, *T. rex* is a surprisingly common fossil in some localities. Sustained field studies by Jack Horner and colleagues (2011) have found that *T. rex* represents, on average, 24 percent of the associated dinosaur remains (that is, specimens represented by more than three bones) in the sediments of the Hell Creek Formation (fig. 5.2D). This makes them the second-most-common type of dinosaur skeleton in this rock unit, and this— along with the general desirability of *Tyrannosaurus* bones—accounts for their significant presence in museums and private collections. Of these, *T. rex* accounts for almost 11 percent (71 of 658) of associated dinosaur

FIGURE 5.1

One tyrant, three biomes. Typically depicted as denizens of lush, swampy forests, *Tyrannosaurus* had a broad geographic range that saw them dwell in a number of distinctive environments, from forests and mountain ranges to vast, open plains.

material from the Hell Creek and Lance formations. Isolated king tyrant teeth and bones add over 1,300 specimens from the same quarries, representing 4.6 percent of fragmentary fossils from these localities (Stein 2019). Most dinosaurs, and many fossil taxa in general, are known from single specimens, so amassing almost 1,500 examples of a single species is exceptional.

As outlined in our anatomical discussions of chapters 3 and 4, no *Tyrannosaurus* skeleton known to science is 100 percent complete. The most complete example yet found is the famous "Sue," which has 73 percent of (approximately) 300 bones preserved (N. Larson 2008). Only some vertebrae of the neck, the tail tip, and some bones of the hands and feet are missing

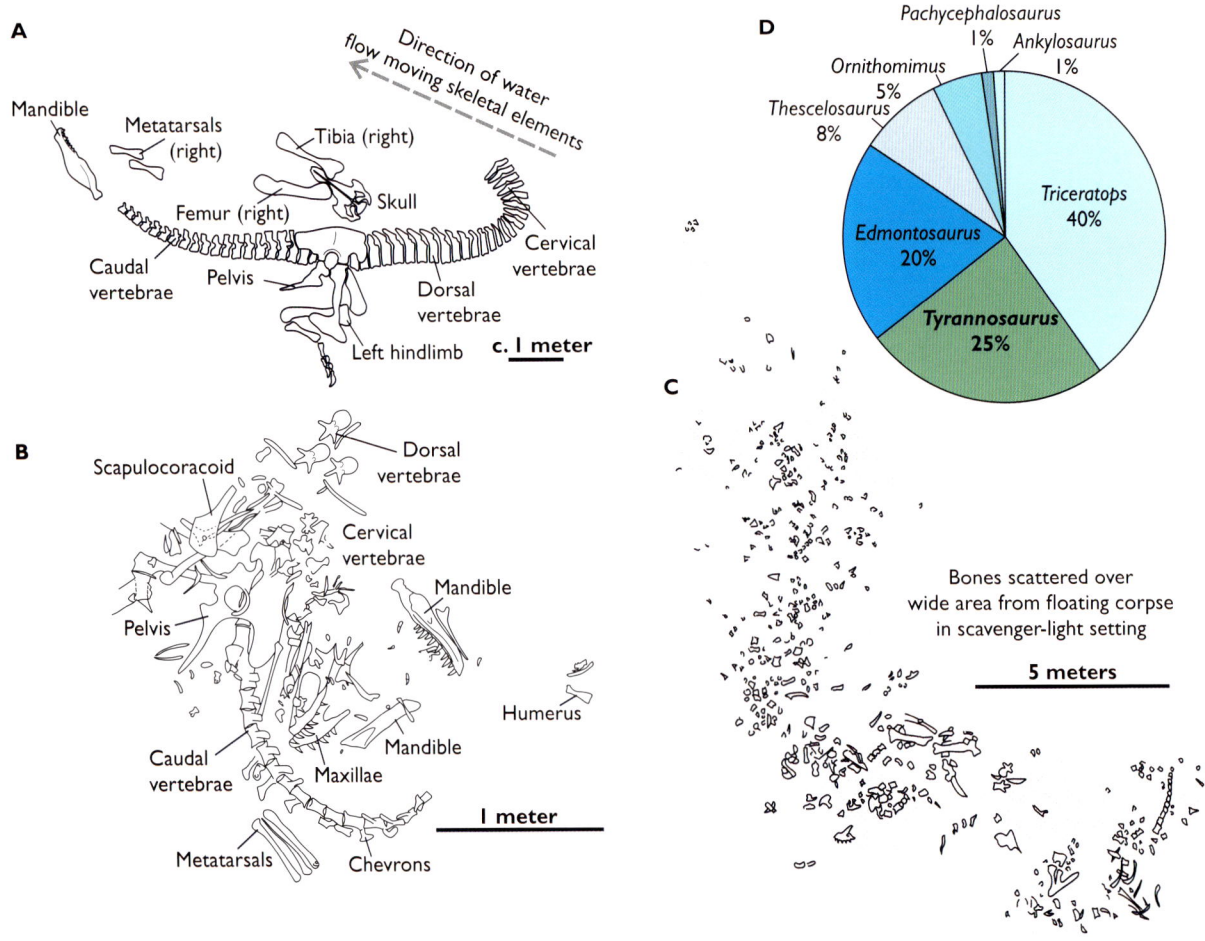

FIGURE 5.2

Particulars of *Tyrannosaurus* fossilization. (A–C) Quarry maps showing the varied preservation styles of *T. rex* skeletons; (A) the mostly articulated, near-complete "Wankel/Nation's *T. rex*" specimen; (B) mostly disarticulated but associated remains of "Jane"; (C) the disarticulated and scattered remains of "Peck's rex"; (D) dinosaur skeleton abundance in the Hell Creek Formation, showing *T. rex* as the second-most-abundant dinosaur. (A) After Horner and Lessem (1993); (B) after Henderson and Harrison (2008); (C) after Derstler and Myers (2008); (D) data from Horner et al. (2011).

(Brochu 2003), which makes it substantially better represented than the sub–50 percent completion rate for most *T. rex* specimens (N. Larson 2008). Near-totally preserved *Tyrannosaurus* skeletons are rare because fossilization requires organisms to be buried, and getting such enormous skeletons into the ground demands specific environmental conditions. The processes that affect organismal remains between death and burial, termed "taphonomy," are mostly destructive: scavenging, decay, physical weathering and transportation (usually via water). They tend to destroy carcasses partly or wholly before they can become fossils, and retarding their effects generally requires burial. Entombing an animal larger than an elephant in sediment is no mean feat, however. It either requires a lot of energy in a short period (i.e., rapid burial in a deluge of sand or mud) or a sustained period with minimal taphonomic disturbance, during which a carcass can progressively be covered by slower sedimentation processes. Both preservation styles are known for *T. rex*, meaning their skeletons are found in varying states of completeness and articulation (fig. 5.2A–C; Horner and Lessem 1993; Derstler and Myers 2008; Henderson and Harrison 2008).

Taphonomic processes may not entirely stop at burial, however, which is why soft tissues, such as skin, organs, muscles, cartilage, and so on, are rarely fossilized: their decay often continues even after they are covered by sediment. Bones may then be altered further by diagenesis, postburial processes that introduce high pressures and temperatures to fossils. During this period, bones can be crushed, deformed, and chemically altered from their original configuration (fig. 3.2). With such a preservational gauntlet to run, it is no fluke that we have so many good specimens of *T. rex*. Factors such as *T. rex* abundance, climatic and environmental conditions, and geological conditions within the Western Interior were evidently optimal for capturing the remains of large dinosaurian predators. The great abundance of outcrops bearing *T. rex* fossils, and the army of people willing to seek and collect them, further explains our substantial king tyrant inventories.

KING TYRANT DISTRIBUTION: GEOGRAPHIC ...

Noting *where Tyrannosaurus* fossils occur is only half of what paleontologists consider when they find a new specimen. They also log which sedimentary rock unit they occur in, a detail essential to revealing where in the geological column their newfound remains come from, thus placing them in the ordered sequence of rocks that make up our geological record. It is from such classifications that we derive the term "formation," a grade of rock unit frequently mentioned in paleontological studies. Geologists categorize geological layers, or strata, based on their size, from beds and laminae (rock units just centimeters or millimeters thick) to enormous, country-spanning groups and supergroups comprising batches of major rock layers. Formations are the "Goldilocks" unit for studies of fossil communities and paleobiogeography. They are large enough to provide substantial samples of fossil life, being composed of many beds (and clusters of beds, known as members) and extending over wide areas (often kilometers), but are not so broad as to span too much geographic space or geological time, coarsening any insights provided by their fossil content. The rocks in formations were also usually created by related environmental processes (for example, a river channel and associated floodplain, or a reef and associated lagoon), such that they capture complementary details about ancient settings and habitats. These can reveal the nature of the environments inhabited by the organisms now preserved as fossils within these rocks, a topic we will return to shortly.

Tyrannosaurus occurs in at least eleven of these formations, spread across the Western Interior sedimentary basin of North America (figs. 5.3–4). Spanning two Canadian provinces and up to eight American states, the north-south extent of *T. rex* measures either almost 1,000 km or nearly 2,000 km, depending on whether tyrannosaurid bones in New Mexico and Texas represent *T. rex* (see chapter 2, also below). Their

LANDS OF THE TYRANTS

northernmost extent occurs in the Willow Creek and Scollard Formations of Alberta and the Frenchman Formation of Saskatchewan. Although not as familiar as certain *Tyrannosaurus* formations from the United States, these units have yielded several important and famous *T. rex* specimens, including the subadult Willow Creek specimen known as "Black Beauty" (N. Larson 2008) and the enormous "Scotty," from the Frenchman Formation (Persons et al. 2020).

Below the Canadian–US border in eastern Montana occurs one of the most famous dinosaur-bearing formations in the world, and a *T. rex* hotspot: the Hell Creek Formation. Some of the first fossils of *Tyrannosaurus* were found in these strata (chapter 1), and it has proved to be an enormously productive *T. rex* unit ever since. Along with the *Tyrannosaurus rex* holotype, Hell Creek localities in southeastern Montana have yielded AMNH 5027, "Jane," "Peck's rex," the "Wankel rex"

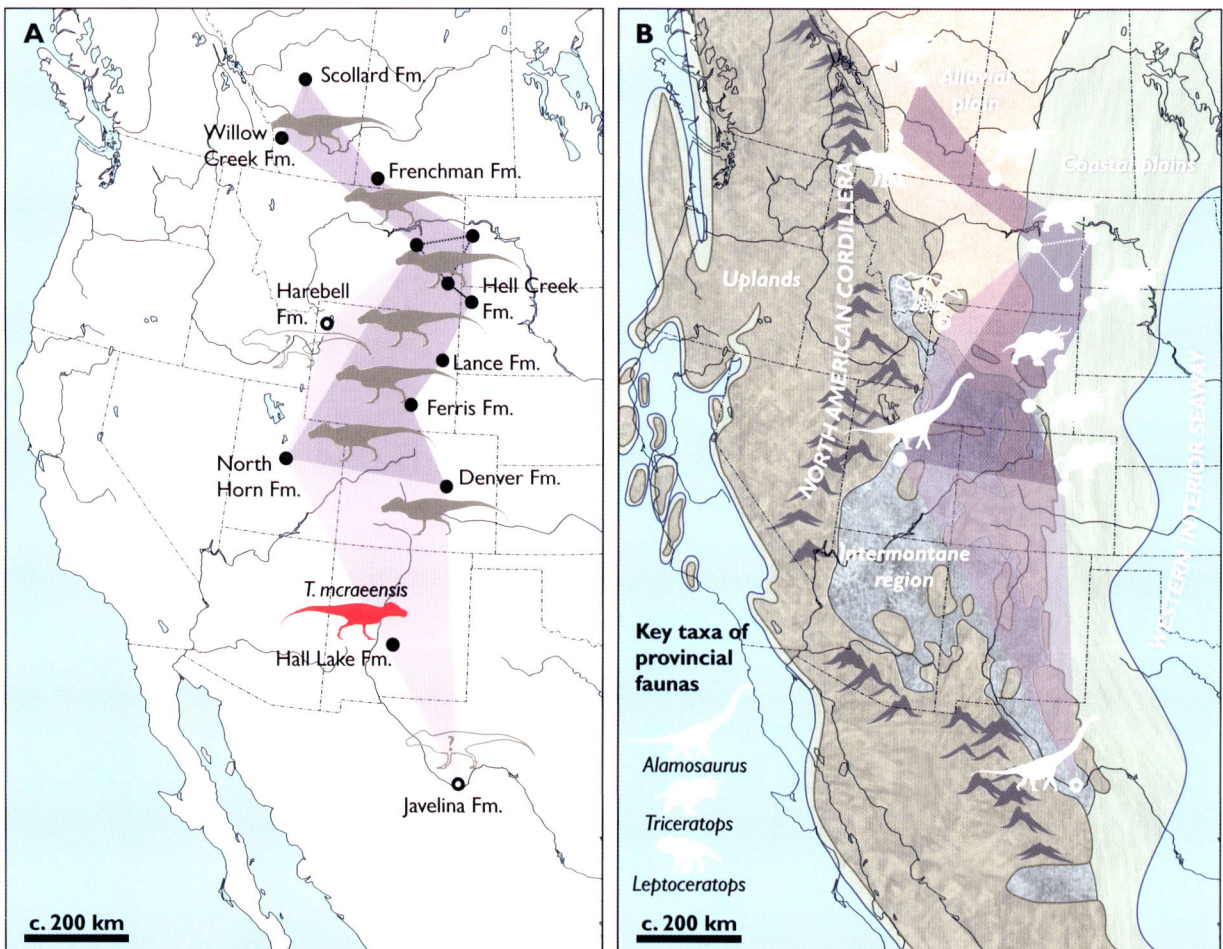

FIGURE 5.3

Geographic distribution of *Tyrannosaurus rex*. (A) Fossil occurrences in North America, with questioned or uncertain examples represented by open circles; (B) reconstructed *T. rex* range across the habitats and dinosaur faunas of Maastrichtian Laramidia. Maps augmented from data presented by Sampson and Loewen (2005).

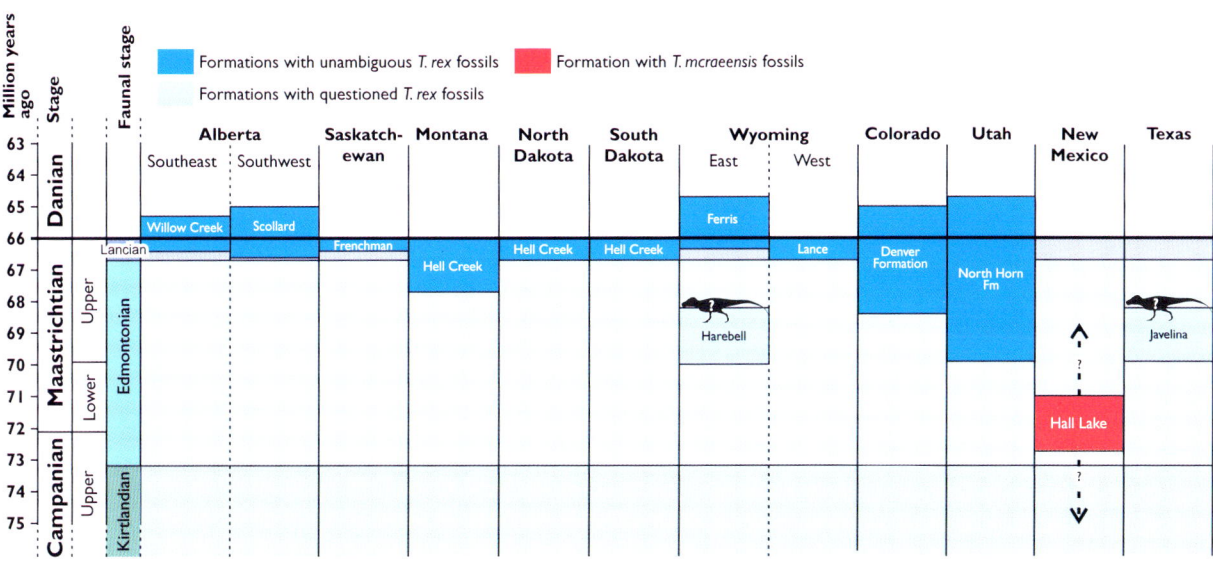

FIGURE 5.4

Stratigraphic ranges of *T. rex*–bearing formations. The geographic relationships among these units are complex and their cataloging here is somewhat simplified: see Fowler (2017) for additional details.

(or "Nation's rex"), the "*Nanotyrannus*" holotype, "Trix," and several more familiar and nicknamed specimens. Hell Creek rocks also crop out in the neighboring states of North and South Dakota, and these provided the famous "Sue" and "Stan" specimens, two of the most studied examples of *Tyrannosaurus*. These are in addition to, as noted above, well over 1,000 Hell Creek *T. rex* specimens, consisting of isolated teeth to near-complete skeletons, in public and private collections. Even if no other strata preserved king tyrant fossils, the Hell Creek Formation would provide an embarrassment of tyrant riches by itself. Lest it be thought that Hell Creek is just "the *T. rex* formation," however, king tyrants are just one component of the Hell Creek biota. These rocks provide a window into the remains of an entire Cretaceous ecosystem and are one of the most important rock units in the world for understanding the final days of the Mesozoic Era (chapter 7).

Further south, in Wyoming, is another geological unit with historic importance: the Lance Formation. Along with providing relatively complete *T. rex* remains

(Dalman 2013), this was the source of Osborn's "*Dynamosaurus*" as well as Marsh's forgotten "*Ornithomimus grandis*" (Breithaupt et al. 2008). A *Tyrannosaurus*-like tooth suggests that *T. rex*, or a close relative, also existed in the northwest of Wyoming (Hodnett et al. 2023), and confirmed king tyrant remains stem from the Ferris Formation in the southern regions of the state (Lillegraven and Eberle 1999).

The northern range of *T. rex*, comprising formations of southern Canada, Hell Creek, and Lance, form the "core" sites of *Tyrannosaurus* discoveries, where their remains are most abundant and well preserved. King tyrant localities become more spread out, their remains sparser, and their identifications more controversial, in more southerly regions (fig. 5.3). The North Horn Formation of Utah has presented fragmentary *T. rex* bones (Sampson and Loewen 2005), while the Coloradan Denver Formation, despite having the accolade of providing the first documented *T. rex* fossil (chapter 1), has since yielded only fragmentary remains and one partial skeleton (Carpenter and Young 2002; Breithaupt et al. 2008).

Further south are two formations with controversial *Tyrannosaurus* content. Partial tyrannosaurid remains have been found in the Hall Lake Formation of New Mexico (this formally being known as the Hall Lake Member of the McRae Formation), either interpreted as *T. rex* (Gillette et al. 1986; Lozinsky et al. 1984; Wolberg et al. 1986; Carr and Williamson 2000) or, more recently, *T. mcraeensis* (chapter 2, Dalman et al. 2024). The only tyrannosaur bones from the Javelina Formation of Texas, a partial caudal vertebra and a maxilla, are also of disputed identity. As noted in chapter 2, some regard them as unquestionably *T. rex* (e.g., Lawson 1976; Carr and Williamson 2000), but others see them as a close relative (e.g., Carpenter 1990; Wick 2014; Brochu 2003; Carr 2020). Uncertainties about the age of the Javelina Formation, and a suggestion that it may be significantly older than other king tyrant strata (fig. 5.4; Fowler 2017), add to these misgivings. The proposal that *T. mcraeensis*, a southern species interpreted as geologically older than *T. rex*, perhaps adds credibility to the Javelina tyrannosaur belonging to a non-*rex* species, but superior remains are needed to make a definitive call.

... AND TEMPORAL

Considering the ages of these *Tyrannosaurus*-bearing units allows us to consider their distribution in a fourth dimension: time. It is common knowledge that *T. rex* lived at the very end of the Cretaceous period, apparently surviving to witness the extinction event that ended the Mesozoic Era, but when did king tyrants first evolve, and how long was their evolutionary history? And did they live in the same regions throughout that entire interval?

Answering these questions requires a more precise address in geological time for *Tyrannosaurus*. Our 4.5-billion-year-old planet has harbored life for about 4 billion years, of which complex, multicellular life has existed for (roughly) the last 600 million. Most of this interval is taken up by the Phanerozoic Eon, which is divided into three eras: the Paleozoic (550–252 million

years ago, or Ma); the Mesozoic (252–66 Ma); and the Cenozoic (66 Ma onward). The Mesozoic is also divided into three periods: the Triassic (250–201 Ma); the Jurassic (201–145 Ma); and the Cretaceous (145–66 Ma) (fig. 2.2). Each of these periods is divided into stages of various durations, their extents being based on the persistence of certain fossil assemblages. The final stage of the Cretaceous, and thus the geochronological home of *T. rex*, is known as the Maastrichtian, which ran from 72 to 66 Ma. It is these last six million years that we have to focus on to learn the evolutionary duration of the *T. rex* lineage, a process that requires knowledge of the geological ages of the rocks that preserve their fossils.

Unfortunately, dating geological formations is not straightforward. The study of sedimentary rock sequences and their temporal correlations is an involved process that makes up the discipline of stratigraphy. Determining stratigraphic relationships between rock units can be complicated because rock layers, like fossils, can become broken, distorted, and confused over geological time. This is especially so in the Western Interior Basin of North America where vast, locally variable geology makes for a complex jigsaw of differently aged strata. Efforts are further hampered by many Western Interior formations not bearing rocks or fossils that allow for reliable assessments of their age. Certain rock deposits, especially those created near active volcanoes, bear zircon crystals that allow for absolute geochemical dating. Zircon crystals chemically alter over time at set rates, allowing the ages of individual crystals to be calculated by analyses of their decayed isotope ratios (archaeological carbon dating uses a similar process, but carbon decays too rapidly for use on geological timescales). Thus, if original, freshly formed zircons can be found in a geological formation (as opposed to older examples eroded in from elsewhere), we can obtain a relatively precise estimate for its age. Fossils can also be used to date rocks if a geochemical date has been determined for that species elsewhere, in that co-occurrence of a taxon across two or more formations implies a similar age for both deposits. A combination of these approaches has allowed stratigraphers to correlate and order the majority of the world's sedimentary rocks, but some, including several Maastrichtian formations in the Western Interior, are resilient against efforts to assess their age. Several formations containing *T. rex* and other late-stage American dinosaurs, therefore, remain of uncertain vintage (Johnson 2008; Fowler 2017; Paul et al. 2022).

This is not to imply that there are no constraints whatsoever on the stratigraphic range of *T. rex* and their associated biota. Northern *T. rex*–bearing formations cover periods up to or even through the end-Cretaceous extinction event (fig. 5.4), preserving both the final sediments of the Mesozoic as well as the first of the overlying Paleogene Period (the initial interval of the mammal-dominated Cenozoic Era). Canadian deposits bearing *T. rex* are no older than 67 million years, and the Hell Creek Formation probably spans the final 1 million years of the Cretaceous, or slightly less (Fowler 2017). This places the bulk of our *T. rex* record in a narrow range of up to one million years, matching the fossil records of dinosaur taxa that stratigraphers use to characterize one of North America's final dinosaur communities: *Triceratops* and *Edmontosaurus annectens* (see below).

In the southern United States, however, are *Tyrannosaurus* fossils from units of less certain age. The North Horn, Denver, and Javelina formations are at least 1 or 2 million years older than *T. rex* units in the north, but how much older they are is unclear (Wick 2014; Fowler 2017; Dalman et al. 2024). The Javelina Formation, which may be at least 68 or 69 million years old, might extend the geological range of king tyrants to 3 million years or more if its tyrannosaurid remains are genuine *T. rex* (Wick 2014). Potentially even older is the Hall Lake Formation, home of *T. mcraeensis*, the lower sediments of which have been dated to 73 million years old (Dalman et al. 2024). No other datable sediments are known for this formation, however, and its *Tyrannosaurus* remains stem from sediments occurring many tens of meters above the time-calibrated horizon. How much time these intervening rocks represent is an unknown quantity at present, but Dalman et al. (2024) use the 150 m of dinosaur-bearing sediments layered over the dated Hall Lake sediments to gauge a rough estimate.

If that 150 m stack represents the entire remaining 7 million years of the Cretaceous and sedimentation rates were continuous, the occurrence of the McRae tyrannosaur 30 m above the dated beds gives an estimated age of 70 million years (Dalman et al. 2024).

That the Hall Lake and Javelina formations are older than the "core" *T. rex* localities is potentially further evidenced by their lack of dinosaur species that characterize the final million years or so of Cretaceous North America (Wick et al. 2013; Lehman et al. 2016; Dalman et al. 2024). The significance of this awaits exploration, but it might imply greater ecological plasticity for *Tyrannosaurus* or *T. rex* than is suggested by the relatively uniform cast of species found in northerly and geologically younger king tyrant formations. Such assessments await the means to more accurately place southern *Tyrannosaurus*-bearing units in geological time, however, as well as further studies that will elucidate the distinctiveness of the Javelina and Hall Lake tyrannosaur species.

Because the lower extent of *Tyrannosaurus* stratigraphic range is somewhat ragged, the geological longevity of *Tyrannosaurus rex* remains an open question. Attempts to estimate the duration of *T. rex* evolution through a variety of means, including methods not strictly based on fossil distribution, point to a longevity of 1.2 to 3.6 million years (Marshall et al. 2021). This lack of resolution hampers our understanding of *Tyrannosaurus* ecology and evolution because we are unable to appreciate their place in the development of North America's latest Cretaceous dinosaur communities. Did *T. rex* appear at the metaphorical eleventh hour, flashing into existence for the final curtain call of the Mesozoic and existing within a single dinosaur community (see below)? Or were they a well-established, "veteran" species that lived through several dinosaur faunas? The latter is not impossible, as the Canadian tyrannosaurid *Gorgosaurus* seems to have persisted for 1.7 million years, while contemporary herbivorous dinosaur species turned over every 300 thousand years (Carr et al. 2022). Greater understanding of Western Interior stratigraphy, and resolving the identities of those southern *Tyrannosaurus* specimens, are key investigations needed to resolve this matter.

WHERE THE REXES ROAMED

ALONG WITH RECORDING where and when fossil organisms lived, geological formations capture details about the environments that their fossils were preserved in. If those remains have not been transported long distances from their owner's point of death, this, in turn, imparts data on the settings where extinct organisms lived. Detailed study of these and their fossil content allows us to reconstruct paleoenvironments: the ancient habitats that were home to extinct animals and plants. It is through these that we can understand something of the larger geographic context that *T. rex* was part of, and to picture the habitats that king tyrants considered home.

Appreciating these environments is aided by learning something of the basic geography and climate of North America during Maastrichtian times (fig. 5.3B). In this interval, North America was joined to Asia, forming the landmass Asiamerica, and was undergoing a transformation from long-standing Cretaceous norms into newer, uncharted conditions. For much of the Cretaceous, the climate had been unusually hot and sea levels were very high, flooding many continents with shallow seas. One such waterway, the Western Interior Seaway, divided North America into two landmasses 100 million years ago: Laramidia to the west, and Appalachia to the east. Warm conditions and shallow, nutrient-rich waters made this a productive sea filled with diverse marine reptiles, fish, and invertebrates, the fossils of which provide an especially complete view into a late-Mesozoic marine ecosystem (Everhart 2017). Just before or during the early Maastrichtian however, global sea levels fell and the Western Interior Seaway began retreating, rejoining Laramidia and Appalachia after 30 million years of separation. In this process, new coastal habitats for terrestrial animals were opened up. *Tyrannosaurus* was among the species that encountered these new landscapes, inhabiting the western, Laramidian region of the newly rejoined North America. They mostly existed within the continental interior,

far away from the retreating seaway, but occasional pulses in sea level sometimes brought marine conditions further inland, making *T. rex* a coastal denizen (e.g., Van Vranken and Boyd 2021). Another major geographic feature occurred in the west, the North American Cordillera, a mountain chain that extended north-south through Laramidia. This was the still-growing progenitor of the mountains and uplands that still exist in western North America today, and they may have been a physical barrier to the spread of *Tyrannosaurus*. Although they are poorly known, the dinosaur communities west of these mountains seemingly have a different fauna to those of the Western Interior (Prieto-Márquez & Wagner, 2015.

It was not only geography that was changing in North America during the Maastrichtian. The climate was also shifting, becoming cooler after hothouse conditions had persisted throughout the majority of the Cretaceous. As a species with a substantial north-south geographic range, *T. rex* inhabited settings with a variety of average air temperatures, from a balmy 20–30°C in the south to a cooler 10–15°C in the north. Rainfall seems to have varied from a semiarid 1–2 mm per day in the south to a semitropical 3–4 mm per day in the north (Amiot et al. 2004; Hunter et al. 2008; Wade et al. 2019). These climatic conditions and local geography contributed to creating three major biomes that were inhabited by *T. rex*, roughly arranged north to south: seasonal lowlands, coastal floodplains, and volcanic highlands (fig. 5.1, 5.3B; Sampson and Loewen 2005).

NORTHWEST: MOUNTAIN-SHADOWED SEASONAL LOWLANDS

Canadian king tyrants lived in a habitat largely distinct from that of their US relatives. The Albertan Willow Creek and Scollard Formations represent alluvial plains:

FIGURE 5.5

The alluvial plains—vast, highland-adjacent landscapes crisscrossed by braided rivers—inhabited by northwestern *Tyrannosaurus*. *T. rex* in some of these environments encountered pronounced seasonality with dry summers and relatively dark winters. Here, a subadult king tyrant combats these trying conditions by excavating a highland *Leptoceratops* from their burrow.

wide, flat riverine landscapes formed from material washed off the North American Cordillera (fig. 5.5; Jerzykiewicz 1997). The Saskatchewan Frenchman Formation, meanwhile, captured floodplain sediments in a broad valley, a transitional landscape between mountain-influenced and coastal biomes (McIver 2002). Each of the environments of these northern formations were distinct from one another, being shaped by their distances from the nearby uplands. They are united, however, in having pronounced seasonality, a phenomenon absent from more southerly parts of the Maastrichtian Western Interior. The climate experienced by northern *T. rex* was warm-temperate and somewhat humid, but seasonal water stress was common (Jerzykiewicz 1997; McIver 2002). This was especially so in the regional southeast where, perhaps in the rain shadow of adjacent mountains, Willow Creek localities bears evidence of desiccated soils and drought-adapted plants (Jerzykiewicz 1997). Reduced water stress is evident in the Scollard and Frenchman formations, which were relatively mesic, with well-developed lakes and rivers. The Frenchman paleoenvironment, however, was prone to wildfires (Bamforth et al. 2014).

Driving seasonal shifts across all three formations were daily light levels. At these latitudes, winter months experienced shortened days and lengthened nights that altered plant communities (McIver 2002). These are expressed in the paleobotanical record of the Frenchman Formation (detailed below) where evergreen vegetation is notably rarer than in the climatically similar, but less seasonal, Lance and Hell Creek formations. In Maastrichtian Canada, it seems, most plants fell into dormancy during winter. The presence of frost-averse palms, cycads, and crocodylians in these settings indicates that winter temperatures were not extreme, however, rarely or never dropping to freezing. Maastrichtian dinosaurs may not have been too bothered by these variations in light levels as dinosaurs living even further north, and experiencing even more pronounced winter conditions, did not migrate to avoid cold and dark months (Druckenmiller et al. 2021). If the dinosaurs of northern Canada were unmoved by

seasonal change, those of milder winters in southern Saskatchewan and Alberta may have toughed out their dim winters as well. This matter ties into questions of whether there is evidence of *Tyrannosaurus* migrating, a topic we will address in the next chapter.

EAST: FORESTED COASTAL PLAINS

The majority of *Tyrannosaurus* are found in sediments representing the sweeping floodplains that separated Laramida's westerly mountains from the shrinking

Western Interior Seaway. These forested lowlands are represented by the most extensively exposed and famous *T. rex* localities, the Hell Creek and Lance formations, as well as the Ferris and Denver formations (Sampson and Loewen 2005). The Frenchman Formation has sufficient environmental similarity with these deposits to be considered a bridge between the alluvial-influenced environments of southern Alberta with the more coastal settings of the central United States. The same may be true of the Denver Formation, a floodplain habitat which had sedimentological influences from nearby southern mountain settings (Carpenter and Young 2002).

Most of our reconstructions and knowledge of *T. rex* habitats stem from these formations, such that these well-vegetated floodplains might be considered the

FIGURE 5.6

Fall season in Hell Creek: a dromaeosaur, *Acheroraptor temertyorum*, surrenders a *Leptoceratops* carcass to a young *Tyrannosaurus*. Within just a few years, *T. rex* juveniles would outgrow all their contemporary dinosaurian predators, making even young king tyrants intimidating presences for other Lancian carnivores in the western interior.

"classic" *Tyrannosaurus* paleoenvironment (fig. 5.6). Intense interest in these deposits, especially Hell Creek, has led to detailed insights into their depositional environments (see Fastovsky and Bercovici 2016 for a review). Climatic conditions were warm-temperate or subtropical, and generally humid. Lacking the seasonality characterizing *T. rex* habitats further north, these coastal plains were marked by a high water table and abundant rivers, lakes, ponds, and swamps. The size of these floodplains fluctuated over time as the Western Interior Seaway pushed inland and pulled away during its gradual retreat, sometimes depositing marine sediments and fossils in otherwise continent-derived sediments. These marine incursions probably had little impact on king tyrant lives, but beach-visiting *T. rex* may have occasionally chanced upon opportunities for seafood.

The Hell Creek, Lance, and Frenchman formations provide excellent paleobotanical data (Johnson 2002; McIver 2002; Bamforth et al. 2014; Fastovsky and Bercovici 2016). Hell Creek, in particular, has a flora of almost 400 plant species from over 150 sites, giving it one of the most extensive botanical records of the entire Mesozoic. Somewhat south of the Frenchman Formation and thus escaping seasonal dimness, the floras of Hell Creek and Lance formed woodlands with a tropical flavor, dominated by small- to medium-sized angiosperm trees. Both diversity and abundance were heavily skewed toward flowering plants such that lineages typically dominating Mesozoic floras, such as ferns, conifers, cycads and maidenhair trees, made up less than 10 percent of diversity and abundance (Johnson 2002). Cycads were rare and had low diversity, but conifers were well represented. They included redwoods *Metasequoia* and *Sequoia*, monkey puzzles *Araucarites* as well as various types of cypress (*Parataxodium*, *Taxodium*, *Fokienia*). Flowering shrubs would have been common, as would have flowering trees. Palms (*Sabalites*) were especially abundant, and woodlands were filled with walnuts (*Juglans*, possibly *Dryophyllum*); grapevines (*Vitis*); katsura (*Trochodendroides*); elm-like species; plane (Platanaceae); and magnolia; among many others (Johnson 2002; Fastovsky and Bercovici

2016). While these floodplains lacked pronounced seasonal changes in climate, many of these trees were deciduous, such that the arrival of fall would have been signified by yellow and red canopies amid evergreen trees (fig. 5.6), as well as the skeletonization of woodlands in winter.

Water-loving plants were also diverse, including willows (*Salix*); aquatic flowers (*Trapago*); *Equisetum* horsetails, some of which grew to enormous sizes (Kiely pers. comm. 2023); as well as various mosses (Fastovsky and Bercovici 2016). Coals hint at the presence of swamps and peats filled with rotten vegetation. Overall, places like Maastrichtian Hell Creek—in spite of its name—seem like climatically stable, pleasant places to live. Environmental concerns may have been limited to occasional floods and wildfires, evidenced by fossil charcoal (Fastovsky and Bercovici 2016). Distant volcanic eruptions contributed ash to some of these communities, but this geological activity was probably insufficiently close to pose significant risk to the plants or animals that inhabited these floodplains.

SOUTHWEST: THE UPLANDS

The latter cannot be said of our last biome. The southern extent of *T. rex* range is perhaps the least familiar, or least reconstructed, of their major habitats, despite the geography of this region probably being the most dramatic. King tyrants of the North Horn formation lived in intermontane regions, dwelling on highland floodplains in river-formed valleys surrounded by mountains (Olsen 1995). These environments formed from eroded mountain sediments in the same way as those of the Willow Creek and Scollard formations, but, rather than wind and rain washing this material onto adjacent lowlands, these rocks accumulated within the upland basins of the growing southern Cordillera.

Even more dramatic was the environment of Hall Lake Formation (Lozinsky et al. 1984), habitat of *T. mcraeensis* or, in some views, another realm of *T. rex*. These maturing mountains caused tectonic activity that brought clichéd vintage artworks of *Tyrannosaurus* alongside volcanoes to life, Hall Lake sediments regularly recording evidence of local pyroclastic volcanism (fig. 5.7; Jahns et al. 1955; Buck and Mack 1995). Leaf litter was geologically captured by volcanic ash falls and low-energy floods to reveal forty to fifty species of ferns, cycads, conifers, and flowering plants. These were of broadly similar composition to the plant life of more northerly formations, although conifers and cycads seem to have been more common (Upchurch

FIGURE 5.7

Together (and scientifically validated) at last: *Tyrannosaurus* and volcanoes! Whether *T. rex* or *T. mcraeensis*, Hall Lake *Tyrannosaurus* lived immediately adjacent to volcanically active mountains, bringing numerous clichéd paleoartworks to life.

and Mack 1998). The presence of substantial trees with trunk diameters of 1.5 m is recorded by trunk stumps preserved in-situ with Hall Lake rocks (Upchurch and Mack 1998).

With a constantly humid, subtropical climate, there is little evidence of strong seasonal variation in this southern region, although some sediments indicate water was scarce on occasion (Buck and Mack 1995). Over time, the North Horn Formation appears to have become slightly wetter, leading to the formation of marshes and swamps on valley floodplains (Lawton et al. 1993). Given the environmental contrast of this bi-ome to the more open habitats to the north and east, it is intriguing to ponder whether mountain *Tyranno-saurus* and other dinosaurs were distinguished from their lowland counterparts in some details of appear-ance and behavior. On such matters, of course, we can only speculate.

MEETING THE NEIGHBORS

FULLY FLESHING OUT *T. rex* habitats requires knowl-edge of the animals that they lived alongside, includ-ing those that they may have seen as prey, those that may have been threats or competitors, and those that formed the background characters of their world. A rich fauna of dinosaurs and other species has been uncov-ered from the sediments of latest Maastrichtian Lara-midia and, with a biota that includes several household names—*Tyrannosaurus, Triceratops, Ankylosaurus, Ed-montosaurus annectens*—this community surely ranks as one of the most famous of all Mesozoic ecosystems. These species are part of the "Lancian" faunal stage of western North American dinosaur evolution and their geological proximity to the end of the Cretaceous pe-riod makes them one of the last non-avian dinosaur assemblages. Continued study of Lancian faunas over the last century has made them especially well under-stood, despite their surprisingly short-lived existence. The taxa that geologically define Lancian rocks, *Tricer-atops* and *Edmontosaurus annectens*, appear in the

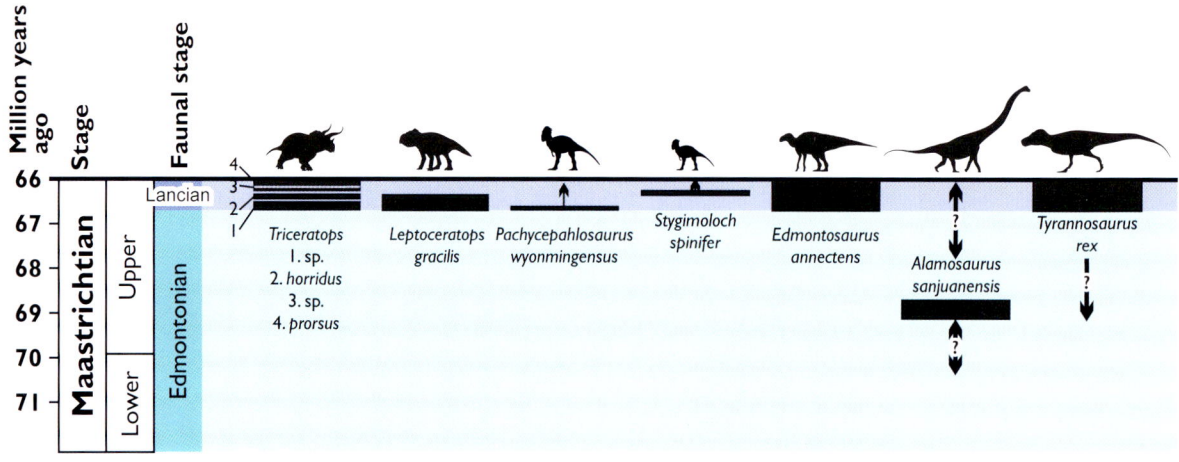

FIGURE 5.8

Stratigraphic distribution of *Tyrannosaurus rex* and key Lancian dinosaur species, based on Fowler (2017). The geological age of *T. rex* is poorly constrained, with them possibly being a relatively short-lived, purely Lancian taxon or, alternatively, a species that spanned several million years.

fossil record about 67 million years ago and existed for a mere 800 thousand years before their extinction at the end of the Cretaceous (fig. 5.8; Fowler 2017). Preceding dinosaur faunal stages enjoyed much longer durations: the Edmontonian lasted about six million years (c. 73 to 67 million years ago), the Kirtlandian ran for two million years (c. 75 to 73 million years ago) and the Judithian spanned five million (c. 80 to 75 million years ago) (Fowler 2017).

Comparing Lancian dinosaurs with those of these older stages highlights a second characteristic of the last North American dinosaurs: that their diversity was relatively low. As we will discuss more in chapter 7, the communities inhabited by *T. rex* bore relatively few species that ranged widely across the Western Interior rather than, as was the case in previous intervals, multiple species segregating into latitude-based geographic provinces (Lehman 2001). Some evidence of provincialism is still apparent in the Lancian (fig. 5.3B), but faunal geographic assemblages are defined more on the abundances of certain taxa rather than, as in the Edmontonian and Judithian, the presence or absence of distinctive species (Lehman 2001). With no geographic barriers dividing northern from southerly dinosaurs in the latest Cretaceous, it is thought that these faunas reflected segregation through dinosaurian preferences for different climates or plant communities, but some researchers use the lack of separating agents to question whether North American dinosaur provincialism actually existed at all, especially in the Maastrichtian (e.g., Carr and Williamson 2000; Lucas et al. 2016; Dean et al. 2020).

Whether representing true faunal segregation or not, some geographic differences in late Maastrichtian dinosaur faunas seem evident (Sampson and Loewen 2005; Wick 2014). *Tyrannosaurus* was a constant across these communities (Carr and Williamson 2000; Sampson and Loewen 2005), which could imply that king tyrants were ecologically versatile and adapted to the different prey species and habitats of each region. It might also be noted however, that likely staples of *T. rex* diets, such as large-horned dinosaurs (*Triceratops*, *Torosaurus*),

hadrosaurids (*Edmontosaurus annectens*), and *Thescelosaurus*, occur in almost every deposit yielding king tyrants. Perhaps *T. rex* were so omnipresent because, in the low-diversity Lancian, familiar food sources were also geographically widespread?

Forging a picture of Lancian biotas is aided by the latest Maastrichtian formations of the Western Interior being so intensely sampled. We thus have a good understanding of the animals that lived in these settings despite a number of preservation biases hampering their fossilization. Small-bodied taxa are especially rare (e.g., Horner et al. 2011), such that we have excellent and numerous remains of many large species, but only fragmentary remains of less substantial ones. Perseverance and careful analysis has seen a number of diminutive Lancian species being named however, fleshing out the diversity of small- and midsized forms that would have lived alongside the more commonly found giants. Tempting as it is to refer to these species as living in the shadow of *Tyrannosaurus*, the journey from hatchling to adult king tyrant would have seen *T. rex* enter the world as an animal no bigger than a midsized dog, during which *they* lived in the shade of much larger beings.

DENIZENS OF WATERWAYS AND THE UNDERSTORY

Some of the most interesting Lancian animals represent species at the opposite end of the size spectrum to *Tyrannosaurus*, and they may have played little direct roles in king tyrant lives outside of their youngest years. As they matured, these small birds, mammals, lizards, and invertebrates probably faded from their notice, but these species were probably far more conspicuous, and perhaps sometimes even dangerous, to young juveniles.

One aspect that makes smaller Lancian species noteworthy is their modernity, with the clams and snails that inhabited the streams, lakes, and coastlines witnessed by *T. rex* belonging to familiar living taxa. The Lancian freshwater clams *Corbicula* and *Plethobasus* are still extant today, and others, such as the marine *Crassostrea* (oysters), are major sources of our food. Similarly, alongside extinct species, Lancian fish include taxa sought by anglers, such as bowfin (*Amia*), gars (*Lepisosteus*), and sturgeon (*Acipenser*) (Brinkman et al. 2014). Sharks and rays swam through the same waters as these fish (Cook et al. 2014), while an array of extant and extinct frog and salamander genera swam and hopped around (Wilson et al. 2014). Turtles, including many soft-shelled forms, were diverse and apparently abundant (Fastovsky and Bercovici 2016).

Lizards and snakes, too, were diverse with almost thirty Lancian lizard taxa and three snake species known (Longrich et al. 2012). Lizards were primarily represented by the extinct polyglyphanodonts, but extant groups such as skinks, iguanas, anguids, as well as relatives of varanoids were also present. Some of these reptiles developed large enough species, such as the 6 kg lizard *Palaeosaniwa* and 3 kg snake *Cerberophis*, to probably pose predatory risk to small dinosaurs (see chapter 6; Longrich et al. 2012), at least before king tyrants grew big enough to turn the tables on them. Some Lancian crocodylians and the croc-convergent *Champsosaurus* were adapted for pursuing fish, but larger, more robust species also represented danger for smaller dinosaurs. While most Lancian crocodylians were within the size range of living examples, older parts of the Javelina Formation contain remains of the enormous alligatoroid *Deinosuchus riograndensis*, a 10 m long species that primarily ate large turtles, but also consumed dinosaurs on occasion (Schwimmer 2002). As noted above, it remains uncertain if *Tyrannosaurus* was present in the Javelina Formation, but, if so, even relatively large tyrants could have been vulnerable to these great reptiles.

A diverse set of mammal fossils are known from Lancian deposits (Sloan and Van Valen 1965; Wilson 2014), but their morphologies and ecologies are poorly understood on account of their fossils mostly comprising teeth and jaws. Mesozoic mammals have been stereotyped since the nineteenth century as tiny, uninteresting animals living in fear of the larger, more spectacular dinosaurs (Panciroli 2021), but this perspective has been overturned as better fossils have revealed their adaptive diversity. Alongside shrew- or possum-like species, Mesozoic mammals also included relatively large predators, gliders, burrowers, and swimmers (Luo 2007; Panciroli 2021).

Our better-known Lancian mammals also did not conform to cliché. The Hell Creek Formation contains two relatively large species, including *Meniscoessus robustus*, a c. 3 kg multituberculate that may have been herbivorous, foraging on communities of newly flourishing flowering plants (Wilson et al. 2012). Even larger was *Didelphodon vorax* (fig. 5.9): a somewhat mustelid-like mammal related to marsupials. Their jaws and teeth were among the strongest of any Mesozoic mammal, indicating a diet of flesh, shelled invertebrates, and vertebrate tissues, possibly including bone (Wilson et al. 2016). Tooth chemistry and some skeletal proportions have been used to interpret *Didelphodon* as semi-aquatic (Longrich 2004; Claytor 2023), although this conclusion is disputed (Fox and Naylor 2006). Reaching 4–6 kg and perhaps over half a meter in snout-vent length, *Didelphodon* may have been large enough to intimidate hatchling *T. rex*, although the possible semiaquatic habits of this mammal may have limited interaction with terrestrial dinosaurs. As with other Lancian mammals, they were likely prey for younger king tyrants but ignored by larger *T. rex* in pursuit of more substantial meals.

Sharing the issues of being evidently diverse, but with limited insight into form and function, are Lancian birds. Fossils of these flying dinosaurs provide one of the most comprehensive insights into the diversity of latest Maastrichtian birds with at least seventeen species known (Longrich et al. 2011), albeit from very fragmentary, isolated, and often broken bones. Curiously, they all seem to represent exclusively Mesozoic lineages, despite evidence of modern bird groups evolving in the Cretaceous (Field et al. 2020). With size estimates of these species ranging from 200 g to 5 kg, they include some of the largest (presumed) flying birds of the Mesozoic. Their ecologies, unfortunately, are unpredictable on account of their scrappy remains,

FIGURE 5.9

A diverse community of mammals lived alongside *T. rex*, the largest and best known of which was *Didelphodon vorax*. A relatively large (for a Mesozoic mammal) 5 kg carnivore with powerful jaws, *Didelphodon* may have intimidated the smallest, youngest king tyrants, although larger juveniles may have seen them, and other Lancian mammals, as food.

although the referral of some bones to the flightless, semiaquatic Hesperornithes predicts the presence of swimming avians in Lancian waterways.

THE COMPETITION ... OR LACK THEREOF

Perhaps surprisingly, non-tyrannosaur Lancian dinosaur predators are not much larger than the species discussed above. This reflects one of the most intriguing facets of *Tyrannosaurus* biology: that a century of collecting Lancian fossils across numerous geological formations has not yet yielded any midsized predatory dinosaurs—excepting, of course, the controversial pygmy tyrant "*Nanotyrannus*" (chapter 2). *T. rex*, it seems, were the sole large predators in their range, and the abundance of their fossils suggests that they were not rare animals, either. Associated *T. rex* remains are found as often, and sometimes more frequently, than the herbivorous *Edmontosaurus* (fig. 5.2D; Horner et al. 2011), tying *Tyrannosaurus* and this hadrosaur for second-most-abundant skeletons in Hell Creek behind the first-place *Triceratops* (Horner et al. 2011). Of course, interpreting fossil assemblages requires teasing reality from the artifice created by preservation biases, and, given Hell Creek's preservation bias toward large taxa, it is unsurprising that the three largest dinosaur species in this unit provide the bulk of associated skeletons. The absence of non-tyrannosaur predators in Hell Creek is compelling, however, when we consider that subadult and juvenile *T. rex* were able to enter its fossil record, at least occasionally. Clearly, Hell Creek's preservation biases were not so exclusive that remains of all small or midsized dinosaur predators were denied passage through deep time. Where, then, are the remains of other carnivorous theropods?

One potential explanation of the absence of small or midsized predators in Lancian deposits is "tyrannosaurid niche assimilation." This posits that the dramatic ontogenetic changes in tyrannosaurid size, form, and biomechanics allowed them to operate as multiple "ecological species" through their lives, monopolizing or assimilating predatory roles that would usually be shared across multiple taxa (e.g., Russell 1970; Bakker et al. 1988; Woodward 2020; Holtz 2021). Typical dinosaur predator ecosystems are speciose, with differently sized and adapted carnivores filling different niches, but tyrannosaurids seem to have filled most of them alone. As we will explore further in chapter 7, it is not unusual for tyrannosaurid ecosystems to lack midtier carnivores (Holtz 2021), so their absence from the Lancian habitats frequented by *T. rex* is not aberrant, at least compared to other North American, late Cretaceous ecosystems. Niche assimilation may also explain the relative abundance of *T. rex* fossils. We might consider a modern mammalian predator ecosystem as "high diversity, low density" where a given area on, for example, an African savannah has a pride of lions, a pack of wild dogs, and a hyena clan all sharing prey resources. In this system, fossil representation of each species would be correspondingly reduced. But in "low diversity, high density" tyrannosaurid predator ecology, there is nothing but differently aged tyrannosaurs to occupy most carnivore ecological space. That, in turn, requires larger populations of tyrannosaurs to operate in each niche

and, in turn, create more abundant fossils. Whatever the explanation for their dominance of Lancian ecosystems, the curious absence of large non–*T. rex* carnivores is a reminder that, while some parts of the Cretaceous would have been familiar, the latest Mesozoic was still an unusual place where the rules of modern ecosystems did not always apply.

While midsized and large *Tyrannosaurus* monopolized upper-tier predatory niches, one ecology that tyrannosaurids never seemed to occupy fully was that of small-bodied predators; a consequence, perhaps, of entering the world as relatively large hatchlings (chapters 4 and 6). We have limited knowledge of the smallest predatory dinosaurs that lived alongside *T. rex*, but they included *Richardoestesia*, *Paronychodon* and *Pectinodon*. These diminutive animals were theropods of some sort, but they are only represented by teeth and scraps of jaw, so their real identities remain enigmatic. Their fossils are sufficiently common across Late Cretaceous dinosaur sites that they are typically regarded as "form taxa": fossils that are distinctive and diagnosable, but probably not "real" taxonomic species (Larson and Currie 2013). *Paronychodon* and *Pectinodon* might pertain to the feathered, talon-clawed theropod groups Dromaeosauridae and Troodontidae, while *Richardoestesia* is a genuinely mysterious taxonomic entity (Larson and Currie 2013). Further tooth discoveries hint at additional dromaeosaur and troodontid species in Hell Creek but, without additional bones, there is little to say about their biology (Gates et al. 2013).

A more confidently identified small Hell Creek theropod that lived alongside *Tyrannosaurus* is the dromaeosaur *Acheroraptor temertyorum* (fig. 5.6; Evans et al. 2013). Only the jaws and teeth of *Acheroraptor* are known, but they allow for predictions of a roughly 2 m long, perhaps 15 kg species similar in body plan to *Velociraptor* or *Deinonychus*. A slightly larger (perhaps 3 m long) dromaeosaur, *Dineobellator notohesperus* from the New Mexican Ojo Alamo Formation, may have lived alongside king tyrants if their southerly distribution is substantiated (Jasinski et al. 2020). Dromaeosaurs were hypercarnivorous predators, but, unlike their movie counterparts, there is no evidence that they ganged up to dispatch prey many times their size (Roach and Brinkman 2007). On the contrary, they are better interpreted as pursuers of small game, their powerful foot claws allowing them to mantle prey like raptorial birds (Fowler et al. 2011b). Dromaeosaurs may have posed risks to the eggs and hatchlings of *Tyrannosaurus*, although juveniles would rapidly surpass them in size and potentially reverse this relationship (fig. 5.6). Before then, speed may have been the best ally for juvenile *T. rex* to evade dromaeosaur attacks. Although popularly depicted as fast and lethal animals, and with the name of their most famous member, *Velociraptor* ("fast thief"), only enhancing this reputation, dromaeosaur hindlimbs were adapted more for powerful gripping than fast running, and they are not regarded as swift theropods by specialists (Fowler et al. 2011b; Persons and Currie 2016).

Another Hell Creek dromaeosaur, *Dakotaraptor steini*, represents the largest known theropod to have lived alongside *T. rex* (DePalma et al. 2015). Touted as a relative giant for their group and estimated to have attained a similar size to the burly dromaeosaur *Utahraptor ostrommaysi*, the existence of a Lancian dromaeosaur reaching or surpassing 250 kg is potentially significant: Perhaps some moderately-sized predators lived alongside *T. rex* after all? However, the validity of *Dakotaraptor* has proven controversial. Represented by an assortment of remains from a bone bed containing a mix of vertebrate fossils (DePalma et al. 2015), it is now clear that the *Dakotaraptor* holotype was chimeric. One bone, an alleged furcula, actually belonged to a turtle (Arbour et al. 2016), and growing, though not yet formally published, concerns from theropod researchers point to other bones pertaining to tyrannosaurids, ornithomimids, and oviraptorosaurs (Cau 2023). Only one *Dakotaraptor* bone, a tail vertebra, is unmistakably dromaeosaurian. Although this is larger than the caudal vertebrae predicted for *Acheroraptor*, it is probably not sufficiently distinct to substantiate the existence of a giant Hell Creek dromaeosaur species on its own. This assessment awaits further investigation, however; for

now, we can only conclude that uncertainty surrounds the prospect of quarter-tonne, talon-footed theropods antagonizing young *T. rex* and other Lancian dinosaurs.

With theropod dinosaurs failing to challenge all but the smallest *T. rex*, it may come as a surprise that another reptile group entirely may have presented the only serious risk to larger king tyrants seeking prey and carrion (excluding, of course, other *Tyrannosaurus*). These were the giant pterosaurs (fig. 5.10), the last of the flying reptiles that shared ancestry with dinosaurs in the Triassic (fig. 2.2) but evolved in parallel throughout the rest of the Mesozoic, developing a huge diversity of species and ecologies. Only one lineage seems to have lived alongside *Tyrannosaurus*: the Azhdarchidae. Thanks to the giant forms *Hatzegopteryx* and *Quetzalcoatlus*, azhdarchids are one of the most famous pterosaur groups. They possessed all the principal features of flying reptiles: elongate (and in this case, toothless) jaws, long necks and limbs, and powerfully developed wings (Witton 2013; Naish and Witton 2017; Andres and Langston 2021). Azhdarchids were generally large pterosaurs with adult wingspans no smaller than 2–3 m, and the largest species reached 10 m or more across the wings (Witton and Habib 2010).

An undoubted azhdarchid cervical vertebra from the Hell Creek Formation leaves no doubt that azhdarchids with c. 5 m wingspans lived alongside *T. rex* (Henderson and Peterson 2006). A giant Maastrichtian species, *Quetzalcoatlus northropi*, may not have been a true *T. rex* contemporary as it dwelt in pre-Lancian Texas (Dalla Vecchia et al. 2013). Such giants are found all over the globe, however, some existing in the last few thousand years of the Maastrichtian (Lawson 1975; Dalla Vecchia et al. 2013). It is thus plausible, if unproven by the patchy pterosaur fossil record, that giant azhdarchids flew over the heads of *T. rex*. If so, these fliers may have been a chief concern for king tyrants, as several lines of evidence point to azhdarchids operating as generalist terrestrial foragers with omnivorous or, at larger sizes, carnivorous diets (Witton and Naish 2008, 2013; Naish and Witton 2017; Padian et al. 2021). Historic views of pterosaurs as inept terrestrial locomotors with poor

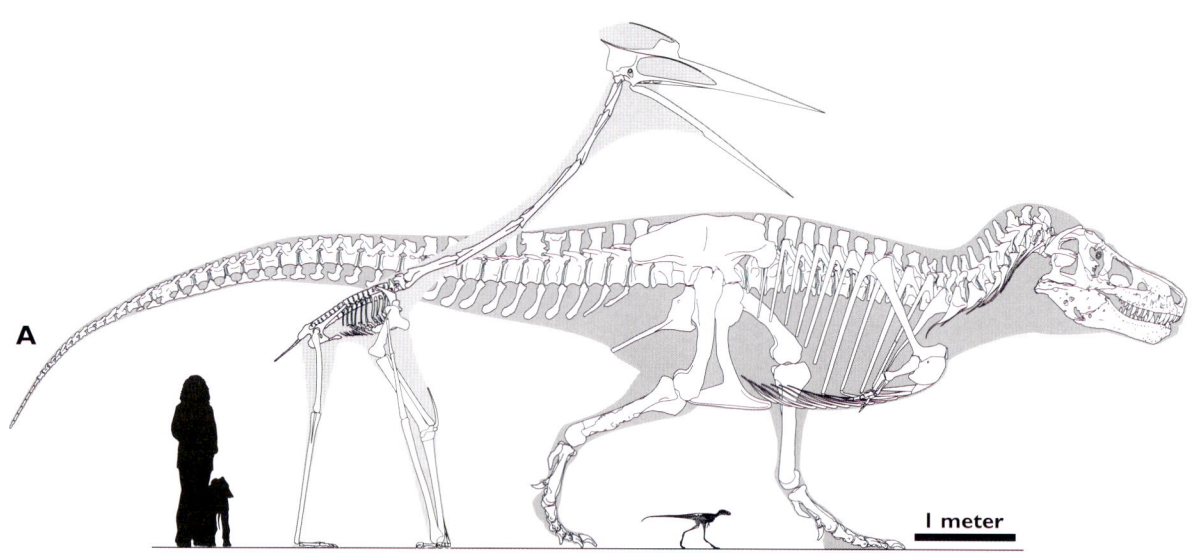

FIGURE 5.10

Other than *Tyrannosaurus*, the largest predators in western North America during the latest Cretaceous were giant azhdarchid pterosaurs. Like giraffe-sized storks, these fliers were proficient ground foragers and probably posed a threat to many small-bodied animals, including juvenile king tyrants. (A) Skeletal restorations of the giant Maastrichtian Texan azhdarchid *Quetzalcoatlus northropi* compared to a *T. rex* adult and hatchling, the latter of which looks like an ideal azhdarchid snack (also see Chapter 6); (B) a large *Tyrannosaurus* is harassed by two giant azhdarchids while trying to enjoy a *Triceratops* meal. Audacious as it might seem for pterosaurs to pilfer food from king tyrants, modern birds can steal food from large mammalian carnivores, and an analogous behavior may have played out in deep time.

flight potential have been overturned in recent decades (see Witton 2013 for an overview), and they are now interpreted as proficient fliers, walkers, and runners. Their takeoff capacity may have been the most impressive example of this mechanic to ever evolve, explaining (in part) their capacity to develop enormous sizes. Launching their lightweight bodies into the air with all four limbs, like a bat, rather than just their legs, like a bird, allowed even the largest flying reptiles to become airborne in seconds, all from a standing start (Habib 2008, 2013).

In this fast and powerful guise, azhdarchid pterosaurs may have represented ecological complications for king tyrants. *T. rex* eggs and hatchlings would have been ideal snacks for giant, ground-stalking pterosaurs (fig. 5.10A), and young *Tyrannosaurus* probably competed with them for small game. Carrion has long been considered a part of azhdarchid diets (Lawson 1975; Witton and Naish 2008), and the largest, giraffe-sized species may have been some of the only animals in Maastrichtian North America that could intimidate adult *T. rex* for scavenging rights (fig. 5.10B). Giant azhdarchids like Q. *northropi* are predicted to have perched their sharp, 2 m long beaks atop necks over 2.5 m in length (Naish and Witton 2017), giving them considerable reach to poke and stab any theropods they found in their way. Outlandish as it may seem for a 200–300 kg pterosaur to have challenged a multi-tonne predator that could bite it in half, modern storks, which are thought to be among the best modern analogues for azhdarchids (see Witton and Naish 2008) and raptorial birds are known to harass or attack larger animals to steal their food. Victims of raptor piracy include goliath herons, shoebill storks, cheetahs, and coyotes, all relatively large species that could easily injure or kill the smaller birds (Schaller 1972; Mock and Mock 1980; Sumba 1989; Jung et al. 2009; John and Lee 2019). In these scenarios, the avian thieves rely on their offensive anatomy or strength of numbers, as well as their fast reflexes and acrobatic capabilities, to bully larger species away from potential meals. Large storks, in addition, are recorded as weaponizing their beaks when agitated by tapirs,

zookeepers, and children, the latter of whom can be killed from the resulting injuries (see Witton and Naish 2013 for review and discussion). We can never know if azhdarchids challenged *T. rex* in this manner, but their anatomy and functionality implies a strength, speed, and offensive capacity for this behavior, and modern animals give us sufficient behavioral precedent to consider it possible.

NEIGHBORING GIANTS, FELLOW ICONS, AND THE HERBIVORE COMMUNITY

The large herbivores that *Tyrannosaurus* lived with, *Triceratops*, *Ankylosaurus*, and *Edmontosaurus*, are dinosaur icons in their own right, but they are just some of the plant-eating dinosaur species that cohabited with *T. rex*. Despite the reduced diversity of Lancian dinosaur communities, many herbivorous lineages remained present in *T. rex* ecosystems, including some of the largest dinosaurs of all time. It is among these herbivores that Lancian provinciality is best evidenced, as their taxonomies are mature enough, and their remains well-enough sampled, that we can provisionally assign significance to their presences and absences in certain regions (fig. 5.3B).

Some herbivores seem to have occurred almost everywhere that *T. rex* did, such as *Thescelosaurus* (fig. 5.11). Several species of this 2.5–4 m long bipedal herbivore are known (Boyd et al. 2009), each bearing low, beaked skulls, relatively short forelimbs of limited reach (Senter and Mackey 2023), and longer, stoutly built hindlimbs. Their widespread distribution occurred despite a lack of obvious defenses against predators, such as offensive structures, armor, or hindlimbs built for running. Instead, *Thescelosaurus* may have been a group-living burrower, a lifestyle suggested by features of their limb anatomy, brain structure, and assemblages of skeletons preserved within burrow structures (Button and Zanno 2023). An enhanced sense of smell may also have alerted *Thescelosaurus* to danger (Button and Zanno 2023).

Larger bipedal herbivores, the hadrosaurs, also seem to have been widely distributed within *T. rex* range, although their fossils are less abundant in southern regions. Most Lancian hadrosaur fossils can be referred to the 12 m long species *Edmontosaurus annectens* (fig. 5.12), a particularly common Western Interior dinosaur (fig. 5.2D). Rare fossils hint at *E. annectens* individuals that surpassed *T. rex* in size, perhaps reaching 15 m long. Thanks to fantastically preserved specimens with copious skin preservation, *E. annectens* are one of the best understood of all Lancian species. (Horner et al. 2011). Colloquially known as "duck-billed" dinosaurs, hadrosaur skeletons are distinguished by their long skulls and wide beaks, batteries of chewing teeth, slender forelimbs, and powerful hindlimbs adapted for fast running (Persons and Currie 2014). *E. annectens* seems to be restricted to Lancian deposits (Fowler 2017) and is distinct from an older *Edmontosaurus* species with a cranial crest, *E. regalis*. Fossils with *T. rex* bite marks indicate that these *E. annectens* may have formed a major food source for king tyrants (fig. 4.24; chapter 6). Along with running away, gregarious behavior, evidenced by extensive bonebeds (e.g., Snyder et al. 2020; Wosik and Evans 2022), may have been a chief hadrosaur defense against predatory acts. Intriguingly, these deposits only comprise large individuals, implying that *E. annectens* formed age-segregated herds where adults lived without juveniles.

FIGURE 5.11

The small neornithischian *Thescelosaurus neglectus*, one of the more abundant and widespread smaller dinosaurian herbivores that lived alongside *T. rex*. Recent research suggests that this inoffensive-looking dinosaur may have had a good sense of smell and dwelled in burrows, both properties that would help avoid predator attention.

FIGURE 5.12

Edmontosaurus annectens watch the sunrise. Represented by thousands of specimens and several examples with extensive skin preservation, *E. annectens* is one of the most completely known of all dinosaurs. They grew to similar sizes, if not larger, than *T. rex*, and fossils show that they were regular components of king tyrant diets.

Other Lancian dinosaurs seem to have been more geographically restricted. The horned dinosaurs, or ceratopsians, characterize distinct dinosaur faunas that lived northeast of the North American Cordillera (Lehman 2001; Sampson and Loewen 2005). The small, 2 m long, and relatively archaic ceratopsian *Leptoceratops gracilis* (fig. 5.13) mostly occurred in the northwestern part of *Tyrannosaurus* range (although also see Ott 2006) and marks a region where larger ceratopsians, such as *Triceratops*, were rare. *Leptoceratops* resembled the famous *Protoceratops* in lacking facial horns, but they were stockier creatures than their Mongolian cousins. Living in burrows (Fowler et al. 2019), excavated by using powerful motions of their heads and strengthened necks (VanBuren 2013), may have helped *Leptoceratops* avoid predators (though see figure 5.5).

Triceratops (figs. 1.22, 5.14) also distinguishes a Lancian dinosaur community, being particularly abundant in northeastern and eastern regions of the Western Interior basin (Lehman 2001; Sampson and Loewen 2005). This "*Triceratops* fauna" is the largest dinosaur province of the region (fig. 5.3B), with Lancian sediments representing ancient coastal floodplains dominated by this famous ceratopsian (Horner et al. 2011). Two *Triceratops* species are recognized from their distinct snout morphologies: the geologically older (fig. 5.8) *T. horridus* has a slightly longer, shallower snout and smaller nasal horn than the younger *T. prorsus* (Scannella et al. 2014). An especially large-horned dinosaur that reached 8 m long and several tonnes in mass, *Triceratops* also sported especially large body scales (Bell et al. 2022), a relatively short and solid neck frill, as well as—famously—three prominent facial horns. The latter have been widely interpreted as possible predator deterrents (on which read on), but, even so, interactions between *Triceratops* and *T. rex* are well evidenced from

FIGURE 5.13

The burrowing herbivore *Leptoceratops gracilis*, a horned dinosaur that inhabited the northwesterly regions of known king tyrant range.

FIGURE 5.14

The other super-famous Lancian dinosaur: *Triceratops horridus*. Fossils show that *T. rex* and *Triceratops* fought during life and that king tyrants consumed *Triceratops* flesh despite the formidable armament adorning *Triceratops* faces.

fossil remains (see chapter 6). Indeed, this genus seems to have been a mainstay of *T. rex* diets.

More widely distributed, although much rarer than *Triceratops*, was another ceratopsid, *Torosaurus latus*. Similar in size and appearance to *Triceratops* except for a longer and perforated cranial frill, some disagreement has occurred over whether *Torosaurus* is a valid genus or merely the mature form of its more famous relative (Scannella and Horner 2010; Longrich and Field 2012; Maiorino et al. 2013; Mallon et al. 2022). Most data currently points to the two being distinct. In addition to inhabiting northern and coastal areas, *Torosaurus* remains are known from the southern, once-mountainous regions of the Western Interior Basin that, to date, have yet to yield *Triceratops*. Among these discoveries are the dubious taxon "*Torosaurus*" *utahensis* (Hunt and Lehman 2008), an Edmontonian species that probably requires removal from the *Torosaurus* genus (Scannella and Horner 2010).

FIGURE 5.15

Feathered theropod contemporaries of *T. rex*, the ostrich dinosaurs *Ornithomimus velox* (left) and the large oviraptorosaur *Anzu wyliei* (right). Two enantiornithine birds, *Avisaurus archibaldi*, rest in the middle distance.

Fossils of herbivorous or omnivorous theropods such as the ostrich-like ornithomimids *Ornithomimus* and *Struthiomimus* occur throughout the northern and central range of *T. rex*, as do the remains of caenagnathid oviraptorosaurs, including the large-bodied species *Anzu wyliei* and the smaller *Eoneophron infernalis* (fig. 5.15; Lamanna et al. 2014; Atkins-Weltman et al. 2024). The taxonomy of western American ornithomimids requires review, and future analyses may identify additional genera in some Lancian deposits (Longrich 2008; Claessens and Loewen 2016), but the occurrence of more than one midsized ornithomimid species with *T. rex* is not in doubt. The elongate hindlimb proportions of these dinosaurs, their convergent development of the arctometatarsal condition (chapters 3 and 4; Holtz 1995) and their middling body size maximized their speed potential to make them among the fastest of all dinosaurs (Russell 1972; Paul 1988b; Carrano, 1999; Dececchi et al. 2020; Rhodes et al. 2021). Chases between these animals and the similarly swift juvenile *T. rex* may have been among the most exciting of any dinosaur predatory acts (fig. 4.19). Caenagnathids, in contrast, are thought to have been faster than the theropod average (Rhodes et al. 2021), but they may have relied on trenchant hand claws or other capabilities to avoid or deter interest from king tyrants. As with ornithomimids,

FIGURE 5.16

Pachycephalosaurus wyomingensis, one of the most distinctive dinosaurs to have lived alongside *Tyrannosaurus*. The thickened skull cap likely served to deliver low-velocity blows to conspecifics during fights.

North American Maastrichtian caenagnathid diversity is probably also underestimated at present (Atkins-Weltman et al. 2024).

The fully herbivorous pachycephalosaurids were similarly distributed to these theropods, overlapping with *Tyrannosaurus* in northern and central latitudes. Several species, including the eponymous *Pachycephalosaurus wyomingensis* (fig. 5.16), *Sphaerotholus buchholtzae*, and *Platytholus clemensi* lived in this region (Woodruff et al. 2021; Horner et al. 2022). Famed for their domed heads that were likely employed in ritualized combat with conspecifics (e.g., Snively and Cox 2008; Peterson et al. 2013; Farke 2014), pachycephalosaurs were also distinctively wide bodied (Paul 2016) and perhaps neither particularly fast nor agile. The same was surely true for the armored ankylosaurs that lived alongside them, such as the ankylosaurid *Ankylosaurus magniventris* (fig. 5.17) and the nodosaur *Denversaurus schlessmani* (Bakker 1988; Burns 2015; Arbour and Mallon 2017).

The hides of both species were covered in blunt bosses and low spikes, and both were large: *Ankylosaurus* especially so at perhaps 8 m long. A club was located on the end of the *Ankylosaurus* tail and injuries on ankylosaurid skeletons show their antagonistic potential (Arbour et al. 2022). *Denversaurus*, though somewhat smaller (c. 6 m long), sported an array of flattened and sharpened osteoderms but, as is typical for nodosaurs, lacked a tail club. These armored species, with their enormous bellies and short limbs, were clearly exploiting the same predatory defenses used by turtles, pangolins, and armadillos today: being so tough and frustrating to subdue that predatory effort outweighed the potential reward.

The third and final faunal province of the Western Interior's Lancian dinosaurs is a southerly region marked by the giant sauropod *Alamosaurus sanjuanensis* (fig. 5.18), the only long-necked dinosaur taxon presently identified from the latest Cretaceous of North America (Gilmore 1946b; Lehman 2001; Sampson and Loewen

FIGURE 5.17

Two large armored dinosaurs, *Ankylosaurus magniventris*, figure out who rules this part of Maastrichtian North America. Emerging data suggests that ankylosaur armor and tail clubs may have developed more for settling intraspecific conflicts than deterring predators, although such features may still have rendered them particularly challenging, complex prey species for king tyrants.

2005). That this sauropod lived alongside *T. rex* in upland, mountainous regions is verified by their co-occurrence in the North Horn Formation of Utah as well as, potentially, the Javelina Formation of Texas, although this is complicated by not only the questioned occurrence of *T. rex* in these deposits but also the taxonomy of *Alamosaurus*. This sauropod has become something of a "wastebasket" taxon comprising any Maastrichtian sauropod fossils from Mexico, New Mexico, Arizona, or Utah, regardless of their form or geological context (Williamson and Weil 2008; D'Emic et al. 2011; Fowler 2017). The validity of *Alamosaurus* as a diagnosable dinosaur taxon is not in any significant doubt (D'Emic et al. 2011), but a growing need for taxonomic housekeeping within this genus prohibits confident statements about their geographic or stratigraphic distribution.

FIGURE 5.18

The largest animal encountered by *T. rex* was among the most gigantic to ever walk the earth: the titanosaur *Alamosaurus sanjuanensis*. Juvenile *Alamosaurus* may have been *T. rex* prey, but adults would have dwarfed, and maybe threatened, even the largest *Tyrannosaurus*.

Long-necked dinosaurs are rare fossils in uppermost Cretaceous sediments of the United States, and the presence of *Alamosaurus* is interpreted as either sauropods recolonizing North America after a disappearance earlier in the Cretaceous (the so-called "sauropod hiatus": see Lucas and Hunt 1989; Williamson and Weil 2008), or else evidence that we have undersampled sauropods from Cretaceous rocks in the Northern Hemisphere (Mannion and Upchurch 2011). In any case, *Alamosaurus* represented a distinctive kind of herbivorous dinosaur in king tyrant communities, one entirely unlike the species in the northern parts of their range. They were, by far, the only Lancian animals significantly bigger than *T. rex* and among the largest of all sauropods: perhaps 26 m long, 5 m to the top of their shoulders, and weighing several tens of tonnes (Fowler and Sullivan 2011; Tykoski and Fiorillo 2017. *Alamosaurus* may have also been armored with osteoderms (Carrano and D'Emic 2015), although we have little idea of the distribution of these structures across titanosaur bodies. We have to wonder whether *Tyrannosaurus* ever attempted to subdue adults of this gigantic species, or instead focused on *Alamosaurus* juveniles as well as the more familiar hadrosaurids and ceratopsians that also made up part of the *Alamosaurus* fauna (Sampson and Loewen 2005).

THE *REX* EFFECT: DID *T. REX* SHAPE THE EVOLUTION OF NORTH AMERICAN DINOSAURS?

For all their size, *Alamosaurus* is not quite the largest known sauropod, whereas species like *E. annectens*, *Triceratops*, and *Ankylosaurus* are among the very biggest representatives of their clades. This, and the well-defended, well-armed nature of *Triceratops* and *Ankylosaurus*, have inspired proposals that these formidable herbivores had specifically evolved to deter an especially large and formidable predator: *Tyrannosaurus* (fig. 5.19; Matthew 1915; Osborn 1917b; Bakker 1986;

Paul 1988a, 2008). Combining these features with Lancian species representing some of the last non-avian dinosaurs has seen the battles between *T. rex* and titanic herbivores portrayed as the crescendo of dinosaur-on-dinosaur violence: the exclamation point at the end of an Mesozoic evolutionary "arms race" between predators and prey (Bakker 1986). Osborn (1917b) saw *Triceratops* as "the climax of ceratopsian defense" that coincided with "the climax of *Tyrannosaurus* offense" (pp. 224–225), while Bakker (1986) noted the poetic quality to their co-occurrence: "It's somehow fitting that these two massive antagonists lived out their co-evolutionary belligerence through the very last days of the very last epoch in the Age of Dinosaurs" (p. 240). Paul, also a fan of the idea that *Triceratops* was the apex defender against theropod aggressors (Paul 1988a), has even proposed that such dangerous prey was a factor in the short lifespans of tyrannosaurids (chapter 6) and echoed Bakker's notion of an arms race: "The upgrading of weaponry in tyrannosaurids and ceratopsids … may represent a Red Queen arms race" (Paul 2008, p. 344; "Red Queen" here refers to a moment in Lewis Carroll's *Through the Looking Glass* where the Red Queen tells Alice, "It takes all the running you can do, to keep in the same place," just as species must adapt to other species' evolutionary modifications to avoid extinction).

The idea that *Tyrannosaurus*, *Triceratops*, *Ankylosaurus*, and other large, offensively geared dinosaurs were products of a long-running battle between predator and prey undoubtedly makes for a compelling natural history narrative, and predatory interactions undoubtedly influenced their evolution. New science on the development of weapons and armor in herbivorous species, both extinct and modern, somewhat derails ideas of dinosaurian arms races and a triumphant Lancian culmination of dinosaur action, however.

Framing this entire discussion is the fact that zoologists are engaged in a long-running conversation about the use of offensive anatomy, such as horns and antlers, as anti-predator devices in modern animals (e.g., Geist 1966; Estes 1991; Roberts 1996; Caro et al. 2003; Loughry and McDonough 2013; Gerstenhaber and Knapp 2022). Counterintuitively, the presence of weapons on prey species can have little overall impact on their predation vulnerability (Schaller 1972). While some species, like African buffalo, aggressively defend themselves from predators with their horns, others, including most antelopes, rarely or never employ their offensive anatomy in this way (Schaller 1972). Similarly, most deer run or hide from predators rather than confronting them with their antlers, despite using these structures to gore and kill rival conspecifics, and giraffes and moose tend to kick predators rather than using their cranial weaponry (Geist 1966, 1999). Fighting predators in any way is rarer than we generally imagine: mostly, horned or antlered prey simply avoid or flee carnivores, and, when caught, some put up surprisingly little resistance (Schaller 1972; Brain et al. 1999).

In part, these observations reflect the horns and antlers of male giraffes, bovids, deer, and other taxa being primarily shaped by sexual selection (Geist 1966; Bro-Jørgensen 2007; Knell et al. 2013), where absorbing, catching, and parrying the blows of rivals during physical intraspecific contests takes adaptive precedence over predator deterrence (Geist 1966; Packer 1983; Bro-Jørgensen 2007). Indeed, the horns and antlers of many male mammals are of such sizes and shapes that they are effectively useless against attacking carnivores (Estes 1991; Roberts 1996). This is not the case for the typically smaller horns and antlers of female mammals, however, the functions of which are more disputed. Some argue that predator defense is their main purpose (Bro-Jørgensen 2007; Stankowich and Caro 2009), implying a wider role for this behavior

FIGURE 5.19

A classic dinosaur interaction: *Tyrannosaurus* takes on *Triceratops*. Since the discovery of *T. rex*, it has been assumed that the horns and frill of *Triceratops* were specifically adapted to counter king tyrant attention; however, new research is framing the evolution of dinosaur horns, spikes, and armor as structures shaped more by intraspecific social factors, and less about predator deterrence.

than has been documented in field studies, but social selection (for instance, to dispute resources with con-specifics, or to mimic males) is also possible, as is non-functionality, where females only have these structures because of genetic conditions related to developing horns in males. The latter may be the case for the fe-males of some species, such as giraffes, which reported-ly never do anything aggressive with their horns.

Similar comments can be made about animal armor where, as with offensive anatomy, intraspecific behav-iors may be an overlooked factor in their development. Nine-banded armadillos, for instance, fight aggressively with one another and bear large scars from the claws of rivals. While some anti-predator utility is an inescap-able quality of being encased in tough skin, it has yet to be demonstrated that this armor reduces armadillo predation risk compared to armorless mammals of sim-ilar size (Loughry and McDonough 2013). The same also appears to be true for lizards, where evidence is growing for intraspecific aggression shaping their ar-mor and osteoderm development more than predation pressures (Arbour et al. 2022). Furthermore, the intu-itive assumption that horned and armored species are more challenging prey than "defenseless" animals is not substantiated among modern animals. On the contrary, some of the most aggressive prey species on the Afri-can savannah are zebras, which bite, kick, and charge at predators despite their lack of horns or other obvious weapons (Kruuk 1972; Schaller 1972).

In sum, while offensive and defensive structures *can* be used as predator deterrents, there are surprisingly few indications of their evolution primarily being driven by predatory pressures among living species. Indeed, whether animals have horns, spikes, or armor seems to matter less to predators than the amount of energy needed to capture and eat a potential prey species (Schaller 1972). Threat of injury is presumably factored into this behavioral calculation, but prey abundance, population density, and nutritional reward against predatory effort seem to trump the presence of offen-sive anatomy. This is expressed by modern predators demonstrably not avoiding aggressive prey armed with large horns: both spotted hyenas and lions habitually attack African buffalo, for instance, one of the most for-midable and aggressive horned mammals in their range (fig. 5.20; Kruuk 1972; Schaller 1972).

There are indications that the above was true for dino-saurs, too. Studies of dinosaur horn and armor evolution are increasingly finding data that shows intraspecific pressures, most likely sexual selection, primarily shaped their configurations, not predators. Since the 1970s re-searchers have noticed that ceratopsian horns and frills are comparable to the sexually selected horns of liv-ing animals in many ways: their exaggerated, complex shapes, their functional peculiarity (i.e., that many seem maladapted for activities other than fighting or signa-ling to conspecifics), their rapid development in later life (fig. 5.21A), their large amount of intraspecific var-iation, and their elevated morphological diversity be-tween species (e.g., Farlow and Dodson 1975; Spassov 1979; Sampson et al. 1997; Horner and Goodwin 2006, 2008; Hone et al. 2011b, 2016; Knell et al. 2013). Fos-sils show that horned dinosaurs also injured each other in ways consistent with ritualized, horn-locking combat, strongly suggesting that they used their horns to settle disputes within their own species (e.g., Farke 2004; Farke et al. 2009; D'Anastasio et al. 2022). Conversely, fossil evidence that their horns were turned on other species remains wanting (Farke 2014).

Similar trends have been identified in ankylosaur ar-mor and tail clubs, which were absent in juveniles and began development later in life (fig. 5.21B; Arbour et al. 2022). Given the ubiquitous targeting of juveniles by predators in modern times, and the likelihood of this behavior also characterizing Mesozoic predation (Hone and Rauhut 2010), the absence of defensive and offensive anatomy in juvenile ankylosaurs is compelling evidence that these structures were not primarily an-ti-predatory in nature. Further support for this hypoth-esis is the large amount of interspecific variation in adult ankylosaur armor and tail weaponry, as well as armored dinosaur fossils with injuries that probably occurred during fights with rivals, not other species (Arbour et al. 2022).

FIGURE 5.20

African lions eat the remains of a cape buffalo. African buffalo are among the most vigorous users of their horns against predators, but they seemingly do little to lower their predation frequency: they remain a chief constituent of lion diets. This gives us food for thought about the relationships between horned dinosaurs and their theropod predators, as well as the evolution of offensive anatomy in herbivorous dinosaur species.

Collectively, these data do not exclude the possibility of *Triceratops*, *Torosaurus*, and *Ankylosaurus* relying on their horns, clubs, and armor when encountering *T. rex*, but we have probably overstated the importance of these structures as predator deterrents, and there is no reason to view Lancian dinosaurs as some sort of climax of dinosaur predator-prey relationships. Accepting the increasingly compelling evidence that the offensive and defensive structures of dinosaur herbivores were largely products of intraspecific selection has several implications, including that only some species used their offensive anatomy to fight off carnivores; that some formidable-looking structures may have been ineffectual predator deterrents; and that some taxa may have lacked the behavioral flexibility to turn their weapons on other species. Our intuitive assumption that horned or armored dinosaurs would be the most challenging prey species for theropods and that "defenseless" dinosaurs like hadrosaurs and sauropods were easier game perhaps plays to stereotypes more than grounded observations of animal behavior. It would be difficult to predict which modern animals are especially dangerous to predators from their bony anatomy alone, and this is surely true for dinosaurs, too. Non-fossilizable factors such as temperament, awareness, physiology, intelligence, and behavioral flexibility are just as important to this property, if not more, than the presence of horns or spikes.

In all, this is a frustrating reminder that the fossil record can reveal what extinct animals were capable of, but not what they were truly like when alive. For all we know, well-armed *Triceratops* and *Ankylosaurus* might have been flighty and nervous, while "defenseless" *Edmontosaurus* and *Alamosaurus* could have been angry, aggressive monsters that charged, mobbed, bit, and trampled marauding *T. rex* at every opportunity (fig. 5.18). These caveats about our capacity to predict the behavior of extinct species are worth considering as we move to our next chapter, addressing the lives and ecology of *Tyrannosaurus*.

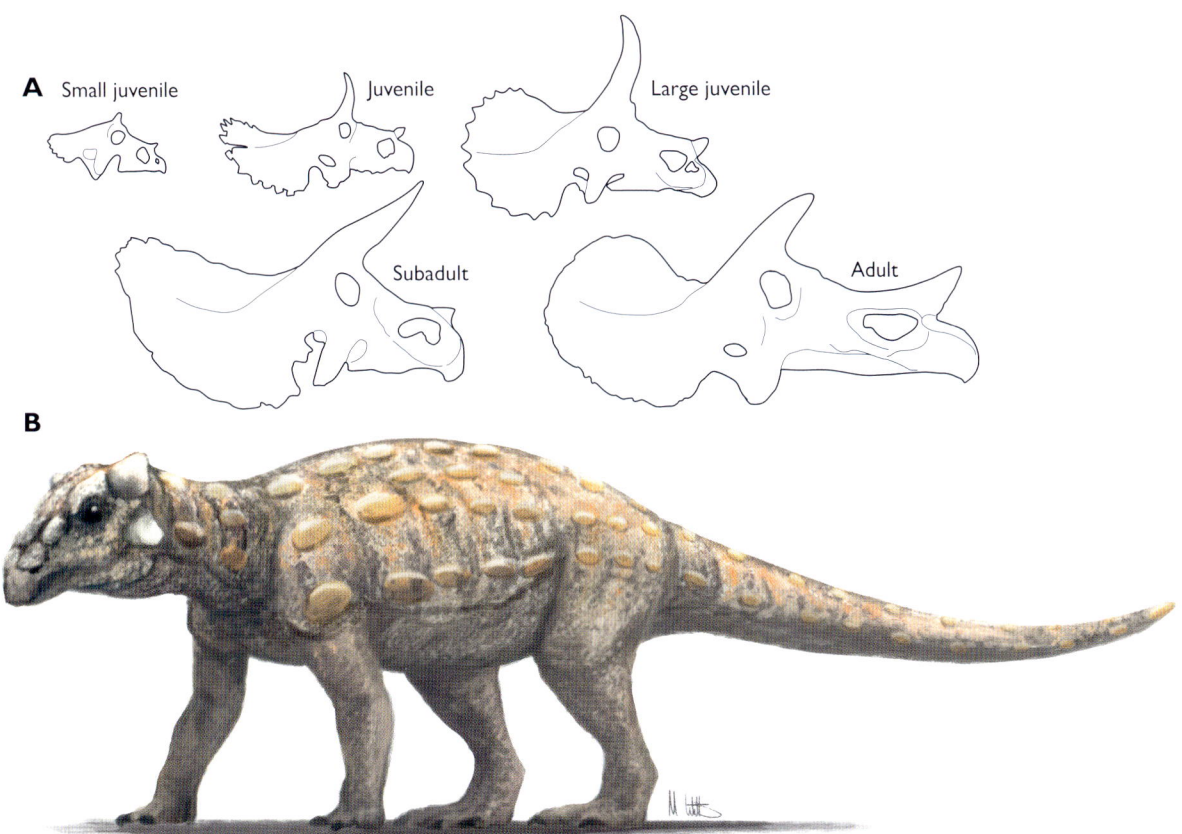

FIGURE 5.21

Horned and armored dinosaurs lack well-developed anti-predator anatomy during their most vulnerable juvenile years, instead developing their armor and horns as they mature. (A) Development of *Triceratops* frills and horns, in which maximum horn-core length and frill dimensions are achieved in adults; (B) restoration of a hatchling *Ankylosaurus* without osteoderm armor or a tail club, a trait suggested by juvenile ankylosaur specimens. The scale pattern of the eventual bony armor was presumably present throughout life, as it is in juvenile armored reptiles. (A) After Horner and Goodwin (2006).

LIFE, FOOD, LOVE, DEATH

6

THE KING TYRANTS of our imaginations are not made of bones, graphs, or equations; they are living animals interacting with other species and other *T. rex*. Researchers have had a lot to say about king tyrant behavior, giving us insights into their social lives, feeding habits, diseases and ailments, as well as their survival prospects. As we shall see, however, some desires to understand dinosaur lifestyles beyond their simplest aspects can push paleontological evidence to, and sometimes beyond, the breaking point. What, if anything, can we confidently say about the lives of *Tyrannosaurus rex*?

THE KING TYRANT SOCIAL SCENE

WE SPOKE IN our first chapter about the intellectual challenges of distinguishing the science of *Tyrannosaurus* from the *T. rex* of our imaginations. Nowhere is this more necessary than in discussions of their life histories and behavior. Often influenced by depictions in popular media, many of us have strong opinions on matters such as *T. rex* predatory habits, their interactions with other *Tyrannosaurus*, their parenting abilities, and so on. A hard truth of paleontology, however, is that the behavior of any extinct animal, even very well-researched ones like *T. rex*, is extremely challenging to reliably interpret from fossils. Aspects of animal lives that we can commonly deduce beyond reasonable doubt mostly pertain to particulars of diet and habitat preferences, elements of reproduction, and some matters concerning injuries or poor health. It takes special fossils and fossil sites, or at least the accumulation of lots of data, to gain insights into questions such as sociality, hunting behavior, courtship rituals, or family dynamics (fig. 6.1).

To that end, the following discussion is not a list of speculations about *T. rex* lives, or a discussion of behaviors that might, or might not, be plausible for *Tyrannosaurus*. Instead, we will specifically focus on elements of *T. rex* life histories that have been investigated and discussed in the last century of research. Some are well evidenced or at least strongly inferable from *T. rex* fossils, others are less compelling. With varying degrees of confidence, we can comment on *Tyrannosaurus* migratory habits, their interactions with other king tyrants, and the possibility of communal living. We also know something about their predatory lifestyles and feeding habits, and can infer some details of how king tyrants entered and exited their Cretaceous world.

FIGURE 6.1

Each no larger than a red fox, a group of hatchling *Tyrannosaurus rex* consider their next steps in a Cretaceous clearing. If they are lucky, three decades of life lay before them, but what, exactly, would their lives involve?

WANDERING NOMADS OR TERRITORIAL SETTLERS?

Territoriality is such a common trait among modern carnivores that we have to wonder whether it was practiced in Mesozoic dinosaurs. But how, possibly, could we glean whether long extinct taxa like *Tyrannosaurus* established territories or not? In recent decades, biomechanical studies and geochemical analysis of dinosaur fossils has permitted exactly this: we are able to deduce whether dinosaurs stayed in one region their whole lives, or migrated between different areas. Thus, it is potentially not beyond modern science to determine whether *T. rex* established territories or was perpetually mobile, perhaps following the movements of prey animals. A definitive conclusion on this matter is probably still several major projects away, but we already have some insights into the roaming habits of king tyrants.

There are several ways to investigate migrating behavior in extinct terrestrial animals. One concerns predictions of physiology and locomotory biomechanics. Modern terrestrial migrators are exclusively tachymetabolic endotherms and have ultraefficient means of locomotion, usually being optimized toward either walking or running, depending on their migratory behavior (Bell and Snively 2008). Their adaptations include especially large elastic ligaments that store and release large volumes of energy with every step, reducing the cost of long-distance travel. Dinosaurs generally lack such structures but may have achieved similar efficiency by simply being large (Bell and Snively 2008; Dececchi et al. 2020). The economy of traveling increases positively with body size, such that multi-tonne dinosaurs may have been able to cross long distances as efficiently as horses, wildebeest, or caribou through brute size rather than finely tuned anatomy. *T. rex* could have been among the champions of long-distance travel for predatory dinosaurs by simply being gigantic and, as noted in chapter 4, having several features of hindlimb anatomy that made it a particularly efficient strider (Dececchi et al. 2020).

Taphonomy can also give us hints about migratory behavior in the form of huge bone beds representing the simultaneous deaths of hundreds or thousands of individuals from a single species (Bell and Snively 2008; Burns et al. 2014). The concentrations of animals in some of these sites are so enormous that they cannot represent typical population densities, instead capturing parts of herds that encountered calamity when traveling between pastures. Vast bonebeds typically comprise herbivorous taxa such as hadrosaurs, ceratopsids, or plant-eating mammals, but carnivores can, on rare occasions, be found in aggregations, too. These tend to have different taphonomic histories to herbivore bone beds however, and also mostly include far fewer animals (see below). To that end, while taphonomy supports notions that some dinosaur herbivores migrated, carnivore bonebeds are less informative on such matters.

Fortunately, extinct animal migration can be less ambiguously recorded through bone isotopes, those variations of atomic composition that capture specifics of physiology and environment throughout bone growth. In chapter 4, we discussed how oxygen isotopes can reveal details about physiology, but other isotope types—such as zinc, carbon, strontium, and calcium—are informative about animal diets, their position in ancient food webs, their habitat preferences, and whether they permanently resided in one region. Strontium isotopes are especially reflective of animal migration as strontium atoms vary with geographic region, their composition reflecting local bedrock and soil conditions. When strontium isotopes are obtained by animals through ingested plants (or ingested prey animals, in the case of carnivores), they are deposited in tooth enamel each day via incremental additions to tooth surfaces. Variation in enamel strontium can thus record animals moving between areas of different geological makeup, and these chemical markers can even be matched to the geology of certain regions, allowing us to not only tell if migration happened, but plot the locations that animals moved between (Fricke et al. 2011; Terrill et al. 2020; Terrill 2021).

Recent work by Joep Schaeffer has applied this technique to *Tyrannosaurus* using teeth from the adult "Trix" specimen (Schaeffer 2016). Across the growth of the tooth (presumably representing about two years—see chapter 3), *T. rex* strontium ratios varied enough to suggest that different soil types were captured by whatever "Trix" ate, and associated oxygen isotope ratios suggest some of this variation reflected changes between wet and dry conditions. This indicates that "Trix" was eating animals that subsisted on vegetation growing in different geological regions, which could mean that *T. rex* roamed around in search of food. However, as a secondary consumer of strontium first ingested by their prey, it is also possible that "Trix's" isotope values reflect consumption of migratory animals passing through a home range (Schaeffer 2016). Isotopic evidence from other studies suggests that likely *Tyrannosaurus* prey species, horned dinosaurs and hadrosaurs, were migrators (Terrill et al. 2020; Terrill 2021), so these interpretations are currently equivocal. Among modern terrestrial carnivores, only a few species follow migrating animals, including spotted hyenas and cheetahs (Bell and Snively 2008; Gnanadesikan et al. 2017). This might suggest we lean toward hypotheses of *T. rex* generally staying in one area year-round, but it is early days for *T. rex* migratory studies. A more pragmatic view may be to wait for future assessments of tooth isotopes in *T. rex* prey species, the strontium ratios of which might be compared to those of *T. rex* for additional insights into who was traveling long distance around Maastrichtian America: predator, prey, or both.

TERRORS OF TYRANNOSAURS

Whether roaming short distances or long range, *T. rex* would have met other king tyrants. It's difficult not to wonder what happened when they did. Were they an irascible species, to the extent of preferring solitary existences, or were they gregarious individuals that sought out the company of others? Might *Tyrannosaurus* have even lived in groups or worked together when

hunting prey? Theropods forming foraging groups or packs is a very popular and exciting concept among many dinosaur fans, and media promotion of certain tyrannosaur discoveries purportedly supporting these ideas has given this concept a strong foothold in *T. rex* hype (Young 2011).

Away from group-living movie dinosaurs and sensationalist press coverage, we're confronted with a hard paleontological reality: identifying group living and sociality in fossil animals is far from straightforward. There are three major hurdles. The first is that the fossil remains of multiple individuals of the same species must be found together in the same locality and sedimentary horizon, thus demonstrating that their remains were genuinely associated at time of burial. The second is demonstrating that these remains pertain to individuals that died together rather than becoming associated through taphonomy, such as separate bodies being transported and bundled together in a flood. Finally, and perhaps most difficult of all, is verifying that this co-occurrence of animal remains captured typical behavior and not a brief concentration of individuals under unusual circumstances. Bone beds of beaked dinosaur herbivores are good examples of how these criteria can be met (e.g., Eberth 2015). Across multiple sites, we see dozens, hundreds, or even thousands of individuals from the same species in one sedimentary horizon; their fossils show identical preservation styles, indicating similar taphonomic histories; and the staggering numbers of individuals imply that they must have existed together in large numbers before catastrophe struck. It can be difficult to make confident deductions about why those dinosaurs were together when they died, or what their social lives were like in detail, but it seems sound to conclude that horned and hadrosaurian dinosaurs lived in large groups, at least some of the time.

No equivalent sites exist for *Tyrannosaurus*. King tyrants are mostly discovered as single specimens, and only on rare occasions will two or more specimens be found in the same quarry. The current record is four king tyrants discovered in the same locality, all recovered from the excavation that begat "Sue" in the early 1990s

(P. Larson 1994; N. Larson 2008; Currie and Eberth 2010). Here, alongside "Sue" itself, were the partial tibia and fibula of a subadult, and solitary skull bones from two juveniles (a small hand claw from the same site, figured by Longrich and Saitta 2024, could represent a fifth individual or additional remains from one of the smaller animals). A second instance of associated *Tyrannosaurus* concerns the discovery of a subadult *T. rex* (the holotype of "*Dinotyrannus megagracilis*," see chapters 2 and 4) in a scrap pile formed during the excavation of a partial skeleton of a large adult (Molnar 1980; P. Larson 1994; N. Larson 2008). Unfortunately, locating this second, smaller *T. rex* among previously disturbed sediments precludes knowing the depositional relationship between these individuals; all that can be inferred is that they were buried in close proximity.

Thus, the "Sue" quarry remains the only confirmed example of a multispecimen *Tyrannosaurus* site. Some have ascribed behavioral significance to this discovery, interpreting the four specimens as a family: the matriarch ("Sue"), a subadult dad, and two offspring (Psihoyos 1994; P. Larson 1994). Contrasting preservation styles argue against such a direct interpretation, however. "Sue," though disarticulated, became fossilized as an almost entire skeleton, while the others were merely broken, solitary bones. This implies differing taphonomic histories for the four tyrants (Currie and Eberth 2010) and, in all likelihood, they did not die together. Moreover, the "Sue" quarry was rich in fossils of all kinds, also yielding parts of ceratopsids, ankylosaurs, ornithopods, non-tyrant theropods, pieces of turtles, amphibians, crocodylians, fishes, lizards, as well as plant fossils of varying types and sizes (N. Larson 2008). The isolated bones of the smaller tyrants better match the preservational style of these scattered, broken fossils more than "Sue," implying that their fossilization probably came via the same processes that captured random pieces of local Hell Creek flora and fauna. The "Sue" quarry, therefore, doesn't clear the first hurdle of demonstrating gregariousness in fossil species, lacking compelling evidence that its four king tyrants were preserved simultaneously.

Away from *Tyrannosaurus*, other localities with associated tyrannosaurids have been proposed as evidence of social behavior in these dinosaurs. Several occur in North America, including a bonebed of at least twenty-two differently aged *Albertosaurus sarcophagus* (Currie 1998; Currie and Eberth 2010; Young 2011). From this, Phillip Currie (1998) envisaged complex hunting habits for tyrannosaurids where fast, agile juveniles drove prey toward larger, more powerful adults. Elsewhere, two sites preserved multiple *Daspletosaurus* (Currie et al. 2005; Currie and Eberth 2010), and another yielded four differently aged, *Teratophoneus*-like tyrannosaurids (Titus et al. 2021). Perhaps the most intriguing are three short trackways of tyrannosaurids, apparently moving together in the same direction (fig. 6.2; McCrea et al. 2014). Richard T. McCrea and colleagues used this site to coin the collective noun for tyrannosaurids: a "terror"—because, what else could it be? Associated tyrants also occur in Asia, with three localities providing two or more *Tarbosaurus* in close association (Currie and Eberth 2010). It's hard to explain all these multispecimen sites as happening through random chance, and, collectively, they start to form a case for tyrannosaurids exhibiting some degree of gregariousness. They existed in tight-enough associations, at least, to fossilize together on occasion. But do they demonstrate communality?

A number of dinosaur paleontologists have voiced opinions on this question, and most are unconvinced (e.g., Roach and Brinkman 2007; quoted responses in Young 2011). The problem is not with the concept of social *Tyrannosaurus* per se, but with the difficulty of appreciating what the associated remains of tyrant dinosaurs truly represent. Critics note that carnivorous animals gather for many non-communal reasons, such as a glut of food or because environmental stresses force them into close quarters. Through the coarseness of fossilization, such behaviors are difficult to distinguish from animals living in tight social groups. Furthermore, as noted by Roach and Brinkman (2007), the formation of clans or packs, and coordinated cooperation among predators, is a specialized behavior that, today, mostly exists among mammals. Prolonged interaction with our

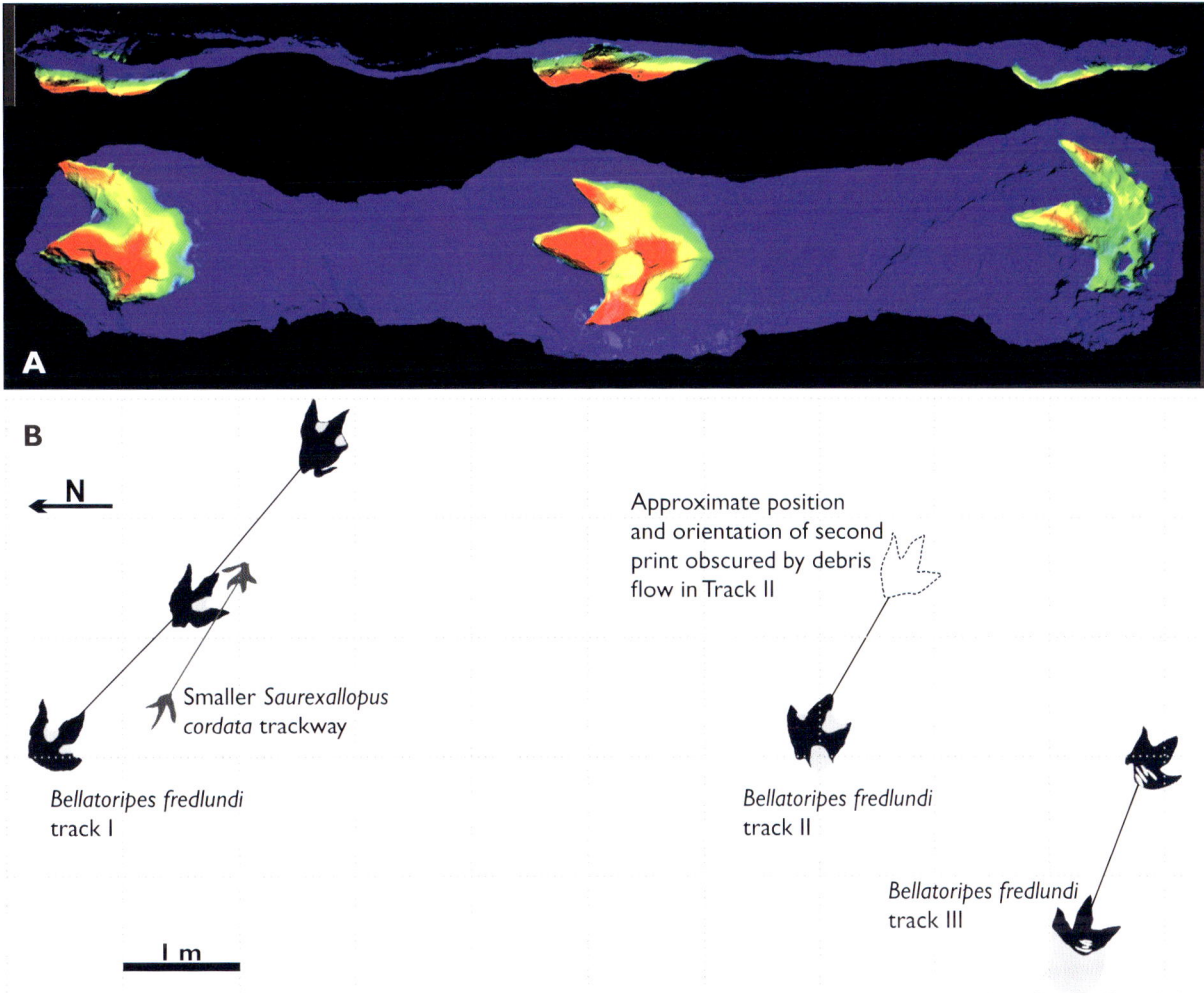

A

B

N

Approximate position
and orientation of second
print obscured by debris
flow in Track II

Smaller *Saurexallopus
cordata* trackway

Bellatoripes fredlundi
track I

Bellatoripes fredlundi
track II

Bellatoripes fredlundi
track III

I m

FIGURE 6.2

Associated tyrannosaurid tracks from British Columbia, Canada, possibly recording three individuals walking together. (A) Photogrammetric detail of the longest (three-print) track in side and plan aspect; (B) tracksite map showing the three closely associated footprint sets oriented in the same direction. The second footprint of track II has been obscured, and its position is only roughly indicated here. Images adapted from McCrea et al. (2014).

mothers and families during the first stages of our lives make mammals predisposed to developing complex, close-knit communities, but comparable behavior is rare on the reptile and bird branch of the evolutionary tree. Birds, crocodylians, and lizards express some group-based behaviors, but they are simpler than those of our own lineage. Most frequently, these are congregations around food, with different degrees of cooperation exhibited when consuming it. Crocodylians can work together to more efficiently dismember prey, but Komodo dragons can attack, kill, and consume smaller individuals competing for carcass access (fig. 6.3; Auffenberg 1981).

LIFE, FOOD, LOVE, DEATH

FIGURE 6.3

What does it mean to discover associated tyrannosaur fossils? It can be difficult to say for certain. As shown here by a gathering of Komodo dragons, nonsocial carnivores can congregate to access food or escape environmental stresses, and such behaviors need to be ruled out before assuming sites with multiple tyrannosaur specimens represent evidence of sustained group living.

Indeed, the largest of these lizards can be dangerous to other conspecifics in general, predating their smaller brethren like they would any other game.

More cooperation is seen among predatory reptiles and birds that forage commensally, where individuals benefit from the presence of others even if they do not specifically coordinate their actions with one another. In these scenarios, prey may flee one animal but rush toward another, making commensal hunting more successful than foraging alone. While such scenes give the impression of organized hunting efforts, they might be better viewed as individuals exploiting one another for their own benefit, and, moreover, they require only limited degrees of cooperation between participants. Possible coordinated

hunting behavior has been documented for crocodylians by Dinets (2015), but, if genuine, it seems to be very rare: over 3,000 hours of observation were required to log seventeen instances of group hunting, only two or three of which may have been truly coordinated events.

This is not to say, however, that routine, advanced cooperative predatory behavior is beyond the reptile-bird line of evolution. A few species of raptorial birds, most notably Harris's hawks, hunt in their mated pairs (fig. 6.4). Even here, however, this behavior is less sophisticated than that developed by mammals. Unlike cooperating dogs, cats, or primates, Harris's hawks do not target prey that would be beyond the means of one individual to subdue. Rather, they seek greater success at catching

FIGURE 6.4

Truly coordinated, cooperative hunting is rare among reptiles and birds, possibly indicating rarity among Mesozoic dinosaurs as well. Harris's hawks are unusual for working together to catch prey, although they maximize their effectiveness at catching the same game targeted by single hawks rather than combining forces against larger quarry, as do groups of carnivorous mammals.

the same prey that they would hunt as individuals. Such behaviors seem generally hardwired into avians as, across hundreds of years of ornithological observations, only a few examples of birds attacking prey beyond the subduing capacity of a single individual have been documented (Roach and Brinkman 2007). It seems that, even when working in teams, birds mostly still think as individuals.

Extrapolating these observations and discoveries to *Tyrannosaurus* perhaps casts doubt on ideas of complex sociality and group hunting. If living reptiles and birds are our analogues, any group hunting by *T. rex* and kin would be commensal and self-serving rather than planned and coordinated. Tackling larger prey in groups and cooperative hunting cannot be ruled out, but this mammal-grade behavior would be an exceptional occurrence

for a species related to living reptiles and birds. Exactly why some tyrannosaurids are preserved in bone beds may never be known, but these sites are more probably temporary aggregations of individuals drawn together by food, water, or environmental stresses than the mass deaths of packs, prides, or "terrors" of tyrant dinosaurs.

BITE CLUB: *T. REX* SOCIAL ETIQUETTE

The social lives of king tyrants may remain mysterious, but evidence that they routinely engaged with one another is directly indicated by large punctures and scars on their skulls. A suite of evidence indicates that these

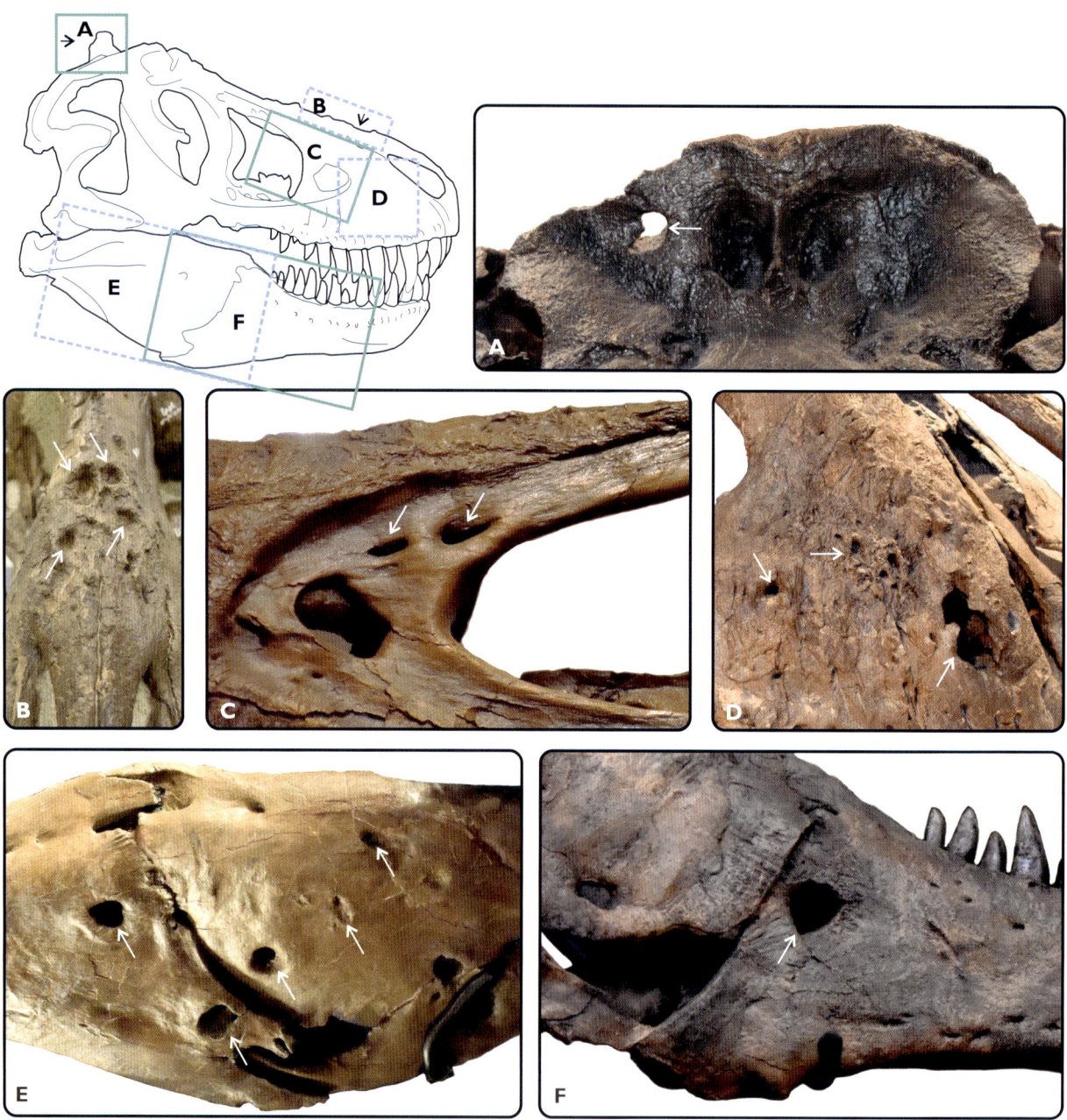

FIGURE 6.5

Evidence of intraspecific facial biting in *T. rex,* with injuries interpreted as possible bite marks arrowed. (A) Rear view of the punctured parietal bone of "Stan"; (B) round depressions in the top of the nasal of American Museum of Natural History specimen 5027; (C) punctures in the maxilla of "Jane"; (D) holes and irregular repair tissue on the right maxilla of "Trix"; (E) the controversial perforations in the lower jaw of "Sue," interpreted as bite marks by some but the results of infections by others; (F) a single large hole in the lower jaw of "Peck's rex".

injuries were probably made from bites to the face incurred during fights with other king tyrants (fig. 6.5; Molnar 1991; Tanke and Currie 1998; P. Larson 2001; Larson and Donnan 2002; Rothschild and Molnar 2008; Rothschild 2013; Bastiaans 2016; Peterson et al. 2009; Brown et al. 2022; Rothschild et al. 2022). We cannot say whether all *T. rex* interactions were violent, but this outcome was surely common as at least fifteen

T. rex skulls record possible wounds from facial biting. Some of these injuries are of disputed causes, either being interpreted as lesions from disease or perhaps inflicted by claws rather than teeth (see Rega and Brochu, 2001; Wolff et al. 2009; Rothschild 2013; Rothschild et al. 2022), but the majority are attributed to *Tyrannosaurus* grabbing each other's heads with their jaws hard enough to penetrate flesh and bone (fig. 6.6). Among

FIGURE 6.6

Social interactions, *Tyrannosaurus*-style. Many theropods appear to have engaged in face biting but the strong jaws of tyrannosaurids make this behavior especially well marked in their fossils, and *T. rex* is no exception. Reconstructed bite trajectories suggest that antagonists mostly stood alongside one another and lunged laterally with their jaws, rather than tackling one another face-on.

living animals, the development of regionalized armored skin is often correlated with intraspecific aggression of this kind (Hieronymus 2009) and the possibility of violent, face-targeted interactions between *Tyrannosaurus* might explain those indications of toughened, armored facial skin discussed in chapter 3.

T. rex was not unique in this behavior. On the contrary, evidence of face biting is so well documented across theropods that it was probably common for the entire group, only diminishing in the maniraptorans as, it is suggested, feather displays eventually took precedence over more direct aggression (Brown et al. 2022). Perhaps because their bites were especially strong and liable to leave wounds, tyrannosaurids have an especially good record of face biting. Brown et al. (2022) reported 324 bite-made lesions in a survey of 202 tyrant dinosaur specimens (not including *T. rex* data), and the recovery of a tyrannosaur tooth tip from one wound leaves no doubt about the cause (Bell and Currie 2010). Injuries likely caused by face biting in *Tyrannosaurus* and other tyrant dinosaurs cover most of the skull (fig. 6.5), with puncture wounds occurring at the back of the lower jaw in a number of specimens (Rothschild and Molnar 2008), on the lateral and upper surface of the snout (Peterson et al. 2009; Rothschild 2013; Bastiaans 2016), as well as behind the orbit and on the top of the occiput (Rothschild 2013). Clustering of bite marks on certain parts of the skull suggest stereotyped or ritualized approaches to fighting, perhaps with tyrants holding their faces alongside one another, attempting to seize their rival's head with lateral lunges of the jaws (Brown et al. 2022).

The behavioral catalysts for these interactions are all but impossible to determine from fossils, but face biting is widespread in living animals, from salamanders and turtles to carnivorous mammals and raptorial birds. In all cases, it dramatically increases in frequency after the onset of sexual maturity (Brown et al. 2022; Rothschild et al. 2022), and we see evidence of this in tyrannosaurids, too. Bite-derived facial injuries in the subadult *T. rex* specimen "Jane" (Peterson et al. 2009) and other half-sized tyrannosaurids (Brown et al. 2022) indicate

that face biting began well before the attainment of maximum size, but around the estimated start of reproductive maturity (chapter 4). Among living animals, males tend to suffer more face biting wounds than females, implying that there's a greater chance of battered and wounded *T. rex* skulls representing males (Brown et al. 2022). However, the skew of bite-mark data toward male animals is not tight enough to make this a reliable sex identifier and, moreover, plenty of species, including some crocodylians, have near-equal or female-biased

wound distributions among the sexes (Brown et al. 2022). These reflect animals of both sexes injuring one another in disputes over food and territory (Schaller 1972; Auffenberg 1981; Grigg and Kirshner 2015), and for all we know, fights over such matters accounted for many *T. rex* face scars. Whatever their cause, *T. rex* facial wounds are of such size to suggest that battles took place between individuals of similar stature, echoing a commonly observed behavior among living animals where fights only break out among evenly matched individuals (Brown et al. 2022). Evidence of healing shows that face biting was, on the whole, nonfatal.

FIGURE 6.7

The flattened ornament of *Tyrannosaurus* may have reduced their showiness, but potentially increased the pushing and shoving capacity of their heads. Head-butting has been proposed as an explanation for the cranial features of a number of theropods, including *T. rex*.

Confrontations between *T. rex* may have not always involved biting. Bakker (1986) and Paul (1988a, 2016) have linked the robust nasal bones and orbital ornaments of tyrannosaurids with another form of intraspecific combat: head-butting (fig. 6.7). Modern researchers are increasingly attuned to the idea of theropods once pushing and shoving one another around with their robust cranial horns and bosses (Sereno and Brusatte 2008; Cau et al. 2013), especially abelisaurids (e.g., Novas 1989; Mazzetta et al. 1998, 2009; Hieronymus 2009; Snively et al. 2011; Delcourt 2018; Cerroni et al. 2021). *Tyrannosaurus* may have been well adapted for such acts. As noted in chapter 3, *T. rex* was an unusually unornamented tyrannosaur: rather than horns above their eyes and spiky scales along their snouts, even very old, well-ossified *Tyrannosaurus* sported low snout scales, flattened lacrimal ornaments, and rounded postorbital bosses. Cornified skin around the eye and snout would make the top of the head especially damage resistant, and the overall gestalt of this configuration is a theropodan bulldozer: a broad, shovel-shaped skull roof to heave and shove other animals with. At present, we can only speculate on whether *T. rex* engaged each other in this way, but studies of their internal skull architecture may elucidate, via searches for shock-absorbing adaptations, the plausibility of head-butting, flank-butting, or head- hoving matches.

TYRANNOSAURUS REX: DINOSAUR HUNTER

FEW TOPICS GET *Tyrannosaurus* fans as excited as how *T. rex* caught and ate their prey, and understandably so: if the thought of these epic, gigantic animals acting out life-or-death chases and deploying their bone-crushing jaws on other giant dinosaurs doesn't excite you *even a little*, you may need to see a doctor. Academics happily opine about *Tyrannosaurus* predation as well, with dinosaur literature containing many deliberations on tyrannosaurid hunting strategies, the biomechanics of predation and feeding, as well as—most famously—discussion over their predilections for scavenged carcasses over live prey. Of all the topics concerning king tyrant behavior, their predatory acts are probably the most vulnerable to contamination from our own intuited ideas and opinions, as well as the "rule of cool" overriding evidence-led hypothesizing. This is a subject to carefully tread through as we review what we know, or have good reason to believe, about *Tyrannosaurus* foraging behavior.

THE PREDATOR VS. SCAVENGER [N]ONTROVERSY

Throughout the 1990s and 2000s, popular discussions of *Tyrannosaurus* ecology were dominated by a single question: Was *T. rex* primarily a scavenger or a predator? This topic has formed the basis of entire television documentaries and exhibits (the skeleton in Figure 1.2 is from such a dedicated display), as well as many heated conversations among *T. rex* fans. Modern interest in this idea can be traced back to a handful of influential and provocative publications by paleontologist Jack Horner, wherein he argued that *T. rex* anatomy was ill-suited to capturing live prey but was ideally adapted for obligate scavenging (Horner and Lessem 1993; Horner 1994). The concept of scavenging tyrant dinosaurs is far older than these works, however, dating back to the earliest research on tyrannosaurids. Like most opinions on dinosaurs from this time, tyrannosaur scavenging has roots in twentieth-century notions of dinosaur sluggishness and biomechanical ineffectuality (Lambe 1917; Halstead 1970; Barsbold 1983; also see chapter 4 for historic interpretations of weak-jawed *T. rex*).

Horner's post–Dinosaur Renaissance scavenger hypothesis brought new data into play but still largely leaned on notions of king tyrants being slow, useless predators (Horner and Lessem 1993; Horner 1994). It proposed that *T. rex* could not move quickly, that their arms were short and useless, and that their jaws and teeth not up to the demands of physically subduing prey. Small eyes were linked to poor eyesight, while a keen sense of smell was argued as well suited for tracking down rotting, stinking carrion (Horner 1994; Horner and Lessem 1993). This proposal attracted tentative support from a few researchers, with Farlow (1994) rationalizing that the great height of *T. rex* would aid visual and olfactory detection of carcasses. Ruxton and Houston (2003) calculated that a modern Serengeti-style ecosystem could support a theropod of *Tyrannosaurus* size through a supply of carrion, assuming it could find them fast enough.

Publicly, the king tyrant scavenging hypothesis has been framed as a true debate between dinosaur experts, a protracted dialogue where arguments are equally weighted in favor of obligate scavenging and more active, predatory lifestyles. However, the real academic legacy of Horner's resurgence of the scavenger hypothesis was a cascade of literature exploring the evidence for tyrannosaurid predation. These projects included examinations of tyrannosaurid anatomy (Farlow and Holtz 2002; Holtz 2008; Paul 2008; Carpenter 2013); the energetics of scavenging lifestyles (Ruxton and Houston 2003, 2004; Carbone et al. 2011); studies of tooth strength and bite force (Erickson et al. 1996; Meers 2002); and the identification of healed bones in prey species bitten by *T. rex* (Carpenter 1998; Happ 2008; DePalma et al. 2013; Siviero et al. 2020). Behind the public hype, very little significant debating took place, and the "discussion" of *Tyrannosaurus* scavenging was really a bolstering of the long-held, better-supported predatory hypothesis.

We don't need to belabor the point that no authors preferring *T. rex* as a predator preclude carrion from having contributed to king tyrant nutritional intake, as it does for most, maybe all, predatory species of modern times, and there is no disagreement about their large size being an intimidating presence to smaller carnivores they encountered at carcasses (fig. 5.6). Nor do we need to rebut each point suggested in favor of strictly scavenging lifestyles, as they have been covered in earlier chapters. Instead, we can simply restate that *T. rex* eyes were not small, and there is no evidence that

their eyesight was poor (chapters 3 and 4); they were not slow for their size (chapter 4); their arms were not useless, especially as juveniles (chapters 3 and 4); their jaws and teeth were exceptionally strong, literally capable of biting through enormous bones (chapter 4, also see below); and an acute sense of smell (chapter 4) is clearly beneficial for both finding live prey and dead remains. Certain anatomies can indeed predict scavenging lifestyles in fossil taxa, as demonstrated by work on fossil vultures and their relatives (e.g., Hertel 1995), but, in this case, the features of king tyrants suggested to dictate exclusively carrion-based feeding habits are directly contradicted by a wealth of science.

Augmenting these observations are studies showing that it may be energetically impossible for a terrestrial animal to secure their dietary intake from scavenging alone. Work by Graeme D. Ruxton and David C. Houston (2003, 2004) showed that, theoretically, the savannahs of modern Africa might supply enough dead animals to fuel a *T. rex* population, but that they would also struggle to reach carcasses before they were consumed by other animals. Flight, it seems, is essential for obligate scavengers, this being the only traveling mechanism swift enough to locate the short-lived feeding opportunities created by dead and dying animals. Chris Carbone and colleagues (2011) arrived at a similar conclusion, noting that large *T. rex* were probably so outnumbered by smaller predators in Lancian ecosystems that carcasses were mostly stripped before they arrived and, in any case, that the vast nutritional requirements of adult *Tyrannosaurus* may never have been satisfied through scavenging half-eaten meals. In this model, *T. rex* were forced to capture live prey to simply ensure that they received enough food. Investigations of this nature not only shed light on king tyrant ecology, but also explain why dedicated scavenging is generally rare among terrestrial carnivores. The question of obligate scavenging in Mesozoic theropods remains open for a minority of researchers, but even those convinced that some large theropods may have been vulture analogues (e.g., Pahl and Ruedas 2021) suspect that *T. rex* was unable to satisfy their energy demands from carcasses alone.

Without doubt, however, the best evidence that *T. rex* sought living prey comes from fossils of adult *Edmontosaurus* and *Triceratops* that apparently survived predatory encounters with king tyrants. Healed tooth marks, one with the tip of a *Tyrannosaurus* tooth left within it, have been found in *Edmontosaurus* tail bones (Carpenter 1998; DePalma et al. 2013; Siviero et al. 2020) as well as on the horns and frill of a *Triceratops* that, remarkably, had its left brow horn crunched off during a *T. rex* attack (Happ 2008). These individuals lived long enough after their tyrant encounters to begin healing their damaged bones. Whether one of these injuries, a missing vertebral spine tip from an *Edmontosaurus* tail (fig. 6.8; Carpenter 1998) can truly be attributed to *T. rex* has been doubted as the unusual frequency of injuries within hadrosaur caudal skeletons has become apparent. For whatever reason, hadrosaur tails seem to have been highly prone to breaks and other trauma, and some of their damaged, infected bones developed tooth mark–like lesions (Tanke and Rothschild 2014). These, not a misfiring *T. rex* bite, may better explain at least one supposed incident of failed *T. rex* predation. Even so, the identification of undoubted healed tooth marks, and the smoking gun of an *Tyrannosaurus* tooth embedded into partly repaired hadrosaur bone, leaves little doubt that king tyrants attacked other dinosaurs, most likely in the hope of consuming them.

CATCHING PREY

Unfortunately, even bones with shed *T. rex* teeth wedged into them don't tell us much about the spe-

cifics of *Tyrannosaurus* predatory behavior. Gregory S. Paul (2008) argued that these *T. rex* predation traces should be disregarded as evidencing typical *T. rex* hunting efforts because they represent failed attempts at subduing other dinosaurs. There must be some truth in this, as the fact those individuals survived their bites suggests that their attacker made a mistake or gave up for some reason, but we have nowhere near enough predatory *T. rex* traces to deduce whether grabbing hadrosaur tails or ceratopsid faces was atypical. We might argue contrarily that the lottery of fossilization is most likely to preserve average, everyday examples of prehistoric life, and that this extends to records of bite marks and injuries. Grasping ceratopsids by their face might seem intuitively unusual, but seizing animal faces is certainly not unheard of among living carnivores (Schaller 1972; Kruuk 1972). Moreover, grabbing *Edmontosaurus* by their tails doesn't seem like an unlikely event during a *T. rex* and hadrosaur interaction. Turning tail and running from an attacking tyrannosaur seems highly sensible to me!

Opinions vary on how confidently we can predict *Tyrannosaurus* predatory behavior. One view is represented by Horner and Lessem (1993) who feel "it's ... hard to imagine how [*T. rex*] caught its dinner" (p. 206), and this can be contrasted with the perspective characterized by Paul (2008): "Some aspects of tyrannosaurid hunting techniques can be restored with substantial confidence" (p. 344). Perhaps the truth lies somewhere between these extremes. After all, the basics of terrestrial vertebrate predation are not infinitely complex (there are only so many ways to approach prey and then bite, claw, or grapple it into submission), predation-relevant

FIGURE 6.8

Bite marks and possible predatory traces made by *Tyrannosaurus*, including evidence of cannibalism. (A) *Edmontosaurus annectens* skeleton with a missing portion of a caudal vertebra, interpreted by some (but not all) as the result of failed *T. rex* predation; (B) one of the most informative specimens regarding *T. rex* feeding habits: a *Triceratops* pelvis riddled with *Tyrannosaurus* tooth punctures, gouges, and chew marks (white arrows); (C) hadrosaur metatarsal with tooth gouges; (D) *Triceratops* frill with *T. rex* tooth marks; (E) *Tyrannosaurus* toe bone bitten by another king tyrant; (F) *Tyrannosaurus* humerus with *T. rex* tooth gouges. (B) From Gignac and Erickson (2017); (C–F) adapted from Longrich et al. (2010).

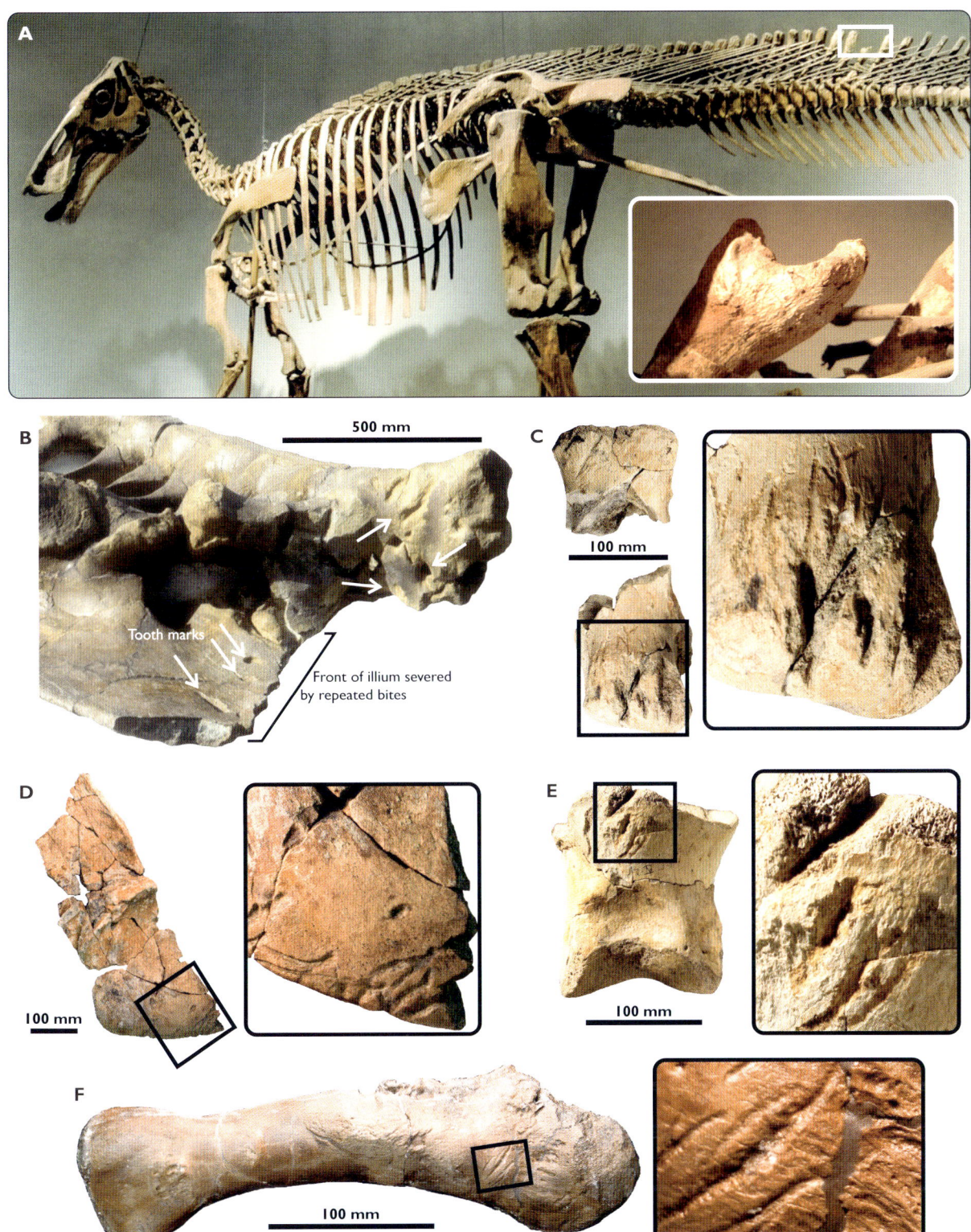

B

500 mm

Tooth marks

Front of illium severed
by repeated bites

C

100 mm

D

100 mm

E

100 mm

F

100 mm

aspects of *T. rex* functional morphology are well studied, and the predatory ecologies of many large modern terrestrial predators are well observed and researched. Certain studies have also directly addressed the predatory significance of king tyrant speed, jaws, claws, and body mass. This must be tempered, however, with acceptance that all behavior is notoriously difficult to predict for fossil animals, and that even basic inferences are difficult to validate through testing.

As the largest terrestrial predators in Maastrichtian Laramidia, even as juveniles, *T. rex* were surely predators of many animals in their respective ecosystems. If the fossil record is anything to go by, the most abundant prey species living alongside *T. rex* were hadrosaurs and horned dinosaurs (fig. 5.2D; Horner et al. 2011). These are also the species that preserve trace evidence of *Tyrannosaurus* feeding (see below), perhaps indicating that they formed a large proportion of *Tyrannosaurus* diets. Presumably, increasingly large and mature individuals were targeted by tyrants of different ages and sizes. Generally, modern predators attack animals of about the same size or somewhat smaller than themselves, mostly ignoring prey that's significantly larger or smaller because they require too much effort to catch or subdue (Schaller 1972). Extrapolating modern predator-to-prey-size ratios to those of Maastrichtian North America, Meers (2002) found that adult *Triceratops* represented ideal prey masses for mature *Tyrannosaurus*, data that align with evidence of healed *T. rex* bites on mature *Triceratops* and *Edmontosaurus* skeletons (Happ 2008; Depalma et al. 2013).

The recovery of *T. rex* in deposits with giant sauropods raises the possibility that king tyrants also attacked these very large herbivores. There is certainly precedent for living predators to occasionally punch above their weight. African lions mostly target animals twice or half their own mass, but sometimes predate adult hippopotamus, rhinoceros, and elephants (Pienaar 1969; Schaller 1972; Brain et al. 1999; Loveridge et al. 2006; Power and Compion 2009), prey individuals that might outweigh their predators by ten to fifteen times

(Power and Compion 2009). Such attacks are uncommon, however, with lions more typically targeting the juveniles of giant animals rather than their parents, and large individuals only becoming game when more manageable prey is scarce (Loveridge et al. 2006; Power and Compion 2009). Conversely, Komodo dragons are known to frequently attack their largest local prey species, water buffalo, whenever possible, supplementing their usual diet of small deer and wild boar (Auffenberg 1981). Both scenarios suggest that *T. rex* attacking large sauropods cannot be ruled out, although we presently lack direct evidence of this interaction. If it happened, *Tyrannosaurus* may not have developed sauropod-specific predation approaches for large prey items, just as living predators do not seemingly need novel strategies to act as giant slayers. Despite the size of elephants, lions bring them down using the same methods deployed against any large species, such as buffalo, giraffes, or rhinos (Schaller 1972; Brain et al. 1999; Power and Compion 2009).

T. rex functional morphology gives some idea what their predatory approaches may have been. Terrestrial vertebrate predators exist on a spectrum between ambushers (hiding or sneaking up on prey to launch an attack at close quarters) and coursers (openly pursuing prey over a long distance). As noted in chapter 4, tyrannosaur hindlimb muscles seem suited for powerful but short-lived chases, and that might point toward a reliance on stalking and ambushing more than coursing (Persons and Currie 2014). Younger, smaller *T. rex* may have been explosive sprinters over short distances, with relatively large hindlimb muscles permitting great acceleration and speed, and their leg proportions conferring enhanced agility (Hutchinson et al. 2011; Snively et al. 2019; Dececchi et al. 2020). At larger sizes and greater masses, *T. rex* are predicted to not only have moved relatively slowly, but also to have lost some capacity for rapid acceleration (Hutchinson et al. 2011; Snively et al. 2019; Decacchi et al. 2020), potentially demanding attacks from closer quarters. Similar changes in locomotion are experienced today by growing Komodo dragons, which transition from rapid pursuit

FIGURE 6.9

The dramatic change in predatory behavior evidenced for *Tyrannosaurus* seems unusual when compared to our modern ecosystems populated by mammals and birds. Among reptilian predators, however, such changes are more commonplace. Komodo dragons undergo a transformation somewhat akin to that imagined for king tyrants, transitioning from fast, tree-climbing, and gracile juveniles (A) to robust, slower, but powerful ambush predators as adults (B).

predators as lithe, agile juveniles to close-range ambushers as large and powerful, but slower adults (fig. 6.9; Auffenberg 1981).

This is not to imply, of course, that *T. rex* became less effective predators as they grew. When thinking about predator speed, *absolute* velocities are less important than *relative* ones: whatever their growth stage, *T. rex* only had to be faster than whatever prey they sought. Even adult *Tyrannosaurus* retained some cursorial adaptations, possessing unusually long legs, elongate feet, and hindlimb musculature adapted for powerful bursts of speed (Holtz 1995, 2008; Hutchinson et al. 2011; Carpenter 2013; Snively et al. 2019). Thus, although capped in performance by the physics of motion at large body size, they outstripped the cursorial adaptations of their prey species, even the running-adapted hadrosaurs. *T. rex* not only had a longer stride than *Edmontosaurus* (fig. 6.10), but were better equipped for acceleration and short-term bursts of speed (Holtz 2008; Carpenter 2013; Persons and Currie 2014).

Hadrosaurs, meanwhile, seem to have had lower top speeds, but greater endurance than their tyrannosaur pursuers (Coombs 1978; Persons and Currie 2014). Specifically, hadrosaur caudofemoralis muscles (those powerful hindlimb retractors that fleshed out the tail base; see Figure 3.11) were comparable in size to those of tyrannosaurids but anchored to a lower point on the thigh (fig. 6.10A). This shortened their strides and correspondingly slowed them down (fig. 6.10B; Persons and Currie 2014), but also imparted greater leverage on the hindlimb itself, making for more efficient limb action and potentially greater endurance. For *Edmontosaurus*, perhaps running longer, not necessarily faster, was crucial to evading king tyrants. The running speeds of ceratopsids are more controversial, with some advocating rhinoceros-like speeds and others suggesting a

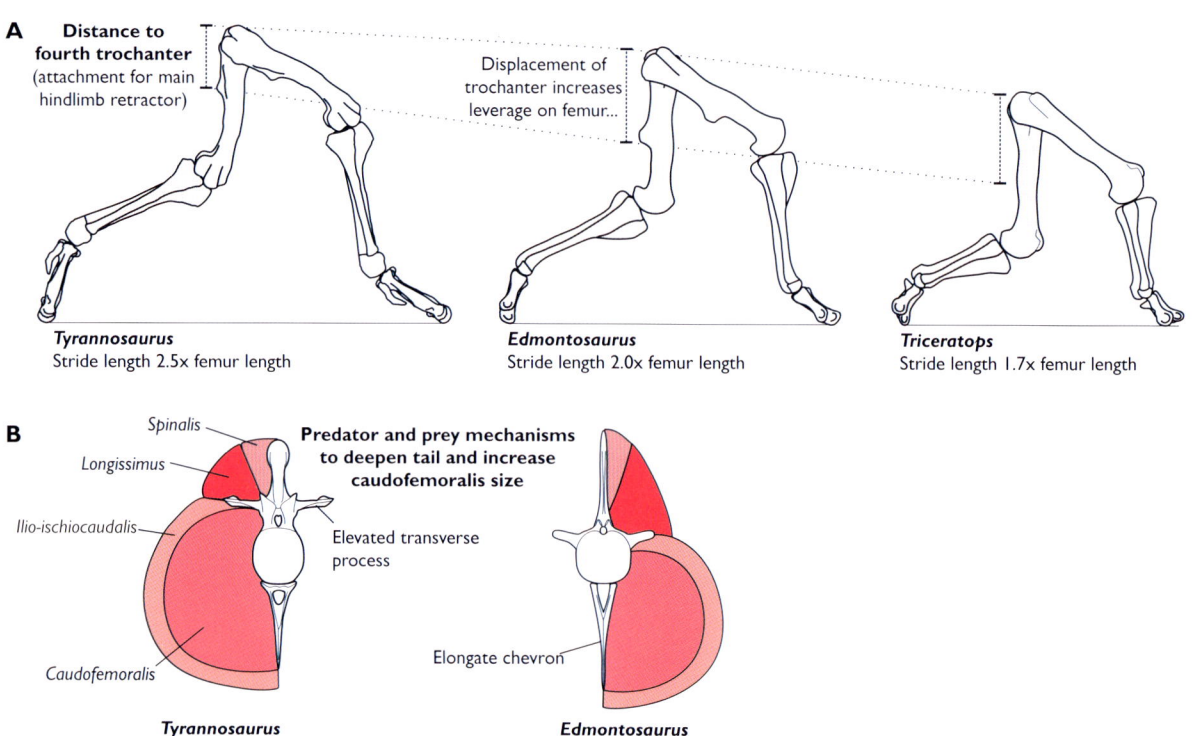

A **Distance to fourth trochanter** (attachment for main hindlimb retractor)

Displacement of trochanter increases leverage on femur...

Tyrannosaurus
Stride length 2.5x femur length

Edmontosaurus
Stride length 2.0x femur length

Triceratops
Stride length 1.7x femur length

B Spinalis

Longissimus

Ilio-ischiocaudalis

Caudofemoralis

Predator and prey mechanisms to deepen tail and increase caudofemoralis size

Elevated transverse process

Elongate chevron

Tyrannosaurus

Edmontosaurus

FIGURE 6.10

The running speed of *Tyrannosaurus* is of great interest to scientists, but, to king tyrants themselves, all that mattered was that they were faster than their prey. (A) Stride length and fourth trochanter positions on adult king tyrants and two known prey taxa, *Edmontosaurus* and *Triceratops*, scaled to the same femur length; (B) the enlarged caudofemoralis volumes in *Tyrannosaurus* and *Edmontosaurus*, where two configurations of vertebral morphology achieved the same goal. (A) *Edmontosaurus* and *Triceratops* limb bones after Paul (2016); (B) after Persons and Currie (2014).

more limited capacity for fast movement, at least compared to hadrosaurs (Coombs 1978; Christiansen and Paul 2001; Maidment et al. 2012, 2014; Persons and Currie 2020). Based on limb proportions alone, however, it stands to reason that, if *T. rex* could outpace *Edmontosaurus*, they could catch *Triceratops* (fig. 6.10A).

The size and power of *Tyrannosaurus* jaws and teeth imply that, like crocodylians, hyenas, dogs, and Komodo dragons, king tyrants were primarily biting predators, rather than cat-like grapplers (Holtz 2008). Bite-based predation against large game can be brutal to witness because most bites, even deep, debilitating ones,

don't end the lives of big animals outright. Grappling predators, particularly cats, have developed behaviors to swiftly dispatch even large individuals with bites to the muzzle or throat, thus making them easier to handle and eat (Schaller 1972), but bite-focused carnivores simply deliver enough damage or exhaust their prey to the point of immobility, after which they start eating it alive. We see this grisly reality across a range of predators. Raptorial birds that don't kill their targets immediately with their talons perch atop them, ripping off strips of flesh with their beaks while pinning them with their feet (Fowler et al. 2011b). Komodo dragons bite at

FIGURE 6.11

The bone-crushing bite of *Tyrannosaurus* is deployed against a fleeing *Torosaurus latus*. Like painted dogs or spotted hyenas, king tyrants may have immobilized prey by wounding or eating their flanks and haunches during pursuits, and their size imbued them with sufficient force to unbalance and topple even large animals. As with modern animal predation, such mechanics do not bring swift, merciful ends to prey individuals, and death via tyrannosaurid attack was probably a terrible way to die.

animal legs to hamstring their prey, or strike at the belly to disembowel them, neither of which are immediately fatal (Auffenberg 1981). Wild dogs and hyenas also aim for disembowelment, usually after a chase where they bite at the legs and flanks, or sometimes topple their targets first by pulling them over with their jaws (Kruuk 1972; Schaller 1972). The recovery of predatory bite marks on hadrosaur tails might record king tyrants also attacking the hindquarters of fleeing dinosaurs, and biomechanical studies have demonstrated the potential for *T. rex* pulling or pushing prey over. Adult *T. rex*, it seems, were sufficiently large to unbalance even large *Triceratops* with only a few seconds of laterally directed force (Krauss and Robinson 2013). Alternatively, a tyrant looming over and pressing down on their prey may have overloaded their target's limbs, these being relatively easy to buckle in large game owing to their already heavily loaded limb muscles (Power and Compion 2009).

As noted in chapter 4, escalations of bite force in growing *Tyrannosaurus* may correlate with the reduction of their arms toward adulthood. In weaker-biting juveniles, longer arms and hands may have helped to restrain or position prey between biting actions, whereas, in powerfully biting adults, the reliance on other appendages to hold prey may have diminished (fig. 4.24). Neck biomechanics indicate that adult *T. rex*, with their limited cervical mobility, did not deploy rapid strikes at prey, but instead bit down and pulled chunks of flesh away in an upward or sideways motion, or else held prey and shook it violently (Snively and Russell 2007a, b). Although fossil evidence shows that attacks from *Tyrannosaurus* were not always fatal, a well-placed king tyrant bite was surely devastating (Paul 1988). Pondering how those jaws worked on animal flesh is almost frightening. Not only would tonnes of force be wrecking any tissues, even large bones, caught between the tyrant's jaws, *not only* would those jaws be so tightly clamped that only their relaxation or the severing of a body part might permit release, but they *also* held their victim to a 6–8 tonne, tyrannosaur-shaped anchor that slowed or outright prohibited escape. We don't need to hype or exaggerate this aspect of *T. rex* paleobiology beyond what our science indicates: a *Tyrannosaurus* attack was surely an awesome, but also perhaps horrifying spectacle of natural history (fig. 6.11).

A TOXIC BITE?

As if king tyrant bites were not severe enough on their own, some scholars have suggested that they were also toxic. This hypothesis was proposed in 1992 by researcher William Abler who, dissatisfied with the conventional interpretation of theropod dental carinae as cutting implements (fig. 3.5), proposed that they

instead stored tiny particles of flesh to promote the growth of malignant bacteria. Drawing specific comparisons to Komodo dragons, he proposed that bites from such teeth would infect wounds and weaken prey, making even a failed tyrannosaur attack potentially lethal. He later suggested that tyrannosaurs bore extensive, *Varanus*-style gums that would occasionally be cut during biting actions, nourishing the lethal microbes within tyrannosaur mouths through the blood they released (Abler 1999). Abler concluded with a gory vision: "The horrific appearance of the feeding tyrannosaur is further exaggerated—their mouths would have run red with their own bloodstained saliva while they dined" (Abler 1999, p. 51).

Abler's hypothesis has not been widely adopted among theropod workers, however, and carinae are still widely regarded as having flesh-cutting roles. This concept is not aided, either, by studies casting doubt on the idea that Komodo dragons exploit bacterial sepsis as a predatory tactic. The mouths of these lizards contain bacteria, but only species expected from their diet and gut flora. Their saliva is thus not considered septic by modern researchers (Fry et al. 2009; Goldstein et al. 2013). Komodo dragons do, however, create a venom that contains compounds linked with paralysis, disorientation, loss of consciousness, cramping, and excessive bleeding. Whether this makes them a truly venomous species is debated because no study has yet demonstrated that their bites deliver potent-enough venom quantities to influence the outcomes of predatory acts. Case studies of extreme dragon wounds on humans provide direct insight into such matters: as terrible as these injuries are, no toxic effects or infections have been reported in humans (Weinstein and White 2015; Boyd et al. 2021). In any case, whether venomous or not, the lack of septic bites in Komodo dragons means they cannot be used to justify toxic tyrannosaur bites. Nor, for that matter, should we entertain ideas of *T. rex* using venom: venom-delivery systems can be identified in fossil reptile jaws (Mitchell et al. 2010), but, to date, no such anatomy has been found in any predatory dinosaur (Gianechini et al. 2011).

FEEDING TIME

A *Tyrannosaurus* behavior we can discuss with relative confidence is their eating habits (fig. 6.12). Predatory animals can leave distinctive marks on bones when feeding on carcasses—sometimes accidentally, sometimes intentionally—and these record their approaches to stripping flesh from skeletons. Gouges indicate teeth being dragged over a bone surface, punctures show bones being gripped, while pulped, broken bones imply gnawing or chewing. Tooth marks on dinosaur bones were largely ignored before the 1990s but paleontologists have now cataloged a great number of instances where theropod dinosaurs left feeding traces on the remains of other animals. Many of these punctured and gouged specimens are isolated, broken bones, explaining why they were overlooked until relatively recently: a disassociated, chewed-up bone is not the sort of specimen sought for exhibition or private collection. It was only post–Dinosaur Renaissance interest in dinosaur biology that encouraged paleontologists to collect these less spectacular, but nevertheless informative, fossils and appreciate their significance (Erickson 1999).

FIGURE 6.12

The "Nation's *T. rex*" (also known as the "Wankel rex") mounted in a feeding pose with *Triceratops* at the Smithsonian National Museum of Natural History. An expansive record of bitten bones gives us insights into how king tyrants fed, among which is evidence that they bit and tugged on the frills of horned dinosaurs, perhaps manipulating the head for access to thick neck musculature (see Figure 6.14; Fowler et al. 2012).

LIFE, FOOD, LOVE, DEATH

Thanks to their powerful bites and occurrences in well-sampled fossil deposits, *Tyrannosaurus* feeding traces have been found on dozens of dinosaur bones (Carpenter 1988; Horner and Lessem 1993; Erickson and Olsen 1996; Happ 2008; Longrich et al. 2010; Fowler et al. 2012; McLain et al. 2018). Theropod bite marks are mostly difficult to assign to a particular species because several, similarly sized and adapted species often coexisted, but the size and shape of their teeth, as well as their status as the only giant theropods in late-Maastrichtian Laramidia (chapter 5), make the bite marks of adult *T. rex* unmistakable. That these traces represent *T. rex* feeding on dying or dead animals is evidenced by their occurrences in places impossible to access in living animals, such as the inner surfaces of joints or the underside of pelvic girdles, as well as a lack of healing tissues around tooth punctures and grooves. Our inventory of *T. rex*–marked bones includes the limb elements, skulls, and pelvic girdles of horned dinosaurs and hadrosaurs, as well as cannibalistically bitten limb bones of other *Tyrannosaurus* (fig. 6.8B–F; Longrich et al. 2010; Fowler et al. 2012). Indeed, *T. rex* bite traces may ultimately prove to be commonplace if we look hard enough: Fowler et al. (2012) report that 8 to 14 percent of a large sample of *Triceratops* bones bear *T. rex* scars.

Each of these bite marks tells us something about what *Tyrannosaurus* ate and how they fed. Most of their feeding traces are found on large dinosaur bones, so we don't have much data on how *T. rex* interacted with smaller bones or, indeed, smaller dinosaurs. These were likely wholly destroyed during feeding (perhaps even swallowed whole; Figure 3.1), or, if any remains survived, they had low chances of fossilization (Erickson and Olsen 1996; Hone and Rauhut 2010; Longrich et al. 2010). The bite marks that have survived to be studied today may also reflect later-stage carcass processing where flesh-stripped bones were nibbled, but not consumed. If *T. rex* were like living species, they ate the fleshiest, most readily digested, and nutrient-rich components of a body before scraping and biting away tissues closer to bone. A fairly reliable pattern of carcass consumption is observed among large living carnivores (Blumenschine 1986) where the hindlimb muscles are eaten first, then the contents of the abdominal cavity and muscles of the shoulders and neck, followed by the fleshier parts of the face. Limb bones are then gnawed before the contents of the skull are extracted toward the end of the sequence. If this pattern was reflected in the Mesozoic, our record of *T. rex* feeding behavior mostly constitutes later-stage feeding (fig. 6.13).

Of all the specimens revealing *T. rex* feeding habits, one of the most enlightening is a large *Triceratops* pelvis riddled with tooth marks (fig. 6.8B; Erickson and Olsen 1996). The posterior end of this limb girdle contains some fifty-eight tooth punctures and gouges and twenty-two possible others, many of which overlap and occur so widely over this gigantic bone that it must have been tumbled about and bitten from different sides. The marks themselves represent different feeding styles. Centimeter-deep, 10 cm long scars imply "puncture and pull" defleshing actions similar, though not identical, to those created by feeding Komodo dragons (D'Amore and Blumenschine 2009). Bone scours left by these lizards are often curved, reflecting arcing movements of their jaws, rather than straight, as with most theropods, implying that *Tyrannosaurus* and kin pulled directly back to remove flesh from corpses (D'Amore and Blumenschine 2009). Strong neck muscles surely assisted in this function, as might backward pulling using the hindlimbs (Snively and Russell 2007a, b).

Puncture marks of various depths (up to 3.7 cm deep), however, are probably not defleshing actions. They perhaps instead record gripping and positioning behavior, or else attempts at breaking through bone. Concentrations of punctures around bone margins are records of this action, and our bite-riddled *Triceratops* pelvis lost a piece of the left ilium and almost one of its massive sacral vertebrae to a gnawing *T. rex*. This latter trace may have reflected a king tyrant severing the pelvis from the rest of the carcass (Erickson and Olsen 1996). The great diameters of most tooth marks implies that the largest teeth were employed in these especially destructive feeding actions, but more delicate feeding traces left by the smaller, incisiform premaxillary teeth record *Tyrannosaurus* nibbling between vertebral processes. Regular bite spacing across the *Triceratops* hip specimen implies a certain regularity of approach, as if the feeding *T. rex* was working its way across the bone systematically rather than chaotically gorging itself (Erickson 1999). *Tyrannosaurus* carcass consumption, it seems, was a blend of brute force and finesse: tumbling and dismembering giant bones one moment, delicately extracting choice tissues the next.

Evidence of forceful feeding habits are also evidenced elsewhere on *Triceratops* specimens. The large, heavy skulls of horned dinosaurs were supported by generous neck musculature that would have been attractive to hungry carnivores, but accessing it would have been complicated by those expansive cranial frills. Horned dinosaurs have some of the largest heads of all land animals and most large carnivores would have struggled to eat their deep neck tissues, the head being too large and heavy to move aside. The enormous skulls of

FIGURE 6.13

King tyrants are some of the only dinosaurs that we know of that habitually bit into and consumed vast quantities of bone. Although they seem to have scraped away and nipped at flesh between skeletal components, they also actively gnawed at and broke off large pieces of bone, perhaps to eat, perhaps to better access parts of carcasses, or perhaps both.

LIFE, FOOD, LOVE, DEATH

FIGURE 6.14

Tyrannosaurus wrecks *Triceratops*. Bite marks from a large sample of *Triceratops* skulls show that *Tyrannosaurus* routinely ate neck muscles only accessible on decapitated horned dinosaur corpses. This potentially explains the presence of tooth gouges and marks on *Triceratops* frills as traces made by king tyrants leveraging these enormous heads from ceratopsid bodies.

Triceratops were no obstacle to *T. rex*, however: if a *Tyrannosaurus* wanted access to that guarded neck meat, they may have simply grabbed the *Triceratops* head and pulled it off the corpse (figs. 6.12, 6.14; Fowler et al. 2012). And yes, that is the most awesome sentence in this book.

Evidence of this behavior includes tooth gouges on several *Triceratops* frills that likely represent carcass manipulation and pulling actions

(fig. 6.8D), there surely being no significant flesh on this part of horned dinosaur skulls to justify biting it for other reasons. Simultaneously, feeding traces on at least two *Triceratops* occipital regions show that *T. rex* ate deep neck flesh only accessible on heads separated from their bodies (Fowler et al. 2012). Other agents may have been at work here—maybe the heads became detached in another way and *T. rex* simply scavenged them?—but the occurrences of these bite marks on a number of specimens suggests they were not chance interactions between species, but a common *T. rex* behavior. Following models of carcass consumption, such acts probably reflected later-stage feeding where a lot of the choice *Triceratops* meat had already been eaten. Even imagined on a partially stripped *Triceratops* carcass, however, the notion of king tyrants pulling such huge dinosaurs apart reminds us of the enormous strength and biomechanical capabilities that giant dinosaurs must have possessed.

LOVE, DEATH, AND *T. REX*

KING TYRANT, QUEEN TYRANT: IDENTIFYING MALE AND FEMALE *TYRANNOSAURUS*

Along with finding suitable places to live, interacting with conspecifics, and sourcing food, a major priority in *Tyrannosaurus* lives would have been reproduction: the passing of tyrant genes onto the next generation. Our knowledge of dinosaur reproductive biology has grown tremendously in the last century, but some aspects remain frustratingly beyond reach. Among them are many seemingly simple lines of inquiry, such as the nature of dinosaur nests and the capacity to distinguish male and female specimens. This latter point is particularly pertinent for king tyrants as their specimen nicknames encourage us to genderize their remains: "Jane," "Petey," "Stan," "Sue," and so on. Mostly, these pet names honor people or circumstances associated with the history of the specimen and have no significance toward the actual sex of the *T. rex* individual. Regardless, they inspire an obvious question: What ability do we have to distinguish male from female *Tyrannosaurus*?

Utilizing a variety of methods and taxa, efforts to determine the sex of dinosaurs has proven a frequent investigative avenue among researchers (see P. Larson 1994 and Mallon 2017 for reviews). Just about every

major dinosaur clade has been interrogated for features that hint at sexual dimorphism, and *T. rex* has featured prominently in these discussions (Carpenter 1990; 1999; Molnar 1991; Larson and Frey 1992; P. Larson 1994, 1997, 2008b; Powell 1998; Larson and Donnan 2002; Carpenter and Smith 2001; Mallon 2017; Hone and Mallon 2017). Among non-avian theropods at least, *Tyrannosaurus* presents one of our best chances of detecting sexual differences because we have a relatively large number of adult, sexually mature *T. rex* specimens.

When variation in *T. rex* specimens was first explored in the 1990s, several potentially sexually dimorphic features were identified using the then-popular idea of "gracile" and "robust" tyrannosaur morphs (fig. 6.15; chapter 2). Features of pelvic and tail osteology that might permit passage of eggs or anchor male reproductive anatomy seemingly allowed identification of these morphs as either male ("gracile" specimens) or female ("robust"). Specifically, Carpenter (1990, 1999) postulated that the backward-projecting pelvic bone, the ischium, was angled at 37° from the horizontal axis of the pelvis in robust "female" individuals compared to 35° in gracile "males." Larson (1994) also proposed that "gracile" specimens had narrower pelvic regions

and that a neighboring set of bones, the tail-suspended chevrons, were both more prominent and located further forward than those of "robust" morphs (P. Larson 1994, 1997). The latter was bolstered by dissections of alligators that purportedly showed a similar sexually distinctive chevron condition (Larson and Frey 1992; P. Larson 1994), promisingly grounding sexual identification of theropod skeletons in modern animals. With these distinctions seemingly established, other skeletal features could be assigned sexual significance (Carpenter 1990; P. Larson 1994, 1997, 2008b). These included the presence and size of cranial ornaments (Molnar 1991; P. Larson 1994) and the shape of *T. rex* humeri, where proposed females bore somewhat bowed upper arm bones, but males had straight humeral shafts (fig. 2.17; P. Larson 1994; Carpenter and Smith 2001).

As neat as it would be to be able to tell male from female *T. rex* apart from simply looking at their skulls, arms, or pelvic regions, none of these ideas have withstood further scrutiny. We mentioned in chapter 2 that the increase of *T. rex* sample sizes over the last twenty years has removed any meaningful distinction between "robust" and "gracile" morphologies, simultaneously dissolving any interpretations of these representing

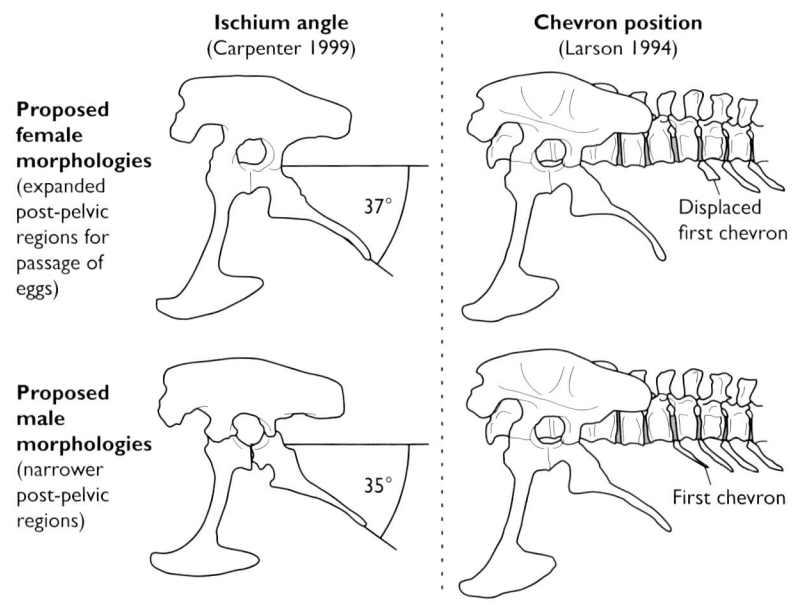

Ischium angle
(Carpenter 1999)

Chevron position
(Larson 1994)

Proposed female morphologies
(expanded post-pelvic regions for passage of eggs)

37°

Displaced first chevron

Proposed male morphologies
(narrower post-pelvic regions)

35°

First chevron

FIGURE 6.15

Proposed sexually dimorphic features in the pelvic region of *T. rex*. The utility of these features for distinguishing king tyrant sexes has proved doubtful, and they are generally not used by modern theropod experts. Images adapted from Carpenter (1999) and Larson (1994).

distinct sexes. Only one relatively recent study (P. Larson 2008b) has attempted to quantify sexual dimorphism in *T. rex*, and, when these data were subjected to statistical analysis, no evidence of sexually distinct body types were found (Mallon 2017). Nor, for that matter, can detailed assessments of *T. rex* growth show a point where skull morphology splits into two forms, as would be expected if king tyrant cranial ornament were dimorphic (Carr 2020). Even the seemingly compelling chevron data has ultimately proven doubtful, as crocodylian tails show a lot of variation in proximal chevron development, but only between individuals, not sexes. Errors in dissection or data recording may have contributed to overinterpretation of this trait in alligators, while a wider examination of tyrannosaurid tails suggests that individual variation, as well as incomplete preservation, factors into their chevron positions (Brochu 2003; Erickson et al. 2005).

Lest it be thought that we are building to a conclusion that *T. rex* was *not* dimorphic, however, a more nuanced assessment is required: we presently lack the data to tell if males were distinct from females. This issue applies to investigations of sexual dimorphism in all dinosaurs, in fact, and the historic arc of sexual determination in *T. rex*—initial promise followed by an eventual dashing of hopes by more careful and quantified analyses—is typical of virtually all studies into dinosaur sexual distinction. The usual expectation that expanding fossil datasets may ultimately provide us with the sample sizes we need to tease out more meaningful insights may, for once, not solve these problems, either. Studies of bone measurements taken from modern sexually dimorphic crocodylians and birds show that their sexual distinctions are masked by growth and individual variation, even when differences between males and females are very marked (Godfrey et al. 1993; Bonnan et al. 2008; Mallon 2017). Simply measuring *T. rex* bones and expecting male and female morphs to cluster in our graphs and data tables is, in light of this, naive.

We should not, however, become totally despondent about our capacity to identify dinosaur sexes. Medullary bone, the distinctive bone type held within dinosaur long bones that was discussed in chapter 4, is exclusively linked to egg production in modern birds and likely identifies gravid females when discovered in fossil reptiles. The identification of such tissues in the leg bones of two king tyrants strongly suggests that they were females (fig. 4.9; Schweitzer et al. 2005b, 2016; Woodward et al. 2020), and they remain, to date, the only relatively confident sex determinations made for any *T. rex*. While these specimens bear no external features indicative of dimorphism, further discoveries of *T. rex* with medullary bone may help us recognize sex-related distinctions in decades to come. Although we struggle to find evidence of dimorphism in bone measurement datasets of living animals, applying some a priori knowledge of specimen sex can help to tease dimorphs apart in such studies by sharpening our "search image" of sex-based traits (Mallon 2017). Thus, our quest to fathom dimorphism in *T. rex* (and other dinosaurs) is not merely about finding more fossils, but finding examples with medullary bone. This will take time because, as noted in chapter 2, medullary bone is both challenging to identify and generally rare, reflecting its absence from female dinosaur bones outside of reproductive cycles. It is probably a job for future generations of paleontologists, with much greater numbers of female *T. rex* bones, to figure out how female tyrants were distinguished from males, if they were at all.

MAKING BABY TYRANTS

Whatever differences existed between male and female king tyrants, producing the next generation required copulation and egg laying. The unusual anatomy and gigantic sizes of dinosaurs has led to a body of speculations about their copulatory behavior (e.g., Halstead 1975; Fritz 1988; Carpenter 1999; Isles 2009). Unsurprisingly, *Tyrannosaurus* is frequently considered in these, even though their body shapes make them far from the most mechanically perplexing species when pondering dinosaur coitus. A slew of artwork and even museum-mounted skeletons (fig. 6.16) present ideas

FIGURE 6.16

Taking the topic of "mounting *Tyrannosaurus* skeletons" to a new level, this display of copulating *T. rex* at the Museo del Jurásico de Asturias (MUJA), Spain, is the ultimate manifestation of our curiosity into dinosaur sex lives. Note that the position of the rearing "male" in this display is implausible, the femora being retracted back well beyond the anthropological limits of the pelvis, and the knee being overextended into a straightened hindlimb. A lower, stooping pose seems more likely (see text and Figure 6.17A).

on how *Tyrannosaurus* mated, many of which violate basic aspects of theropod functional anatomy. Popular culture tends to imply that dinosaur mating is a total enigma to scientists, but understanding of their skeletal function and the reproductive behavior of extant reptiles allow for some sensible insights.

Unless they differed from all living reptiles and birds, dinosaur genitals were contained within a cloaca: a single opening that deposited eggs, excreted waste, and transferred sperm between mating dinosaurs. Reptile cloacae are located behind the pelvis at the base of the tail and were likely relatively small in dinosaurs, about the same diameters as their eggs (Fritz 1988). The cloacae of male lizards, snakes, turtles, crocodylians, and many birds contain a wholly retractable penis-like phallus. Male dinosaurs almost certainly possessed these as well, their absence in some birds reflecting secondary loss of this ancestral feature (Carpenter 1999; Isles

2009). Lizard and snake phalluses are paired structures that alternately enter the female cloaca to deliver sperm via grooves along their top surfaces, but those of archosaurs are unpaired, transmitting sperm along a single channel (Carpenter 1999; Isles 2009). The likelihood of a grooved phallus in male dinosaurs suggests that they probably didn't mate using "cloacal kissing," touching inverted cloacae in the manner of most birds.

Having a basic model of dinosaur cloacal anatomy helps to constrain their mating postures. We can identify, for instance, that the location of the cloaca under a wide, heavy tail, and the limited twisting capabilities of dinosaur bodies, means they could not mate like lizards and snakes, which are able to bring their cloacae close by rotating their hips and tails into position (Frey 1995). We see the impact of these stiffened torsos in other archosaurs, crocodylians, which almost exclusively mate while floating, the male depressing their tails below the female far enough for their phalluses to reach the female's cloaca (Isles 2009). Regardless of trunk rigidity, both lizard and crocodylian males close the distance to female cloacas by placing one leg over the back of their partners, suggesting this might be a strategy used by all reptiles with long, powerful tails (Isles 2009).

Factoring these observations into large theropod anatomy allows us to predict how they might have mated (Carpenter 1999; Isles 2009). Contrary to some illustrations (e.g., Fritz 1988), king tyrant copulation probably did not involve animals clambering over one another with limbs and tails projecting at awkward angles. It probably took place low to the ground, the female leaning forward to elevate her cloaca and flexing her tail sideways. In this pose, a stooping male could half-step over her, his tail tip resting on the ground to bring the tail base into proximity with his partner's, leaving the phallus only a short distance to reach the cloaca and deliver sperm (fig. 6.17A; Carpenter 1999; Isles 2009). This pose would be relatively stable, an important factor if *Tyrannosaurus* reproductive acts were long-winded (copulation can last for hours in some reptiles) and in alleviating toppling risks. Further stability may have come from males grasping the females with

their forelimbs (Osborn 1906) or jaws, both of which would be viable for a male hunched over a female. Biting the backs or necks of females is a common practice among mating animals, and is a possible explanation for bite marks on the back of some *T. rex* skulls (e.g., the parietal bone of "Stan"; Figure 6.5A).

Successful *T. rex* copulation led to the next stage of reproduction: the formation of eggs. Our record of dinosaur eggs, egg clutches, eggshells, and nests is impressive (e.g., Carpenter 1999; Norell et al. 2020) and some diversity in egg morphology and nesting practices are apparent from these data. A few influential discoveries and interpretations have crafted stereotypes about how dinosaurs created nests and reared their young, however, such that a lot of this diversity, and our state of knowledge about dinosaur eggs and nests in general, is misconstrued in popular culture. Rimmed nests scraped into the ground with partially buried eggs, perhaps with a pile of rotting vegetation or a feathered dinosaur perched atop to keep them warm, are common components of modern dinosaur media, as are hatchlings being looked after by their parents. The dinosaur egg record contains many biases and issues that complicate this picture, however (Carpenter 1999): most of our eggs stem from Cretaceous dinosaurs, and we are only rarely able to assign eggs to certain clades or species. Genuine fossil nests, rather than isolated clutches of eggs or eggshell fragments, are also very rare. Accordingly, dinosaur nest structures and egg treatment (for instance, their degree of aerial exposure, their burial under sediment or vegetation, their arrangements in nesting sites) and related topics, such as parental care, are controversial subjects among researchers.

Uncertainty surrounds dinosaur nesting habits because we are still learning fundamental truths about dinosaur eggs, some details of which challenge previously held, long-term assumptions. For instance, dinosaur eggs are now considered so porous that most or all probably required entombment in vegetation mounds or sediment to prevent desiccation (e.g., Deeming 2006; Tanaka et al. 2015; note that some authors prefer a partially exposed egg model for some

FIGURE 6.17

Making baby king tyrants. (A) *Tyrannosaurus* copulation was probably not as complex as some sources suggest, likely involving poses not dissimilar to those used by living reptiles; (B) egg clutches laid by large predatory theropods tentatively imply that they were buried in sediment rather than in open nests or piles of vegetation; (C) speculative king tyrant egg shapes based on the predicted sizes of hatchlings (see Chapter 4) and the eggs of other large theropods: *Prismatoolithus* (round shape) and *Macroelongatoolithus* (oval). Note the scale bar: neither egg is particularly large compared to the size of even a young adult *T. rex*; (D) hatching day for baby king tyrants, their first task perhaps being to dig themselves free from their sediment scrape.

bird-like dinosaurs: see, for example; Varricchio et al. 1999; Mainwaring et al. 2023). Even more dramatically, it seems that dinosaur eggshell integrity was variable, with the first dinosaurs producing thin-shelled soft eggs, like those of lizards and turtles, rather than hard eggs, like those of crocodylians and birds (Norell et al. 2020, though also see Legendre et al. 2020). Ornithischians, sauropodomorphs, and theropods apparently developed hard-shelled eggs independently of one another, tremendously complicating our understanding of their reproductive evolution. Now more than ever, it is challenging to make any sweeping statements about how dinosaurs, as a group, approached egg production and nesting.

Tyrannosaurids, meanwhile, have stubbornly refused to give any significant insight into their eggs, nests, or embryos, despite the general abundance of their adult bones. Why they remain so elusive is genuinely puzzling. Recent discoveries of the first perinatal (perhaps one year old or less) tyrannosaurid bones in Campanian/Maastrichtian deposits of western North America are an important milestone in this quest for data, but only deepen the mystery in several respects (Funston et al. 2021). These tiny tyrannosaur fossils stem from the same deposits that yield some of the best hadrosaur and troodontid egg and embryo material from North America, suggesting that tyrant neonates were present in the same far inland, relatively arid environments as these nesting dinosaurs. And yet, while tyrannosaur eggs and young juveniles were apparently present in the right sedimentary basins for their fossils to preserve, they remain extremely rare. As large predators, tyrannosaur populations were probably lower than those of smaller carnivores and herbivores, so some degree of rarity is expected for their eggs and embryos, but their near total absence suggests additional complicating factors. Did tyrannosaurs lay their eggs in upland areas that do not generally enter the geological record? Were their eggs or nests easily destroyed? Did they lay relatively few eggs? Further interrogation of the fossil beds that have provided our first perinatal tyrant bones may yet give some clues on these matters.

Without any tyrannosaur-specific egg or nest data to guide us, we can only make broad predictions about what *T. rex* eggs and nests were like. They likely laid hard eggs with well-mineralized shells, rather than soft, leathery ones, because the development of hard-shelled eggs seems to be ancestral to tetanuran theropods (Norell et al. 2020; Funston et al. 2021). Their eggs were likely much smaller than we might intuitively assume, as they are for dinosaurs in general (Ruxton et al. 2014). Among large predatory theropods, eggs tentatively attributed to the 8–9 m long *Allosaurus* are a mere 150 by 80 mm; those of the midsized *Lourinhanosaurus* were similar, 130 by 94 mm (Carpenter 1999; Deeming 2006); and badly crushed eggs attributed to the 11 m long *Torvosaurus* were probably not significantly bigger (Araújo et al. 2013). The largest non-avian theropod eggs, laid by gigantic oviraptorosaurs, were 400–450 mm by 150 mm (Pu et al. 2017). While king tyrant embryos would easily be accommodated within eggs of these dimensions (Funston et al. 2021), such (relatively) large eggs contrast with the smaller eggs assigned to other large tetanuran predators, and it remains to be seen which egg type tyrannosauroids used. Possible evidence that *T. rex* eggs were comparatively small are their coprolites, which suggest a narrow cloaca from their 15 cm width (chapter 3). Efforts to crudely estimate egg size by scrunching a hypothetical, meter-long (Russell 1970; Currie 2003b; Funston et al. 2021) tyrannosaur skeleton into the oval or circular shapes typifying theropod eggs suggest that *T. rex* eggs could have been 120–200 mm wide and 300–350 mm long (fig. 6.17C); large for a theropod, but still remarkably small for an elephant-sized animal.

Along with egg dimensions, it is difficult to predict how many eggs *T rex* laid. Some large theropod egg clusters contain twenty-six or even thirty-four eggs (Carpenter 1999), and a correlation between body size and fecundity in living reptiles (Carpenter 1999) implies that *Tyrannosaurus* clutches might have been of similar numbers. Dinosaur egg clutches were never huge, however, with total clutch masses often being far lower, proportionate to body size, than those of crocodylians and birds (Ruxton et al. 2014). Despite their large body size, *T. rex* probably laid a few dozen eggs in their nests at most, not hundreds.

To keep their eggs warm, prevent desiccation, and avert predation, *T. rex* eggs were probably covered or buried (fig. 6.17B, D). Eggs possibly laid by *Allosaurus*, as well as more confidently identified examples from the mysterious theropod *Lourinhanosaurus* and *Torvosaurus* (Carrano et al. 2013; Araújo et al. 2013; Tanaka et al. 2015), have highly porous eggshells adapted for water conductance in low-oxygen, high-carbon-dioxide settings, and would probably have desiccated rapidly if exposed to open air (Deeming 2006; Araújo et al. 2013; Tanaka et al. 2015). Their survival likely depended on being encased in extremely humid conditions provided by rotting vegetation or damp ground. It's not clear from fossils, alas, which strategy most theropods used, but details of a seemingly undisturbed *Torvosaurus* egg clutch suggest they were deposited in a hole, recalling the nests of sea turtles (Araújo et al. 2013). Many sauropod egg sites also suggest burial in sediment (Dhiman et al. 2023 and references therein), and this may have been a more common strategy for dinosaur nesting than the stereotypical rimmed nest. We can only guess at *T. rex* egg development times, but large dinosaur eggs are estimated to have incubated for several months before hatching (Ruxton et al. 2014).

WERE KING TYRANTS ATTENTIVE PARENTS?

With no eggs, no embryos, no substantial remains of very young animals, and no unambiguous associations between adult and juvenile tyrant fossils, we have limited insights into *Tyrannosaurus* parenting. And yet, this topic is another popular, often romanticized aspect of dinosaur media, appearing often in novels, films, and documentaries. From the *Jurassic Park* sequel novel and movie *The Lost World* (1997) to documentaries like *Walking with Dinosaurs* and *Prehistoric Planet*, we often show *T. rex* as attentive parents, guiding their off-

spring through at least the early years of their lives. The likelihood of these scenarios is uncertain as our data on tyrant parenting, and perhaps dinosaur parenting in general, is not extensive.

One element we have some insight into concerns the independence of dinosaur offspring. Species represented by growth series that include juvenile stages can reveal whether young individuals were altricial (that is, entering the world in a relatively helpless state and relying on their parents for all aspects of their survival) or precocial (starting life with sufficiently developed anatomy to survive without, or with only very little, parental assistance). Altriciality is rarer among vertebrates than precociality (Carrier 1996; Dial and Carrier 2012) and can be inferred when fossil juveniles have short, functionally limited limbs that grow rapidly once a certain developmental stage is reached. Precocial species demonstrate the opposite: their limbs are long and well-developed from the moment they enter the world, and they "grow into" these oversize extremities as they get older. Scaling equations and rare fossils show that juvenile tyrannosaurids had long legs and arms compared to adults, features that better match precociality than altriciality (fig. 4.4; Russell 1970; Currie 2003b; Williams et al. 2010; Funston et al. 2021). Furthermore, the presence of characteristic tyrannosaurid dentition (cutting maxillary teeth and "D"-shaped premaxillary teeth) and expanded housing for jaw musculature in young juveniles points to key predatory adaptations existing throughout *T. rex* lives (Carr and Williamson 2004; Funston et al. 2021). These data suggest that young tyrants were not helpless and reliant on their parents for basic survival. They were, at least, independently mobile and had the right anatomy to catch their own food.

The possibility that king tyrants exhibited some basic parenting is suggested by living archosaurs, however. Both birds and crocodylians look after their precocial offspring to some degree, with crocodylians (mainly mothers) guarding their nests and protecting their newly hatched progeny for their first year (Grigg and Kirshner 2015). Our most ancient modern bird lineages, the ratites, gamebirds, and waterfowl, exhibit similar behavior (Varricchio et al. 2008). The recovery of maniraptoran dinosaurs sat atop their nests, probably in guarding rather than brooding postures, given the possible necessity of egg burial to prevent desiccation (Deeming 2006), suggests that parenting instincts were present in some non-avian dinosaurs. Properties of their nests and anatomy might indicate paternal-dominant care (Varricchio et al. 2008). Making sweeping comments about parenting strategies in fossil groups is unwise because they are so varied across living animals, but these consistent indications of nest guarding and hatchling protection in living and extinct archosaurs predicts similar behavior for *Tyrannosaurus*. Conversely, evidence that large king tyrants occasionally ate smaller *Tyrannosaurus* suggests that, as with crocodylians and lizards, not all adults were friendly toward minors (Mclain et al. 2018).

Given all the above, what might we imagine life was like for newly hatched *Tyrannosaurus*? If parental care was largely protective, they might have been fairly independent animals capable of finding and killing their own food (fig. 6.18). Hatching at a meter or so in length, and already equipped with robust teeth, powerful bites, and the longest arms they would ever possess relative to body size (Currie 2003b; Williams et al. 2010), baby tyrants may have focused on taking vertebrate prey from their earliest days: perhaps small mammals, reptiles, and other baby dinosaurs, supplemented by larger invertebrates, eggs and carrion (fig. 6.18A). Fueling an active lifestyle and their accelerated growth rate

FIGURE 6.18

The three major pillars of life as a small juvenile king tyrant: (A) Eat plenty of nutritious food to fuel rapid growth; (B) use those long legs to pursue prey; (C) try to avoid becoming a meal for something else, such as the large lizard *Palaeosaniwa canadensis*. Most hatchling *T. rex* probably ended up with a variant on option "C."

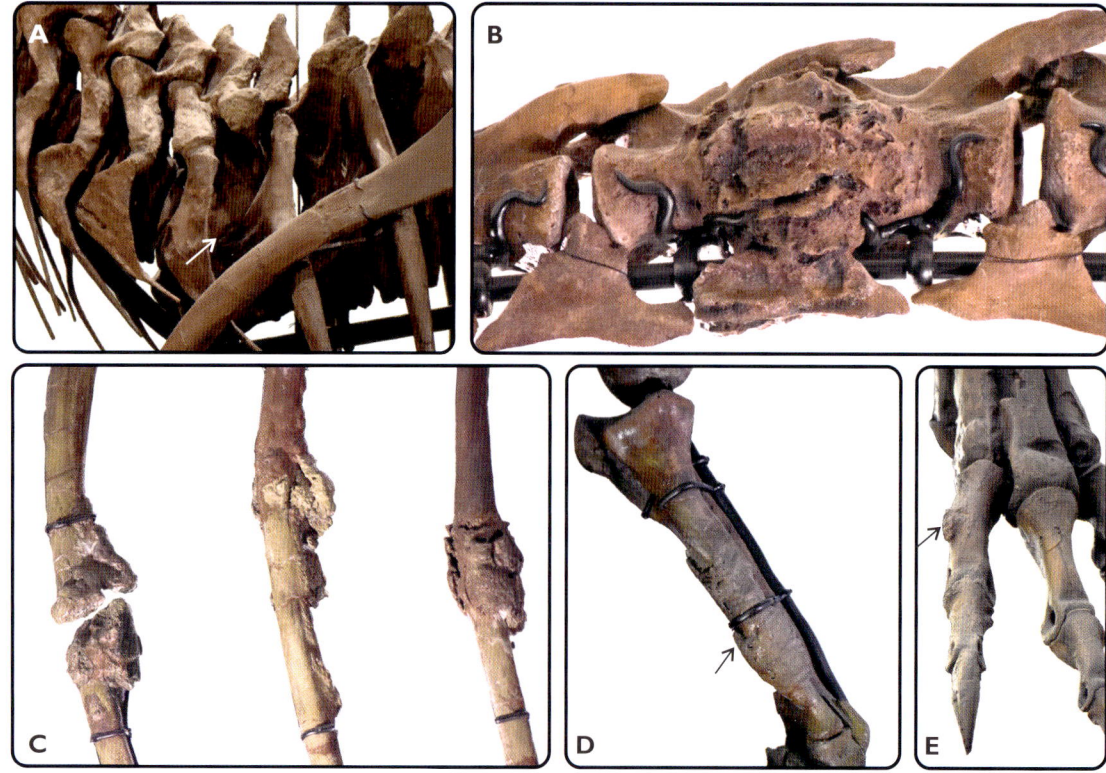

FIGURE 6.19

Perhaps the most terrifying sight for a young king tyrant: the sinister outline of a stalking azhdarchid pterosaur. Fast on the ground as well as in the air, the energy-intensive flighted lifestyles of pterosaurs probably made them perpetually hungry, and the long limbs and necks of azhdarchids gave them an elevated perspective for seeking small prey. A baby tyrant this close to a foraging azhdarchid, especially in open ground, was probably in trouble.

must have made baby king tyrants ravenous for food, and their long legs surely helped chase it down (fig. 6.18B). Their speed was also surely their main strategy to avoid danger; their lightweight, gangly bodies being far from the dinosaur wrestling frames of their parents. Most baby *T. rex* probably lived short lives with violent ends, often ending up as prey items for other predators (figs. 6.18C, 6.19). Assuming modern laws of population dynamics and juvenile mortality, of the dozens or hundreds of eggs produced by a pair of king tyrants in their lifetimes, only one or two would survive to adulthood.

INJURIES, DISEASES, AND INFECTIONS

Transitioning from juvenility to maturity evidently took its toll on *T. rex* individuals. Hallmarks of injuries and disease, known as "pathologies," are well represented on their skeletons in the form of damaged, broken, and infected bones (figs. 6.5, 6.20). Many of these occur on the famous "Sue" specimen, thanks to which we have a particularly in-depth glimpse into the ailments of a large predatory dinosaur (Rothschild and Molnar 2008). This is not to imply that *T. rex* was unusually clumsy or prone to medical complaints; most *T. rex* specimens have few or no pathologies, and "Sue", perhaps on account of their advanced age and the "snowballing" effects of one

pathology leading to others, was exceptionally compromised compared to other *T. rex*. This trend is replicated in other well-known theropod species where inventories of mostly "healthy" fossils skeletons are contrasted against a few highly pathological individuals (e.g., Molnar 2001; Rothschild and Tanke 2005; Foth et al. 2015; Gutherz et al. 2020).

Ranging from minor ailments to serious injury, the pathological bones of *T. rex* are sometimes fully healed, evidencing successful recovery from whatever incidents or diseases caused them. Many are not, however, inviting speculation about the condition of some king tyrants during their final days. Particularly in the popular press, *Tyrannosaurus* pathologies are often reflexively attributed to battles with other species or conspecifics. Some injuries, such as those related to face biting, almost certainly resulted from such acts (fig. 6.5), but the causes of their injured or diseased bones are mostly unidentifiable. Paleontologists with knowledge of medical trauma can offer valuable insights into fossil pathologies, but broken, infected, or deformed bones often provide little causal information on their own. Accordingly, some popular interpretations of *T. rex* injuries, which include attributing smashed limbs to blows from ankylosaur tails, or fused tail vertebrae to mating injuries, are speculations that have not been borne out under further study (e.g., Psihoyos 1994; P. Larson 1997, 2001; Bastiaans 2016 vs. Molnar 2001; Brochu 2003).

FIGURE 6.20

Tyrannosaurus injuries and pathologies. (A) Fused vertebrae at the neck-torso junction (arrowed), bones that may have never separated during embryology; (B) fused caudal vertebrae reflecting injury or infection (or both); (C) broken and partly healed dorsal ribs; (D) a fractured or infected fibula (swollen region indicated by arrow); (E) bone growth on a pedal phalanx, more likely resulting from a blood infection rather than direct injury.

Among the most frequent injuries on *T. rex* skeletons are punctures and gouges attributed to the teeth and claws of conspecifics, the likes of which we discussed earlier in this chapter. As with bites delivered to *T. rex* prey, injuries to living king tyrants, rather than cannibalism of dying or dead individuals, are evidenced by healed bone textures. Some of these, such as the lesions on the mandibles of "Sue" and several other *T. rex* specimens (fig. 6.5E), are controversial, interpreted by some as bite marks (e.g., P. Larson 1997, 2001; Rothschild and Molnar 2008; Rothschild et al. 2022) but abscesses from microbial infections by others (e.g., Brochu 2003; Rega and Brochu 2003; Wolff et al. 2009). Proponents of the latter argue that the positioning of the wounds at the back of the jaw is unusual for a facial bite, and that the distribution of holes does not correspond to rows of teeth. Similarities have instead been drawn between these openings and those found in birds infected by *Trichomonosis*, a fatal protozoan that erodes the bones of the lower jaw and necrotizes soft tissues of the throat (Wolff et al. 2009). In this hypothesis, face biting could still contribute to the formation of these pathologies because face-to-face contact can spread infections between individuals, much like the face-biting of Tasmanian devils transmits deadly oral cancers (Wolff et al. 2009).

Elsewhere on king tyrant bodies, broken teeth and bones are found in several specimens (fig. 6.20C–D). Their teeth sometimes fused together, became heavily worn, or adopted mangled, warped shapes (Molnar 2001; Brochu 2003), presumably resulting from injuries to the jaw causing abnormal growth. Fractures, often marked by swelling of the surrounding bone, have been documented in ribs (Brochu 2003; Rothschild and Molnar 2008; Bastiaans 2016) and sometimes in hindlimb bones. Early reports of a fractured rib in the "Sue" specimen mentioned a tooth embedded within the pathology, seemingly providing evidence of a *T. rex* biting another's flank (P. Psihoyos 1994; P. Larson 2001; Rothschild and Molnar 2008). This "tooth," however, was later interpreted as broken chunks of rib, leaving the cause of this injury uncertain (Brochu

2003). The possible fracture and infection of "Sue's" left fibula (fig. 6.20D) has proven of particular interest, with P. Larson (1997, 2001) speculating that this injury would have crippled the animal sufficiently that it only survived thanks to conspecifics providing food. Brochu (2003) disagreed, noting that theropods bore most of their weight on the larger lower hindlimb bone, the tibia, and doubted any scenario where "Sue" was rendered immobile.

Pathologies occur among king tyrant vertebrae, too. Some *T. rex* specimens have vertebrae that knitted together, sometimes from unknown causes (fig. 6.20B; Brochu 2003), but other times because they probably failed to fully separate during embryonic development (fig. 6.20A; Rothschild and Molnar 2008). Perhaps the most dramatic king tyrant vertebral pathology has been found in the tail base of the "Wyrex" specimen, where much of the tail was apparently lost following severe trauma (Anné et al. 2023b). Popular sources—of course—attribute this to a severing bite from another *T. rex*, but this claim requires evaluation in peer-reviewed literature at the time of writing. Whatever the cause, this pathology was not immediately fatal as the affected region shows signs of bony reaction and remodeling. There is no sign that the tail was re-growing, however, as occurs to varying extents among living reptiles with lost tails (Anné et al. 2023b).

Infections are also common in *T. rex* fossils (Rothschild and Molnar 2008). Represented by irregular swellings and excavations of bone surfaces, they occur across *Tyrannosaurus* bodies, from their skulls and humeri to fibulae and even their toes (fig. 6.20E; Brochu 2003; Vittore and Henderson 2013; Bastiaans 2016). Some of these are suspected to have resulted from infections or necrosis of neighboring soft tissues spreading into bones. A large pit in the right humerus of "Sue" is one such pathology of suspected infected origin, and it is associated with a number of other injuries to the shoulder and adjacent ribs. This may have been a consequence of major trauma to this area resulting in tendon avulsion: a portion of bone being dislocated by an overstressed tendon (Brochu 2003). Similar avulsive acts may explain infections on the tail of "Trix" (Bastiaans 2016). Conversely, the swollen, infected toe bone of "Jane" (fig. 6.20E) is an isolated injury without associated trauma. An attack on its foot by another animal may have been the cause, but infections of this nature are often the result of blood-borne bacterial infections that can be obtained through even minor cuts or abrasions (Vittore and Henderson 2013). Though surely causing pain and lameness, these infected bones probably did not directly kill their tyrannosaurs, at least initially. Left untreated, bone infections can spread and eventually prove fatal, however (Vittore and Henderson 2013).

Along with bearing various injuries and infections, "Sue" and "Trix" possess pathologies likely indicative of advanced age (Rothschild et al. 1997; Rothschild and Molner 2008; Bastiaans 2016). In "Sue", this includes dorsal vertebrae that exhibit evidence of Spondylosis deformans: degradation of the joints between vertebrae leading to the formation of bony spurs between vertebral contacts. "Trix" shows similar, as yet undiagnosed, vertebral degradation (Bastiaans 2016). "Sue" also suffered from gout, a form of arthritis. Revealed by characteristic erosions on the right hand (Rothschild et al. 1997), gout causes pain and swelling around joints and develops from concentrations of uric acid in the bloodstream, these often resulting from dehydration, kidney failure, or a diet heavy in purine (an organic compound concentrated in animal flesh and internal organs; Rothschild et al. 1997). "Sue's" gout was not an especially extreme case, but it, and other degradations of skeletal health, provide insight into the lives of older *T. rex*. Apparently, not even *Tyrannosaurus* was immune to the ravages of advancing age.

LIVE FAST, DIE YOUNG?

As evidenced by their battered bodies, *T. rex* lives could be tough: an existence of tackling large prey, fighting other *T. rex*, avoiding cannibal conspecifics, and battling disease. Our relatively high numbers of tyrannosaurid specimens have allowed for estimates of mortality rates for different growth stages (fig. 6.21; Erickson et al. 2006), which in turn reveal the chances of king tyrants making it to old age. Using data from the 22 specimen-strong *Albertosaurus* bonebed mentioned earlier, Gregory Erickson and colleagues (2006) were able to calculate tyrannosaurid population structure and figure out which percentages constituted juveniles, subadults, and adults. These were then compared to "composite" populations derived from museum collections of tyrant dinosaur bones, to which our large collections of *T. rex* were well suited. For all taxa, the survivorship of young juveniles was difficult to calculate because of the scarcity of their fossils, so modern mortality rates among reptiles, birds, and mammals were used: these predict that 60 percent of tyrannosaurids died within their first few years of life (Erickson et al. 2006). Survivorship improved markedly thereafter, with both the bonebed data and artificial tyrant populations showing more animals surviving than dying until midlife. After this, chances of death continually increased with age, forming a slowly plunging curve on survivability plots (fig. 6.21).

Such mortality patterns are typical for large, fast-growing animals, such as bigger-bodied birds and mammals, and suggest comparable life histories for *T. rex* (Erickson et al. 2006). The most survivable period of tyrannosaur existence seems to reflect a time in which young animals had outgrown predation risk from smaller predators but had not yet reached sexual maturity, allowing them to focus on staying well fed and healthy while, presumably, avoiding encounters with larger tyrannosaurs. This lessened mortality rate may have contributed to the rarity of juvenile and early-subadult tyrannosaur fossils, their greater survivability limiting the number of remains that could enter the fossil record (Erickon et al. 2006). A midlife rise in mortality rate correlates with the presumed onset of adolescence, presumably reflecting sexual behavior promoting dangerous interactions with conspecifics and placing greater physiological strains on females. Finally, in later life, advancing years probably lessened their abilities to recover from injuries and heightened their susceptibility to disease. It seems that only 2 percent of *Tyrannosaurus* made it

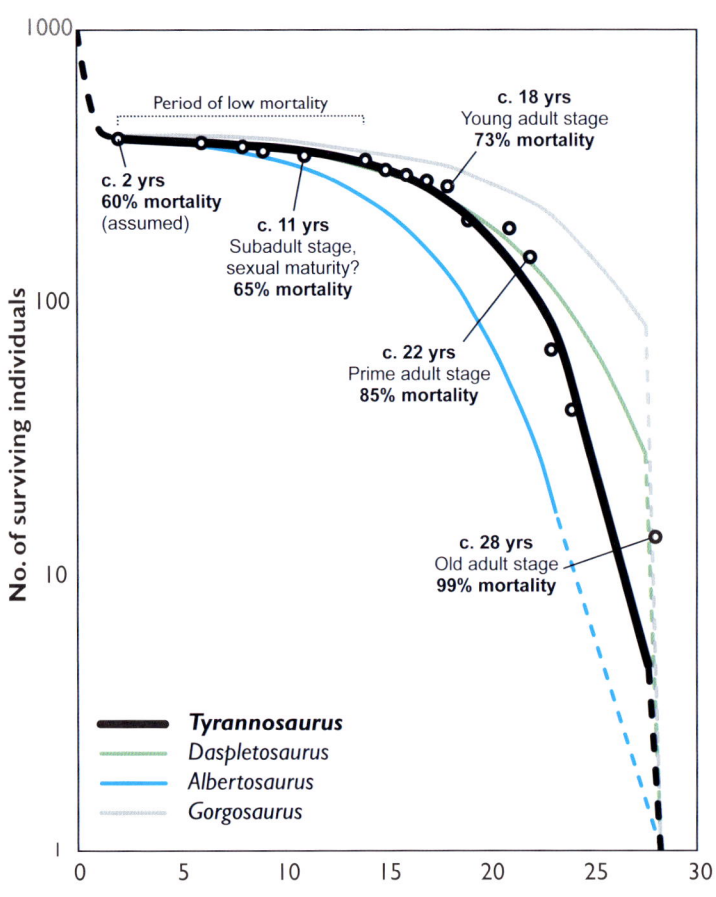

FIGURE 6.21

Estimated survivorship rates for tyrannosaurids, calculated from the age and size distribution of associated tyrannosaurid specimens (*Albertosaurus*) and composite "populations" for *Tyrannosaurus*, *Daspletosaurus*, and *Gorgosaurus*. After Erickson et al. (2006).

to their greatest size and oldest ages, of around thirty years (Erickson et al. 2004, 2006; Erickson 2014; Cullen et al. 2020).

Dying close to a thirtieth birthday might seem very young, but this seems to have been long-lived for a tyrannosaurid. Estimates of dinosaur ages show that most tyrant dinosaur species struggled to reach their late twenties (Erickson et al. 2004), contrasting markedly with the longevity of today's multi-tonne animals. Hippopotamus and elephants can live for more than sixty years, while large whales may live for well over a century (Szekely et al. 2015). Responding to this, Paul (2008) argued that *T. rex* and their relatives led "such dangerous lives that they had to reach sexual maturity and breed rapidly before they died" (Paul 2008, p. 308). For this explanation to be valid, however, it has to be demonstrated that *Tyrannosaurus* and their kin lived

truly short lives compared to other Mesozoic dinosaurs. Comparing like-for-like here is critical because longevity is sensitive to different aspects of animal reproduction and locomotion. The long lives of large mammals, for instance, are necessary to facilitate reproductive cycles that can last years, the likes of which did not trouble large egg-laying dinosaurs. Similarly, flighted animals also seem to live longer lives than terrestrial species, perhaps because flight capabilities reduce predation risk (Szekely et al. 2015).

Efforts to compare tyrannosaur longevity against other multi-tonne Mesozoic dinosaur species are hampered by our small datasets of dinosaur ages, but early signals suggest that dinosaurs had variable lifespans. LAG data shows that our oldest c. 1.5 tonne hadrosaurs reached maturity after thirteen years or so, and 2.5 tonnne horned ceratopsids may have survived until around eighteen (Erickson 2014). Sauropods seem to have been relatively long-lived, although conclusions about their longevity vary. Some model gigantic sauropods like *Alamosaurus* as taking at least forty-five years to reach full size, but others posit longer life spans (e.g., seventy years for *Apatosaurus* to attain maturity) (Lehman and Woodward 2008; Erickson 2014). Among theropods, immature specimens of the large (c. 1.5 tonne) theropod *Allosaurus* died at eighteen years old, suggesting that this genus gained maturity at twenty-two to twenty-eight years old (Bybee et al. 2006). Other allosauroids seem to have lived much longer, with a slow-growing, unnamed carcharodontosaurid reaching thirty-nine to fifty-three years (Cullen et al. 2020). These limited comparisons suggest that tyrannosaurid lifespans were not unusual for Mesozoic dinosaurs, and, indeed, roughly thirty-year-old *T. rex* are among the oldest known theropod specimens. Whether they were long- or short-lived, pathologies indicative of geriatric life stages, such as arthritis and degrading vertebral cartilage, suggest that these individuals were not cut down in their prime, but had lived to an age where their bodies were past peak health.

Larger sampling of dinosaur ages may ultimately paint a different picture of *T. rex* longevity in future, but evidence that they did not live on a perpetual knife-edge between survival and death is provided by data suggesting that tyrannosaurs enjoyed relatively good chances of day-to-day survival for at least part of their lives, and that their projected mortality rates were not unusual compared to large birds and mammals (Erickson et al. 2006). Perhaps we do not need to view tyrannosaurs as having especially short, dangerous existences and that, contrary to their popular (and sometimes scientific) reputation, *Tyrannosaurus* possessed typical instincts for self-preservation when confronted with challenging situations. These were put to their ultimate test when king tyrants faced their reckoning at the end of the Cretaceous. Having learned much about how *Tyrannosaurus* lived, it is now time to consider how they met their end.

7

DEATH OF A SPECIES, END OF AN ERA

THE STORY OF the Mesozoic Era is ultimately a tragedy: just as we are beginning to understand and admire the character of king tyrants, it is time to kill them off. *Tyrannosaurus* were among the dinosaurs that experienced the asteroid-induced extinction that drew the curtain on the Age of Reptiles. Controversy has surrounded this event for decades, but careful analysis has allowed us to establish that king tyrants were almost certainly victims of the end-Cretaceous extinction, as well as revealing the horrific fate that befell them and other North American dinosaurs.

APPROACHING THE FINISH LINE

TYRANNOSAURUS IS AN unusual fossil species because, along with other organisms found in the very latest rocks of the Cretaceous, we have some understanding of what caused their extinction. The majority of fossil taxa enter and disappear from the rock record without fanfare, their range across geological time possibly constituting an accurate representation of their true evolutionary span, or possibly not. It is only when species occur immediately before widespread, well-marked disruptive events in the fossil record that we have hopes of identifying the agent of their demise. Many such events are known from the last several hundred million years and a few truly monumental examples, known as "The Big Five," are considered "mass extinctions": planet-wide catastrophes that wiped out huge percentages of contemporary lifeforms.

The end-Cretaceous extinction, often abbreviated as the "K/Pg extinction" (using abbreviations for *Kreide*, the German word for "Cretaceous," and Paleogene, the following geological period), of 66 million years ago ranks among the hardest hitting of these episodes, dooming about 75 percent of species alive at that time. Widely considered to have been caused by a cataclysmic asteroid impact into what is now the Chicxulub region of Mexico (fig. 7.1), the K/Pg extinction marked the end of many great Mesozoic lineages. All dinosaurs except birds, all marine reptiles except turtles and palaeophid snakes, all pterosaurs, and all ammonites were wiped out. Few clades made it through without reductions in diversity. This was the biggest upset to life since the even more dramatic Permian extinction of over 180 million years prior, and resulted in

FIGURE 7.1

The end of the Mesozoic Era and the final days of the non-avian dinosaurs were marked by the collision of a mountain-sized asteroid into what is now the Mexican Yucatán Peninsula. North American dinosaurs would have experienced some of the worst and most immediate effects of this planetary event: nothing on this continent was safe, even king tyrants.

a drastically different course for life in the Cenozoic Era that followed.

Undoubtedly, the K/Pg event ties into the mythology established around *Tyrannosaurus*. Could we invent a more fitting end to an animal built up as the most gigantic and macho of all dinosaurian predators than one where a giant space rock hurtled into the Earth? If *T. rex* had to go, they had to go with a bang—and nothing less than one of the largest "bangs" in the history of our planet would do. As with everything surrounding *T. rex* hype, however, we have to be careful to separate popular mythologizing from paleontological facts. In reality, the nature and causes of the end-Cretaceous extinction remain a sufficiently controversial and contested topic that many details surrounding the fate of *T. rex* and other non-avian dinosaurs are uncertain. The question of whether dinosaurs were ravaged by an extraterrestrial impact is the proverbial tip of the iceberg when it comes to their extinction: debates continue about the health of dinosaur communities in the millions of years preceding their exit from existence, and, based on fossil distribution, some have doubted that non-avian dinosaurs even made it to the very end of the Cretaceous. As just one species among many that went extinct during the K/Pg event, *T. rex* has never been the sole focus of a Cretaceous extinction study. However, in being the sole large theropod of Lancian ecosystems, it is an important taxon to consider when attempting to contextualize the final days of the Mesozoic dinosaurs.

EARLY CONSIDERATIONS: KING TYRANTS AS WORN-OUT, SEXLESS SUPER-FREAKS

Dinosaur extinction has not always been a major focus of paleontologists. The realization that an "Age of Reptiles" preceded the "Age of Mammals" dates back to the 1830s, but nineteenth-century paleontologists were not intrigued by the circumstances that instigated this dramatic faunal turnover. Rather, it was early

twentieth-century researchers that first viewed dinosaur extinction as a worthy research subject, although the earliest contributions to this discussion are more remembered for their naive, unscientific, or downright fanciful notions than their astute insights (see Benton 1990 for a comprehensive review). Some of these ideas are noteworthy for their inversion of perceived dinosaur grandeur, taking properties that were celebrated about the great reptiles for promotional purposes by museums and movies and linking them to their demise. Viewed as one of the largest and last of the predatory dinosaurs, such thoughts cast the qualities that made *T. rex* famous as their Achilles' heel.

Among these first concepts of dinosaurian extinction was "racial senility," a sustained period of relaxed evolutionary pressures resulting in a degradation of their genetic line and the development of exorbitant body forms ill-suited to meet changing conditions (e.g., Woodward 1910; Swinton 1934). This viewed the evolutionary history of the entire dinosaur lineage almost like the life of a single organism, in which a healthy, vigorous youth allows for adaptability against new challenges, a settled and steady midlife permits sustainable existence for a long duration, after which an inevitable decline in health and the onset of senility in later years establishes vulnerabilities against new conditions. Symptoms of lineages that had reached this "senile" stage included "over-elaboration of skeletal structure" (Swinton 1934, p. 180), such as the development of spikes and horns, toothlessness, as well as great size. These applied, of course, to a great many dinosaurs, including the gigantic *Tyrannosaurus*. No catastrophic scenario was required to push Late Cretaceous dinosaurs over the edge in this proposal, as they had become so adaptively indulgent that even minor shifts in environmental conditions brought the curtain down on the whole group.

Not all early twentieth-century authors agreed with the concept of racial senility, but the large sizes of dinosaurs were still identified as agents in their extinction. Nopcsa (1917) attributed their great stature to physical abnormalities in their brains. He speculated that overactive pituitary glands caused gigantism, leading to terrific dinosaurian size, thickened facial bones, and unusually long limbs. Once again, *Tyrannosaurus* fit this diagnosis perfectly. Nopcsa (1917) further linked malfunctioning pituitary glands to a lessened reproductive drive, casting *T. rex* and other large dinosaurs as hormonally unbalanced, sick, and distorted animals with little interest in perpetuating their own existence. Raymond (1939), meanwhile, also thought that dinosaur brains led to their downfall, but laid the blame on their assumed limited intelligence. "Even the Jovian *Tyrannosaurus*, greatest of all the hunters," he wrote, "had a remarkably small braincase" (Raymond 1939, p. 148). "Obviously, animals with such small brains, which showed little improvement during the ages of the Mesozoic, were not particularly well fitted for survival" (Raymond 1939, p. 153).

The physicality of dinosaurs was their downfall in each of these concepts, and, as the greatest embodiments of theropod power and size, *T. rex* were clearly doomed even as they were regarded by other scholars as relatively "advanced" dinosaurs. As might be expected, ideas such as racial senility, malfunctioning pituitary glands, and dimwittedness are no longer considered viable factors in dinosaur extinction; they tell us much more about early twentieth-century attitudes toward prehistoric animals than anything about the last days of the Cretaceous (Benton 1990).

A FRUSTRATED MODERN PERSPECTIVE

The K/Pg extinction was addressed more scientifically from the 1970s onward. Here, scientists began to form testable hypotheses that could explain the widespread decimation of life at the end of the Mesozoic. The most important breakthrough in this interval was the discovery of evidence for a cataclysmic end-Cretaceous asteroid impact (Alvarez et al. 1980) and the identification of a similarly dated asteroid impact site at Chicxulub, on the coast of the Yucatán Peninsula (Hildebrand et al. 1991). The significance of the K/Pg impact event in

the Cretaceous extinction is still debated (see below), but most paleontologists are satisfied that, whatever else was happening at the end of the Mesozoic, a huge asteroid striking the Gulf of Mexico was the killer blow.

Although considering the K/Pg extinction from a more holistic perspective has proven fruitful, the charisma of non-avian dinosaurs means that their disappearance continues to draw particular attention. Today, a few major themes frame this discussion, with perhaps the largest controversy concerning the diversity of the last dinosaur communities. Were they already in decline and merely finished off by the asteroid, or were they in good health and snuffed out in their prime? Related to this, experts have questioned whether the fossil record truly documents dinosaurs persisting until the last moments of the Mesozoic, or whether they died out immediately before the asteroid struck. Both are key questions if we wish to understand the demise of *Tyrannosaurus*, as well as how we might view the king tyrant world. Were *T. rex* parts of struggling, dying ecosystems? Was their status as such a widespread predator across North America reflective of diminishing dinosaur diversity? And did *T. rex* live to witness the cataclysm that ended the Age of Reptiles?

Investigating these matters requires meticulous observations of the rocks and fossils deposited in the Western Interior Basin at the end of the Cretaceous. To assess how dinosaurs like *T. rex* went extinct, we need field sites representing several million years of unbroken, carefully dated, dinosaur-bearing latest Cretaceous sediments that extend right up to the K/Pg boundary. Ideally, these will span not just the Maastrichtian, but also the previous geological stage, the Campanian (82–72 Ma), to give greater context to changes in the last few million years of the Cretaceous. Geologists can identify the end of the Mesozoic (or, technically, the start of the Cenozoic) via a globally distributed sediment bed rich with the metal iridium: a rare element on Earth, but relatively abundant in asteroids. It was the discovery of this layer that first evidenced an extraterrestrial impact at the end of the Mesozoic, and its distribution across the planet makes it a distinctive horizon that marks the Cretaceous-Paleogene transition worldwide. Only western North America, however, is currently recognized as having rocks with a sufficiently long and continuous pre-impact terrestrial fossil record to inform us about the final days of the dinosaurs (Brusatte et al. 2015). This is good news for anyone wishing to understand the extinction of Laramidian dinosaurs like *T. rex*, but means that little is known of what was happening to dinosaurs outside of North America as the Cretaceous drew to a close.

MAASTRICHTIAN LARAMIDIA: A CRETACEOUS SUCCESS STORY, OR MEASURE OF DECAY?

It is, in part, this lack of data on the changes experienced by latest Cretaceous dinosaur communities that has given rise to the long-running discussion about their pre-extinction diversity. Mesozoic dinosaur diversity fluctuated over time (fig. 7.2) and some researchers posit that pre-impact dinosaur communities were losing species richness, perhaps terminally, making them vulnerable to extinction whether an asteroid collision was imminent or not (e.g., Sloan and Rigby 1986). The last few million years of the Mesozoic were fraught with change and challenges, all of which could have put pressure on dinosaur faunas. Global temperatures were cooling, climate-altering mass volcanism was occurring in India (forming the famous Deccan Traps), floral communities were transforming thanks to the proliferation of flowering plants, and rapid falls in sea levels were lessening the amount of productive coastal communities (e.g., Lloyd et al. 2008; Archibald et al. 2010; Mitchell et al. 2012). The final days of the dinosaurs may have seen them struggling to cope with rapidly changing conditions. This raises an interesting set of questions regarding the status of king tyrants: Was their role as the sole large predator of western North America a consequence of reduced dinosaur diversity? And how atypical was the diversity of the Lancian dinosaur communities inhabited by *T. rex*?

Taking these questions in turn, we might first consider whether the absence of large predators living alongside *T. rex* was unusual and, perhaps, a sign that dinosaur communities were in trouble. Setting aside controversies over the existence of additional, problematic *Tyrannosaurus* species (chapter 2), it is mostly undoubted that *T. rex* was the sole large predator in Lancian ecosystems. As some of the most intensively studied dinosaur-bearing rocks on the planet, the chance we have overlooked a major second large predator, or a "mid-tier" carnivorous theropod in the latest Maastrichtian rocks of western North America is close to zero (Holtz 2021). This condition, with one arch-predator spread over a wide geographic range, contrasts markedly with Early Cretaceous and Jurassic dinosaur communities that bore multiple large-bodied carnivores, often from different clades (Holtz 2021). Were these older dinosaur systems simply more productive or less stressed than those of the Maastrichtian, permitting their carriage of several big carnivore species?

A broader look at tyrannosaurid evolution suggests not. As discussed in chapter 5, tyrannosaurids are thought to have monopolized mid- and large-sized predator niches in Campanian and Maastrichtian Asiamerica, with multiple dinosaur communities having just one or two tyrant species that absorbed the roles occupied by multiple large predators in the Jurassic and Early Cretaceous (Holtz 2021). The catalyst behind the Late Cretaceous proliferation of tyrannosaurids and their capacity to "assimilate" predatory niches (Holtz 2021) remains uncertain, but, for whatever reason, Campanian and Maastrichtian Asiamerican dinosaur communities tend to contain one large predator. In a survey of twenty-nine tyrant-dominated assemblages, Holtz (2021) noted only two formations with more than one midsized or larger predatory species: the early Maastrichtian Nemegt Formation of Mongolia, and the Campanian Dinosaur Park Formation of Alberta, Canada. We cannot, therefore, see the status of *Tyrannosaurus* as a solo predator in Lancian ecosystems as unusual. On the contrary, such a condition seems ecologically typical for the last 15 million years of tyrannosaurid evolution.

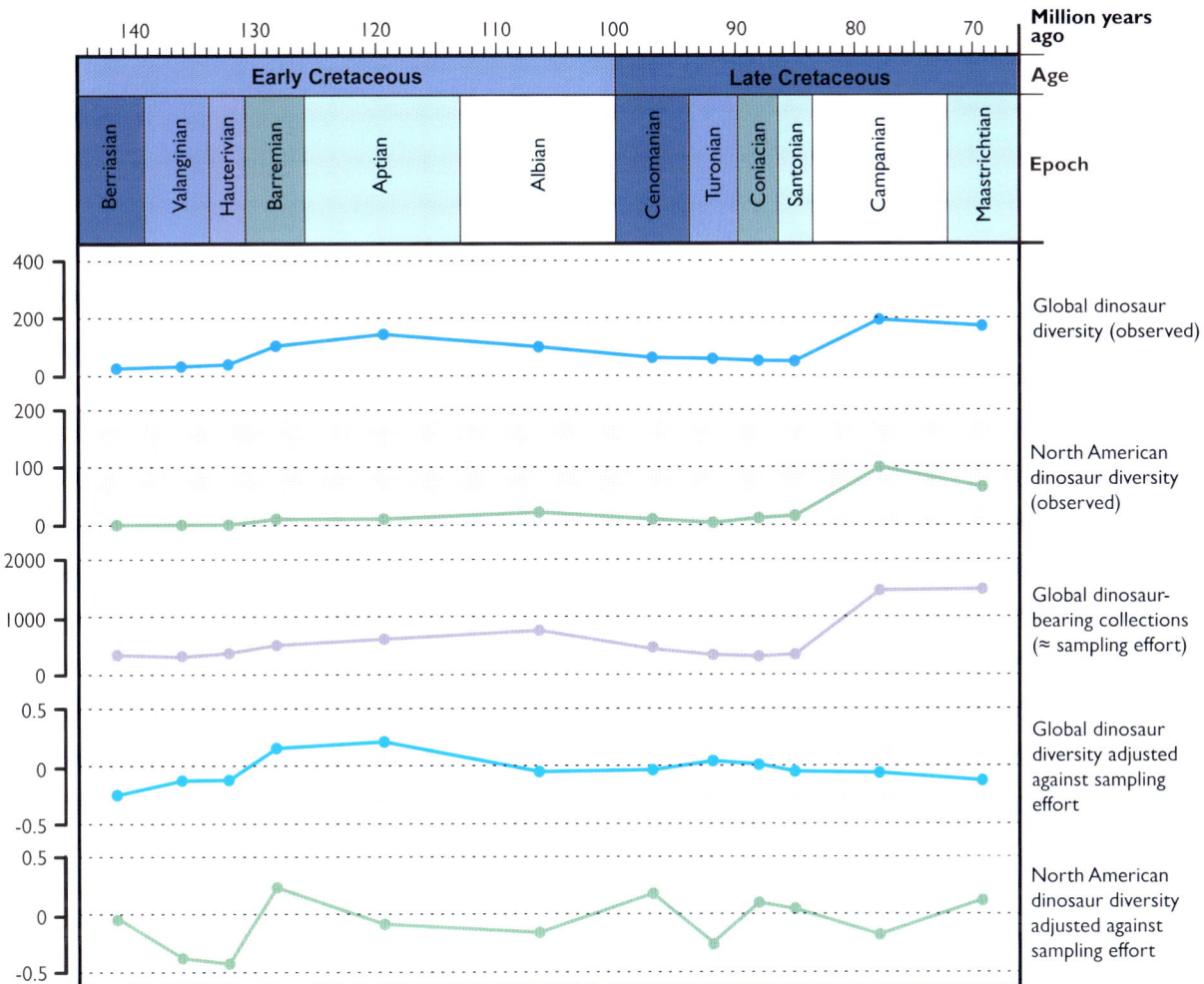

FIGURE 7.2

Cretaceous dinosaur diversity in context. At face value, North American dinosaur species richness seems to climax in the Campanian, before declining in the Maastrichtian. This fall tracks a lessening of dinosaur sampling, however, and efforts to correct this bias suggests that Maastrichtian diversity may not have been unusually low. Not all researchers agree with this assessment, however, and debates about the health of the final dinosaur communities continue. Data adapted and simplified from Brusatte et al. (2015).

As much as *Tyrannosaurus* was more extreme than some tyrannosaurids in aspects of growth, body size, and perceived ecological niche shifts (e.g., Cullen et al. 2020; Jevnikar 2021; Voris et al. 2021; Therrien et al. 2021), they represented a variation of typical tyrannosaurid ecologies, not something altogether novel.

Perhaps, then, tyrannosaur ecological monopolization reflects another cause: Did Campanian-Maastrichtian dinosaur communities of North America and Asia possess fewer species than those of earlier Mesozoic ecosystems, limiting the number of carnivore niches and, by extension, taxonomic diversity? It seems not. Holtz (2021) found that most tyrant-dominated Asiamerican dinosaur communities have comparable diversity to those of earlier Cretaceous and Jurassic formations, suggesting that prey diversity was no different despite the reduced number of predators. Indeed, tyrannosaurids existed as a solitary large predatory species even during the Campanian, the most diverse dinosaur interval of Cretaceous North America on record. This would argue, it seems, in favor of individual tyrant species being more ecologically adaptable than other large-bodied dinosaur predators, suggesting the solitary nature of *T. rex* in Lancian dinosaur fossil assemblages is probably not a symptom of dinosaur decline.

Theropods are just one dinosaur group, of course, and the question of Maastrichtian dinosaur diversity reduction has, traditionally, focused on non-avian dinosaur diversity as a whole (Russell 1984; Sloan and Rigby 1986; Sheehan et al. 1991; Brusatte et al. 2015; Sakamato et al. 2016; Chiarenza et al. 2019; Bonsor et al. 2020; Condamine et al. 2021). Over time, researchers have polarized into "gradualists," who interpret the Cretaceous impact as finishing off a group that was already in decline, and "catastrophists," who regard dinosaurs as maintaining typical levels of species richness

until they were suddenly wiped out by the K/Pg event. Opinions on this matter are divided because there are several ways to interpret Late Cretaceous dinosaur diversity, and differing analytic techniques can produce conflicting results. Understanding how *T. rex* fit into Mesozoic communities relies, in part, on finding the truth of this matter.

The most straightforward approach to this conundrum is to simply count dinosaur species from the Maastrichtian and compare them with that of the Campanian. At face value, this certainly suggests a loss of species richness toward the end of the Mesozoic (fig. 7.3). Campanian rocks of western North America contain an especially rich dinosaur fauna comprising over forty types of ankylosaur, hadrosaur, theropods (including several tyrannosaurids), and ceratopsids, compared to something like twenty-five to thirty-three species in the Maastrichtian Hell Creek Formation (Brusatte et al. 2015). This implies a 20–40 percent decline of dinosaur species richness in the last 15 million years of the Mesozoic, and large herbivores represent the bulk of this loss. Where dozens of large hadrosaur and ceratopsid taxa roamed the Campanian, only three were present at the time when dinosaurs met their final extinction. This decrease in large herbivore diversity is a rare point of

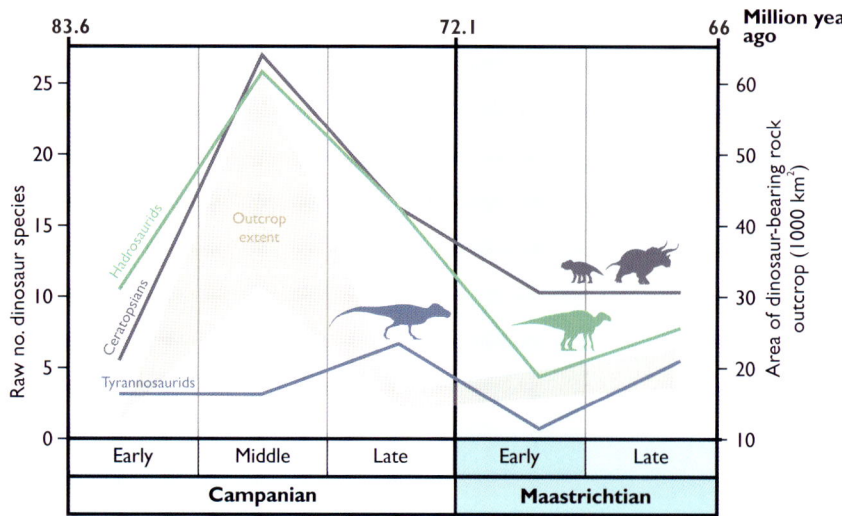

FIGURE 7.3

Was *Tyrannosaurus* part of an impoverished dinosaur fauna? Raw species counts of Campanian and Maastrichtian North American dinosaurs suggest so, but their diversity also mirrors the area of fossiliferous dinosaur strata from the latest Cretaceous. Perhaps sampling biases are skewing our perception? Redrawn from Chiarenza et al. (2019).

consensus among modern workers on dinosaur extinction (e.g., Brusatte et al. 2015; Bonsor et al. 2020; Condamine et al. 2021), although disagreements occur over how to interpret it. Are we picking up on a real biological signal, or are sampling issues and fossil preservation biases to blame? Comparing Maastrichtian dinosaur diversity against fossiliferous outcrop extent and sampling effort suggest that our species richness signal *is* biased, specifically by the area of available fossils (figs. 7.2–3). It seems unlikely that another dozen or so species of large dinosaur herbivores exist in the intensively sampled Hell Creek Formation, but other Maastrichtian strata in North America have received far less research attention. Learning more about the dinosaur communities in these sites is required to shed light on the reality of this apparent diversity loss (Bonsor et al. 2020). Furthermore, as noted in Chapter 5, the taxonomy of several Lancian dinosaur taxa, including ornithomimids, oviraptorosaurs and *Alamosaurus*, await reappraisal, it being suspected that their fossil inventories contain more species than is currently appreciated.

With these caveats in mind, the question of whether *T. rex* was an inhabitant of a declining dinosaur community remains open. At face value, it seems that yes, *T. rex* may have lived at a time when some North American dinosaur communities were waning, at least within some parts of their range. It is not clear, however, if this reflected limited, local loss of species richness or a widespread, global decline in dinosaur diversity. Either way, we might wonder if some of the unusual anatomy of *T. rex* was shaped by a diminishing pool of prey species. Given the evidence that *T. rex* routinely ate two of the only abundant giant herbivore species it lived alongside, might their particularly large adult size and powerful bites be specializations for predating *Edmontosaurus* and *Triceratops*? Perhaps the smaller sizes, higher speeds, and (somewhat) lessened bite strengths of earlier tyrant dinosaurs permitted a wider prey choice in more diverse ecosystems, while *T. rex* evolved a body plan specifically able to tackle the only abundant, large-bodied prey available to them. The unique ecological circumstances of *Tyrannosaurus*

might go some way to explaining some of their unusual features, but this idea might be difficult to test, given that the evolution of *T. rex* and their prey was cut short by a dramatic extinction event.

DID *T. REX* MAKE IT TO THE END?

Alongside debates about the real or artificial nature of dinosaur diversity loss are discussions over whether non-avian dinosaurs reached the final days of the Mesozoic. Some proponents of "gradualist" dinosaur extinction have argued that dinosaurs were extinct, or so rare as to be "functionally extinct," thousands of years before the end of the Cretaceous, pointing to a paucity of dinosaur fossils in the uppermost layers of the Hell Creek Formation as evidence of a pre-impact disappearance (e.g., Clemens and Archibald 1980; Clemens et al. 1981; Williams 1994). The extensive outcrops, preservation of the iridium layer, and abundant dinosaur fossils make the Hell Creek Formation one of the best geological units to study the final days of the dinosaurs, and the absence of fossils in its uppermost, youngest sediments is not controversial (fig. 7.4): the 3 meters beneath the K/Pg boundary genuinely has relatively few vertebrate fossils. This zone is known as the "3 m gap" (e.g., Alvarez 1983; Pearson et al. 2002; Lyson et al. 2011). *T. rex* is among those species with a final occurrence around 3 m beneath the iridium layer, concurrent with the youngest remains of most other Hell Creek dinosaurs (Williams 1994; Pearson et al. 2002).

This sharp exclusion of most dinosaur remains from the final stages of Hell Creek deposition requires explanation. Some have interpreted it as an honest signal: that dinosaur fossils vanish before the end-Cretaceous marker horizon because dinosaurs did too (e.g., Clemens et al. 1981; Williams 1994). Whether they were *totally* extinct before the K/Pg impact is disputed because dinosaur bones can occur, rarely, in the 3 m gap, suggesting some populations were still extant. Alternatively, these bones could be explained by the phenomenon of fossil reworking, where ancient streams eroded rocks

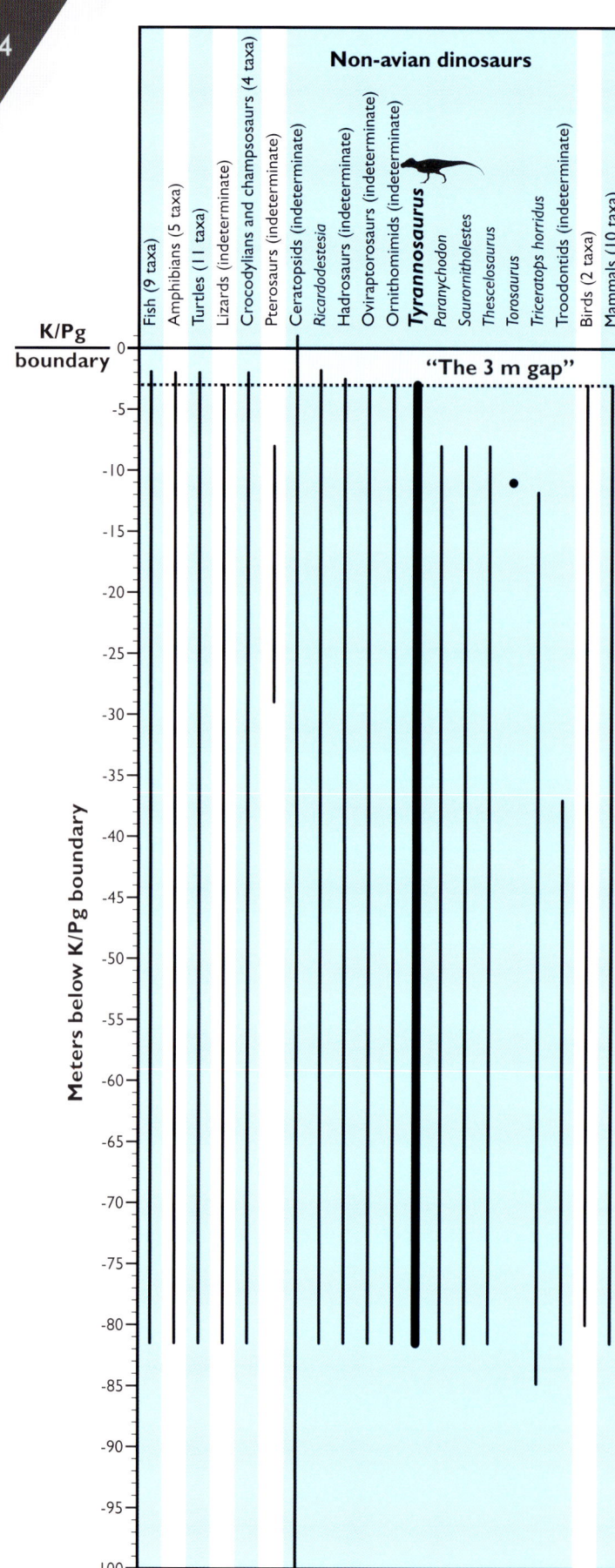

FIGURE 7.4

Hell Creek's mysterious "3 m gap," the relatively non-fossiliferous sediments immediately beneath the K/Pg boundary. Interpreted by some as representing a pre-K/Pg decline in dinosaurs, surveys of Hell Creek fossils show that remains of almost all vertebrates are rare in these beds, and that preservational conditions, not ancient ecological factors, are probably behind this phenomenon. Adapted and somewhat simplified from Pearson et al. (2002).

and soils containing bony material to redeposit them in younger sediments. Reworked fossils of these kinds are common geological phenomena and can transport much older remains, sometimes even millions of years older, into younger rocks. That this phenomenon occurred in Hell Creek is widely accepted, and it bolsters the possibility that the final non-avian dinosaur fossils of the Mesozoic are mere ghosts of geologically older individuals.

Taking the 3 m gap at face value is probably oversimplistic, however. Statistical studies modeling the frequency of dinosaur finds through Hell Creek have concluded that their absence in the last few meters of Cretaceous strata is not indicative of a reduction in dinosaur abundance (Sheehan et al. 1991, 2000; Pearson et al. 2002). The Hell Creek Formation is a substantial sedimentary unit, about 100 m thick, and several meters can exist between fossiliferous horizons. The absence of dinosaurs in one 3 m thick band is therefore interesting, but it is not statistically significant. Indeed, statisticians have argued that finding abundant dinosaur remains at the very top of the formation is an unrealistic expectation if we assume a normal distribution of fossils throughout Hell Creek (Alvarez 1983). This is a fundamental paleontological principle known as the Signor-Lipps effect, which posits that the uppermost geological occurrence of a fossil species almost certainly never represents their final extinction (Signor and Lipps 1982). This cautions against reading the last stratigraphic occurrence of a fossil species as their extinction marker as this can give false readings of declines in diversity before a true mass extinction event, an observation that originated amid discussion of how to interpret the decline of species approaching the K/Pg boundary (Signor and Lipps 1982).

The 3 m gap has not only been criticized on theoretical grounds. A number of researchers have sought dinosaur specimens that might narrow or close the gap altogether, and have had success in doing so. Scouring Hell Creek strata immediately below the iridium layer has yielded dinosaur fossils at a frequency comparable to lower parts of the formation in some areas (Sheehan et al. 2000), and at least some of these are demonstrably not reworked from older rocks. Of particular note is a *Triceratops* horn from just 13 cm below the K/Pg boundary (Lyson et al. 2011) which, in being deposited in a poorly rooted ancient soil rather than a stream bed, could not have been eroded and re-deposited from older sediments. This fossil almost certainly represents a horned dinosaur that lived and died just thousands of years before the Cretaceous impact event.

Some paleontologists have also investigated whether the scarcity of vertebrate fossils below the iridium layer has a geological, rather than biological, explanation. These studies are among the most important and persuasive counterarguments against face-value interpretations of the 3 m gap because, in looking beyond dinosaurs, they bring other, critical data to the debate. The reality of the 3 m gap is that *all* Hell Creek vertebrates suddenly become much rarer below the iridium layer, including fish, amphibians, turtles, crocodylians, and lizards (fig. 7.4; Pearson et al. 2002). This is suspicious because, if this gap represents a true biological event, it implies that almost the entire Hell Creek vertebrate fauna died out before the K/Pg impact despite their radically different ecologies and lifestyles. This seems highly improbable, and a more plausible explanation is a relatively simple, mundane one: that a change in sedimentological conditions at the top of the Hell Creek Formation lessened the likelihood of preserving bone. Sedimentological analysis of the 3 m gap shows that it contains no calcareous minerals, indicating a degree of elevated acidity and geological processes that were hostile to bone preservation (Retallack et al. 1987). The "3 m gap" is therefore almost certainly an artifact caused by geological biases and not a record of early dinosaur extinction in the Hell Creek region. Thus, while the geologically youngest *T. rex* bones stem from 3 m or so below the iridium layer, the wealth of statistical, paleontological and geological data give no compelling reason to think that *Tyrannosaurus*, other non-avian dinosaurs, or the rest of the Hell Creek fauna, did not make it to the bitter end of the Mesozoic.

GOODBYE, *T. REX*

THE FATE OF North America's non-bird dinosaurs was set in motion somewhere in our solar system long before their final days on Earth. While *Tyrannosaurus* and other dinosaurs lived out their Late Cretaceous existences, an enormous asteroid was on a collision course with the southernmost tip of North America (fig. 7.1). The resulting impact into what is now the Chicxulub region of the Yucatán Peninsula was the abrupt end of the entire Mesozoic Era, and the sediments deposited from its fallout mark the very first moments of the Cenozoic. This means that, technically, non-avian dinosaur populations died in the Paleogene, not the Mesozoic, but their existence for hours, days, weeks or even a few years after the impact is imperceptible against the grand scale of geological time. It is not inconceivable that small dinosaur populations hung on somewhere in the Paleogene for a short while, and dinosaur fossils have even been reported from post-Cretaceous rocks (e.g., Sloan and Rigby 1986; Fassett 2009). These are widely regarded as either reworked or incorrectly dated, however (e.g., Brusatte et al. 2015), and in any case, some short-term survival doesn't change how we interpret that geologically instant transition from Mesozoic to Cenozoic. The time of the non-avian dinosaurs ended 66 million years ago, and the king tyrants disappeared with them.

We could, theoretically, end our discussion there. A giant rock from space killed all the *Tyrannosaurus*—conversation over. But to do so would ignore the detailed picture emerging about the final moments of North America's dinosaurs, and the frankly terrifying circumstances that *Tyrannosaurus* and their contemporaries witnessed at their end. King tyrants are frequently characterized as the biggest, scariest monsters of the Mesozoic, but the effects of the K/Pg impact were far, far more dreadful than any dinosaur could ever be, even in fiction. It's difficult to conceive of the scale, speed, and energy of the Chicxulub event. The 150–200 km-wide crater (Morgan et al. 2022) is either the second- or third-largest impact trace on the planet, second only to the 2 billion-year-old, 180–300 km wide Vredefort impact structure of South Africa and, maybe, the 500 km wide Deniliquin multiple-ring feature of Australia, recently proposed as the remnant of a massive impact during the early Paleozoic (Glikson and Yeates 2022; Glikson 2023). The outcomes of these rare "mega-impacts" are beyond anything witnessed by a human, reducing the explosive output of even our largest atomic bombs to a rounding error. Geological data and impact simulations suggest that they triggered continent-sized fireballs, hurricane-force shockwaves, and blankets of superheated ejecta, the scale of which are difficult for us to imagine. Considering these data against where we know *Tyrannosaurus* lived gives us a ground-level, almost personal insight into the final days of king tyrants.

Over 350 sites recording the K/Pg impact are known around the globe (Schulte et al. 2010), giving us a detailed view of how events unfolded 66 million years ago. Studies of seasonally deposited growth lines in the bones of fish killed during the impact suggest that

FIGURE 7.5

The ravages of the Chicxulub impact on the western interior of North America and the range of *Tyrannosaurus*. (A) The effects of the impact at increasing distance. *T. rex* escaped the instant death that befell species within a 1,000 km radius of the impact site, but the majority of their known population areas were still close enough to experience some of the worst effects. (B) Did *Tyrannosaurus* witness the Chicxulub impact? Modeling the estimated height of the impact plume and ejecta curtain against the curvature of the Earth suggests many *T. rex* could have observed the event that ended the Mesozoic Era, despite their distance from the collision itself. Simulated impact details in B adapted from Morgan et al. (2022).

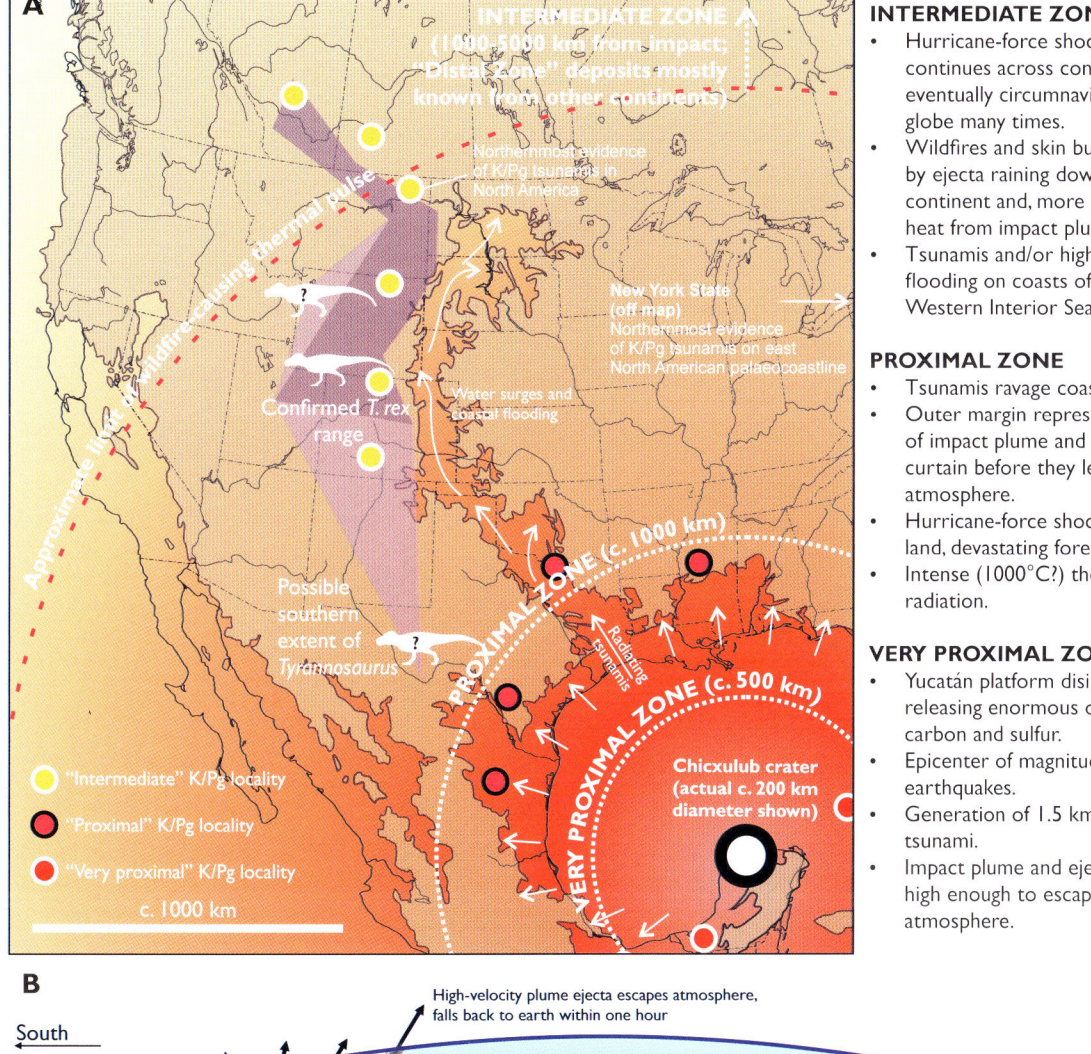

A

INTERMEDIATE ZONE
(1000-5000 km from impact;
"Distal Zone" deposits mostly
known from other continents)

Northernmost evidence
of K/Pg tsunami in
North America

New York State
(off map)
Northernmost evidence
of K/Pg tsunami on east
North American palaeocoastline

Confirmed *T. rex*
range

Water surges and
coastal flooding

Approximate limit of wildfire-causing thermal pulse

?

?

Possible
southern
extent of
Tyrannosaurus

PROXIMAL ZONE (c. 1000 km)

Radiating
tsunamis

VERY PROXIMAL ZONE (c. 500 km)

Chicxulub crater
(actual c. 200 km
diameter shown)

⬤ "Intermediate" K/Pg locality

⬤ "Proximal" K/Pg locality

⬤ "Very proximal" K/Pg locality

c. 1000 km

INTERMEDIATE ZONE

- Hurricane-force shockwave continues across continent, eventually circumnavigating the globe many times.
- Wildfires and skin burns caused by ejecta raining down onto continent and, more proximally, heat from impact plume.
- Tsunamis and/or high-energy flooding on coasts of remnant Western Interior Seaway.

PROXIMAL ZONE

- Tsunamis ravage coastlines.
- Outer margin represents extent of impact plume and ejecta curtain before they leave atmosphere.
- Hurricane-force shockwave hits land, devastating forests.
- Intense (1000°C?) thermal radiation.

VERY PROXIMAL ZONE

- Yucatán platform disintegrates, releasing enormous quantities of carbon and sulfur.
- Epicenter of magnitude 10 earthquakes.
- Generation of 1.5 km tall tsunami.
- Impact plume and ejecta rise high enough to escape atmosphere.

B

South

High-velocity plume ejecta escapes atmosphere,
falls back to earth within one hour

North

Top of atmosphere

Ejecta cloud
(Soot, dust, sulfate
aerosols etc.)

Closest possible occurrence
of *Tyrannosaurus* to impact
(Texas-Mexico border)

Closest southeastern
occurrence of *T. rex*
to impact (Colorado)

Hell Creek

Impact plume
(Gas and fine debris
expanding from impact
at high velocity)

Ejecta curtain

Horizon line

Top of stratosphere

Earth's surface

450 km

Chicxulub impact site
(Yucatán Peninsula)

c. 1200 km

c. 1700 km

c. 2000 km

1000 km

it occurred in spring (During et al. 2022), striking the Chicxulub region at a 45–60° angle, perhaps from the southwest (Artemieva and Morgan 2009; Gulick et al. 2019). As might be expected, how animals and plants experienced the events that followed was related to their proximity to this Cretaceous ground zero (fig. 7.5). Most of our global K/Pg impact sites are considered "distal"-grade deposits, represented primarily by the iridium layer and particles deposited by the globally dispersed ejecta cloud. Those of North America and a good portion of South America were in an "intermediate" effect zone, a region 1,000–5,000 km from Chicxulub that experienced more pronounced fallout from the ejecta as well as floods caused by the impact. Parts of the southern United States and Middle America are rated as "proximal" (500–1,000 km from the impact) or "very proximal" (0–500 km). These experienced 360 kph tsunamis that were perhaps 1.5 km tall (Range et al. 2022) as well as the full force of the deadly asteroid shockwave. Some models posit that the impact plume (the mix of rapidly expanding gas, vaporized rock and other debris; essentially the impact explosion) expanded to cover the entire 2,000 km diameter of these two zones as it rose hundreds of kilometers into, and eventually out of, the atmosphere (Morgan et al. 2022). In this scheme, the recorded geographic range of *T. rex* lies within the "intermediate" fallout zone (fig. 7.5), removing them from the instant death that surely befell anything living within the "proximal" or "very proximal" zones, but not sparing them from other catastrophic, violent, and very immediate impact effects. We can divide these into circumstances occurring seconds, minutes, hours, and days after the asteroid's arrival.

Despite their 1,700–2,500 km distance from the Yucatán Peninsula, king tyrants may have witnessed and would have felt the asteroid's collision (figs. 7.5B, 7.6). In these first few seconds, the impact plume, modeled to have risen hundreds of kilometers into and beyond the atmosphere (Morgan et al. 2022), would have been visible over the horizon. It was probably literally blinding to observe, being far brighter than even the midday sun. Eye damage was of secondary concern to the thermal pulse from the blast, however. Intense heat of several hundred or perhaps even over a thousand degrees centigrade radiated from the impact plume, igniting vegetation and burning animals up to 2,400 km away (Morgan et al. 2013; Santa Catharina et al. 2022).

GOODBYE, *T. REX*

FIGURE 7.6

A split-second witness to the end of the world, a mother king tyrant observes the expansion of the impact plume thousands of kilometers away, over the horizon. Any animals that observed this were likely blinded by the light, and, even at this distance, the heat was enough to sear flesh and ignite fires.

FIGURE 7.7

The south coast of North America was, within minutes of the K/Pg event, inundated by enormous tidal waves. Their size is debated, but they were likely measurable in many tens of meters, dwarfing even the mighty *Alamosaurus*. Having already ripped up shallow-marine environments, these waves would now batter and flood inland settings.

FIGURE 7.8

Forests of North America minutes after impact. Wildfires, ignited by thermal radiation from the impact explosion and falling debris from the rapidly expanding ejecta curtain, are joined by floodwaters from tsunamis that forced their way northward through the remnant of the Western Interior Seaway.

This distance covers the bulk of *Tyrannosaurus* geographic range. Against this sudden intense radiation, animals already sheltered before the impact had the best chance of survival, but multi-tonne animals adapted to life in open habitats were probably especially vulnerable. We have to wonder if a large proportion of the southern and mid-latitude *T. rex* population were killed or horrifically burned in the first few seconds of the impact.

Blinding light and flesh-searing heat were only the start. The seas, the earth, and the skies rapidly brought their own catastrophes to western North America. Magnitude 10 earthquakes arrived in *T. rex* habitats several minutes after impact, toppling standing animals and triggering tsunamis, compounding those formed from the impact itself (fig. 7.7; Mittal et al. 2022; Range et al. 2022). Eventually joined by the original tidal wave a few hours later, which was still tens of meters tall, these waves surged through the remnant Western Interior Seaway, battering, scouring, and flooding coastal communities. Undoubtedly, many *Tyrannosaurus* were swept along and drowned in these surging waters.

But perhaps worst of all was the scenario unfolding in the skies. The ejecta plume, a cloud of pulverized rock, dust, and debris from the impact, had initially been launched hundreds of kilometers into space, but was now collapsing downward (Morgan et al. 2022). Traveling at 2–4 km per second and dense enough to block out the sun, this cloud would have arrived in *T. rex* habitats ten to fifteen minutes after the impact, raining debris as it spread (Morgan et al. 2013). As with any particles entering Earth's atmosphere, the contents of this cloud burned as they plunged back to Earth, directing another wave of intense heat radiation on the plants and animals below (fig. 7.8). The severity of this firestorm is debated because the evidence for widespread wildfires at the K/Pg boundary is limited, even in North America (Morgan et al. 2022). However, such radiation likely ignited dry vegetation and desiccated previously lush landscapes, making them prone to burning later (Morgan et al. 2013; Schulte et al. 2010). Any immediate fires may have been put out by the next impact

effect: shockwave-derived, hurricane-force winds. About two hours after the impact, 180 kph winds would have torn into southern *T. rex* habitats (Morgan et al. 2013), leveling trees and throwing debris, causing further animal fatalities. Even in their northern range, *T. rex* would have experienced winds surpassing 100 kph, enough to strip leaves from trees (Morgan et al. 2013).

Eventually, as hours turned to days after the impact, the tsunami waters receded, winds died down, and wildfires burned themselves out. The environments they left behind had been wrecked. Casualties among mid- and large-sized dinosaurs were undoubtedly tremendous, and their burned, drowned, and smoke-choked bodies must have been strewn everywhere, a scene of biological carnage that we can barely fathom. *Tyrannosaurus* living in the northernmost United States and Canada stood the best chance of weathering these events, having avoided the worst of the tsunamis and shockwave. But any that lived through that apocalyptic day would have found themselves in a wasteland. They were entering the final and longest-lasting phase of the K/Pg event: the impact winter. By now, the ejecta cloud had dispersed across the entire planet and would remain in place for years (Morgan et al. 2022), blocking sunlight and limiting the photosynthesis essential to reestablishing and rebuilding ravaged plant communities and the ecosystems they supported. The chemistry of the impact cloud, derived from the rocks vaporized at Chicxulub, added further complications, cooling the Earth's surface by 10–26°C, lowering precipitation rates, and reducing ozone levels, increasing exposure to ultraviolet radiation (Schulte et al. 2015; Morgan et al. 2022).

Once a rich, productive environment capable of sustaining complex communities of animal life, the Western Interior of North America was now a barren wilderness that could only support detritus-based food webs. The forests and woodlands that nourished *Tyrannosaurus* prey were burnt and blackened, or else slowly dying from lack of sunlight. Only ferns, adapted to growing in dim light, and fungi, nourished by the decaying remains of Cretaceous life, could thrive

FIGURE 7.9

A survivor of the K/Pg event is compelled to resume the springtime mating calls they were performing mere days before in lush woodlands, before their world ended. With an impact taking place in spring, perhaps the final days of king tyrants were not only spent starving amid the ruins of the Cretaceous, but also acutely aware of their unusual solitude.

(Morgan et al. 2022). For any remaining *T. rex*, this was the end: they were multi-tonne predators in a world without the capacity to support animals much above 5–10 kg. The factors explaining animal survival across the K/Pg boundary are still discussed, but it seems that only small-bodied species with unfussy eating habits endured this gloomy period in Earth's history. Avian dinosaurs may have survived because some had adapted to eating seeds and bugs rather than flesh by the end of the Cretaceous (D. Larson et al. 2016). Perhaps some small dinosaur carnivores made an effort at living in the Paleogene, scavenging what they could or subsisting on the diminutive species that carried life into the Cenozoic. In these broken ecosystems, infant *T. rex* may have survived longer than their larger, more powerful parents, being potential predators of small game (chapter 6). Indeed, with no larger predators to bother them, some may have even fared well on a diet of small-bodied vertebrates. But, ultimately starvation awaited any *Tyrannosaurus* that lived to see the other side of the K/Pg impact, for even resourceful juveniles would eventually grow to sizes demanding nourishment from now extinct larger prey. Having avoided death by fire, flood, and physical shock, any remaining king tyrants lived a sorry, starving existence until their inevitable ends came (fig. 7.9).

AFTERLIFE

WHEN THE LAST *Tyrannosaurus* perished in the fallout of the K/Pg impact, the bones of previous *T. rex* generations were already safely buried in the ground, some for a million years or more. Over 66 million years later, these *T. rex* individuals would be exhumed from their sediments and displayed around the world, their anatomy, physiology, and life history subjected to intense study by inquisitive scientists, and their tissues revivified and reanimated by countless artists and filmmakers (fig. 7.10). More than most fossil species, *T. rex* have not been allowed to die.

So strong is our desire for *Tyrannosaurus* to have endured past their end-Mesozoic fate that artists and scientists have attempted to imagine where their evolutionary history would have progressed had their lineage not been suddenly wiped out. From some perspectives, this concept is somewhat perverse: Mesozoic dinosaurs *had* to suffer near total annihilation for our own mammalian evolution to occur, and yet we seem somehow dissatisfied with the fact that they were unable to continue their evolutionary story. *Tyrannosaurus* has frequently survived in isolated corners of the planet in adventure fiction, from 1933's *King Kong* to various filmic reimaginings of Conan Doyle's *The Lost World*. The *T. rex* of these narratives are portrayed without modification from their Mesozoic guise, in essence showing them as "living fossils": organisms unchanged from their geologically distant state in the fossil record. Many animals today are considered "living fossils," including the coelacanth, the tuatara, and crocodylians. "Lost world" fictions play into this idea that animals can, by virtue of suppressed evolution, bring archaic, prehistoric biology into modern times for human explorers to find and, inevitably, conquer. The idea that animals can survive for tens of millions of years without modification is actually heavily flawed, however: species like coelacanths, turtles, and crocodylians may seem "prehistoric," but they have amassed just as many evolutionary and adaptive miles as any other species, and are often highly modified and unusual compared to their fossil cousins. If the *T. rex* line had endured past the Cretaceous, it almost certainly would not have remained unchanged for long.

It was in this spirit that one of the more influential speculative works on post-Mesozoic dinosaurs was presented: Dougal Dixon's 1988 book *The New Dinosaurs: An Alternative Evolution*. Dixon is famous for speculative evolution, especially for his 1981 book *After Man*, which imagines what life will evolve on Earth following the extinction of humanity. Taking the same approach

FIGURE 7.10

"Thomas," a young adult *T. rex* skeleton displayed at the Natural History Museum, Los Angeles, is one of the geologically youngest substantial king tyrant specimens. By the time this animal lived and died, generations of other king tyrants had already entered the rock record as fossils, awaiting resurrection as objects of scientific study and cultural interest at the hands of intelligent primates 66 million years later.

to Mesozoic dinosaurs, Dixon's *New Dinosaurs* explored the evolutionary history of an Earth where most dinosaur lines had not been destroyed by the Chicxulub impact. Among the products of this alternative natural history was "*Ganeosaurus tardus*," also known as

the "gourmand": a colossal, 17 m long South American tyrannosaur adapted for scavenging (fig. 7.11). Sporting short legs, robust feet, and no arms, Dixon cast his *T. rex* descendant as a specialist scavenger on account of their sluggish locomotion. Distensible jaws permitted the consumption of whole carcasses at once, and an armored back allowed gourmands to rest where they fed, protected from predators (Dixon 1988).

Dixon's *New Dinosaurs* has inspired several speculative tyrannosaurs. The gourmand seems to be directly referenced, or is at least strongly converged upon by, the similarly imagined "*Megapubis acheirus*" from the Japanese book アノ古生物、進化し続けるとどうなるの?("If Extinct Organisms Continued to Evolve—The History of Life on Earth"). As its name implies, this "armless one with a large pubis" is characterized by a huge pelvis; so large, in fact, that it almost reaches the ground even in a standing animal. This is imagined as providing a seat for these enormous fictional scavengers to rest on while they wait for surrounding animals to die, recalling fishing humans sleepily waiting for a bite on their line. How this scenario is squared against the needs of scavenging animals to be especially wide ranging (chapter 6) is not clear.

The legacy of the gourmand is also seen in Gregory S. Paul's work. In a critical 1990 review of Dixon's *New Dinosaurs*, Paul chastised their lack of speed (see chapter 4 for Paulian views on *T. rex* locomotion), relegation to scavenging despite the predatory adaptations of tyrannosaurs, as well as the implausibility of large terrestrial animals to sustain themselves on carrion alone. In an alternative imagined post-Mesozoic dinosaur fauna, Paul posited that tyrannosaurids would continue to exist within North American dinosaur ecosystems with little change from their latest Cretaceous forms, even tens of millions of years later: "How could such fast, powerful predators be improved?" (Paul 1990, p. 313). He proposed that only tyrannosaur arms would change, becoming tiny, fingerless nubbins, that their skull openings would close further, and that their stereoscopic vision would be augmented. Indeed, in contrast to Dixon's post-Mesozoic Earth, where dinosaurs had diversified

FIGURE 7.11

Dougal Dixon's gourmand *"Ganeosaurus tardus,"* a speculative post-Mesozoic tyrannosaur adapted for scavenging that was invented for his 1988 book *The New Dinosaurs: An Alternative Evolution.* The gourmand has influenced a number of other speculative Cenozoic tyrannosaurs.

into niches and forms unrepresented from their true fossil record, the Cenozoic dinosaurs inhabiting Paul's vision were broadly similar in ecology and anatomy to real Lancian species. He attributed this to "adaptive entrenchment," where clades are constrained in anatomy and genetics by elements of their original physical and genetic makeup.

Dixon's work has inspired filmmakers, too. His *New Dinosaurs* was namechecked by the movie producers of the 2005 remake of *King Kong*, which imagined a modern, "evolved" tyrannosaur in the guise of *"Vastatosaurus rex"*: a gigantic *T. rex* in all but name with a distinctly crocodylian design motif, including lipless jaws and an armored back. In a fun nod to the scientific history of *Tyrannosaurus*, *Vastatosaurus* also featured three fingers, but a need to retain a basic tyrannosaur shape and ecology for narrative purposes (that is, being a formidable foe to a giant ape) limited the creative potential for imagining a modern-day tyrannosaur

descendant. In promotional media, the team behind 2005's *King Kong* admitted that their intent was to make an especially monstrous version of *Tyrannosaurus* (*Recreating the Eighth Wonder: The Making of King Kong* 2006).

It is, of course, difficult to say if any of these speculations are plausible. Theropod evolution shows that some body plans can exist for tens of millions of years without major modifications, while other clades are defined by near-constant evolutionary novelty. We can't even be certain that the oft-suggested loss of tyrannosaur arms would eventually play out, given that there is no strong consensus on their function. If, as discussed in chapter 4, they retained some use despite their size, they may well have been retained in future generations even if Dollo's Law—the observation that evolution is generally nonreversible—implies that Cenozoic tyrants may have struggled to redevelop additional digits. Similarly, it also does not seem certain that the Cenozoic children of *T. rex* would remain huge, or even grow larger, than their Cretaceous ancestor. Most Mesozoic theropods were considerably smaller than *Tyrannosaurus*, suggesting that only particular ecological conditions promoted the development of giant dinosaur predators. Greater adaptive opportunities seem to have generally been found at smaller sizes, and, on grounds of probability, Cenozoic tyrannosaurs may have been smaller than the king tyrants that begat them.

Nor, for that matter, should we regard tyrannosaur anatomy as unimprovable. The success of tyrannosaurids demonstrates that their body plans were well optimized for predatory roles in Late Cretaceous terrestrial ecosystems, but they may have been forced to adapt as the natural world moved on. Even within the Mesozoic, conditions were already changing in ways that could have shaped *T. rex* descendants. How, for instance, may the apparent reappearance of sauropods in North America altered tyrannosaur evolution after millions of years of targeting hadrosaur and ceratopsid prey? And, for that matter, how might tyrannosaurs have adapted to the last 66 million years of challenges and changes in Earth's biosphere: shifting climates, evolving plant communities, and the introduction of newly evolved animal species? As fond as we are of tyrannosaur form, tyrant dinosaurs were not eternal, unchanging Olympians. Their anatomy was a response to specific adaptive circumstances, and as prone to modification through natural selection as that of any other species.

It is undeniably fun to imagine what the future of tyrannosaurid evolution may have held (fig. 7.12), but our suggestions can only echo our personal and cultural reflections on *T. rex* and their relatives. We see this expressed in the various speculative *Tyrannosaurus* descendants discussed above: Dixon's gourmand is a projection of older, scavenger-based hypotheses of tyrannosaur lifestyles, just as Paul's armless sprinter is the embodiment of his perception of tyrant dinosaurs as powerful superpredators. No matter how much science we attempt to inject into our speculations, they ultimately owe more to individual perspectives than they do to the realities of adaptive evolution.

A

B

FIGURE 7.12

We can only imagine what the future may have held for descendants of *Tyrannosaurus rex* had they not become extinct 66 million years ago. If their lineage had survived until relatively recent times, and our planetary environment followed largely the same course as we know today, Tyrannosauridae may have included long-legged, small-bodied species adapted for hunting small game in open grasslands (A) or compactly built, cold-adapted species living at high latitudes (B). Imagining such speculative animals is a fun exercise, but their plausibility is impossible to test.

THE VALUE OF *TYRANNOSAURUS*

As we find ourselves at an interface between fossils, science, and culture, our discussion of *Tyrannosaurus rex* has run full circle: we are, once again, viewing king tyrants through a subjective and emotional human eye rather than the forensic lens of science. If we take a leaf from the above discussion and indulge in some speculative history, however, we might conclude our king tyrant journey with a question that sheds light on both our continued obsession with *T. rex*, as well as their scientific importance: Would *Tyrannosaurus* have made anywhere near as large an impact on science and culture if it had been discovered in modern times, and not over a century ago?

It is difficult to see this answer as anything other than a probable "yes." Even if we knew of other tyrannosaurids, finding *T. rex* today would still push the boundaries of what we know this group, and predatory dinosaurs in general, were capable of anatomically and functionally. The abundance of their fossils would still make them the best-known giant theropod, quickly eclipsing our slim knowledge of giant carcharodontosaurids and spinosaurids, and the potential for determining details of their growth, life history, and ecology would still be ranked as among the most comprehensive for any dinosaur species. Indeed, had dinosaur paleontology progressed for the last century without *T. rex*, their discovery might be heralded as a game-changer: a species that finally answered questions framed by other dinosaur discoveries. Indeed, a case could be made that this scenario has played out somewhat in reality. We may have known about king tyrants since the early 1900s, but their attainment of "model organism" status has only emerged in the last few decades. It is only recently, therefore, that *T. rex* has helped to address dinosaur topics that were previously poorly understood or controversial. Amid the hype around *T. rex*, it is easy to forget that their fossil record is unusually good, and that our seemingly inexhaustible interest in them is fueled partly by their capacity to keep providing answers.

It is harder to say whether *Tyrannosaurus* would have developed the same pop-culture appeal if they had been discovered today, in part because it's difficult to imagine what modern dinosaur culture would be without them. From a popularization of science perspective, the discovery and promotion of *T. rex* by the AMNH was proverbial lightning in a bottle. Exceptional individuals, from Brown's uncanny ability to discover fantastic dinosaur specimens to Osborn's savvy eye for establishing academic and public kudos—to say nothing of coining the note-perfect name *Tyrannosaurus rex*—conspired to unleash *T. rex* on a public primed for spectacular dinosaurs. Today's paleontological culture is far more complex than that of a century ago, and it is harder for new species to establish themselves in the roster of household dinosaur names. Even so, the giddy thrill of standing next to a mounted *T. rex*, their recognizability, their status as one of the largest carnivores to ever walk the Earth, and the inevitable media interest in new king tyrant research, would have given them a good chance of making a considerable cultural splash. It is not difficult to imagine king tyrants adopting commanding positions in museums, movies, and paleoartworks even if they had been discovered in the last few years, not 100 years ago.

Throughout this book we have seen how *Tyrannosaurus* has been analyzed and probed from every angle, and how the resultant data and interpretations of their anatomy, functionality, and evolution are now supporting one another to form robust, realistic hypotheses about *T. rex* biology. We have also seen the strain between the popular version of *T. rex* and our scientific understanding, with the hyped, sensationalized variant often being very different to what paleontologists interpret from fossil bones. And yet, to admit that the human history of *T. rex* might have played out in similar fashion if they had been discovered today is to concede that *Tyrannosaurus* might have some special qualities. They combine a checklist that paleontologists desire in a fossil taxon—an abundance of excellently preserved fossils in relatively accessible locations—with characteristics that many of us, including scientists, find intuitively intriguing and captivating: large size, fascinating anatomy, and a hint of faded predatory danger. It's no mystery, in truth, why *T. rex* remains one of the most popular dinosaurs with paleontologists and the public alike (fig. 7.13).

It's perhaps this quality that always has, and probably always will, define *Tyrannosaurus*. King tyrants are all things to all people, so they will always be exciting, infuriating, awe-inspiring, and frustrating, in equal measure. They represent a benchmark for paleobiological research, but draw attention away from studies of other species. Their fossils are remarkable natural objects to see and handle, but their increasing desirability and monetary value strains the relationship between commercial and academic paleontology. Their appearance in artwork and other media is familiar to the point of cliché, but their reconstructed anatomy and biology undeniably inspires our creativity. Our personal position in this sea of contradicting viewpoints determines the version of *Tyrannosaurus* that we carry with us. In all the noise and bustle surrounding *T. rex*, my preferred location is with an eye on their fossils and the data they continue to provide, processing new studies with an open mind. The king tyrants we see crafted by storytellers, advertising agencies, and Hollywood special effects artists are sometimes incredible, but have nothing on what science tells us about the reality of *Tyrannosaurus rex*.

FIGURE 7.13

Scenes such as these at the American Museum of Natural History, with thick crowds surrounding a *Tyrannosaurus* display that has existed in various capacities for over a century, are perhaps the most visual reminder of the continuing fascination that king tyrants hold for us. Few other animals, living or extinct, have so many different values—research subject, fantasy monster, financial asset—to so many people. Perhaps it is this that has become the ultimate legacy of *T. rex*, and the culture we have developed around them.

REFERENCES

Abler, W. L. (1992). The serrated teeth of tyrannosaurid dinosaurs, and biting structures in other animals. Paleobiology, 18, 161–183.

Abler, W. L. (1999). The teeth of the tyrannosaurs. Scientific American, 281, 50–51.

Alexander, R. M. (1976). Estimates of speeds of dinosaurs. Nature, 261, 129–130.

Alexander, R. M. (1985). Mechanics of posture and gait of some large dinosaurs. Zoological Journal of the Linnean Society, 83, 1–25.

Alvarez, L. W. (1983). Experimental evidence that an asteroid impact led to the extinction of many species 65 million years ago. Proceedings of the National Academy of Sciences, 80, 627–642.

Alvarez, L. W., Alvarez, W., Asaro, F., & Michel, H. V. (1980). Extraterrestrial cause for the Cretaceous-Tertiary extinction. Science, 208, 1095–1108.

Amiot, R., Lécuyer, C., Buffetaut, E., Fluteau, F., Legendre, S., & Martineau, F. (2004). Latitudinal temperature gradient during the Cretaceous Upper Campanian–Middle Maastrichtian: δ¹⁸O record of continental vertebrates. Earth and Planetary Science Letters, 226, 255–272.

Anderson, J. F., Hall-Martin, A., & Russell, D. A. (1985). Long-bone circumference and weight in mammals, birds and dinosaurs. Journal of Zoology, 207, 53–61.

Andres, B., & Langston, W., Jr, (2021). Morphology and taxonomy of Quetzalcoatlus Lawson 1975 (Pterodactyloidea: Azhdarchoidea). Journal of Vertebrate Paleontology, 41, 46–202.

Anné, J., Canoville, A., Edwards, N. P., Schweitzer, M. H., & Zanno, L. E. (2023a). Independent evidence for the preservation of endogenous bone biochemistry in a specimen of Tyrannosaurus rex. Biology, 12, 264.

Anné, J., Whitney, M., Brocklehurst, R., Donnelly, K., & Rothschild, B. (2023b). Unusual lesions seen in the caudals of the hadrosaur, Edmontosaurus annectens. The Anatomical Record, 306, 594–606.

Apesteguía, S., Smith, N. D., Valieri, R. J., & Makovicky, P. J. (2016). An unusual new theropod with a didactyl manus from the Upper Cretaceous of Patagonia, Argentina. PLOS One, 11, e0157793.

Aranciaga Rolando, A. M., Motta, M. J., Agnolín, F. L., Manabe, M., Tsuihiji, T., & Novas, F. E. (2022). A large Megaraptoridae (Theropoda: Coelurosauria) from Upper Cretaceous (Maastrichtian) of Patagonia, Argentina. Scientific Reports, 12, 6318.

Aranciaga Rolando, A. M., Novas, F. E., & Agnolín, F. L. (2019). A reanalysis of Murusraptor barrosaensis Coria & Currie (2016) affords new evidence about the phylogenetic relationships of Megaraptora. Cretaceous Research, 99, 104–127.

Araújo, R., Castanhinha, R., Martins, R. M., Mateus, O., Hendrickx, C., Beckmann, F., ... & Alves, L. C. (2013). Filling the gaps of dinosaur eggshell phylogeny: Late Jurassic theropod clutch with embryos from Portugal. Scientific Reports, 3, 1924.

Arbour, V. M., & Mallon, J. C. (2017). Unusual cranial and postcranial anatomy in the archetypal ankylosaur Ankylosaurus magniventris. Facets, 2, 764–794.

Arbour, V. M., Zanno, L. E., & Evans, D. C. (2022). Palaeopathological evidence for intraspecific combat in ankylosaurid dinosaurs. Biology Letters, 18, 20220404.

Arbour, V. M., Zanno, L. E., Larson, D. W., Evans, D. C., & Sues, H. D. (2016). The furculae of the dromaeosaurid dinosaur Dakotaraptor steini are trionychid turtle entoplastra. PeerJ, 4, e1691.

Archibald, J. D., Clemens, W. A., Padian, K., Rowe, T., Macleod, N., Barrett, P. M., ... & Sahni, A. (2010). Cretaceous extinctions: Multiple causes. Science, 328, 973–973.

Artemieva, N., & Morgan, J. (2009). Modeling the formation of the K–Pg boundary layer. Icarus, 201, 768–780.

Associated Press (2012, April 18). Dino castings dispute ends; key issue unresolved. Auction Central News. https://www.liveauctioneers.com/news/top-news/naturalhistory/dino-castings-dispute-ends-key-issue-unresolved/

Atkins-Weltman, K. L., Simon, D. J., Woodward, H. N., Funston, G. F., & Snively, E. (2024). A new oviraptorosaur (Dinosauria: Theropoda) from the end-Maastrichtian Hell Creek Formation of North America. PLOS One, 19, e0294901.

Auffenberg, W. (1981). The behavioral ecology of the Komodo monitor. University Presses of Florida.

Augusta, J., & Burian, Z., 1958. Prehistoric animals. Spring Books.

Aureliano, T., Ghilardi, A. M., Guilherme, E., Souza-Filho, J. P., Cavalcanti, M., & Riff, D. (2015). Morphometry, bite-force, and paleobiology of the Late Miocene Caiman Purussaurus brasiliensis. PLOS One, 10, e0117944.

Bakker, R. T. (1972). Anatomical and ecological evidence of endothermy in dinosaurs. Nature, 238, 81–85.

Bakker, R. T. (1975). Dinosaur renaissance. Scientific American, 232, 58–79.

Bakker, R. T. (1986). The dinosaur heresies: New theories unlocking the mystery of the dinosaurs and their extinction. William Morrow.

Bakker, R. T. (1988). Review of the Late Cretaceous nodosauroid Dinosauria: Denversaurus schlessmani, a new armor-plated dinosaur from the Latest Cretaceous of South Dakota, the last survivor of the nodosaurians, with comments on stegosaur-nodosaur relationships. Hunteria 1, 1–23.

Bakker, R. T., Williams, M., & Currie, P. J. (1988). Nanotyrannus, a new genus of pygmy tyrannosaur, from the latest Cretaceous of Montana. Hunteria, 1, 1–30.

Bamforth, E. L., Button, C. L., & Larsson, H. C. (2014). Paleoclimate estimates and fire ecology immediately prior to the end-Cretaceous mass extinction in the Frenchman Formation (66 Ma), Saskatchewan, Canada. Palaeogeography, Palaeoclimatology, Palaeoecology, 401, 96–110.

Barker, C. T., Naish, D., Newham, E., Katsamenis, O. L., & Dyke, G. (2017). Complex neuroanatomy in the rostrum of the Isle of Wight theropod Neovenator salerii. Scientific Reports, 7, 3749.

Baron, M. G., Norman, D. B., & Barrett, P. M. (2017). A new hypothesis of dinosaur relationships and early dinosaur evolution. Nature, 543, 501–506.

Barrick, R. E., & Showers, W. J. (1994). Thermophysiology of Tyrannosaurus rex: Evidence from oxygen isotopes. Science, 265, 222–224.

Barsbold, R. (1983). Carnivorous dinosaurs from the Cretaceous of Mongolia. Joint Soviet-Mongolian Paleontological Expedition, Transactions. 19, 5–120. (In Russian.)

REFERENCES

Bastiaans, D. (2016). What was wrong with Trix? In: Besselink, M. (Ed.) *Trix: The grand old lady.* Naturalis Biodiversity Center, Leiden. pp. 46–47.

Bates, K. T., & Falkingham, P. L. (2012). Estimating maximum bite performance in *Tyrannosaurus rex* using multi-body dynamics. Biology Letters, 8, 660–664.

Bates, K. T., & Falkingham, P. L. (2018). The importance of muscle architecture in biomechanical reconstructions of extinct animals: A case study using *Tyrannosaurus rex.* Journal of Anatomy, 233, 625–635.

Bates, K. T., Manning, P. L., Hodgetts, D., & Sellers, W. I. (2009). Estimating mass properties of dinosaurs using laser imaging and 3D computer modelling. PLOS One, 4, e4532.

Begum, S. (2022, November 22). Cancelled T-Rex auction sparks calls for respect of copyright, more transparency on "real bones." The Straits Times. https://www.straitstimes.com/singapore/cancelled-t-rex-auction-sparks-calls-for-respect-of-copyright-more-transparency-on-real-bones

Bell, P. R., & Currie, P. J. (2010). A tyrannosaur jaw bitten by a confamilial: Scavenging or fatal agonism? Lethaia, 43, 278–281.

Bell, P. R., & Snively, E. (2008). Polar dinosaurs on parade: A review of dinosaur migration. Alcheringa, 32, 271–284.

Bell, P. R., Campione, N. E., Persons IV, W. S., Currie, P. J., Larson, P. L., Tanke, D. H., & Bakker, R. T. (2017). Tyrannosauroid integument reveals conflicting patterns of gigantism and feather evolution. Biology Letters, 13, 20170092.

Bell, P. R., Hendrickx, C., Pittman, M., Kaye, T. G., & Mayr, G. (2022). The exquisitely preserved integument of *Psittacosaurus* and the scaly skin of ceratopsian dinosaurs. Communications Biology, 5, 809.

Benson, R. B. (2008). New information on *Stokesosaurus*, a tyrannosauroid (Dinosauria: Theropoda) from North America and the United Kingdom. Journal of Vertebrate Paleontology, 28, 732–750.

Benson, R. B., Butler, R. J., Carrano, M. T., & O'Connor, P. M. (2012). Air-filled postcranial bones in theropod dinosaurs: Physiological implications and the "reptile"–bird transition. Biological Reviews, 87, 168–193.

Benson, R. B., Carrano, M. T., & Brusatte, S. L. (2010). A new clade of archaic large-bodied predatory dinosaurs (Theropoda: Allosauroidea) that survived to the latest Mesozoic. Naturwissenschaften, 97, 71–78.

Benton, M. J. (1990). Scientific methodologies in collision: The history of the study of the extinction of the dinosaurs. Evolutionary Biology, 24, 371–400.

Benton, M. J. (2000). A brief history of dinosaur paleontology. In: Paul, G. S. (Ed.) *The Scientific American book of dinosaurs.* St. Martin's Press. pp. 10–44.

Benton, M. J. (2004). Origin and relationships of Dinosauria. In: Weishampel, D. B., Dodson, P., and Osmólska, H. (Eds.) *The Dinosauria* (second edition). University of California Press. pp. 7–19.

Benton, M. J. (2021). The origin of endothermy in synapsids and archosaurs and arms races in the Triassic. Gondwana Research, 100, 261–289.

Bern, M., Phinney, B. S., & Goldberg, D. (2009). Reanalysis of *Tyrannosaurus rex* mass spectra. Journal of Proteome Research, 8, 4328–4332.

Bever, G. S., Brusatte, S. L., Balanoff, A. M., & Norell, M. A. (2011). Variation, variability, and the origin of the avian endocranium: Insights from the anatomy of *Alioramus altai* (Theropoda: Tyrannosauroidea). PLOS One, 6, e23393.

Bever, G. S., Brusatte, S., Carr, T. D., Xu, X., Balanoff, A. M., Norell, M., ... & Mongolian-American Museum Paleontological Project. (2013). The braincase anatomy of the late Cretaceous dinosaur *Alioramus* (Theropoda, Tyrannosauroidea). Bulletin of the American Museum of Natural History, 376, 1–72.

Bjil, S. (2016). The growth and age of Trix. In: Besselink, M. (Ed.) *Trix: The grand old lady.* Naturalis Biodiversity Center, Leiden. pp. 74–75.

Black, R. (2022, April 2). Give *T. rex* a Rest! Slate. https://slate.com/technology/2022/04/tyrannosaurus-rex-popularity-tired-paleontology.html

Blumenschine, R. J. (1986). Carcass consumption sequences and the archaeological distinction of scavenging and hunting. Journal of Human Evolution, 15, 639–659.

Boatman, E. M., Goodwin, M. B., Holman, H. Y. N., Fakra, S., Zheng, W., Gronsky, R., & Schweitzer, M. H. (2019). Mechanisms of soft tissue and protein preservation in *Tyrannosaurus rex.* Scientific Reports, 9, 15678.

Bonnan, M. F., Farlow, J. O., & Masters, S. L. (2008). Using linear and geometric morphometrics to detect intraspecific variability and sexual dimorphism in femoral shape in *Alligator mississippiensis* and its implications for sexing fossil archosaurs. Journal of Vertebrate Paleontology, 28, 422–431.

Bonsor, J. A., Barrett, P. M., Raven, T. J., & Cooper, N. (2020). Dinosaur diversification rates were not in decline prior to the K-Pg boundary. Royal Society Open Science, 7, 201195.

Bouabdellah, F., Lessner, E., & Benoit, J. (2022). The rostral neurovascular system of *Tyrannosaurus rex.* Palaeontologia Electronica, 25, 1–20.

Boyd, B. S., Colon, F., Doty, J. F., & Sanders, K. C. (2021). Beware of the dragon: A case report of a Komodo dragon attack. Foot & Ankle Orthopaedics, 6. doi:10.1177/24730114211015623.

Boyd, C. A., Brown, C. M., Scheetz, R. D., & Clarke, J. A. (2009). Taxonomic revision of the basal neornithischian taxa *Thescelosaurus* and *Bugenasaura.* Journal of Vertebrate Paleontology, 29, 758–770.

Bradbury, R. (2003). *Dinosaur tales.* ibooks.

Brain, C., Forge, O., & Erb, P. (1999). Lion predation on black rhinoceros (*Diceros bicornis*). African Journal of Ecology, 37, 107–109.

Brazaitis, P., & Watanabe, M. E. (2011). Crocodilian behaviour: A window to dinosaur behaviour? Historical Biology, 23, 73–90.

Breithaupt, B. H., Southwell, E., & Matthews, N. (2008). Wyoming's *Dynamosaurus imperiosus* and other early discoveries of *Tyrannosaurus rex* in the Rocky Mountain West. In: Larson, P., & Carpenter, K. (Eds.) Tyrannosaurus rex, *the tyrant king.* Indiana University Press. pp. 57–61.

Brett-Surman, M. K., & Farlow, J. O. (1997). Some irreverent thoughts about dinosaur metabolic physiology: Jurisphagous food consumption rates of *Tyrannosaurus rex.* In: Farlow, J. O., & Brett-Surman, M. K. (Eds.) *The complete dinosaur.* Indiana University Press. pp. 350–351.

Brinkman, D. B., Newbrey, M. G., & Neuman, A.G. (2014). Diversity and paleoecology of actinopterygian fish from vertebrate microfossil localities of the Maastrichtian Hell Creek Formation of Montana. In: Wilson, G. P., Clemens, W. A., Horner, J. R., & Hartman, J. H. (Eds.) *Through the end of the Cretaceous in the type locality of the Hell Creek Formation in Montana and adjacent areas.* Geological Society of America Special Paper, 503, 247–270.

REFERENCES

Brinkman, P. D. (2010). *The second Jurassic dinosaur rush: Museums and paleontology in America at the turn of the twentieth century.* University of Chicago Press.

Brochu, C. A. (2000). A digitally-rendered endocast for *Tyrannosaurus rex*. Journal of Vertebrate Paleontology, 20, 1–6.

Brochu, C. A. (2003). Osteology of *Tyrannosaurus rex*: Insights from a nearly complete skeleton and high-resolution computed tomographic analysis of the skull. Journal of Vertebrate Paleontology, 22, 1–138.

Bro-Jørgensen, J. (2007). The intensity of sexual selection predicts weapon size in male bovids. Evolution, 61, 1316–1326.

Brown, B. (1915). *Tyrannosaurus*, a Cretaceous carnivorous dinosaur: The largest flesh eater that ever lived. Scientific American, 113, 322–323.

Brown, C. M., Currie, P. J., & Therrien, F. (2022). Intraspecific facial bite marks in tyrannosaurids provide insight into sexual maturity and evolution of bird-like intersexual display. Paleobiology, 48, 12–43.

Brown, C. M., VanBuren, C. S., Larson, D. W., Brink, K. S., Campione, N. E., Vavrek, M. J., & Evans, D. C. (2015). Tooth counts through growth in diapsid reptiles: Implications for interpreting individual and size-related variation in the fossil record. Journal of Anatomy, 226, 322–333.

Brownstein, C. D. (2021). Dinosaurs from the Santonian–Campanian Atlantic coastline substantiate phylogenetic signatures of vicariance in Cretaceous North America. Royal Society Open Science, 8, 210127.

Brusatte, S. (2018). *The rise and fall of the dinosaurs: The untold story of a lost world*. Macmillan.

Brusatte, S. L. (2010). Tyrannosaurus rex, the tyrant king. Journal of Vertebrate Paleontology, 30, 304–305.

Brusatte, S. L., & Benson, R. B. (2013). The systematics of Late Jurassic tyrannosauroid theropods from Europe and North America. Acta Palaeontologica Polonica, 58, 47–54.

Brusatte, S. L., & Carr, T. D. (2016). The phylogeny and evolutionary history of tyrannosauroid dinosaurs. Scientific Reports, 6, 20252.

Brusatte, S. L., Benson, R. B., & Norell, M. A. (2011). The anatomy of *Dryptosaurus aquilunguis* (Dinosauria: Theropoda) and a review of its tyrannosauroid affinities. American Museum Novitates, 2011, 1–53.

Brusatte, S. L., Butler, R. J., Barrett, P. M., Carrano, M. T., Evans, D. C., Lloyd, G. T., ... & Williamson, T. E. (2015). The extinction of the dinosaurs. Biological Reviews, 90, 628–642.

Brusatte, S. L., Carr, T. D., & Norell, M. A. (2012). The osteology of *Alioramus*, a gracile and long-snouted tyrannosaurid (Dinosauria: Theropoda) from the Late Cretaceous of Mongolia. Bulletin of the American Museum of Natural History, 366, 1–197.

Brusatte, S. L., Carr, T. D., Erickson, G. M., Bever, G. S., & Norell, M. A. (2009). A long-snouted, multihorned tyrannosaurid from the Late Cretaceous of Mongolia. Proceedings of the National Academy of Sciences, 106, 17261–17266.

Brusatte, S. L., Carr, T. D., Williamson, T. E., Holtz, T. R., Jr., Hone, D. W. E., & Williams, S. A. (2016). Dentary groove morphology does not distinguish "*Nanotyrannus*" as a valid taxon of tyrannosauroid dinosaur. Comment on: "Distribution of the dentary groove of theropod dinosaurs: Implications for theropod phylogeny and the validity of the genus *Nanotyrannus* Bakker et al., 1988." Cretaceous Research, 65, 232–237.

Brusatte, S. L., Nesbitt, S. J., Irmis, R. B., Butler, R. J., Benton, M. J., & Norell, M. A. (2010a). The origin and early radiation of dinosaurs. Earth-Science Reviews, 101, 68–100.

Brusatte, S. L., Norell, M. A., Carr, T. D., Erickson, G. M., Hutchinson, J. R., Balanoff, A. M., ... & Xu, X. (2010b). Tyrannosaur paleobiology: New research on ancient exemplar organisms. Science, 329, 1481–1485.

Buck, B. J., & Mack, G. H. (1995). Latest Cretaceous (Maastrichtian) aridity indicated by paleosols in the McRae Formation, south-central New Mexico. Cretaceous Research, 16, 559–572.

Buckley, M., Warwood, S., van Dongen, B., Kitchener, A. C., & Manning, P. L. (2017). A fossil protein chimera; difficulties in discriminating dinosaur peptide sequences from modern cross-contamination. Proceedings of the Royal Society. Series B: Biological Sciences, 284, 20170544.

Buffetaut, E. (1987). *A short history of vertebrate paleontology*. Croom Helm.

Burns, M. E. (2015). Intraspecific variation in Late Cretaceous nodosaurids (Ankylosauria: Dinosauria). Journal of Vertebrate Paleontology, SVP Program and Abstracts Book 2015, 99–100.

Burns, M. E., Coy, C., Arbour, V. M., Currie, P. J., & Koppelhus, E. B. (2014). The Danek *Edmontosaurus* Bonebed: New insights on the systematics, biogeography, and palaeoecology of Late Cretaceous dinosaur communities. Canadian Journal of Earth Sciences, 51, v–vii.

Butler, R. J., Barrett, P. M., & Gower, D. J. (2012). Reassessment of the evidence for postcranial skeletal pneumaticity in Triassic archosaurs, and the early evolution of the avian respiratory system. PLOS One, 7, e34094.

Button, D. J., & Zanno, L. E. (2023). Neuroanatomy of the late Cretaceous *Thescelosaurus neglectus* (Neornithischia: Thescelosauridae) reveals novel ecological specialisations within Dinosauria. Scientific Reports, 13, 19224.

Bybee, P. J., Lee, A. H., & Lamm, E. T. (2006). Sizing the Jurassic theropod dinosaur *Allosaurus*: Assessing growth strategy and evolution of ontogenetic scaling of limbs. Journal of Morphology, 267, 347–359.

Calvo, J. O., & Coria, R. A. (1998). New specimen of *Giganotosaurus carolinii* (Coria & Salgado, 1995), supports it as the largest theropod ever found. Gaia, 15, 117–122.

Campbell, K. E., & Marcus, L. (1992). The relationship of hindlimb bone dimensions to body weight in birds. Natural History Museum of Los Angeles County Science Series, 36, 395–412.

Campione, N. E., & Evans, D. C. (2012). A universal scaling relationship between body mass and proximal limb bone dimensions in quadrupedal terrestrial tetrapods. BMC Biology, 10, 1–22.

Campione, N. E., & Evans, D. C. (2020). The accuracy and precision of body mass estimation in non-avian dinosaurs. Biological Reviews, 95, 1759–1797.

Campione, N. E., Barrett, P. M., & Evans, D. C. (2020). On the ancestry of feathers in Mesozoic dinosaurs. In: Foth, C. Rauhut, O. W. M. (Eds.) *The evolution of feathers: From their origin to the present*. Springer. pp. 213–243.

Campione, N. E., Evans, D. C., Brown, C. M., & Carrano, M. T. (2014). Body mass estimation in non-avian bipeds using a theoretical conversion to quadruped stylopodial proportions. Methods in Ecology and Evolution, 5, 913–923.

Canale, J. I., Apesteguía, S., Gallina, P. A., Mitchell, J., Smith, N. D., Cullen, T. M., ... & Makovicky, P. J. (2022). New giant carnivorous

REFERENCES

dinosaur reveals convergent evolutionary trends in theropod arm reduction. Current Biology, 32, 3195–3202.

Candeiro, C. R. D. A., Brusatte, S. L., Vidal, L., & Pereira, P. V. L. G. D. C. (2018). Paleobiogeographic evolution and distribution of Charcharodontosauridae (Dinosauria, Theropoda) during the middle Cretaceous of North Africa. Papéis Avulsos de Zoologia, 58, e20185829.

Cantlay, J. C., Martin, G. R., McClelland, S. C., Potier, S., O'Brien, M. F., Fernández-Juricic, E., ... & Portugal, S. J. (2023). Binocular vision and foraging in ducks, geese and swans (Anatidae). Proceedings of the Royal Society. Series B, 290, 20231213.

Carbone, C., Turvey, S. T., & Bielby, J. (2011). Intra-guild competition and its implications for one of the biggest terrestrial predators, Tyrannosaurus rex. Proceedings of the Royal Society. Series B: Biological Sciences, 278, 2682–2690.

Caro, T. I. M. (2005). The adaptive significance of coloration in mammals. BioScience, 55, 125–136.

Caro, T. I. M. (2013). The colours of extant mammals. In: Hofreiter, M. (Ed.) Genetic basis and evolutionary causes of colour variation in vertebrates. Seminars in Cell & Developmental Biology, Academic Press, 24, 542–552.

Caro, T. M., Graham, C. M., Stoner, C. J., & Flores, M. M. (2003). Correlates of horn and antler shape in bovids and cervids. Behavioral Ecology and Sociobiology, 55, 32–41.

Carpenter, K. (1988). Evidence of predatory behavior by Tyrannosaurus. In: Horner, J. R. (Ed.) International symposium on vertebrate behavior as derived from the fossil record. Museum of the Rockies, Montana State University. (unpaginated)

Carpenter, K. (1990). Variation in Tyrannosaurus rex. In: Carpenter, K., & Currie, P. Dinosaur systematics: Perspectives and approaches. Cambridge University Press. pp. 141–145.

Carpenter, K. (1992). Tyrannosaurids (Dinosauria) of Asia and North America. In: Mateer, N., & Chen, P. (Eds.) Aspects of nonmarine Cretaceous geology. China Ocean Press. pp. 250–268.

Carpenter, K. (1998). Evidence of predatory behavior by carnivorous dinosaurs. Gaia, 15, 135–144.

Carpenter, K. (1999). Eggs, nests, and baby dinosaurs: A look at dinosaur reproduction. Indiana University Press.

Carpenter, K. (2002). Forelimb biomechanics of nonavian theropod dinosaurs in predation. Senckenbergiana lethaea, 82, 59–75.

Carpenter, K., (2013). A closer look at the hypothesis of scavenging versus predation by Tyrannosaurus rex. In: Parrish, J. M., Molnar, R. E., Currie, P. J., & Koppelhus, E. B. (Eds.) Tyrannosaurid paleobiology. Indiana University Press. pp. 265–277.

Carpenter, K., & Smith, M. (2001). Forelimb osteology and biomechanics of Tyrannosaurus rex. In: Tanke, D. H., & Carpenter, K. (Eds.) Mesozoic vertebrate life. Indiana University Press. pp. 90–116.

Carpenter, K., & Young, D. B. (2002). Late Cretaceous dinosaurs from the Denver Basin, Colorado. Rocky Mountain Geology, 37, 237–254.

Carr, T. D. (1999). Craniofacial ontogeny in tyrannosauridae (Dinosauria, Coelurosauria). Journal of Vertebrate Paleontology, 19, 497–520.

Carr, T. D. (2020). A high-resolution growth series of Tyrannosaurus rex obtained from multiple lines of evidence. PeerJ, 8, e9192.

Carr, T. D. (2021). Tyrannosaurus rex: An endangered species. Journal of Vertebrate Paleontology, SVP Program and Abstracts Book 2021, 78–79.

Carr, T. D. (2023). A reappraisal of tyrannosauroid fossils from the Iren Dabasu Formation (Coniacian–Campanian), Inner Mongolia, People's Republic of China. Journal of Vertebrate Paleontology, 42, e2199817.

Carr, T. D., & Williamson, T. E. (2000). A review of Tyrannosauridae (Dinosauria, Coelurosauria) from New Mexico. New Mexico Museum of Natural History and Science Bulletin, 17, 113–145.

Carr, T. D., & Williamson, T. E. (2004). Diversity of late Maastrichtian Tyrannosauridae (Dinosauria: Theropoda) from western North America. Zoological Journal of the Linnean Society, 142, 479–523.

Carr, T. D., & Williamson, T. E. (2010). Bistahieversor sealeyi, gen. et sp. nov., a new tyrannosauroid from New Mexico and the origin of deep snouts in Tyrannosauroidea. Journal of Vertebrate Paleontology, 30, 1–16.

Carr, T. D., Napoli, J. G., Brusatte, S. L., Holtz, T. R., Hone, D. W. E., Williamson, T. E., & Zanno, L. E. (2022). Insufficient evidence for multiple species of Tyrannosaurus in the latest Cretaceous of North America: A comment on "The tyrant lizard king, queen and emperor: Multiple lines of morphological and stratigraphic evidence support subtle evolution and probable speciation within the North American genus Tyrannosaurus." Evolutionary Biology, 49, 327–341.

Carr, T. D., Varricchio, D. J., Sedlmayr, J. C., Roberts, E. M., & Moore, J. R. (2017). A new tyrannosaur with evidence for anagenesis and crocodile-like facial sensory system. Scientific Reports, 7, 44942.

Carr, T. D., Williamson, T. E., & Schwimmer, D. R. (2005). A new genus and species of tyrannosauroid from the Late Cretaceous (Middle Campanian) Demopolis Formation of Alabama. Journal of Vertebrate Paleontology, 25, 119–143.

Carr, T. D., Williamson, T. E., Britt, B. B., & Stadtman, K. (2011). Evidence for high taxonomic and morphologic tyrannosauroid diversity in the Late Cretaceous (Late Campanian) of the American Southwest and a new short-skulled tyrannosaurid from the Kaiparowits formation of Utah. Naturwissenschaften, 98, 241–246.

Carrano, M. T. (1999). What, if anything, is a cursor? Categories versus continua for determining locomotor habit in mammals and dinosaurs. Journal of Zoology, 247, 29–42.

Carrano, M. T., & D'Emic, M. D. (2015). Osteoderms of the titanosaur sauropod dinosaur Alamosaurus sanjuanensis Gilmore, 1922. Journal of Vertebrate Paleontology, 35, e901334.

Carrano, M. T., & Hutchinson, J. R. (2002). Pelvic and hindlimb musculature of Tyrannosaurus rex (Dinosauria: Theropoda). Journal of Morphology, 253, 207–228.

Carrano, M. T., & Sampson, S. D. (2008). The phylogeny of Ceratosauria (Dinosauria: Theropoda). Journal of Systematic Palaeontology, 6, 183–236.

Carrano, M. T., Benson, R. B., & Sampson, S. D. (2012). The phylogeny of Tetanurae (Dinosauria: Theropoda). Journal of Systematic Palaeontology, 10, 211–300.

Carrano, M. T., Mateus, O., & Mitchell J. (2013). First definitive association between embryonic Allosaurus bones and Prismatoolithus eggs in the Morrison Formation (Upper Jurassic, Wyoming, USA). Journal of Vertebrate Paleontology, SVP Program and Abstracts Book 2013, 101.

Carrier, D. R. (1996). Ontogenetic limits on locomotor performance. Physiological zoology, 69, 467–488.

Caspar, K. R., Gutiérrez-Ibáñez, C., Bertrand, O. C., Carr, T., Colbourne, J., Erb, A., ... & Hurlburt, G. R. (2024). How smart was T. rex? Testing claims of exceptional cognition in dinosaurs and the application of neuron count estimates in palaeontological research. The Anatomical Record, 1–32.

Cau, A. (2023, September 23). *Dakotaraptor* non esiste. Theropoda. https://theropoda.blogspot.com/2023/09/dakotaraptor-non-esiste.html

Cau, A., Dalla Vecchia, F. M., & Fabbri, M. (2013). A thick-skulled theropod (Dinosauria, Saurischia) from the Upper Cretaceous of Morocco with implications for carcharodontosaurid cranial evolution. Cretaceous Research, 40, 251–260.

Cerroni, M. A., Canale, J. I., & Novas, F. E. (2021). The skull of *Carnotaurus sastrei* Bonaparte 1985 revisited: Insights from craniofacial bones, palate and lower jaw. Historical Biology, 33, 2444–2485.

Chatterjee, S. (1985). *Postosuchus*, a new thecodontian reptile from the Triassic of Texas and the origin of tyrannosaurs. Philosophical Transactions of the Royal Society of London. Series B: Biological Sciences, 309, 395–460.

Cherry, M. A. (2005). A *Tyrannosaurus*-Rex aptly named Sue: Using a disputed dinosaur to teach contract defenses. North Dakota Law Review, 81, 295.

Chiarenza, A. A., Mannion, P. D., Lunt, D. J., Farnsworth, A., Jones, L. A., Kelland, S. J., & Allison, P. A. (2019). Ecological niche modelling does not support climatically-driven dinosaur diversity decline before the Cretaceous/Paleogene mass extinction. Nature Communications, 10, 1091.

Chin, K., Tokaryk, T. T., Erickson, G. M., & Calk, L. C. (1998). A king-sized theropod coprolite. Nature, 393, 680–682.

Christiansen, P. (1998). Strength indicator values of theropod long bones, with comments on limb proportions and cursorial potential. Gaia, 15, 241–255.

Christiansen, P., & Fariña, R. A. (2004). Mass prediction in theropod dinosaurs. Historical Biology, 16, 85–92.

Christiansen, P., & Paul, G. S. (2001). Limb bone scaling, limb proportions, and bone strength in neoceratopsian dinosaurs. Gaia, 16, 13–29.

Chure, D. J. (1995). A reassessment of the gigantic theropod *Saurophagus maximus* from the Morrison Formation (Upper Jurassic) of Oklahoma, USA. In: Sun, A., and Wang, T. (Eds.) *Sixth symposium on Mesozoic terrestrial ecosystems and biota*. China Ocean Press. pp. 103–106.

Claessens, L. P., & Loewen, M. A. (2016). A redescription of *Ornithomimus velox* Marsh, 1890 (Dinosauria, Theropoda). Journal of Vertebrate Paleontology, 36, e1034593.

Claessens, L. P., O'Connor, P. M., & Unwin, D. M. (2009). Respiratory evolution facilitated the origin of pterosaur flight and aerial gigantism. PLOS One, 4, e4497.

Clarke, J. A., Chatterjee, S., Li, Z., Riede, T., Agnolin, F., Goller, F., ... & Novas, F. E. (2016). Fossil evidence of the avian vocal organ from the Mesozoic. Nature, 538, 502–505.

Claytor, J. R. (2023). Insights into the evolution and ecology of mammals from the Hell Creek region of northeastern Montana. Unpublished doctoral dissertation, University of Washington.

Clemens, W. A., & Archibald, J. D. (1980). Evolution of terrestrial faunas during the Cretaceous-Tertiary transition. Mémoires de la Société Géologique de France, Nouvelle Série, 139, 67–74.

Clemens, W. A., Archibald, J. D., & Hickey, L. J. (1981). Out with a whimper not a bang. Paleobiology, 7, 293–298.

Codron, D., Carbone, C., & Clauss, M. (2013). Ecological interactions in dinosaur communities: Influences of small offspring and complex ontogenetic life histories. PLOS One, 8, e77110.

Colbert, E. H. (1962). The weights of dinosaurs. American Museum Novitates, 2076, 1–16.

Colbert, E. H. (1965). *The age of reptiles*. W. W. Norton.

Colbert, E. H., Gillette, D. D., & Molnar, R. E. (2012). North American dinosaur hunters. In: Brett-Surman, M. K., Holtz, T. R., Jr., & Farlow, J. O. (Eds.) *The complete dinosaur* (second edition). Indiana University Press. pp. 61–72.

Condamine, F. L., Guinot, G., Benton, M. J., & Currie, P. J. (2021). Dinosaur biodiversity declined well before the asteroid impact, influenced by ecological and environmental pressures. Nature Communications, 12, 3833.

Cook, T. D., Newbrey, M. G., Brinkman, D. B., Kirkland, J. I., Wilson, G. P., Clemens, W. A., ... & Hartman, J. H. (2014). Euselachians from the freshwater deposits of the Hell Creek Formation of Montana. In: Wilson, G. P., Clemens, W. A., Horner, J. R., & Hartman, J. H. (Eds.) *Through the end of the Cretaceous in the type locality of the Hell Creek Formation in Montana and adjacent areas*. Geological Society of America Special Paper, 503, 229–246.

Coombs, W. P., Jr. (1978). Theoretical aspects of cursorial adaptations in dinosaurs. The Quarterly Review of Biology, 53, 393–418.

Cope, E. D. (1866). Remarks on the remains of a gigantic extinct dinosaur from the Cretaceous Greensand of New Jersey. Proceedings of the Academy of Natural Sciences of Philadelphia, 18, 275–279.

Cope, E. D. (1868). On some Cretaceous Reptilia. Proceedings of the Academy of Natural Sciences of Philadelphia, 30, 233–242.

Cope, E. D. (1892). Fourth note on the Dinosauria of the Laramie. The American Naturalist, 26, 756–758.

Coria, R. A., & Currie, P. J. (2016). A new megaraptoran dinosaur (Dinosauria, Theropoda, Megaraptoridae) from the Late Cretaceous of Patagonia. PLOS One, 11, e0157973.

Coria, R. A., & Salgado, L. (1995). A new giant carnivorous dinosaur from the Cretaceous of Patagonia. Nature, 377, 224–226.

Cornwall, W. (2020, May 22). Court rules "Dueling Dinos" belong to landowners, in a win for science. Science. https://www.science.org/content/article/court-rules-dueling-dinos-belong-landowners-win-science

Cost, I. N., Middleton, K. M., Sellers, K. C., Echols, M. S., Witmer, L. M., Davis, J. L., & Holliday, C. M. (2020). Palatal biomechanics and its significance for cranial kinesis in *Tyrannosaurus rex*. The Anatomical Record, 303, 999–1017.

Cuadrado, M., Martín, J., & López, P. (2001). Camouflage and escape decisions in the common chameleon *Chamaeleo chamaeleon*. Biological Journal of the Linnean Society, 72, 547–554.

Cuff, A. R., Demuth, O. E., Michel, K., Otero, A., Pintore, R., Polet, D. T., ... & Hutchinson, J. R. (2022). Walking—and running and jumping—with dinosaurs and their cousins, viewed through the lens of evolutionary biomechanics. Integrative and Comparative Biology, 62, 1281–1305.

Culhane, J. (1987). *Walt Disney's* Fantasia. Harry N. Abrams.

Cullen, T. M., Canale, J. I., Apesteguía, S., Smith, N. D., Hu, D., & Makovicky, P. J. (2020). Osteohistological analyses reveal diverse strategies of theropod dinosaur body-size evolution. Proceedings of the Royal Society. Series: B, 287, 20202258.

Cullen, T. M., Larson, D. W., Witton, M. P., Scott, D. Maho, T. Brink, K. S., Evans, D. C. and Reisz, R. (2023). Theropod dinosaur facial reconstruction and the importance of soft tissues in paleobiology. Science, 379, 1348–1352.

Currey, J. D. (2006). *Bones: Structure and mechanics*. Princeton University Press.

Currie, P. J. (1998). Possible evidence of gregarious behavior in tyrannosaurids. Gaia, 15, 271–277.

Currie, P. J. (2003a). Cranial anatomy of tyrannosaurid dinosaurs from the Late Cretaceous of Alberta, Canada. Acta Palaeontologica Polonica, 48, 191–226.

Currie, P. J. (2003b). Allometric growth in tyrannosaurids (Dinosauria: Theropoda) from the upper Cretaceous of North America and Asia. Canadian Journal of Earth Sciences, 40, 651–665.

Currie, P. J. (2023). Celebrating dinosaurs: Their behaviour, evolution, growth, and physiology. Canadian Journal of Earth Sciences, 60, 263–293.

Currie, P. J., & Eberth, D. A. (2010). On gregarious behavior in *Albertosaurus*. Canadian Journal of Earth Sciences, 47, 1277–1290.

Currie, P. J., Trexler, D., Koppelhus, E. B., Wicks, K., & Murphy, N. (2005). An unusual multi-individual tyrannosaurid bonebed in the Two Medicine Formation (Late Cretaceous, Campanian) of Montana (USA). In: Carpenter, K. (Ed.) *The carnivorous dinosaurs*. Indiana University Press. pp. 313–324.

Czaczkes, T. J. (2022). Advanced cognition in ants. Myrmecological News, 32, 51–64.

Dalla Vecchia, F. M., Riera, V., Oms, J. O., Dinarès-Turell, J., Gaete, R., & Galobart, A. (2013). The last pterosaurs: First record from the uppermost Maastrichtian of the Tremp Syncline (Northern Spain). Acta Geologica Sinica-English Edition, 87, 1198–1227.

Dalman, S. G. (2013). New examples of *Tyrannosaurus rex* from the Lance formation of Wyoming, United States. Bulletin of the Peabody Museum of Natural History, 54, 241–254.

Dalman, S. G., Loewen, M. A., Pyron, R. A., Jasinski, S. E., Malinzak, D. E., Lucas, S. G., ... & Longrich, N. R. (2024). A giant tyrannosaur from the Campanian–Maastrichtian of southern North America and the evolution of tyrannosaurid gigantism. Scientific Reports, 14(1), 22124.

Dal Sasso, C., Maganuco, S., Buffetaut, E., & Mendez, M. A. (2005). New information on the skull of the enigmatic theropod *Spinosaurus*, with remarks on its size and affinities. Journal of Vertebrate Paleontology, 25, 888–896.

D'Amore, D. C., & Blumensehine, R. J. (2009). Komodo monitor (*Varanus komodoensis*) feeding behavior and dental function reflected through tooth marks on bone surfaces, and the application to ziphodont paleobiology. Paleobiology, 35, 525–552.

D'Anastasio, R., Cilli, J., Bacchia, F., Fanti, F., Gobbo, G., & Capasso, L. (2022). Histological and chemical diagnosis of a combat lesion in *Triceratops*. Scientific Reports, 12, 3941.

Darwin, C. (1871). *The descent of man, and selection in relation to sex*. John Murray.

Davey, C. (2019). *The American Museum of Natural History and how it got that way*. Fordham University Press.

Dawson, T. J., & Maloney, S. K. (2004). Fur versus feathers: The different roles of red kangaroo fur and emu feathers in thermoregulation in the Australian arid zone. Australian Mammalogy, 26, 145–151.

Dean, C. D., Chiarenza, A. A., & Maidment, S. C. (2020). Formation binning: A new method for increased temporal resolution in regional studies, applied to the Late Cretaceous dinosaur fossil record of North America. Palaeontology, 63, 881–901.

Dececchi, T. A., Mloszewska, A. M., Holtz, T. R., Jr., Habib, M. B., & Larsson, H. C. (2020). The fast and the frugal: Divergent locomotory strategies drive limb lengthening in theropod dinosaurs. PLOS One, 15, e0223698.

Deeming, D. C. (2006). Ultrastructural and functional morphology of eggshells supports the idea that dinosaur eggs were incubated buried in a substrate. Palaeontology, 49, 171–185.

Delcourt, R. (2018). Ceratosaur palaeobiology: New insights on evolution and ecology of the southern rulers. Scientific Reports, 8, 1–12.

Delcourt, R., & Grillo, O. N. (2018). Tyrannosauroids from the Southern Hemisphere: Implications for biogeography, evolution, and taxonomy. Palaeogeography, Palaeoclimatology, Palaeoecology, 511, 379–387.

D'Emic, M. D., O'Connor, P. M., Pascucci, T. R., Gavras, J. N., Mardakhayava, E., & Lund, E. K. (2019). Evolution of high tooth replacement rates in theropod dinosaurs. PLOS One, 14, e0224734.

D'Emic, M. D., O'Connor, P. M., Sombathy, R. S., Cerda, I., Pascucci, T. R., Varricchio, D., ... & Curry Rogers, K. A. (2023). Developmental strategies underlying gigantism and miniaturization in non-avialan theropod dinosaurs. Science, 379, 811–814.

D'Emic, M. D., Wilson, J. A., & Williamson, T. E. (2011). A sauropod dinosaur pes from the latest Cretaceous of North America and the validity of *Alamosaurus sanjuanensis* (Sauropoda, Titanosauria). Journal of Vertebrate Paleontology, 31, 1072–1079.

DePalma, R. A., Burnham, D. A., Martin, L. D., Larson, P. L., & Bakker, R. T. (2015). The first giant raptor (Theropoda: Dromaeosauridae) from the Hell Creek formation. Paleontological Contributions, 14, 1–16.

DePalma, R. A., Burnham, D. A., Martin, L. D., Rothschild, B. M., & Larson, P. L. (2013). Physical evidence of predatory behavior in *Tyrannosaurus rex*. Proceedings of the National Academy of Sciences, 110, 12560–12564.

de Ricqles, A. J. (1974). Evolution of endothermy: Histological evidence. Evolutionary Theory, 1, 51–80.

Derstler, K. L., & Myers, J. M. (2008). Taphonomy of the *Tyrannosaurus rex* Peck's rex from the Hell Creek Formation of Montana. In: Larson, P., and Carpenter, K. (Eds.) Tyrannosaurus rex, *the tyrant king*. Indiana University Press. pp. 74–81.

Designing Godzilla. (2006) *Godzilla* (DVD), BFI.

Desmond, A. J. (1975). *The hot-blooded dinosaurs: A revolution in palaeontology*. Dial Press.

Dhiman, H., Verma, V., Singh, L. R., Miglani, V., Jha, D. K., Sanyal, P., ... & Prasad, G. V. (2023). New Late Cretaceous titanosaur sauropod dinosaur egg clutches from lower Narmada valley, India: Palaeobiology and taphonomy. PLOS One, 18, e0278242.

Dial, T. R., & Carrier, D. R. (2012). Precocial hindlimbs and altricial forelimbs: Partitioning ontogenetic strategies in mallards (*Anas platyrhynchos*). Journal of Experimental Biology, 215, 3703–3710.

Dinets, V. (2015). Apparent coordination and collaboration in cooperatively hunting crocodilians. Ethology Ecology & Evolution, 27(2), 244–250.

Dingus, L., & Norell, M. (2010). *Barnum Brown: The man who discovered* Tyrannosaurus rex. University of California Press.

Dixon, D. (1988). *The new dinosaurs: An alternative evolution*. Salem House.

Dong, Z. M. 1979. Cretaceous dinosaur fossils in south China. In: IVPP Academia Sinica, Nanjing Institute of Geology and Paleontology,

REFERENCES

Academia Sinica (Eds.) *Memoirs on Mesozoic red beds of south China*, Science Press. pp. 342–350. [in Chinese]

Druckenmiller, P. S., Erickson, G. M., Brinkman, D., Brown, C. M., & Eberle, J. J. (2021). Nesting at extreme polar latitudes by non-avian dinosaurs. Current Biology, 31, 3469–3478.

Dunham, W. (2024, January 11). Scientists conclude New Mexico fossil is new *Tyrannosaurus* species. Reuters. https://www.reuters.com/science/scientists-conclude-new-mexico-fossil-is-new-tyrannosaurus-species-2024-01-11/

During, M. A., Smit, J., Voeten, D. F., Berruyer, C., Tafforeau, P., Sanchez, S., ... & van der Lubbe, J. H. (2022). The Mesozoic terminated in boreal spring. Nature, 603, 91–94.

Eberth, D. A. (2015). Origins of dinosaur bonebeds in the Cretaceous of Alberta, Canada. Canadian Journal of Earth Sciences, 52, 655–681.

Elbein, A. (2024a, January 3). Study aims to bring a tinier tyrannosaur back from oblivion. New York Times. https://www.nytimes.com/2024/01/03/science/tyrannosaurus-rex-nanotyrannus-fossils.html

Elbein, A. (2024b, January 11). New origin story for *Tyrannosaurus rex* suggested by fossil. New York Times. https://www.nytimes.com/2024/01/11/science/new-tyrannosaur-species-fossil.html

Erickson, G. M. (1996). Incremental lines of von Ebner in dinosaurs and the assessment of tooth replacement rates using growth line counts. Proceedings of the National Academy of Sciences, 93, 14623–14627.

Erickson, G. M. (1999). Breathing life into *Tyrannosaurus rex*. Scientific American, 281, 42–49.

Erickson, G. M. (2014). On dinosaur growth. Annual Review of Earth and Planetary Sciences, 42, 675–697.

Erickson, G. M., & Olson, K. H. (1996). Bite marks attributable to *Tyrannosaurus rex*: Preliminary description and implications. Journal of Vertebrate Paleontology, 16, 175–178.

Erickson, G. M., Currie, P. J., Inouye, B. D., & Winn, A. A. (2006). Tyrannosaur life tables: An example of nonavian dinosaur population biology. Science, 313, 213–217.

Erickson, G. M., Gignac, P. M., Steppan, S. J., Lappin, A. K., Vliet, K. A., Brueggen, J. D., ... & Webb, G. J. (2012). Insights into the ecology and evolutionary success of crocodilians revealed through bite-force and tooth-pressure experimentation. PLOS One, 7, e31781.

Erickson, G. M., Kirk, S. D. V., Su, J., Levenston, M. E., Caler, W. E., & Carter, D. R. (1996). Bite-force estimation for *Tyrannosaurus rex* from tooth-marked bones. Nature, 382, 706–708.

Erickson, G. M., Lappin, A. K., & Larson, P. (2005). Androgynous rex–the utility of chevrons for determining the sex of crocodilians and non-avian dinosaurs. Zoology, 108, 277–286.

Erickson, G. M., Makovicky, P. J., Currie, P. J., Norell, M. A., Yerby, S. A., & Brochu, C. A. (2004). Gigantism and comparative life-history parameters of tyrannosaurid dinosaurs. Nature, 430, 772–775.

Erickson, G. M., Rogers, K. C., & Yerby, S. A. (2001). Dinosaurian growth patterns and rapid avian growth rates. Nature, 412, 429–433.

Estes, R. D. (1991). The significance of horns and other male secondary sexual characters in female bovids. Applied Animal Behaviour Science, 29, 403–451.

Evans, D. C., Larson, D. W., & Currie, P. J. (2013). A new dromaeosaurid (Dinosauria: Theropoda) with Asian affinities from the latest Cretaceous of North America. Naturwissenschaften, 100, 1041–1049.

Everhart, M. J. (2017). *Oceans of Kansas: A natural history of the Western Interior Sea*. Indiana University Press.

Evers, S. W., Rauhut, O. W., Milner, A. C., McFeeters, B., & Allain, R. (2015). A reappraisal of the morphology and systematic position of the theropod dinosaur *Sigilmassasaurus* from the "middle" Cretaceous of Morocco. PeerJ, 3, e1323.

Ezcurra, M. D., Butler, R. J., & Gower, D. J. (2013). "Proterosuchia": The origin and early history of Archosauriformes. In: Nesbitt, S. J., Desojo, J. B., & Irmis, R. B. (Eds.) *Anatomy, phylogeny and palaeobiology of early archosaurs and their kin*. Geological Society, London, Special Publications, 379, 9–33.

Fabbri, M., Navalón, G., Benson, R. B., Pol, D., O'Connor, J., Bhullar, B. A. S., ... & Ibrahim, N. (2022). Subaqueous foraging among carnivorous dinosaurs. Nature, 603, 852–857.

Farke, A. (2009). Review of Tyrannosaurus rex: *The tyrant king*. Palaeontologia Electronica, 12, R2; 2pp.; http://palaeo-electronica.org/2009_1/index.html

Farke, A. A. (2004). Horn use in *Triceratops* (Dinosauria: Ceratopsidae): Testing behavioral hypotheses using scale models. Palaeontologia Electronica, 7, 10p.

Farke, A. A. (2014). Evaluating combat in ornithischian dinosaurs. Journal of Zoology, 292, 242–249.

Farke, A. A., Wolff, E. D., & Tanke, D. H. (2009). Evidence of combat in *Triceratops*. PLOS One, 4, e4252.

Farlow, J. O. (1993). On the rareness of big, fierce animals; speculations about the body sizes, population densities, and geographic ranges of predatory mammals and large carnivorous dinosaurs. American Journal of Science, 293A, 167–199.

Farlow, J. O. (1994). Speculations about the carrion-locating ability of tyrannosaurs. Historical Biology, 7, 159–165.

Farlow, J. O., & Dodson, P. (1975). The behavioral significance of frill and horn morphology in ceratopsian dinosaurs. Evolution, 29, 353–361.

Farlow, J. O., & Holtz, T. R., Jr. (2002). The fossil record of predation in dinosaurs. The Paleontological Society Papers, 8, 251–266.

Farlow, J. O., Brinkman, D. L., Abler, W. L., & Currie, P. J. (1991). Size, shape, and serration density of theropod dinosaur lateral teeth. Modern Geology, 16, 161–198.

Farlow, J. O., Gatesy, S. M., Holtz, T. R., Jr., Hutchinson, J. R., & Robinson, J. M. (2000). Theropod locomotion. American Zoologist, 40, 640–663.

Farlow, J. O., Smith, M. B., & Robinson, J. M. (1995). Body mass, bone "strength indicator," and cursorial potential of *Tyrannosaurus rex*. Journal of Vertebrate Paleontology, 15, 713–725.

Fassett, J. E. (2009). New geochronologic and stratigraphic evidence confirms the Paleocene age of the dinosaur-bearing Ojo Alamo Sandstone and Animas Formation in the San Juan Basin, New Mexico and Colorado. Palaeontologia Electronica, 12, 146p.

Fastovsky, D. E., & Bercovici, A. (2016). The Hell Creek Formation and its contribution to the Cretaceous–Paleogene extinction: A short primer. Cretaceous Research, 57, 368–390.

Field, D. J., Benito, J., Chen, A., Jagt, J. W., & Ksepka, D. T. (2020). Late Cretaceous neornithine from Europe illuminates the origins of crown birds. Nature, 579, 397–401.

Fiorillo, A. R., & Tykoski, R. S. (2014). A diminutive new tyrannosaur from the top of the world. PLOS One, 9, e91287.

Foffa, D., Cuff, A. R., Sassoon, J., Rayfield, E. J., Mavrogordato, M. N., & Benton, M. J. (2014). Functional anatomy and feeding biomechan-

REFERENCES

ics of a giant Upper Jurassic pliosaur (Reptilia: Sauropterygia) from Weymouth Bay, Dorset, UK. Journal of Anatomy, 225, 209–219.

Ford, T. L. (1997) Did theropods have lizard lips? Southwest Paleontological Symposium—Proceedings, 1997, 65–78.

Foster, W., Brusatte, S. L., Carr, T. D., Williamson, T. E., Yi, L., & Lü, J. (2022). The cranial anatomy of the long-snouted tyrannosaurid dinosaur *Qianzhousaurus sinensis* from the Upper Cretaceous of China. Journal of Vertebrate Paleontology, 41, e1999251.

Foth, C., Evers, S. W., Pabst, B., Mateus, O., Flisch, A., Patthey, M., & Rauhut, O. W. (2015). New insights into the lifestyle of *Allosaurus* (Dinosauria: Theropoda) based on another specimen with multiple pathologies. PeerJ, 3, e940.

Fowler, D. W. (2017). Revised geochronology, correlation, and dinosaur stratigraphic ranges of the Santonian-Maastrichtian (Late Cretaceous) formations of the Western Interior of North America. PLOS One, 12, e0188426.

Fowler, D. W., & Sullivan, R. M. (2011). The first giant titanosaurian sauropod from the Upper Cretaceous of North America. Acta Palaeontologica Polonica, 56, 685–690.

Fowler, D. W., Freedman, E. A., Scannella, J. B., & Kambic, R. E. (2011b). The predatory ecology of *Deinonychus* and the origin of flapping in birds. PLOS One, 6, e28964.

Fowler, D. W., Scannella, J. B., Goodwin, M. G., & Horner, J. R. (2012) How to eat a *Triceratops*: Large sample of toothmarks provides new insight into the feeding behavior of *Tyrannosaurus*. Journal of Vertebrate Paleontology, 32, 96A.

Fowler, D. W., Wilson, J. P., Freedman Fowler, E. A., & Horner, J. R. (2019). The horned dinosaur *Leptoceratops* (Ornithischia: Neoceratopsia) raised its young in communal nesting burrows: Evidence from three new bonebeds in the Hell Creek Formation (Maastrichtian, Late Cretaceous), Montana. Abstracts volume of Cretaceous and Beyond, Paleontology of the Western Interior, Publications of North Dakota Geological Survey, Miscellaneous Series, 94, 20.

Fowler, D. W., Woodward, H. N., Freedman, E. A., Larson, P. L., & Horner, J. R. (2011a). Reanalysis of "*Raptorex kriegsteini*": A juvenile tyrannosaurid dinosaur from Mongolia. PLOS One, 6, e21376.

Fox, R. C., & Naylor, B. G. (2006). Stagodontid marsupials from the Late Cretaceous of Canada and their systematic and functional implications. Acta Palaeontologica Polonica, 51, 13–36.

Frey, R. (1995). Der Zusammenhang zwischen Begattungsstellung, Lokomotionsweise und Kopulationsorgan bei Vertebrata, mit Ausnahme der Mammalia. Eine vergleichende Betrachtung. Journal of Zoological Systematics and Evolutionary Research, 33, 17–31.

Fricke, H. C., Hencecroth, J., & Hoerner, M. E. (2011). Lowland–upland migration of sauropod dinosaurs during the Late Jurassic epoch. Nature, 480, 513–515.

Fritz, S. (1988) *Tyrannosaurus sex*: A love tail. Omni, 10, 64–69.

Fry, B. G., Wroe, S., Teeuwisse, W., van Osch, M. J., Moreno, K., Ingle, J., ... & Norman, J. A. (2009). A central role for venom in predation by *Varanus komodoensis* (Komodo Dragon) and the extinct giant *Varanus* (*Megalania*) *priscus*. Proceedings of the National Academy of Sciences, 106, 8969–8974.

Funston, G. F., Powers, M. J., Whitebone, S. A., Brusatte, S. L., Scannella, J. B., Horner, J. R., & Currie, P. J. (2021). Baby tyrannosaurid bones and teeth from the Late Cretaceous of western North America. Canadian Journal of Earth Sciences, 58, 756–777.

Galton, P. M. (1970). The posture of hadrosaurian dinosaurs. Journal of Paleontology, 44, 464–473.

Galton, P. M. (1973). The cheeks of ornithischian dinosaurs. Lethaia, 6, 67–89.

Garrick, L. D., Lang, J. W., & Herzog, H. A. (1978). Social signals of adult American alligators. Bulletin of the American Museum of Natural History, 160, 153–192.

Gates, T. A., Zanno, L. E., & Makovicky, P. J. (2013). Theropod teeth from the upper Maastrichtian Hell Creek Formation "Sue" Quarry: New morphotypes and faunal comparisons. Acta Palaeontologica Polonica, 60, 131–139.

Gatesy, S. M., Bäker, M., & Hutchinson, J. R. (2009). Constraint-based exclusion of limb poses for reconstructing theropod dinosaur locomotion. Journal of Vertebrate Paleontology, 29, 535–544.

Gauthier, J. (1986). Saurischian monophyly and the origin of birds. Memoirs of the California Academy of Sciences, 8, 1–55.

Geist, V. (1966). The evolution of horn-like organs. Behaviour, 27, 175–214.

Geist, V. (1999). *Deer of the world*. Swan Hill Press.

Gerstenhaber, C., & Knapp, A. (2022). Sexual selection leads to positive allometry but not sexual dimorphism in the expression of horn shape in the blue wildebeest, *Connochaetes taurinus*. BMC Ecology and Evolution, 22, 107.

Gianechini, F. A., Agnolín, F. L., & Ezcurra, M. D. (2011). A reassessment of the purported venom delivery system of the bird-like raptor *Sinornithosaurus*. Paläontologische Zeitschrift, 85, 103–107.

Gignac, P. M., & Erickson, G. M. (2017). The biomechanics behind extreme osteophagy in *Tyrannosaurus rex*. Scientific Reports, 7, 2012.

Gillette, D., Wolberg, D., & Hunt, A. (1986). *Tyrannosaurus rex* from the McRae Formation (Lancian, Upper Cretaceous), Elephant Butte Reservoir, Sierra County, New Mexico. New Mexico Geological Society Guidebook, 37, 235–238.

Gilmore, C. W. (1920). Osteology of the carnivorous Dinosauria in the United State National museum: With special reference to the genera *Antrodemus* (*Allosaurus*) and *Ceratosaurus*. *Bulletin of the United States National Museum* (No. 110). US Government Printing Office.

Gilmore, C. W. (1946a). New carnivorous dinosaur from the Lance Formation of Montana. Smithsonian Miscellaneous Collections 106, 1–19.

Gilmore, C. W. (1946b). Reptilian fauna of the North Horn Formation of central Utah. United States Geological Survey Professional Paper, 210C, 1–52.

Glikson, A. Y. (2023). An asteroid impact origin of the Hirnantian (end-Ordovician) glaciation and mass extinction. Gondwana Research, 118, 153–159.

Glikson, A. Y., & Yeates, A. N. (2022). Geophysics and origin of the Deniliquin multiple-ring feature, Southeast Australia. Tectonophysics, 837, 229454.

Glut, D. F. (1997). *Dinosaurs: The encyclopedia*. McFarland and Company.

Gnanadesikan, G. E., Pearse, W. D., & Shaw, A. K. (2017). Evolution of mammalian migrations for refuge, breeding, and food. Ecology and Evolution, 7, 5891–5900.

Godefroit, P., Sinitsa, S. M., Dhouailly, D., Bolotsky, Y. L., Sizov, A. V., McNamara, M. E., ... & Spagna, P. (2014). A Jurassic ornithischian dinosaur from Siberia with both feathers and scales. Science, 345, 451–455.

REFERENCES

Godfrey, L. R., Lyon, S. K., & Sutherland, M. R. (1993). Sexual dimorphism in large-bodied primates: The case of the subfossil lemurs. American Journal of Physical Anthropology, 90, 315–334.

Gold, M. E. L., Brusatte, S. L., & Norell, M. A. (2013). The cranial pneumatic sinuses of the tyrannosaurid Alioramus (Dinosauria: Theropoda) and the evolution of cranial pneumaticity in theropod dinosaurs. American Museum Novitates, 2013, 1–46.

Goldner, O., & Turner, G. (1976). The making of King Kong: The story behind a film classic. Ballantine Books.

Goldstein, E. J., Tyrrell, K. L., Citron, D. M., Cox, C. R., Recchio, I. M., Okimoto, B., ... & Fry, B. G. (2013). Anaerobic and aerobic bacteriology of the saliva and gingiva from 16 captive Komodo dragons (Varanus komodoensis): New implications for the "bacteria as venom" model. Journal of Zoo and Wildlife Medicine, 262–272.

Gould, S. J. (1995). Dinosaur in a haystack: Reflections in natural history. Harvard University Press.

Gower, D. J. (2000). Rauisuchian archosaurs (Reptilia, Diapsida): An overview. Neues Jahrbuch für Geologie und Paläontologie, Abhandlungen, 218, 447–488.

Grigg, G., & Kirshner, D. (2015). Biology and evolution of crocodylians. CSIRO Publishing.

Guinard, G. (2015). Introduction to evolutionary teratology, with an application to the forelimbs of Tyrannosauridae and Carnotaurinae (Dinosauria: Theropoda). Evolutionary Biology, 42, 20–41.

Guinness World Records (2020). Guinness World Records 2021. Guinness World Records Limited.

Gulick, S. P. S., Morgan, J. V., & IODP-ICDP Expedition Scientists. (2019). Scientific drilling into the K-Pg Chicxulub impact crater: Discoveries from IODP-ICDP expedition 364. Geological Society of America Abstracts with Programs, 51, 5. doi: 10.1130/abs/2019AM-338465.

Gutherz, S. B., Groenke, J. R., Sertich, J. J., Burch, S. H., & O'Connor, P. M. (2020). Paleopathology in a nearly complete skeleton of Majungasaurus crenatissimus (Theropoda: Abelisauridae). Cretaceous Research, 115, 104553.

Greshko, M. (2020, October 12). "Stan" the T. rex just sold for $31.8 million—and scientists are furious. National Geographic. https://www.nationalgeographic.com/science/article/stan-tyrannosaurus-rex-sold-at-auction-paleontologists-are-furious

Habib, M. (2013). Constraining the air giants: Limits on size in flying animals as an example of constraint-based biomechanical theories of form. Biological Theory, 8, 245–252.

Habib, M. B. (2008). Comparative evidence for quadrupedal launch in pterosaurs. Zitteliana, B28, 159–166.

Halstead, L. B. (1975). The evolution and ecology of the dinosaurs. Book Club Associates, London.

Happ, J. (2008). An analysis of predator-prey behavior in a head-to-head encounter between Tyrannosaurus rex and Triceratops. In: Larson, P., & Carpenter, K. (Eds.) Tyrannosaurus rex, the tyrant king. Indiana University Press. pp. 355–370.

Harryhausen, R., & Dalton, T. (2010). Ray Harryhausen: An animated life. Aurum Press.

Hartman, S. (2013, July 7). Mass estimates: North vs South redux. Dr. Scott Hartman's Skeletal Drawing.com. https://www.skeletaldrawing.com/home/mass-estimates-north-vs-south-redux772013

Hartman, S. (2019, April 2). The lip post. Dr. Scott Hartman's Skeletal Drawing.com. https://www.skeletaldrawing.com/home/the-lip-post1

Hartman, S. (2020, November 28). Road to Spinosaurus IV: Not your father's JP3-osaurus. Dr. Scott Hartman's Skeletal Drawing.com. https://www.skeletaldrawing.com/home/road-to-spinosaurus-iv-not-your-fathers-jp3-osaurus11282020

Hartman, S. A., Lovelace, D. M., Linzmeier, B. J., Mathewson, P. D., & Porter, W. P. (2022). Mechanistic thermal modeling of Late Triassic terrestrial amniotes predicts biogeographic distribution. Diversity, 14, 973.

Hay, O. P. (1910). On the manner of locomotion of the dinosaurs especially Diplodocus, with remarks on the origin of the birds. Proceedings of the Washington Academy of Sciences, 12, 1–25.

Haynes, G. (1991). Mammoths, mastodons, and elephants: Biology, behavior and the fossil record. Cambridge University Press.

Heilmann, G. (1927). The origin of birds. HF & G. Witherby.

Helmore, E. (2019, February 24). Dinosaur fossil collectors "price museums out of the market." The Observer. https://www.theguardian.com/science/2019/feb/24/dinosaur-fossils-collectors-museums-price-sale

Henderson, D. M. (1999). Estimating the masses and centers of mass of extinct animals by 3-D mathematical slicing. Paleobiology, 25, 88–106.

Henderson, D. M. (2003). The eyes have it: The sizes, shapes, and orientations of theropod orbits as indicators of skull strength and bite force. Journal of Vertebrate Paleontology, 22, 766–778.

Henderson, D. M. (2013). Sauropod necks: Are they really for heat loss? PLOS One, 8, e77108.

Henderson, D. M. (2018). A buoyancy, balance and stability challenge to the hypothesis of a semi-aquatic Spinosaurus Stromer, 1915 (Dinosauria: Theropoda). PeerJ, 6, e5409.

Henderson, D. M., & Snively, E. (2004). Tyrannosaurus en pointe: Allometry minimized rotational inertia of large carnivorous dinosaurs. Proceedings of the Royal Society of London. Series B: Biological Sciences, 271, S57-S60.

Henderson, M. D., & Harrison, W. H. (2008). Taphonomy and environment of deposition of juvenile tyrannosaurid skeleton from the Hell Creek Formation (latest Maastrichtian) of southeastern Montana. In: Larson, P., and Carpenter, K. (Eds.) Tyrannosaurus rex, the tyrant king. Indiana University Press. pp. 82–90.

Henderson, M. D., & Peterson, J. E. (2006). An azhdarchid pterosaur cervical vertebra from the Hell Creek Formation (Maastrichtian) of southeastern Montana. Journal of Vertebrate Paleontology, 26, 192–195.

Herculano-Houzel, S. (2023). Theropod dinosaurs had primate-like numbers of telencephalic neurons. Journal of Comparative Neurology, 531, 962–974.

Hertel, F. (1995). Ecomorphological indicators of feeding behavior in recent and fossil raptors. The Auk, 112, 890–903.

Hieronymus, T. L. (2009). Osteological correlates of cephalic skin structures in Amniota: Documenting the evolution of display and feeding structures with fossil data. Doctoral dissertation, Ohio University.

Hieronymus, T. L., & Witmer, L. M. (2010). Homology and evolution of avian compound rhamphothecae. The Auk, 127, 590–604.

Hieronymus, T. L., Witmer, L. M., Tanke, D. H., & Currie, P. J. (2009). The facial integument of centrosaurine ceratopsids: Morphological

REFERENCES

and histological correlates of novel skin structures. The Anatomical Record, 292, 1370–1396.

Hildebrand, A. R., Penfield, G. T., Kring, D. A., Pilkington, M., Camargo Z, A., Jacobsen, S. B., & Boynton, W. V. (1991). Chicxulub crater: A possible Cretaceous/Tertiary boundary impact crater on the Yucatan Peninsula, Mexico. Geology, 19, 867–871.

Hippensteel, S., & Condliffe, S. (2013). Profiting from the past: Are fossils a sound investment? GSA Today, 23, 27–29.

Hirt, M. R., Jetz, W., Rall, B. C., & Brose, U. (2017). A general scaling law reveals why the largest animals are not the fastest. Nature Ecology & Evolution, 1, 1116–1122.

Hodnett, J. P. M, Carrano, M. T., Santucci, V. L., Tweet, J. S., Visaggi, C. C. (2023). A tyrannosaur (Dinosauria; Theropoda; Tyrannosauridae) from the Late Cretaceous (Maastrichtian) Harebell Formation of Yellowstone National Park, Wyoming. New Mexico Museum of Natural History and Science Bulletin, 94, 233–238.

Holliday, C. M., & Witmer, L. M. (2008). Cranial kinesis in dinosaurs: Intracranial joints, protractor muscles, and their significance for cranial evolution and function in diapsids. Journal of Vertebrate Paleontology, 28, 1073–1088.

Holliday, C. M., Porter, W. R., Vliet, K. A., & Witmer, L. M. (2020). The frontoparietal fossa and dorsotemporal fenestra of archosaurs and their significance for interpretations of vascular and muscular anatomy in dinosaurs. The Anatomical Record, 303, 1060–1074.

Holtz, T. R., Jr. (1994). The phylogenetic position of the Tyrannosauridae: Implications for theropod systematics. Journal of Paleontology, 68, 1100–1117.

Holtz, T. R., Jr. (1995). The arctometatarsalian pes, an unusual structure of the metatarsus of Cretaceous Theropoda (Dinosauria: Saurischia). Journal of Vertebrate Paleontology, 14, 480–519.

Holtz, T. R., Jr. (2000). A new phylogeny of the carnivorous dinosaurs. Gaia, 15, 5–61.

Holtz, T. R., Jr. (2001). The phylogeny and taxonomy of the Tyrannosauridae. In: Tanke, D. H., & Carpenter, K. (Eds.) Mesozoic vertebrate life. Indiana University Press. pp. 64–83

Holtz, T. R., Jr. (2004). Tyrannosauroidea. In: Weishampel, D. B., Dodson, P., and Osmólska, H. (Eds.) The Dinosauria (second edition). University of California Press. pp. 111–136.

Holtz, T.R., Jr. (2007). Dinosaurs: The most complete, up-to-date encyclopedia for dinosaur lovers of all ages. Random House.

Holtz, T. R., Jr. (2008). A critical reappraisal of the obligate scavenging hypothesis for Tyrannosaurus rex and other tyrant dinosaurs. In: Larson, P., & Carpenter, K. (Eds.) Tyrannosaurus rex, the tyrant king. Indiana University Press. pp. 371–396.

Holtz, T. R., Jr. (2021). Theropod guild structure and the tyrannosaurid niche assimilation hypothesis: Implications for predatory dinosaur macroecology and ontogeny in later Late Cretaceous Asiamerica. Canadian Journal of Earth Sciences, 58, 778–795.

Hone, D. W. E. (2012). Variation in the tail length of non-avian dinosaurs. Journal of Vertebrate Paleontology, 32, 1082–1089.

Hone, D. W. E. (2016). The tyrannosaur chronicles: The biology of the tyrant dinosaurs. Bloomsbury.

Hone, D. W. E., & Holtz, T. R., Jr. (2017). A century of spinosaurs—a review and revision of the Spinosauridae with comments on their ecology. Acta Geologica Sinica-English Edition, 91, 1120–1132.

Hone, D. W. E., & Holtz, T. R., Jr. (2021). Evaluating the ecology of Spinosaurus: Shoreline generalist or aquatic pursuit specialist? Palaeontologia Electronica, 24, 1–28.

Hone, D. W. E., & Mallon, J. C. (2017). Protracted growth impedes the detection of sexual dimorphism in non-avian dinosaurs. Palaeontology, 60, 535–545.

Hone, D. W. E., & Rauhut, O. W. (2010). Feeding behaviour and bone utilization by theropod dinosaurs. Lethaia, 43, 232–244.

Hone, D. W. E., Naish, D., & Cuthill, I. C. (2011b). Does mutual sexual selection explain the evolution of head crests in pterosaurs and dinosaurs? Lethaia, 45, 139–156.

Hone, D. W. E., Wang, K., Sullivan, C., Zhao, X., Chen, S., Li, D., ... & Xu, X. (2011a). A new, large tyrannosaurine theropod from the Upper Cretaceous of China. Cretaceous Research, 32, 495–503.

Hone, D. W. E., Wood, D., & Knell, R. J. (2016). Positive allometry for exaggerated structures in the ceratopsian dinosaur Protoceratops andrewsi supports socio-sexual signaling. Palaeontologia Electronica, 19, 1–13.

Horner, J. R. (1994). Steak knives, beady eyes, and tiny little arms (a portrait of T. rex as a scavenger). The Paleontological Society Special Publications, 7, 157–164.

Horner, J. R., & Goodwin, M. B. (2006). Major cranial changes during Triceratops ontogeny. Proceedings of the Royal Society. Series B: Biological Sciences, 273, 2757–2761.

Horner, J. R., & Goodwin, M. B. (2008). Ontogeny of cranial epi-ossifications in Triceratops. Journal of Vertebrate Paleontology, 28, 134–144.

Horner, J. R., & Lessem, D. (1993). The complete T. rex: How stunning new discoveries are changing our understanding of the world's most famous dinosaur. Simon and Schuster.

Horner, J. R., & Padian, K. (2004). Age and growth dynamics of Tyrannosaurus rex. Proceedings of the Royal Society of London. Series B: Biological Sciences, 271, 1875–1880.

Horner, J. R., Goodwin, M. B., & Evans, D. C. (2022). A new pachycephalosaurid from the Hell Creek Formation, Garfield County, Montana, USA. Journal of Vertebrate Paleontology, 42, e2190369.

Horner, J. R., Goodwin, M. B., & Myhrvold, N. (2011). Dinosaur census reveals abundant Tyrannosaurus and rare ontogenetic stages in the Upper Cretaceous Hell Creek Formation (Maastrichtian), Montana, USA. PLOS One, 6, e16574.

Hu, C. Z., Cheng, Z.-W., Pang, Q. Q., & Fang, X.-S. (2001). Shantungosaurus giganteus. Geological Publishing House.

Hübner, T. R. (2012). Bone histology in Dysalotosaurus lettowvorbecki (Ornithischia: Iguanodontia)–variation, growth, and implications. PLOS One, 7, e29958.

Hunt, R. K., & Lehman, T. M. (2008). Attributes of the ceratopsian dinosaur Torosaurus, and new material from the Javelina Formation (Maastrichtian) of Texas. Journal of Paleontology, 82, 1127–1138.

Hunter, S. J., Valdes, P. J., Haywood, A. M., & Markwick, P. J. (2008). Modelling Maastrichtian climate: Investigating the role of geography, atmospheric CO2 and vegetation. Climate of the Past Discussions, 4, 981–1019.

Hurlburt, G. R., Ridgely, R. C., & Witmer, L. M. (2013). Relative size of brain and cerebrum in tyrannosaurid dinosaurs: An analysis using brain-endocast quantitative relationships in extant alligators. In: Parrish, J. M., Molnar, R. E., Currie, P. J., & Koppelhus, E. B. (Eds.) Tyrannosaurid paleobiology. Indiana University Press. pp. 135–154.

Hurum, J. H., & Sabath, K. (2003). Giant theropod dinosaurs from Asia and North America: Skulls of Tarbosaurus bataar and Tyrannosaurus rex compared. Acta Palaeontologica Polonica, 48, 161–190.

REFERENCES

Hurum, J. R. H., & Currie, P. J. (2000). The crushing bite of tyrannosaurids. Journal of Vertebrate Paleontology, 20, 619–621.

Hutchinson, J. R. (2004). Biomechanical modeling and sensitivity analysis of bipedal running ability. II. Extinct taxa. Journal of Morphology, 262, 441–461.

Hutchinson, J. R., & Allen, V. (2009). The evolutionary continuum of limb function from early theropods to birds. Naturwissenschaften, 96, 423–448.

Hutchinson, J. R., & Garcia, M. (2002). Tyrannosaurus was not a fast runner. Nature, 415, 1018–1021.

Hutchinson, J. R., Anderson, F. C., Blemker, S. S., & Delp, S. L. (2005). Analysis of hindlimb muscle moment arms in Tyrannosaurus rex using a three-dimensional musculoskeletal computer model: Implications for stance, gait, and speed. Paleobiology, 31, 676–701.

Hutchinson, J. R., Bates, K. T., Molnar, J., Allen, V., & Makovicky, P. J. (2011). A computational analysis of limb and body dimensions in Tyrannosaurus rex with implications for locomotion, ontogeny, and growth. PLOS One, 6, e26037.

Hutchinson, J. R., Famini, D., Lair, R., & Kram, R. (2003). Are fast-moving elephants really running? Nature, 422, 493–494.

Hutchinson, J. R., Ng-Thow-Hing, V., & Anderson, F. C. (2007). A 3D interactive method for estimating body segmental parameters in animals: Application to the turning and running performance of Tyrannosaurus rex. Journal of Theoretical Biology, 246, 660–680.

Ibrahim, N., Maganuco, S., Dal Sasso, C., Fabbri, M., Auditore, M., Bindellini, G., ... & Pierce, S. E. (2020b). Tail-propelled aquatic locomotion in a theropod dinosaur. Nature, 581, 67–70.

Ibrahim, N., Sereno, P. C., Dal Sasso, C., Maganuco, S., Fabbri, M., Martill, D. M., ... & Iurino, D. A. (2014). Semiaquatic adaptations in a giant predatory dinosaur. Science, 345, 1613–1616.

Ibrahim, N., Sereno, P. C., Varricchio, D. J., Martill, D. M., Dutheil, D. B., Unwin, D. M., ... & Kaoukaya, A. (2020a). Geology and paleontology of the upper cretaceous Kem Kem group of eastern Morocco. ZooKeys, 928, 1–216.

Isles, T. E. (2009). The socio-sexual behaviour of extant archosaurs: Implications for understanding dinosaur behaviour. Historical Biology, 21, 139–214.

Jahns, R. H., Kottlowski, F. E., & Kuellmer, F. J. (1955). Volcanic rocks of south-central New Mexico. In: New Mexico Geological Society, 6th Field Conference. pp. 92–95.

Jasinski, S. E., Sullivan, R. M., & Dodson, P. (2020). New dromaeosaurid dinosaur (Theropoda, Dromaeosauridae) from New Mexico and biodiversity of dromaeosaurids at the end of the Cretaceous. Scientific Reports, 10, 5105.

Jerison, H. J. (1973). Evolution of the brain and intelligence. Academic Press.

Jerzykiewicz, T. (1997). Stratigraphic framework of the uppermost Cretaceous to Paleocene strata of the Alberta Basin. Geological Survey of Canada, Bulletin, 510, 1–121.

Jevnikar, E. M. (2021). A comprehensive ontogenetic analysis of Tarbosaurus bataar provides insight into intraskeletal and individual variation of tyrannosaurid growth. Master's thesis, North Carolina State University.

Ji, Q., Ji, S.A., & Zhang, L. J. (2009). First large tyrannosauroid theropod from the Early Cretaceous Jehol Biota in northeastern China. Geological Bulletin of China. 28, 1369–1374.

John, J., & Lee, W. (2019). Kleptoparasitism of Shoebills Balaeniceps rex by African Fish Eagles Haliaeetus vocifer in western Tanzania. Tanzania Journal of Science, 45, 131–143.

Johnson, K. (2008). How old is T. rex? Challenges with the dating of terrestrial strata deposited during the Maastrichtian stage of the Cretaceous period. In: Larson, P., & Carpenter, K. (Eds.) Tyrannosaurus rex, the tyrant king. Indiana University Press. pp. 63–65.

Johnson, K. R. (2002). Megaflora of the Hell Creek and lower Fort Union Formations in the western Dakotas: Vegetational response to climate change, the Cretaceous-Tertiary boundary event, and rapid marine transgression. In: Hartman, J. H., Johnson, K. R., & Nichols, D. J., (Eds.) The Hell Creek Formation and the Cretaceous-Tertiary boundary in the northern Great Plains: An integrated continental record of the end of the Cretaceous. Geological Society of America Special Paper, 361, 329–391.

Johnson-Ransom, E., Li, F., Xu, X., Ramos, R., Midzuk, A. J., Thon, U., ... & Snively, E. (2024). Comparative cranial biomechanics reveal that Late Cretaceous tyrannosaurids exerted relatively greater bite force than in early-diverging tyrannosauroids. The Anatomical Record, 307, 1897–1917.

Jones, E. D. (2020). Assumptions of authority: The story of Sue the T-rex and controversy over access to fossils. History and Philosophy of the Life Sciences, 42, 2.

Jung, T. S., Everatt, K. T., & Andresen-Everatt, L. M. (2009). Kleptoparasitism of a coyote (Canis latrans) by a golden eagle (Aquila chrysaetos) in northwestern Canada. Northwestern Naturalist, 90, 53–55.

Kawabe, S., & Hattori, S. (2022). Complex neurovascular system in the dentary of Tyrannosaurus. Historical Biology, 34, 1137–1145.

Kaye, T. G., Gaugler, G., & Sawlowicz, Z. (2008). Dinosaurian soft tissues interpreted as bacterial biofilms. PLOS One, 3, e2808.

Keillor, T. M. (2013). Jane, in the flesh. In: Parrish, J. M., Molnar, R. E., Currie, P. J., & Koppelhus, E. B. (Eds.) Tyrannosaurid paleobiology. Indiana University Press. pp. 156–174.

Kingma, B. R., Frijns, A. J., Schellen, L., & van Marken Lichtenbelt, W. D. (2014). Beyond the classic thermoneutral zone: Including thermal comfort. Temperature, 1, 142–149.

Kingsley, E. P., Eliason, C. M., Riede, T., Li, Z., Hiscock, T. W., Farnsworth, M., ... & Clarke, J. A. (2018). Identity and novelty in the avian syrinx. Proceedings of the National Academy of Sciences, 115, 10209–10217.

Kjærgaard, P. C. (2012). The fossil trade: Paying a price for human origins. Isis, 103, 340–355.

Knell, R. J., Naish, D., Tomkins, J. L., & Hone, D. W. E. (2013). Sexual selection in prehistoric animals: Detection and implications. Trends in Ecology & Evolution, 28, 38–47.

Knoll, F. (2008). Buccal soft anatomy in Lesothosaurus (Dinosauria: Ornithischia). Neues Jahrbuch für Geologie und Paläontologie-Abhandlungen, 248, 355–364.

Krauss, D., & Robinson, J. (2013). The biomechanics of a plausible hunting strategy for Tyrannosaurus rex. In: Parrish, J. M., Molnar, R. E., Currie, P. J., & Koppelhus, E. B. (Eds.) Tyrannosaurid paleobiology. Indiana University Press. pp. 251–264.

Kruuk, H. (1972). The spotted hyena: A study of predation and social behavior. University of Chicago Press.

Ksepka, D. T., Balanoff, A. M., Smith, N. A., Bever, G. S., Bhullar, B. A. S., Bourdon, E., ... & Smaers, J. B. (2020). Tempo and pattern of avian brain size evolution. Current Biology, 30, 2026–2036.

REFERENCES

Kundrát, M., Xu, X., Hančová, M., Gajdoš, A., Guo, Y., & Chen, D. (2018). Evolutionary disparity in the endoneurocranial configuration between small and gigantic tyrannosauroids. Historical Biology, 32, 1–15.

Lamanna, M. C., Sues, H. D., Schachner, E. R., & Lyson, T. R. (2014). A new large-bodied oviraptorosaurian theropod dinosaur from the latest Cretaceous of western North America. PLOS One, 9, e92022.

Lambe, L. M. (1917). The Cretaceous theropodous dinosaur *Gorgosaurus*. Government Printing Bureau.

Lambert, H., Carder, G., & D'Cruze, N. (2019). Given the cold shoulder: A review of the scientific literature for evidence of reptile sentience. Animals, 9, 821.

Langer, M. C., Ezcurra, M. D., Rauhut, O. W., Benton, M. J., Knoll, F., McPhee, B. W., … & Brusatte, S. L. (2017). Untangling the dinosaur family tree. Nature, 551, E1–E3.

Larramendi, A. (2016). Shoulder height, body mass, and shape of proboscideans. Acta Palaeontologica Polonica, 61, 537–574.

Larramendi, A., Paul, G. S., & Hsu, S. Y. (2021). A review and reappraisal of the specific gravities of present and past multicellular organisms, with an emphasis on tetrapods. The Anatomical Record, 304, 1833–1888.

Larson, D. W., & Currie, P. J. (2013). Multivariate analyses of small theropod dinosaur teeth and implications for paleoecological turnover through time. PLOS One, 8, e54329.

Larson, D. W., Brown, C. M., & Evans, D. C. (2016). Dental disparity and ecological stability in bird-like dinosaurs prior to the end-Cretaceous mass extinction. Current Biology, 26, 1325–1333.

Larson, N. (2008). One-hundred years of *Tyrannosaurus rex*: The skeletons. In: Larson, P., and Carpenter, K. (Eds.) Tyrannosaurus rex, the tyrant king. Indiana University Press. pp. 1–55.

Larson, P. (1997). The king's new clothes: A fresh look at *Tyrannosaurus rex*. In: Wolberg, D. L., Stump, E., & Rosenberg, G. D. (Eds.) DinoFest international proceedings. Academy of Natural Sciences. pp. 65–71.

Larson, P. (2001). Pathologies in *Tyrannosaurus rex*: Snapshots of a killer's life. Journal of Vertebrate Paleontology, 21, 71A–72A.

Larson, P. (2008a). Atlas of the skull bones of *Tyrannosaurus rex*. In: Larson, P., & Carpenter, K. (Eds.) Tyrannosaurus rex, the tyrant king. Indiana University Press. pp 233–243.

Larson, P. (2008b). Variation and sexual dimorphism in *Tyrannosaurus rex*. In: Larson, P., & Carpenter, K. (Eds.) Tyrannosaurus rex, the tyrant king. Indiana University Press. pp. 103–128.

Larson, P. (2013). The case for *Nanotyrannus*. In: Parrish, J. M., Molnar, R. E., Currie, P. J., & Koppelhus, E. B. (Eds.) Tyrannosaurid paleobiology. Indiana University Press. pp. 15–53.

Larson, P., & Donnan, K. (2002). *Rex appeal: The amazing story of Sue, the dinosaur that changed science, the law, and my life*. Invisible Cities Press.

Larson, P. L. (1994). *Tyrannosaurus* sex. The Paleontological Society Special Publications, 7, 139–156.

Larson, P. L., & Frey, E. (1992). Sexual dimorphism in the abundant upper Cretaceous theropod *Tyrannosaurus rex*. Journal of Vertebrate Paleontology, 12, 38A.

Larsson, H. C. (2008). Palatal kinesis of *Tyrannosaurus rex*. In: Larson, P., & Carpenter, K. (Eds.) Tyrannosaurus rex, the tyrant king. Indiana University Press. pp. 245–252.

Lawson, D. A. (1975). Pterosaur from the latest Cretaceous of West Texas: Discovery of the largest flying creature. Science, 187, 947–948.

Lawson, D. A. (1976). *Tyrannosaurus* and *Torosaurus*, Maestrichtian dinosaurs from Trans-Pecos, Texas. Journal of Paleontology, 50, 158–164.

Lawton, T. F., Talling, P. J., Hobbs, R. S., Trexler Jr, J. H., Weiss, M. P., & Burbank, D. W. (1993). Structure and stratigraphy of Upper Cretaceous and Paleogene strata (North Horn Formation), eastern San Pitch Mountains, Utah—Sedimentation at the front of the Sevier orogenic belt. US Geological Survey Bulletin 1787-II, II1–II33.

Lee, A. H., & Werning, S. (2008). Sexual maturity in growing dinosaurs does not fit reptilian growth models. Proceedings of the National Academy of Sciences, 105, 582–587.

Legendre, L. J., Rubilar-Rogers, D., Vargas, A. O., & Clarke, J. A. (2020). The first dinosaur egg remains a mystery. bioRxiv, 2020.12.10.406678.

Lehman, T. M. (2001). Late Cretaceous dinosaur provinciality. In: Tanke, D. H., & Carpenter, K. (Eds.) *Mesozoic vertebrate life*. Indiana University Press. pp. 310–328.

Lehman, T. M., & Carpenter, K. (1990). A partial skeleton of the tyrannosaurid dinosaur *Aublysodon* from the Upper Cretaceous of New Mexico. Journal of Paleontology, 64(6), 1026–1032.

Lehman, T. M., & Woodward, H. N. (2008). Modeling growth rates for sauropod dinosaurs. Paleobiology, 34, 264–281.

Lehman, T. M., Wick, S. L., & Wagner, J. R. (2016). Hadrosaurian dinosaurs from the Maastrichtian Javelina Formation, Big Bend National Park, Texas. Journal of Paleontology, 90, 333–356.

Leitch, D. B., & Catania, K. C. (2012). Structure, innervation and response properties of integumentary sensory organs in crocodilians. Journal of Experimental Biology, 215, 4217–4230.

Lescaze, Z. (2017). *Paleoart: Visions of the prehistoric past*. Taschen.

Li, D., Norell, M. A., Gao, K. Q., Smith, N. D., & Makovicky, P. J. (2009). A longirostrine tyrannosauroid from the Early Cretaceous of China. Proceedings of the Royal Society. Series B: Biological Sciences, 277, 183–190.

Li, Z., Zhou, Z., & Clarke, J. A. (2018). Convergent evolution of a mobile bony tongue in flighted dinosaurs and pterosaurs. PLOS One, 13, e0198078.

Lillegraven, J. A., & Eberle, J. J. (1999). Vertebrate faunal changes through Lancian and Puercan time in southern Wyoming. Journal of Paleontology, 73, 691–710.

Lindgren, J., Uvdal, P., Engdahl, A., Lee, A. H., Alwmark, C., Bergquist, K. E., … & Jacobs, L. L. (2011). Microspectroscopic evidence of Cretaceous bone proteins. PLOS One, 6, e19445.

Lipkin, C., & Carpenter, K. (2008). Looking again at the forelimb of *Tyrannosaurus rex*. In: Larson, P., & Carpenter, K. (Eds.) Tyrannosaurus rex, the tyrant king. Indiana University Press. pp. 166–190.

Lipkin, C., Sereno, P. C., & Horner, J. R. (2007). The furcula in *Suchomimus tenerensis* and *Tyrannosaurus rex* (Dinosauria: Theropoda: Tetanurae). Journal of Paleontology, 81, 1523–1527.

Lloyd, G. T., Davis, K. E., Pisani, D., Tarver, J. E., Ruta, M., Sakamoto, M., … & Benton, M. J. (2008). Dinosaurs and the Cretaceous terrestrial revolution. Proceedings of the Royal Society. Series B: Biological Sciences, 275, 2483–2490.

Lockley, M., Janke, P. R., & Triebold, M. (2011). Tracking *Tyrannosaurus*: Notes on purported *T. rex* tracks. Ichnos, 18, 172–175.

Lockley, M., Kukihara, R., & Mitchell, L. (2008). Why *Tyrannosaurus rex* had puny arms: An integral morphodynamic solution to a simple puzzle in theropod paleobiology. In: Larson, P., & Carpenter,

REFERENCES

K. (Eds.) Tyrannosaurus rex, *the tyrant king*. Indiana University Press. pp. 131–164.

Lockley, M. G., & Hunt, A. P. (1994). A track of the giant theropod dinosaur *Tyrannosaurus* from close to the Cretaceous/Tertiary boundary, northern New Mexico. Ichnos, 3, 213–218.

Loewen, M. A., Irmis, R. B., Sertich, J. J., Currie, P. J., & Sampson, S. D. (2013). Tyrant dinosaur evolution tracks the rise and fall of Late Cretaceous oceans. PLOS One, 8, e79420.

Longrich, N. R. (2004). Aquatic specialization in mammals from the Late Cretaceous of North America. Journal of Vertebrate Paleontology, 24, 84A.

Longrich, N. R. (2008). A new, large ornithomimid from the Cretaceous Dinosaur Park Formation of Alberta, Canada: Implications for the study of dissociated dinosaur remains. Palaeontology, 51, 983–997.

Longrich, N. R., & Field, D. J. (2012). *Torosaurus* is not *Triceratops*: Ontogeny in chasmosaurine ceratopsids as a case study in dinosaur taxonomy. PLOS One, 7, e32623.

Longrich, N. R., & Saitta, E. T. (2024). Taxonomic status of *Nanotyrannus lancensis* (Dinosauria: Tyrannosauroidea)—a distinct taxon of small-bodied tyrannosaur. Fossil Studies, 2, 1–65.

Longrich, N. R., Bhullar, B. A. S., & Gauthier, J. A. (2012). Mass extinction of lizards and snakes at the Cretaceous–Paleogene boundary. Proceedings of the National Academy of Sciences, 109, 21396–21401.

Longrich, N. R., Horner, J. R., Erickson, G. M., & Currie, P. J. (2010). Cannibalism in *Tyrannosaurus rex*. PLOS One, 5, e13419.

Longrich, N. R., Tokaryk, T., & Field, D. J. (2011). Mass extinction of birds at the Cretaceous–Paleogene (K–Pg) boundary. Proceedings of the National Academy of Sciences, 108, 15253–15257.

Loughry, W. J., & McDonough, C. M. (2013). *The nine-banded armadillo: A natural history*. University of Oklahoma Press.

Loveridge, A. J., Hunt, J. E., Murindagomo, F., & Macdonald, D. W. (2006). Influence of drought on predation of elephant (*Loxodonta africana*) calves by lions (*Panthera leo*) in an African wooded savannah. Journal of Zoology, 270, 523–530.

Lozinsky, R. P., Hunt, A. P., Wolberg, D. L., & Lucas, S. G. (1984). Late Cretaceous (Lancian) dinosaurs from the McRae Formation, Sierra County, New Mexico. New Mexico Geology, 6, 72–77.

Lü, J., Yi, L., Brusatte, S. L., Yang, L., Li, H., & Chen, L. (2014). A new clade of Asian Late Cretaceous long-snouted tyrannosaurids. Nature Communications, 5, 3788.

Lucas, S. G., & Hunt, A. P. (1989). *Alamosaurus* and the sauropod hiatus in the Cretaceous. In: Farlow, J. O. (Ed.), *Paleobiology of the dinosaurs*. Geological Society of America Special Paper, 238. pp. 75–85.

Lucas, S. G., Sullivan, R. M., Lichtig, A. J., Dalman, S. G., & Jasinski, S. E. (2016). Late Cretaceous dinosaur biogeography and endemism in the Western Interior Basin, North America: A critical re-evaluation. New Mexico Museum of Natural History and Science Bulletin, 71, 195–213.

Luo, Z. X. (2007). Transformation and diversification in early mammal evolution. Nature, 450, 1011–1019.

Lyson, T. R., Bercovici, A., Chester, S. G., Sargis, E. J., Pearson, D., & Joyce, W. G. (2011). Dinosaur extinction: Closing the "3 m gap." Biology Letters, 7, 925–928.

Macdonald, D. W. (Ed.) 2009. *The encyclopedia of mammals*. Oxford University Press.

Maidment, S. C., Bates, K. T., Falkingham, P. L., VanBuren, C., Arbour, V., & Barrett, P. M. (2014). Locomotion in ornithischian dinosaurs: An assessment using three-dimensional computational modelling. Biological Reviews, 89, 588–617.

Maidment, S. C., Linton, D. H., Upchurch, P., & Barrett, P. M. (2012). Limb-bone scaling indicates diverse stance and gait in quadrupedal ornithischian dinosaurs. PLOS One, 7, e36904.

Mainwaring, M. C., Medina, I., Tobalske, B. W., Hartley, I. R., Varricchio, D. J., & Hauber, M. E. (2023). The evolution of nest site use and nest architecture in modern birds and their ancestors. Philosophical Transactions of the Royal Society. Series B, 378, 20220143.

Maiorino, L., Farke, A. A., Kotsakis, T., & Piras, P. (2013). Is *Torosaurus Triceratops*? Geometric morphometric evidence of late Maastrichtian ceratopsid dinosaurs. PLOS One, 8, e81608.

Maleev, E. A. (1955a). Giant carnivorous dinosaurs of Mongolia. Doklady Akademii Nauk SSSR. 104, 634–637.

Maleev, E. A. (1955b). New carnivorous dinosaurs from the Upper Cretaceous of Mongolia. Doklady Akademii Nauk SSSR. 104, 779–783.

Mallon, J. C. (2017). Recognizing sexual dimorphism in the fossil record: Lessons from nonavian dinosaurs. Paleobiology, 43, 495–507.

Mallon, J. C., & Hone, D. W. E. (2024). Estimation of maximum body size in fossil species: A case study using *Tyrannosaurus rex*. Ecology and Evolution, 14, e11658.

Mallon, J. C., Holmes, R. B., Bamforth, E. L., & Schumann, D. (2022). The record of *Torosaurus* (Ornithischia: Ceratopsidae) in Canada and its taxonomic implications. Zoological Journal of the Linnean Society, 195, 157–171.

Maloney, S. K. (2008). Thermoregulation in ratites: A review. Australian Journal of Experimental Agriculture, 48, 1293–1301.

Mannion, P. D., & Upchurch, P. (2011). A re-evaluation of the 'mid-Cretaceous sauropod hiatus' and the impact of uneven sampling of the fossil record on patterns of regional dinosaur extinction. Palaeogeography, Palaeoclimatology, Palaeoecology, 299, 529–540.

Marsh, O. C. (1884). Principal characters of American Jurassic dinosaurs; Part VIII, the order Theropoda. American Journal of Science, 3, 329–340.

Marsh, O. C. (1890). Description of new dinosaurian reptiles. American Journal of Science, 3, 81–86.

Marsh, O. C. (1896). *The dinosaurs of North America*. U.S. Government Printing Office.

Marshall, C. R., Latorre, D. V., Wilson, C. J., Frank, T. M., Magoulick, K. M., Zimmt, J. B., & Poust, A. W. (2021). Absolute abundance and preservation rate of *Tyrannosaurus rex*. Science, 372, 284–287.

Martin, G. R. (2007). Visual fields and their functions in birds. Journal of Ornithology, 148, 547–562.

Martin, G. R. (2017). What drives bird vision? Bill control and predator detection overshadow flight. Frontiers in neuroscience, 11, 278042.

Matthew, W. D. (1915). *Dinosaurs with special reference to the American Museum: Collections (No. 5)*. American Museum of Natural History.

Mattison, R., & Griffin, E. (1989). Limb use and disuse in ratites and tyrannosaurids. Journal of Vertebrate Paleontology Abstracts, 9, 32A.

REFERENCES

Mazzetta, G. V., Cisilino, A. P., Blanco, R. E., & Calvo, N. (2009). Cranial mechanics and functional interpretation of the horned carnivorous dinosaur *Carnotaurus sastrei*. Journal of Vertebrate Paleontology, 29, 822–830.

Mazzetta, G. V., Fariña, R. A., & Vizcaíno, S. F. (1998). On the palaeobiology of the South American horned theropod *Carnotaurus sastrei* Bonaparte. Gaia, 15, 192.

McCrea, R. T., Buckley, L. G., Farlow, J. O., Lockley, M. G., Currie, P. J., Matthews, N. A., & Pemberton, S. G. (2014). A 'terror of tyrannosaurs': The first trackways of tyrannosaurids and evidence of gregariousness and pathology in Tyrannosauridae. PLOS One, 9, e103613.

McDonald, A. T., Wolfe, D. G., & Dooley Jr, A. C. (2018). A new tyrannosaurid (Dinosauria: Theropoda) from the Upper Cretaceous Menefee Formation of New Mexico. PeerJ, 6, e5749.

McIver, E. E. (2002). The paleoenvironment of *Tyrannosaurus rex* from southwestern Saskatchewan, Canada. Canadian Journal of Earth Sciences, 39, 207–221.

McKeown, M., Brusatte, S. L., Williamson, T. E., Schwab, J. A., Carr, T. D., Butler, I. B., ... & Vogel, S. C. (2020). Neurosensory and sinus evolution as tyrannosauroid dinosaurs developed giant size: Insight from the endocranial anatomy of *Bistahieversor sealeyi*. The Anatomical Record, 303, 1043–1059.

McLain, M. A., Nelsen, D., Snyder, K., Griffin, C. T., Siviero, B., Brand, L. R., & Chadwick, A. V. (2018). Tyrannosaur cannibalism: A case of a tooth-traced tyrannosaurid bone in the Lance Formation (Maastrichtian), Wyoming. Palaios, 33, 164–173.

McPhee, R. (2022, April 13). Sir David Attenborough's BBC1 dinosaur show presents softer "woke" version of the T-Rex. The Sun. https://www.thesun.co.uk/tv/18257993/david-attenborough-woke-t-rex/

Meers, M. B. (2002). Maximum bite force and prey size of *Tyrannosaurus rex* and their relationships to the inference of feeding behavior. Historical Biology, 16, 1–12.

Middleton, K. M., & Gatesy, S. M. (2000). Theropod forelimb design and evolution. Zoological Journal of the Linnean Society, 128, 149–187.

Milner, A. R., & Lockley, M. G. (2016). Dinosaur swim track assemblages: Characteristics, contexts, and ichnofacies implications. In: Falkingham, P. L., Marty, D., & Richter, A. (Eds.) *Dinosaur tracks: The next steps*. Indiana University Press. pp 152–80.

Milner, R. (2012). *Charles R. Knight: The artist who saw through time*. Abrams.

Mitchell, J. S., Heckert, A. B., & Sues, H. D. (2010). Grooves to tubes: Evolution of the venom delivery system in a Late Triassic "reptile." Naturwissenschaften, 97, 1117–1121.

Mitchell, J. S., Roopnarine, P. D., & Angielczyk, K. D. (2012). Late Cretaceous restructuring of terrestrial communities facilitated the end-Cretaceous mass extinction in North America. Proceedings of the National Academy of Sciences, 109, 18857–18861.

Mitchell, W. T. (1998). *The last dinosaur book: The life and times of a cultural icon*. University of Chicago Press.

Mittal, T., Sprain, C. J., Renne, P. R., & Richards, M .A. (2022), Deccan volcanism at K-Pg time. In: Koeberl, C., Claeys, P., & Montanari, A. (Eds.) *From the Guajira Desert to the Apennines, and from Mediterranean microplates to the Mexican killer asteroid: Honoring the career of Walter Alvarez*. Geological Society of America Special Paper, 557, 471–496,

Mock, D. W., & Mock, K. C. (1980). Feeding behavior and ecology of the Goliath Heron. The Auk, 97, 433–448.

Molina-Pérez, R., & Larramendi, A. (2019). *Dinosaur facts and figures: The theropods and other Dinosauriformes*. Princeton University Press.

Molnar, R. E. (1978). A new theropod dinosaur from the Upper Cretaceous of central Montana. Journal of Paleontology, 52, 73–82.

Molnar, R. E. (1980). An albertosaur from the Hell Creek formation of Montana. Journal of Paleontology, 54, 102–108.

Molnar, R. E. (1991). The cranial morphology of *Tyrannosaurus rex*. Palaeontographica. Abteilung A, Paläozoologie, Stratigraphie, 217, 137–176.

Molnar, R. E. (1998). Mechanical factors in the design of the skull of *Tyrannosaurus rex* (Osborn, 1905). Gaia, 15, 193–218.

Molnar, R. E. (2001). Theropod paleopathology: A literature survey. In: Tanke, D.H., & Carpenter, K. (Eds.) *Mesozoic vertebrate life*, Indiana University Press. pp. 337–363

Molnar, R. E. (2008). Reconstruction of the jaw musculature of *Tyrannosaurus rex*. In: Larson, P., & Carpenter, K. (Eds.) *Tyrannosaurus rex, the tyrant king*. Indiana University Press. pp. 254–81.

Molnar, R. E. (2013). A comparative analysis of reconstructed jaw musculature and mechanics of some large theropods. In: Parrish, J. M., Molnar, R. E., Currie, P. J., & Koppelhus, E. B. (Eds.) *Tyrannosaurid paleobiology*. Indiana University Press. pp. 176–193.

Moody, R. T. J., & Naish, D. (2010). Alan Jack Charig (1927–1997): An overview of his academic accomplishments and role in the world of fossil reptile research. In: Moody, R. T. J., Buffetaut, E., Naish, D., & Martill, D. M. (Eds.) *Dinosaurs and other extinct saurians: A historical perspective*. Geological Society, London, Special Publications, 343, 89–109.

Morgan, J., Artemieva, N., & Goldin, T. (2013). Revisiting wildfires at the K-Pg boundary. Journal of Geophysical Research: Biogeosciences, 118, 1508–1520.

Morgan, J. V., Bralower, T. J., Brugger, J., & Wünnemann, K. (2022). The Chicxulub impact and its environmental consequences. Nature Reviews Earth & Environment, 3, 338–354.

Morgan, K. (1998). Thermoneutral zone and critical temperatures of horses. Journal of Thermal Biology, 23, 59–61.

Morhardt, A. C. (2009). Dinosaur smiles: Do the texture and morphology of the premaxilla, maxilla, and dentary bones of sauropsids provide osteological correlates for inferring extra-oral structures reliably in dinosaurs? Masters thesis, Western Illinois University.

Morhardt, A. C. (2016). Gross anatomical brain region approximation (GABRA): Assessing brain size, structure, and evolution in extinct archosaurs. Doctoral dissertation, Ohio University.

Müller, O., Záruba, B., Košťák, M., & Walica, R. Rostislav. (2023). *Pravěký svět Zdeňka Buriana—Kniha 1: Od vzniku Země po zánik dinosaurů*. Albatros Media.

Müller, R. T., & Garcia, M. S. (2020). A paraphyletic "Silesauridae" as an alternative hypothesis for the initial radiation of ornithischian dinosaurs. Biology letters, 16, 20200417.

Murali, G., & Kodandaramaiah, U. (2016). Deceived by stripes: Conspicuous patterning on vital anterior body parts can redirect predatory strikes to expendable posterior organs. Royal Society Open Science, 3, 160057.

Myhrvold, C. L., Stone, H. A., & Bou-Zeid, E. (2012). What is the use of elephant hair? PLOS One, 7, e47018.

REFERENCES

Myhrvold, N. P. (2013). Revisiting the estimation of dinosaur growth rates. PLOS One, 8(12), e81917.

Myhrvold, N. P. (2016). Dinosaur metabolism and the allometry of maximum growth rate. PLOS One, 11, e0163205.

Nabavizadeh, A. (2020). New reconstruction of cranial musculature in ornithischian dinosaurs: Implications for feeding mechanisms and buccal anatomy. The Anatomical Record, 303, 347–362.

Naish, D. (2021). Dinopedia: A brief compendium of dinosaur lore. Princeton University Press.

Naish, D., & Cau, A. (2022). The osteology and affinities of Eotyrannus lengi, a tyrannosauroid theropod from the Wealden Supergroup of southern England. PeerJ, 10, e12727.

Naish, D., & Witton, M. P. (2017). Neck biomechanics indicate that giant Transylvanian azhdarchid pterosaurs were short-necked arch predators. PeerJ, 5, e2908.

National Research Council. (1981). Effect of environment on nutrient requirements of domestic animals. National Academies Press.

Neate, R. (2022, November 21). Christie's cancels T rex skeleton auction after doubts raised. The Guardian. https://www.theguardian.com/science/2022/nov/21/christies-cancels-t-rex-skeleton-auction-after-doubts-raised

Nesbitt, S. J. (2011). The early evolution of archosaurs: Relationships and the origin of major clades. Bulletin of the American Museum of Natural History, 2011, 1–292.

Nesbitt, S. J., Barrett, P. M., Werning, S., Sidor, C. A., & Charig, A. J. (2013b). The oldest dinosaur? A Middle Triassic dinosauriform from Tanzania. Biology Letters, 9, 20120949.

Nesbitt, S. J., Brusatte, S. L., Desojo, J. B., Liparini, A., De França, M. A., Weinbaum, J. C., & Gower, D. J. (2013a). Rauisuchia. In: Nesbitt, S. J., Desojo, J. B., & Irmis, R. B. (Eds.) Anatomy, phylogeny and palaeobiology of early archosaurs and their kin. Geological Society, London, Special Publications, 379, 241–274.

Nesbitt, S. J., Denton, R. K., Jr., Loewen, M. A., Brusatte, S. L., Smith, N. D., Turner, A. H., ... & Wolfe, D. G. (2019). A mid-Cretaceous tyrannosauroid and the origin of North American end-Cretaceous dinosaur assemblages. Nature Ecology & Evolution, 3, 892–899.

Newman, B. H. (1970). Stance and gait in the flesh-eating dinosaur Tyrannosaurus. Biological Journal of the Linnean Society, 2, 119–123.

Nicholls, C. (2024, July 18). Stegosaurus skeleton sets auction record, selling for $44.6 million. CNN. https://edition.cnn.com/2024/07/18/science/stegosaurus-fossil-record-intl-scli-scn/index.html

Nieuwland, I. (2019). American dinosaur abroad: A cultural history of Carnegie's plaster Diplodocus. University of Pittsburgh Press.

Nopcsa, F. (1917). Über Dinosaurier. Centralblatt für Mineralogie, Geologie und Paläontologie, 1917, 332–351.

Norell, M., Gaffney, E. S., & Dingus, L. (1995). Discovering dinosaurs in the American Museum of Natural History. Alfred A. Knopf.

Norell, M. A., Wiemann, J., Fabbri, M., Yu, C., Marsicano, C. A., Moore-Nall, A., ... & Zelenitsky, D. K. (2020). The first dinosaur egg was soft. Nature, 583, 406–410.

Norman, D. B. (1985). The illustrated encyclopedia of dinosaurs. Salamander Books.

Norman, D. B., Baron, M. G., Garcia, M. S., & Müller, R. T. (2022). Taxonomic, palaeobiological and evolutionary implications of a phylogenetic hypothesis for Ornithischia (Archosauria: Dinosauria). Zoological Journal of the Linnean Society, 196(4), 1273–1309.

Nothdurft, W., & Smith, J. (2002). The lost dinosaurs of Egypt. Random House.

Novas, F. E. (1989). Los dinosaurios carnívoros de la Argentina. Doctoral thesis, La Plata: Universidad Nacional de La Plata.

Novas, F. E., Aranciaga Rolando, A. M., & Agnolín, F. L. (2016). Phylogenetic relationships of the Cretaceous Gondwanan theropods Megaraptor and Australovenator: The evidence afforded by their manual anatomy. Memoirs of Museum Victoria, 74, 49–61.

O'Connor, P. M. (2004). Pulmonary pneumaticity in the postcranial skeleton of extant Aves: A case study examining Anseriformes. Journal of Morphology, 261, 141–161.

O'Connor, P. M. (2006). Postcranial pneumaticity: An evaluation of soft-tissue influences on the postcranial skeleton and the reconstruction of pulmonary anatomy in archosaurs. Journal of Morphology, 267, 1199–1226.

Olkowicz, S., Kocourek, M., Lučan, R. K., Porteš, M., Fitch, W. T., Herculano-Houzel, S., & Němec, P. (2016). Birds have primate-like numbers of neurons in the forebrain. Proceedings of the National Academy of Sciences, 113, 7255–7260.

Olsen, T. (1995). Fluvial and fluvio-lacustrine facies and depositional environments of the Maastrichtian to Paleocene North Horn Formation, Price Canyon, Utah. The Mountain Geologist, 32, 27–44.

Olshevsky, G. (1995). The origin and evolution of the Tyrannosauridae, part 1. Dino Frontline, 9, 92–119.

Olsson, M., Stuart-Fox, D., & Ballen, C. (2013). Genetics and evolution of colour patterns in reptiles. In: Hofreiter, M. (Ed.) Genetic basis and evolutionary causes of colour variation in vertebrates. Seminars in Cell & Developmental Biology, Academic Press, 24, 529–541.

Osborn, H. F. (1905). Tyrannosaurus and other Cretaceous carnivorous dinosaurs. Bulletin of the AMNH, 21, 259–265.

Osborn, H. F. (1906). Tyrannosaurus, Upper Cretaceous carnivorous dinosaur. (second communication). Bulletin of the AMNH, 22, 281–296.

Osborn, H. F. (1912). Crania of Tyrannosaurus and Allosaurus. Memoirs of the American Museum of Natural History, 1, 3–30.

Osborn, H. F. (1913). Tyrannosaurus: Restoration and model of the skeleton. Bulletin of the AMNH, 32, 91–92.

Osborn, H. F. (1917a). Skeletal adaptations of Ornitholestes, Struthiomimus, Tyrannosaurus. Bulletin of the AMNH, 35, 733–771.

Osborn, H. F. (1917b). The origin and evolution of life: On the theory of action, reaction and interaction of energy. Charles Scribner's Sons, New York.

Ostrom, J. H. (1969). Deinonychus antirrhopus, an unusual theropod from the Early Cretaceous of Montana. Yale Peabody Museum Bulletin, 30, 1–165.

Ostrom, J. H. (1974). Archaeopteryx and the origin of flight. The Quarterly Review of Biology, 49, 27–47.

Ott, C. J. (2006). Cranial anatomy and biogeography of the first Leptoceratops gracilis (Dinosauria: Ornithischia) specimens from the Hell Creek Formation, Southeast. In: Carpenter, K. (Ed.) Horns and beaks: Ceratopsian and ornithopod dinosaurs. Indiana University Press. pp. 213–233.

Packer, C. (1983). Sexual dimorphism: The horns of African antelopes. Science, 221, 1191–1193.

Padian, K. (2022). Why tyrannosaurid forelimbs were so short: An integrative hypothesis. Acta Palaeontologica Polonica, 67, 63–76.

REFERENCES

Padian, K., Cunningham, J. R., Langston, W., Jr., & Conway, J. (2021). Functional morphology of *Quetzalcoatlus* Lawson 1975 (Pterodactyloidea: Azhdarchoidea). Journal of Vertebrate Paleontology, 41, 218–251.

Padian, K., Hutchinson, J. R., & Holtz, T. R., Jr. (1999). Phylogenetic definitions and nomenclature of the major taxonomic categories of the carnivorous Dinosauria (Theropoda). Journal of Vertebrate Paleontology, 19. 69–80.

Pahl, C. C., & Ruedas, L. A. (2021). Carnosaurs as apex scavengers: Agent-based simulations reveal possible vulture analogues in late Jurassic Dinosaurs. Ecological Modelling, 458, 109706.

Panciroli, E. (2021). *Beasts before us: The untold story of mammal origins and evolution*. Bloomsbury.

Pantuso, P. (2019, July 17). Perhaps the best dinosaur fossil ever discovered. So why has hardly anyone seen it? The Guardian. https://www.theguardian.com/science/2019/jul/17/montana-fossilized-dueling-dinosaurs-skeletons-dino-cowboy

Pare, S. (2024, January 3). *Nanotyrannus* vs. *T. rex* saga continues: Controversial study "doesn't settle the question at all." Live Science. https://www.livescience.com/animals/dinosaurs/nanotyrannus-vs-t-rex-saga-continues-controversial-study-doesnt-settle-the-question-at-all

Paul, G. S. (1988a). *Predatory dinosaurs of the world: A complete illustrated guide*. Simon & Schuster.

Paul, G. S. (1988b). Limb design, function and running performance in ostrich-mimics and tyrannosaurs. Gaia, 15, 257.

Paul, G. S. (1990). An improbable view of Tertiary dinosaurs. Evolutionary Theory, 9, 309–315.

Paul, G. S. (1997) Dinosaur models: The good, the bad, and using them to estimate the mass of dinosaurs. In: Wolberg, D. L., Stump, E., & Rosenberg, G. D. (Eds.) *DinoFest international proceedings*. Academy of Natural Sciences. pp. 39–45.

Paul, G. S. (2008). The extreme lifestyles and habits of the gigantic tyrannosaurid superpredators of the late Cretaceous of North America and Asia. In: Larson P., Carpenter K. (Eds.) Tyrannosaurus rex, *the tyrant king*. Indiana University Press. pp. 307–354.

Paul, G. S. (2016). *The Princeton field guide to dinosaurs*. Princeton University Press.

Paul, G. S. (2019). Non-ornithischian dinosaurs probably had lips. Here's why. Prehistoric Times, 127, 44–49.

Paul, G. S., Persons, W. S., & Van Raalte, J. (2022). The tyrant lizard king, queen and emperor: Multiple lines of morphological and stratigraphic evidence support subtle evolution and probable speciation within the North American genus *Tyrannosaurus*. Evolutionary Biology, 49, 156–179.

Pearson, D. A., Schaefer, T., Johnson, K. R., Nichols, D. J., and Hunter, J. P., 2002, Vertebrate biostratigraphy of the Hell Creek Formation in southwestern North Dakota and northwestern South Dakota. In: Hartman, J. H., Johnson, K. R., and Nichols, D. J., (Eds.) *The Hell Creek Formation and the Cretaceous-Tertiary boundary in the northern Great Plains: An integrated continental record of the end of the Cretaceous: Boulder, Colorado*, Geological Society of America Special Paper, 361, 145–167.

Pembury Smith, M. Q., & Ruxton, G. D. (2020). Camouflage in predators. Biological Reviews, 95, 1325–1340.

Persons, W. S., & Currie, P. J. (2014). Duckbills on the run: The cursorial abilities of hadrosaurs and implications for tyrannosaur-avoid-ance strategies. In: Eberth, D. A & Evans, D. C. (Eds.) *Hadrosaurs*. Indiana University Press. pp. 449–458.

Persons, W. S., & Currie, P. J. (2020). The anatomical and functional evolution of the femoral fourth trochanter in ornithischian dinosaurs. The Anatomical Record, 303, 1146–1157.

Persons, W. S., IV, & Currie, P. J. (2011). The tail of *Tyrannosaurus*: Reassessing the size and locomotive importance of the M. caudofemoralis in non-avian theropods. The Anatomical Record, 294, 119–131.

Persons, W. S., IV, & Currie, P. J. (2016). An approach to scoring cursorial limb proportions in carnivorous dinosaurs and an attempt to account for allometry. Scientific Reports, 6, 19828.

Persons, W. S., IV, Currie, P. J., & Erickson, G. M. (2020). An older and exceptionally large adult specimen of *Tyrannosaurus rex*. The Anatomical Record, 303, 656–672.

Peterson, J. E., & Daus, K. N. (2019). Feeding traces attributable to juvenile *Tyrannosaurus rex* offer insight into ontogenetic dietary trends. PeerJ, 7, e6573.

Peterson, J. E., Dischler, C., & Longrich, N. R. (2013). Distributions of cranial pathologies provide evidence for head-butting in dome-headed dinosaurs (Pachycephalosauridae). PLOS One, 8, e68620.

Peterson, J. E., Henderson, M. D., Scherer, R. P., & Vittore, C. P. (2009). Face biting on a juvenile tyrannosaurid and behavioral implications. Palaios, 24, 780–784.

Peterson, J. E., Lenczewski, M. E., & Scherer, R. P. (2010). Influence of microbial biofilms on the preservation of primary soft tissue in fossil and extant archosaurs. PLOS One, 5, e13334.

Peterson, J. E., Tseng, Z. J., & Brink, S. (2021). Bite force estimates in juvenile *Tyrannosaurus rex* based on simulated puncture marks. PeerJ, 9, e11450.

Pienaar, U. D. V. (1969). Predator-prey relationships amongst the larger mammals of the Kruger National Park. Koedoe, 12, 108–176.

Porfiri, J. D., Novas, F. E., Calvo, J. O., Agnolín, F. L., Ezcurra, M. D., & Cerda, I. A. (2014). Juvenile specimen of *Megaraptor* (Dinosauria, Theropoda) sheds light about tyrannosauroid radiation. Cretaceous Research, 51, 35–55.

Porfiri, J. D., Valieri, R. D. J., Santos, D. D., & Lamanna, M. C. (2018). A new megaraptoran theropod dinosaur from the Upper Cretaceous Bajo de la Carpa Formation of northwestern Patagonia. Cretaceous Research, 89, 302–319.

Porter, W. P., & Kearney, M. (2009). Size, shape, and the thermal niche of endotherms. Proceedings of the National Academy of Sciences, 106, 19666–19672.

Porter, W. R., & Witmer, L. M. (2020). Vascular patterns in the heads of dinosaurs: Evidence for blood vessels, sites of thermal exchange, and their role in physiological thermoregulatory strategies. The Anatomical Record, 303, 1075–1103.

Powell, J. S. 1998. Sexual dimorphism in archosaurs: Testing the limits of extant phylogenetic bracket method. Journal of Vertebrate Paleontology 18, 70A.

Power, R. J., & Compion, R. X. S. (2009). Lion predation on elephants in the Savuti, Chobe National Park, Botswana. African Zoology, 44, 36–44.

Preston, D. (1986). *Dinosaurs in the attic*. St. Martin's Press.

Prieto-Márquez, A., Wagner, J. R., Bell, P. R., & Chiappe, L. M. (2015). The late-surviving "duck-billed" dinosaur *Augustynolophus* from

REFERENCES

the upper Maastrichtian of western North America and crest evolution in Saurolophini. Geological Magazine, 152, 225–241.

Psihoyos, L. (1994). *Hunting dinosaurs*. Cassell.

Pu, H., Zelenitsky, D. K., Lü, J., Currie, P. J., Carpenter, K., Xu, L., ... & Shen, C. (2017). Perinate and eggs of a giant caenagnathid dinosaur from the Late Cretaceous of central China. Nature Communications, 8, 14952.

Range, M. M., Arbic, B. K., Johnson, B. C., Moore, T. C., Titov, V., Adcroft, A. J., ... & Wang, H. (2022). The Chicxulub impact produced a powerful global tsunami. AGU Advances, 3, e2021AV000627.

Rauhut, O. W. M. (2003). The interrelationships and evolution of basal theropod dinosaurs. Special Papers in Palaeontology, 69, 1–213.

Rauhut, O. W., & Pol, D. (2019). Probable basal allosauroid from the early Middle Jurassic Cañadón Asfalto Formation of Argentina highlights phylogenetic uncertainty in tetanuran theropod dinosaurs. Scientific Reports, 9, 18826.

Rauhut, O. W., Milner, A. C., & Moore-Fay, S. (2010). Cranial osteology and phylogenetic position of the theropod dinosaur *Proceratosaurus bradleyi* (Woodward, 1910) from the Middle Jurassic of England. Zoological Journal of the Linnean Society, 158, 155–195.

Ray, G. E. (1941), Big for his day. Natural History, 48, 36–39.

Rayfield, E. J. (2004). Cranial mechanics and feeding in *Tyrannosaurus rex*. Proceedings of the Royal Society of London. Series B: Biological Sciences, 271, 1451–1459.

Rayfield, E. J. (2005a). Using finite-element analysis to investigate suture morphology: A case study using large carnivorous dinosaurs. The Anatomical Record, 283, 349–365.

Rayfield, E. J. (2005b). Aspects of comparative cranial mechanics in the theropod dinosaurs *Coelophysis, Allosaurus* and *Tyrannosaurus*. Zoological Journal of the Linnean Society, 144, 309–316.

Raymond, P. E. (1939). *Prehistoric life*. Harvard University Press.

Recreating the eighth wonder: The making of King Kong (2006). *King Kong* (DVD), Universal.

Rega, E. A., & Brochu, C. A. 2001. Paleopathology of a mature *Tyrannosaurus rex* skeleton. Journal of Vertebrate Paleontology, 21, 92A.

Regal, B. (2018). *Henry Fairfield Osborn: Race and the search for the origins of man*. Routledge.

Reiner, A. (2023). Could theropod dinosaurs have evolved to a human level of intelligence? Journal of Comparative Neurology, 531(9), 975–1006.

Reolid, M., Cardenal, F. J., & Reolid, J. (2021). Digital 3D models of theropods for approaching body-mass distribution and volume. Journal of Iberian Geology, 47, 599–624.

Retallack, G. J., Leahy, G. D., & Spoon, M. D. (1987). Evidence from paleosols for ecosystem changes across the Cretaceous/Tertiary boundary in eastern Montana. Geology, 15, 1090–1093.

Reuters. (2015, December 22). Nicolas Cage returns stolen Mongolian dinosaur skull he bought at gallery. The Guardian. https://www.theguardian.com/film/2015/dec/22/nicolas-cage-returns-stolen-mongolian-dinosaur-skull-he-bought-at-gallery

Reynolds, M. (2018, June 21). The dinosaur trade: How celebrity collectors and glitzy auctions could be damaging science. Wired. https://www.wired.co.uk/article/dinosaur-t-rex-auction-sale-private-fossil-trade

Rhodes, M. M., Henderson, D. M., & Currie, P. J. (2021). Maniraptoran pelvic musculature highlights evolutionary patterns in theropod locomotion on the line to birds. PeerJ, 9, e10855.

Riede, T., Eliason, C. M., Miller, E. H., Goller, F., & Clarke, J. A. (2016). Coos, booms, and hoots: The evolution of closed-mouth vocal behavior in birds. Evolution, 70, 1734-1746.

Riede, T., Li, Z., Tokuda, I. T., & Farmer, C. G. (2015). Functional morphology of the *Alligator mississippiensis* larynx with implications for vocal production. Journal of Experimental Biology, 218, 991–998.

Rieppel, L. (2019). *Assembling the dinosaur: Fossil hunters, tycoons, and the making of a spectacle*. Harvard University Press.

Roach, B. T., & Brinkman, D. L. (2007). A reevaluation of cooperative pack hunting and gregariousness in *Deinonychus antirrhopus* and other nonavian theropod dinosaurs. Bulletin of the Peabody Museum of Natural History, 48, 103–138.

Roberts, S. C. (1996). The evolution of hornedness in female ruminants. Behaviour, 133, 399–442.

Rohleder, A. (2001, August 1). Collecting Dinosaur Bones. Forbes. http://www.forbes.com/2001/08/01/0801connguide.html

Romer, A. S. (1959). Vertebrate Paleontology, 1908–1958. Journal of Paleontology, 33, 915–925.

Rose, C., & Reiss, J. O. (1993). Metamorphosis and the vertebrate skull: Ontogenetic patterns and developmental mechanisms. In: Hanken, H., & Hall, B. K. (Eds.) *The skull development*. University of Chicago Press, 1, 289–346.

Ross, R. M., Duggan-Haas, D., & Allmon, W. D. (2013). The posture of *Tyrannosaurus rex*: Why do student views lag behind the science? Journal of Geoscience Education, 61, 145–160.

Rothschild, B., & Tanke, D.H. (2005). Theropod paleopathology: State-of-the-art review. In: Carpenter, K. (Ed.), *The carnivorous dinosaurs*, Indiana University Press. pp. 351–365.

Rothschild, B., O'Connor, J., & Lozado, M. C. (2022). Closer examination does not support infection as cause for enigmatic *Tyrannosaurus rex* mandibular pathologies. Cretaceous Research, 140, 105353.

Rothschild, B. M. (2013). Clawing their way to the top: Tyrannosaurid pathology and lifestyle. In: Parrish, J. M., Molnar, R. E., Currie, P. J., & Koppelhus, E. B. (Eds.) *Tyrannosaurid paleobiology*. Indiana University Press. pp. 210–221.

Rothschild, B. M. (2015). Unexpected behavior in the Cretaceous: Tooth-marked bones attributable to tyrannosaur play. Ethology Ecology & Evolution, 27, 325–334.

Rothschild, B. M., & Molnar, R. E. (2008). Tyrannosaurid pathologies as clues to nature and nurture in the Cretaceous. In: Larson, P., & Carpenter, K. (Eds.) Tyrannosaurus rex, *the tyrant king*. Indiana University Press. pp. 287–304.

Rothschild, B. M., Tanke, D., & Carpenter, K. (1997). Tyrannosaurs suffered from gout. Nature, 387, 357–357.

Rowe, A. J., & Rayfield, E. J. (2024). Morphological evolution and functional consequences of giantism in tyrannosauroid dinosaurs. ISCIENCE, 27, 110679.

Rowe, A. J., & Snively, E. (2021). Biomechanics of juvenile tyrannosaurid mandibles and their implications for bite force: Evolutionary biology. The Anatomical Record, 305, 373–392.

Rowe, M. F., Bakken, G. S., Ratliff, J. J., & Langman, V. A. (2013). Heat storage in Asian elephants during submaximal exercise: Behavioral regulation of thermoregulatory constraints on activity in endothermic gigantotherms. Journal of Experimental Biology, 216, 1774–1785.

Rozhdestvensky, A. K. (1965). Growth changes in Asian dinosaurs and some problems of their taxonomy. Paleontologičeskij žurnal, 3, 95–109.

Russell, A. P., & Bauer, A. M. (2021). Vocalization by extant nonavian reptiles: a synthetic overview of phonation and the vocal apparatus. The Anatomical Record, 304, 1478–1528.

Russell, D. A. (1970). Tyrannosaurs from the Late Cretaceous of western Canada. National Museum of Natural Sciences, Publications, in Paleontology, 1, 1–34.

Russell, D. A. (1972). Ostrich dinosaurs from the Late Cretaceous of western Canada. Canadian Journal of Earth Sciences, 9, 375–402.

Russell, D. A. (1984). The gradual decline of the dinosaurs—fact or fallacy? Nature, 307, 360–361.

Ruxton, G. D., & Houston, D. C. (2003). Could Tyrannosaurus rex have been a scavenger rather than a predator? An energetics approach. Proceedings of the Royal Society of London. Series B: Biological Sciences, 270, 731–733.

Ruxton, G. D., & Houston, D. C. (2004). Obligate vertebrate scavengers must be large soaring fliers. Journal of Theoretical Biology, 228, 431–436.

Ruxton, G. D., Birchard, G. F., & Deeming, D. C. (2014). Incubation time as an important influence on egg production and distribution into clutches for sauropod dinosaurs. Paleobiology, 40, 323–330.

Sakamoto, M., Benton, M. J., & Venditti, C. (2016). Dinosaurs in decline tens of millions of years before their final extinction. Proceedings of the National Academy of Sciences, 113, 5036–5040.

Sakamoto, M., Ruta, M., & Venditti, C. (2019). Extreme and rapid bursts of functional adaptations shape bite force in amniotes. Proceedings of the Royal Society. Series: B, 286, 20181932.

Samman, T. (2013). Tyrannosaurid craniocervical mobility: A preliminary qualitative assessment. In: Parrish, J. M., Molnar, R. E., Currie, P. J., & Koppelhus, E. B. (Eds.) Tyrannosaurid paleobiology. Indiana University Press. pp. 195–210.

Sampson, S. D., & Loewen, M. A. (2005). Tyrannosaurus rex from the Upper Cretaceous (Maastrichtian) North Horn Formation of Utah: Biogeographic and paleoecologic implications. Journal of Vertebrate Paleontology, 25, 469–472.

Sampson, S. D., Ryan, M. J., & Tanke, D. H. (1997). Craniofacial ontogeny in centrosaurine dinosaurs (Ornithischia: Ceratopsidae): Taxonomic and behavioral implications. Zoological Journal of the Linnean Society, 121, 293–337.

Santa Catharina, A., Kneller, B. C., Marques, J. C., Mcarthur, A. D., Cevallos-Ferriz, S. R. S., Theurer, T., ... & Muirhead, D. (2022). Timing and causes of forest fire at the K–Pg boundary. Scientific Reports, 12, 13006.

Santangelo Law Offices (2015, February 11). Tyrannosaurus rex lawsuit settled. https://idea-asset.com/tyrannosaurus-rex-lawsuit-settled/

Saveliev, S. V., & Alifanov, V. R. (2007). A new study of the brain of the predatory dinosaur Tarbosaurus bataar (Theropoda, Tyrannosauridae). Paleontological Journal, 41, 281–289.

Scannella, J. B., & Horner, J. R. (2010). Torosaurus Marsh, 1891, is Triceratops Marsh, 1889 (Ceratopsidae: Chasmosaurinae): Synonymy through ontogeny. Journal of Vertebrate Paleontology, 30, 1157–1168.

Scannella, J. B., Fowler, D. W., Goodwin, M. B., & Horner, J. R. (2014). Evolutionary trends in Triceratops from the Hell Creek formation, Montana. Proceedings of the National Academy of Sciences, 111, 10245–10250.

Schaeffer, J. (2016). The deeds and dealings of Trix. In: Besselink, M. (Ed.) Trix: the grand old lady. Naturalis Biodiversity Center, Leiden. pp. 90–91.

Schaller, G. B. (1972). The Serengeti lion: A study of predator-prey relations. University of Chicago Press.

Scherer, C. R., & Voiculescu-Holvad, C. (2023). Re-analysis of a dataset refutes claims of anagenesis within Tyrannosaurus-line tyrannosaurines (Theropoda, Tyrannosauridae). Cretaceous Research, 155, 105780.

Schmerge, J. D., & Rothschild, B. M. (2016a). Distribution of the dentary groove of theropod dinosaurs: Implications for theropod phylogeny and the validity of the genus Nanotyrannus Bakker et al., 1988. Cretaceous Research, 61, 26–33.

Schmerge, J. D., & Rothschild, B. M. (2016b). When a groove is not a groove: Clarification of the appearance of the dentary groove in tyrannosauroid theropods and the distinction between Nanotyrannus and Tyrannosaurus. Reply to Comment on: "Distribution of the dentary groove of theropod dinosaurs: Implications for theropod phylogeny and the validity of the genus Nanotyrannus Bakker et al., 1988." Cretaceous Research, 65, 238–243.

Schmidt-Nielsen, K., Kanwisher, J., Lasiewski, R. C., Cohn, J. E., & Bretz, W. L. (1969). Temperature regulation and respiration in the ostrich. The Condor, 71, 341–352.

Schroeder, K., Lyons, S. K., & Smith, F. A. (2021). The influence of juvenile dinosaurs on community structure and diversity. Science, 371, 941–944.

Schulte, P., Alegret, L., Arenillas, I., Arz, J. A., Barton, P. J., Bown, P. R., ... & Willumsen, P. S. (2010). The Chicxulub asteroid impact and mass extinction at the Cretaceous-Paleogene boundary. Science, 327, 1214–1218.

Schweitzer, M. H., Wittmeyer, J. L., & Horner, J. R. (2005b). Gender-specific reproductive tissue in ratites and Tyrannosaurus rex. Science, 308, 1456–1460.

Schweitzer, M. H., Wittmeyer, J. L., & Horner, J. R. (2007). Soft tissue and cellular preservation in vertebrate skeletal elements from the Cretaceous to the present. Proceedings of the Royal Society. Series B: Biological Sciences, 274, 183–197.

Schweitzer, M. H., Wittmeyer, J. L., Horner, J. R., & Toporski, J. K. (2005a). Soft-tissue vessels and cellular preservation in Tyrannosaurus rex. Science, 307, 1952–1955.

Schweitzer, M. H., Zheng, W., Organ, C. L., Avci, R., Suo, Z., Freimark, L. M., ... & Asara, J. M. (2009). Biomolecular characterization and protein sequences of the Campanian hadrosaur B. canadensis. science, 324, 626–631.

Schweitzer, M. H., Zheng, W., Zanno, L., Werning, S., & Sugiyama, T. (2016). Chemistry supports the identification of gender-specific reproductive tissue in Tyrannosaurus rex. Scientific Reports, 6, 1–10.

Schwimmer, D. R. (2002). King of the crocodylians: The paleobiology of Deinosuchus. Indiana University Press.

Seebacher, F. (2001). A new method to calculate allometric length-mass relationships of dinosaurs. Journal of Vertebrate Paleontology, 21, 51–60.

Sellers, W. I., & Manning, P. L. (2007). Estimating dinosaur maximum running speeds using evolutionary robotics. Proceedings of the Royal Society. Series B: Biological Sciences, 274, 2711–2716.

REFERENCES

Sellers, W. I., Pond, S. B., Brassey, C. A., Manning, P. L., & Bates, K. T. (2017). Investigating the running abilities of *Tyrannosaurus rex* using stress-constrained multibody dynamic analysis. PeerJ, 5, e3420.

Senter, P. (2007). A new look at the phylogeny of Coelurosauria (Dinosauria: Theropoda). Journal of Systematic Palaeontology, 5, 429–463.

Senter, P. (2008). Voices of the past: A review of Paleozoic and Mesozoic animal sounds. Historical Biology, 20, 255–287.

Senter, P., & Parrish, J. M. (2006). Forelimb function in the theropod dinosaur *Carnotaurus sastrei*, and its behavioral implications. PaleoBios, 26, 7–17.

Senter, P., & Robins, J. H. (2005). Range of motion in the forelimb of the theropod dinosaur *Acrocanthosaurus atokensis*, and implications for predatory behaviour. Journal of Zoology, 266, 307–318.

Senter, P. J., & Mackey, J. J. (2023). Forelimb motion and orientation in the ornithischian dinosaurs *Styracosaurus* and *Thescelosaurus*, and its implications for locomotion and other behavior. Palaeontologia Electronica, 26, 1–19.

Sereno, P. C., & Brusatte, S. L. (2008). Basal abelisaurid and carcharodontosaurid theropods from the Lower Cretaceous Elrhaz Formation of Niger. Acta Palaeontologica Polonica, 53, 15–46.

Sereno, P. C., Dutheil, D. B., Iarochene, M., Larsson, H. C., Lyon, G. H., Magwene, P. M., ... & Wilson, J. A. (1996). Predatory dinosaurs from the Sahara and Late Cretaceous faunal differentiation. Science, 272, 986–991.

Sereno, P. C., Myhrvold, N., Henderson, D. M., Fish, F. E., Vidal, D., Baumgart, S. L., ... & Conroy, L. L. (2022). *Spinosaurus* is not an aquatic dinosaur. Elife, 11, e80092.

Sereno, P. C., Tan, L., Brusatte, S. L., Kriegstein, H. J., Zhao, X., & Cloward, K. (2009). Tyrannosaurid skeletal design first evolved at small body size. Science, 326, 418–422.

Shay, D., & Duncan, J. (1993). *The making of* Jurassic Park. Ballantine Books.

Sheehan, P. M., Fastovsky, D. E., Barreto, C., & Hoffmann, R. G. (2000). Dinosaur abundance was not declining in a "3 m gap" at the top of the Hell Creek Formation, Montana and North Dakota. Geology, 28, 523–526.

Sheehan, P. M., Fastovsky, D. E., Hoffmann, R. G., Berghaus, C. B., & Gabriel, D. L. (1991). Sudden extinction of the dinosaurs: Latest Cretaceous, upper Great Plains, USA. Science, 254, 835–839.

Shimizu, T., Patton, T. B., Szafranski, G., & Butler, A. B. (2009). Evolution of the visual system in reptiles and birds. In: Binder, M. D., Hirokawa, N. Windhorst, U., & Hirsch, M. C. (Eds.) Encyclopedia of neuroscience, Springer, 161, 5–24.

Shuonan, Z, Baiming, Z, Mateer, N. J., & Lucas, S. G. (1985) The Mesozoic reptiles of China. In: Lucas, S. G., & Mateer, N. J. (Eds). *Studies of Chinese fossil vertebrates*. Bulletin of the Geological Institutions of the University of Uppsala, N.S., Vol. II. pp. 133–150.

Signor, P. W. & Lipps, J. H. (1982). Sampling bias, gradual extinction patterns, and catastrophes in the fossil record. In: Leon T. Silver, L. T., & Schultz, P. H. (Eds.) *Geological implications of impacts of large asteroids and comets on the Earth*, Geological Society of America Special Paper, 190, 291–296.

Siviero, B. C., Rega, E., Hayes, W. K., Cooper, A. M., Brand, L. R., & Chadwick, A. V. (2020). Skeletal trauma with implications for intratail mobility in *Edmontosaurus annectens* from a monodominant bonebed, Lance Formation (Maastrichtian), Wyoming USA. Palaios, 35, 201–214.

Sloan, R. E., & Rigby, J. K., Jr. (1986). Response: Cretaceous-Tertiary dinosaur extinction. Science, 234, 1173–1175.

Sloan, R. E., & Van Valen, L. (1965). Cretaceous mammals from Montana. Science, 148, 220–227.

Smith, J. B. (2005). Heterodonty in *Tyrannosaurus rex*: Implications for the taxonomic and systematic utility of theropod dentitions. Journal of Vertebrate Paleontology, 25, 865–887.

Smith, J. B., Lamanna, M. C., Mayr, H., & Lacovara, K. J. (2006). New information regarding the holotype of *Spinosaurus aegyptiacus* Stromer, 1915. Journal of Paleontology, 80, 400–406.

Smith, N. D., Makovicky, P. J., Agnolin, F. L., Ezcurra, M. D., Pais, D. F., & Salisbury, S. W. (2008). A *Megaraptor*-like theropod (Dinosauria: Tetanurae) in Australia: Support for faunal exchange across eastern and western Gondwana in the Mid-Cretaceous. Proceedings of the Royal Society. Series B: Biological Sciences, 275, 2085–2093.

Snively, E., & Cox, A. (2008). Structural mechanics of pachycephalosaur crania permitted head-butting behavior. Palaeontologia Electronica, 11, 3A.

Snively, E., & Russell, A. (2002). The tyrannosaurid metatarsus: Bone strain and inferred ligament function. Senckenbergiana lethaea, 82, 35–42.

Snively, E., & Russell, A. P. (2007a). Craniocervical feeding dynamics of *Tyrannosaurus rex*. Paleobiology, 33, 610–638.

Snively, E., & Russell, A. P. (2007b). Functional variation of neck muscles and their relation to feeding style in Tyrannosauridae and other large theropod dinosaurs. The Anatomical Record, 290, 934–957.

Snively, E., & Samman, T. (2015). Unexpected behavior in the Cretaceous: tooth-marked bones attributable to tyrannosaur play. A comment by E. Snively & T. Samman. Ethology Ecology & Evolution, 27, 422–427.

Snively, E., Cotton, J. R., Witmer, L., Ridgely, R., & Theodor, J. (2011). Finite element comparison of cranial sinus function in the dinosaur *Majungasaurus* and head-clubbing giraffes. American Society of Mechanical Engineers; Summer Bioengineering Conference, 54587, 1075–1076.

Snively, E., Henderson, D. M., & Phillips, D. S. (2006). Fused and vaulted nasals of tyrannosaurid dinosaurs: Implications for cranial strength and feeding mechanics. Acta Palaeontologica Polonica, 51, 435–454.

Snively, E., O'Brien, H., Henderson, D. M., Mallison, H., Surring, L. A., Burns, M. E., ... & Cotton, J. R. (2019). Lower rotational inertia and larger leg muscles indicate more rapid turns in tyrannosaurids than in other large theropods. PeerJ, 7, e6432.

Snively, E., Russell, A. P., & Powell, G. L. (2004). Evolutionary morphology of the coelurosaurian arctometatarsus: Descriptive, morphometric and phylogenetic approaches. Zoological Journal of the Linnean Society, 142, 525–553.

Snyder, K., McLain, M., Wood, J., & Chadwick, A. (2020). Over 13,000 elements from a single bonebed help elucidate disarticulation and transport of an *Edmontosaurus thanatocoenosis*. PLOS One, 15, e0233182.

Soares, D. (2002). An ancient sensory organ in crocodilians. Nature, 417, 241–242.

Spassov, N. B. (1979). Sexual selection and the evolution of horn-like structures of ceratopsian dinosaurs. Palaeontology, Stratigraphy and Lithology, 11, 37–48.

REFERENCES

Stankowich, T., & Blumstein, D. T. (2005). Fear in animals: a meta-analysis and review of risk assessment. Proceedings of the Royal Society. Series B: Biological Sciences, 272, 2627–2634.

Stankowich, T., & Caro, T. (2009). Evolution of weaponry in female bovids. Proceedings of the Royal Society. Series B: Biological Sciences, 276, 4329–4334.

Stein, W. W. (2019). Taking count: A census of dinosaur fossils recovered from the Hell Creek and Lance Formations (Maastrichtian). Journal of Paleontological Sciences, 8, 1–42.

Stevens, K. A. (2006). Binocular vision in theropod dinosaurs. Journal of Vertebrate Paleontology, 26, 321–330.

Stevens, K. A., Larson, P., Wills, E. D., & Anderson, A. (2008). Rex, sit: Digital modeling of Tyrannosaurus rex at rest. In: Larson, P., & Carpenter, K. (Eds.) Tyrannosaurus rex, the tyrant king. Indiana University Press. pp. 193–204.

Stovall, J. W., & Langston, W. (1950). Acrocanthosaurus atokensis, a new genus and species of Lower Cretaceous Theropoda from Oklahoma. The American Midland Naturalist, 43, 696–728.

Stromer, E. 1936. Ergebnisse der Forschungsreisen Prof. E. Stromers in den Wüsten Ägyptens. VII. Baharije-Kessel und -Stufe mit deren Fauna und Flora. Eine ergänzende Zusammenfassung. Abhandlungen der Bayerischen Akademie der Wissenschaften, Mathematisch-naturwissenschaftliche Abteilung n. F., 33, 1–102.

Sumba, S. J. A. (1989). Food procurement through piracy and scavenging in the African fish eagle in Queen Elizabeth National Park, Uganda. African Journal of Ecology, 27, 111–118.

Swinton, W. E. (1934). The dinosaurs. T. Murby.

Szekely, P., Korem, Y., Moran, U., Mayo, A., & Alon, U. (2015). The mass-longevity triangle: Pareto optimality and the geometry of life-history trait space. PLOS Computational Biology, 11, e1004524.

Tanaka, K., Zelenitsky, D. K., & Therrien, F. (2015). Eggshell porosity provides insight on evolution of nesting in dinosaurs. PLOS One, 10, e0142829.

Tanke, D. H., & Currie, P. J. (1998). Head-biting behavior in theropod dinosaurs: paleopathological evidence. Gaia, 15, 167–184.

Tanke, D. H., & Rothschild, B. M. (2014). Paleopathology in Late Cretaceous Hadrosauridae from Alberta, Canada with comments on a putative Tyrannosaurus bite injury on an Edmontosaurus tail. In: Eberth, D. A., & Evans, D. C. (Eds.) Hadrosaurs. Indiana University Press. pp. 540–572.

Tattersdill, W., & Witton, M. P. (2025). The "spin" in Spinosaurus: inventing a modern dinosaur superstar. In: Manias, C. Palaeontology in public: popular science, lost creatures and deep time. UCL Press. pp. 79–108.

Terrill, D. (2021). Applications of strontium concentrations and isotopes preserved in vertebrate enamel and enamel-like tissues to paleoecology studies of Silurian conodonts from Gotland, Sweden, and Cretaceous dinosaurs and plesiosaurs from Alberta, Canada. Doctoral thesis, University of Calgary, Calgary, Canada.

Terrill, D. F., Henderson, C. M., & Anderson, J. S. (2020). New application of strontium isotopes reveals evidence of limited migratory behaviour in Late Cretaceous hadrosaurs. Biology Letters, 16, 20190930.

Therrien, F., & Henderson, D. M. (2007). My theropod is bigger than yours ... or not: estimating body size from skull length in theropods. Journal of Vertebrate Paleontology, 27, 108–115.

Therrien, F., Zelenitsky, D. K., Tanaka, K., Voris, J. T., Erickson, G. M., Currie, P. J., ... & Kobayashi, Y. (2023). Exceptionally preserved stomach contents of a young tyrannosaurid reveal an ontogenetic dietary shift in an iconic extinct predator. Science Advances, 9, eadi0505.

Therrien, F., Zelenitsky, D. K., Voris, J. T., & Tanaka, K. (2021). Mandibular force profiles and tooth morphology in growth series of Albertosaurus sarcophagus and Gorgosaurus libratus (Tyrannosauridae: Albertosaurinae) provide evidence for an ontogenetic dietary shift in tyrannosaurids. Canadian Journal of Earth Sciences, 58, 812–828.

Titus, A. L., Knoll, K., Sertich, J. J., Yamamura, D., Suarez, C. A., Glasspool, I. J., ... & Roberts, E. M. (2021). Geology and taphonomy of a unique tyrannosaurid bonebed from the upper Campanian Kaiparowits Formation of southern Utah: implications for tyrannosaurid gregariousness. PeerJ, 9, e11013.

Tschopp, E., Mateus, O., & Benson, R. B. (2015). A specimen-level phylogenetic analysis and taxonomic revision of Diplodocidae (Dinosauria, Sauropoda). PeerJ, 3, e857.

Tsuihiji, T., Watabe, M., Tsogtbaatar, K., Tsubamoto, T., Barsbold, R., Suzuki, S., ... & Witmer, L. M. (2011). Cranial osteology of a juvenile specimen of Tarbosaurus bataar (Theropoda, Tyrannosauridae) from the Nemegt Formation (Upper Cretaceous) of Bugin Tsav, Mongolia. Journal of Vertebrate Paleontology, 31, 497–517.

Turner, A. H., Makovicky, P. J., & Norell, M. A. (2007). Feather quill knobs in the dinosaur Velociraptor. Science, 317, 1721–1721.

Tykoski, R. S., & Fiorillo, A. R. (2017). An articulated cervical series of Alamosaurus sanjuanensis Gilmore, 1922 (Dinosauria, Sauropoda) from Texas: New perspective on the relationships of North America's last giant sauropod. Journal of Systematic Palaeontology, 15, 339–364.

Ullmann, P. V., Macauley, K., Ash, R. D., Shoup, B., & Scannella, J. B. (2021). Taphonomic and diagenetic pathways to protein preservation, part I: The case of Tyrannosaurus rex specimen MOR 1125. Biology, 10, 1193.

Upchurch, G. R., & Mack, G. H. (1998). Latest Cretaceous leaf megafloras from the Jose Creek Member, McRae Formation of New Mexico. New Mexico Geological Society Guidebook, 49, 209–222.

Van Bijlert, P. A., van Soest, A. K., & Schulp, A. S. (2021). Natural Frequency Method: estimating the preferred walking speed of Tyrannosaurus rex based on tail natural frequency. Royal Society Open Science, 8, 201441.

VanBuren, C. S. (2013). The function and evolution of the syncervical in ceratopsian dinosaurs with a review of cervical fusion in tetrapods. Master's thesis, University of Toronto.

Van der Reest, A. J., Wolfe, A. P., & Currie, P. J. (2016). A densely feathered ornithomimid (Dinosauria: Theropoda) from the Upper Cretaceous Dinosaur Park Formation, Alberta, Canada. Cretaceous Research, 58, 108–117.

Van Vranken, N. E., & Boyd, C. A. (2021). The first in situ collection of a mosasaurine from the marine Breien Member of the Hell Creek Formation in south-central North Dakota, USA. PaleoBios, 38, 1–11.

Varricchio, D. J. (2001). Gut contents from a Cretaceous tyrannosaurid: Implications for theropod dinosaur digestive tracts. Journal of Paleontology, 75, 401–406.

Varricchio, D. J. (2011). A distinct dinosaur life history? Historical Biology, 23, 91–107.

REFERENCES

Varricchio, D. J., Jackson, F., & Trueman, C. N. (1999). A nesting trace with eggs for the Cretaceous theropod dinosaur *Troodon formosus*. Journal of Vertebrate Paleontology, 19, 91–100.

Varricchio, D. J., Moore, J. R., Erickson, G. M., Norell, M. A., Jackson, F. D., & Borkowski, J. J. (2008). Avian paternal care had dinosaur origin. Science, 322, 1826–1828.

Vinther, J. (2015). A guide to the field of palaeo colour: Melanin and other pigments can fossilise: reconstructing colour patterns from ancient organisms can give new insights to ecology and behaviour. BioEssays, 37, 643–656.

Vittore, C. P., & Henderson, M. D. (2013). Brodie abscess involving a tyrannosaur phalanx: Imaging and implications. In: Parrish, J. M., Molnar, R. E., Currie, P. J., & Koppelhus, E. B. (Eds.) *Tyrannosaurid paleobiology*. Indiana University Press. pp. 223–236.

Vogel, G. (2020, October 7). Stan the *T. rex* sells for record $32 million at auction. Science. https://www.science.org/content/article/stan-t-rex-sells-record-32-million-auction

Volpe, R. (Ed.). (2007). *The Age of Reptiles: The art and science of Rudolph Zallinger's great dinosaur mural at Yale*. Peabody Museum of Natural History, Yale University.

Voris, J. T., Therrien, F., Zelenitsky, D. K., & Brown, C. M. (2020). A new tyrannosaurine (Theropoda: Tyrannosauridae) from the Campanian Foremost Formation of Alberta, Canada, provides insight into the evolution and biogeography of tyrannosaurids. Cretaceous Research, 110, 104388.

Voris, J. T., Zelenitsky, D. K., Therrien, F., Ridgely, R. C., Currie, P. J., & Witmer, L. M. (2021). Two exceptionally preserved juvenile specimens of *Gorgosaurus libratus* (Tyrannosauridae, Albertosaurinae) provide new insight into the timing of ontogenetic changes in tyrannosaurids. Journal of Vertebrate Paleontology, 41, e2041651.

Wade, D. C., Abraham, N. L., Farnsworth, A., Valdes, P. J., Bragg, F., & Archibald, A. T. (2019). Simulating the climate response to atmospheric oxygen variability in the Phanerozoic: A focus on the Holocene, Cretaceous and Permian. Climate of the Past, 15, 1463–1483.

Walsberg, G. E. (1983). Coat color and solar heat gain in animals. BioScience, 33, 88–91.

Warshaw, E. A., & Fowler, D. W. (2022). A transitional species of *Daspletosaurus* Russell, 1970 from the Judith River Formation of eastern Montana. PeerJ, 10, e14461.

Wedel, M. J. (2003). The evolution of vertebral pneumaticity in sauropod dinosaurs. Journal of Vertebrate Paleontology, 23, 344–357.

Wedel, M. J. (2004). Skeletal pneumaticity in saurischian dinosaurs and its implications for mass estimates. Journal of Vertebrate Paleontology 24, 127A.

Wedel, M. J. (2009). Evidence for bird-like air sacs in saurischian dinosaurs. Journal of Experimental Zoology, Part A: Ecological Genetics and Physiology, 311, 611–628.

Weinbaum, J. C. (2011). The skull of *Postosuchus kirkpatricki* (Archosauria: Paracrocodyliformes) from the upper Triassic of the United States. PaleoBios, 30, 18–44.

Weinstein, S. A., & White, J. (2015). In response to how not to train your dragon: A case of Komodo dragon bite, by Borek and Charlton. Wilderness & Environmental Medicine, 26, 572–573.

Wells, H. G. (1922). *A short history of the world*. Cassell & Company.

White, M. A., Bell, P. R., Cook, A. G., Barnes, D. G., Tischler, T. R., Bassam, B. J., & Elliott, D. A. (2015b). Forearm range of motion in *Australovenator wintonensis* (Theropoda, Megaraptoridae). PLOS One, 10, e0137709.

White, M. A., Bell, P. R., Cook, A. G., Poropat, S. F., & Elliott, D. A. (2015a). The dentary of *Australovenator wintonensis* (Theropoda, Megaraptoridae); implications for megaraptorid dentition. PeerJ, 3, e1512.

Wick, S. L. (2014). New evidence for the possible occurrence of *Tyrannosaurus* in west Texas, and discussion of Maastrichtian tyrannosaurid dinosaurs from Big Bend National Park. Cretaceous Research, 50, 52–58.

Wick, S. L., & Lehman, T. M. (2013). A new ceratopsian dinosaur from the Javelina Formation (Maastrichtian) of west Texas and implications for chasmosaurine phylogeny. Naturwissenschaften, 100, 667–682.

Williams, M. E. (1994). Catastrophic versus noncatastrophic extinction of the dinosaurs: Testing, falsifiability, and the burden of proof. Journal of Paleontology, 68, 183–190.

Williams, P. (2018). *The dinosaur artist: Obsession, betrayal, and the quest for Earth's ultimate trophy*. Hachette Books.

Williams, S., Brusatte, S., Mathews, J., Currie, P. (2010). A new juvenile *Tyrannosaurus* and a reassessment of ontogenetic and phylogenetic changes in tyrannosauroid forelimb proportions. Journal of Vertebrate Paleontology, 30, 187A.

Williamson, T. E., & Weil, A. (2008). Stratigraphic distribution of sauropods in the Upper Cretaceous of the San Juan Basin, New Mexico, with comments on North America's Cretaceous "sauropod hiatus." Journal of Vertebrate Paleontology, 28, 1218–1223.

Wilson, G. P. (2014). Mammalian extinction, survival, and recovery dynamics across the Cretaceous-Paleogene boundary in northeastern Montana, USA. In: Wilson, G. P., Clemens, W. A., Horner, J. R., & Hartman, J. H. (Eds.) *Through the end of the Cretaceous in the type locality of the Hell Creek Formation in Montana and adjacent areas*. Geological Society of America Special Paper, 503, 365–392.

Wilson, G. P., DeMar, D. G., Jr., & Carter, G. (2014). Extinction and survival of salamander and salamander-like amphibians across the Cretaceous-Paleogene boundary in northeastern Montana, USA. In: Wilson, G. P., Clemens, W. A., Horner, J. R., & Hartman, J. H. (Eds.) *Through the end of the Cretaceous in the type locality of the Hell Creek Formation in Montana and adjacent areas*. Geological Society of America Special Paper, 503, 271–297.

Wilson, G. P., Ekdale, E. G., Hoganson, J. W., Calede, J. J., & Vander Linden, A. (2016). A large carnivorous mammal from the Late Cretaceous and the North American origin of marsupials. Nature Communications, 7, 13734.

Wilson, G. P., Evans, A. R., Corfe, I. J., Smits, P. D., Fortelius, M., & Jernvall, J. (2012). Adaptive radiation of multituberculate mammals before the extinction of dinosaurs. Nature, 483, 457–460.

Wilson, J. A. (1990). The Society of Vertebrate Paleontology 1940–1990, a fifty-year retrospective. Journal of Vertebrate Paleontology, 10, 1–39.

Witmer, L. M. (1997). The evolution of the antorbital cavity of archosaurs: A study in soft-tissue reconstruction in the fossil record with an analysis of the function of pneumaticity. Journal of Vertebrate Paleontology, 17, 1–76.

Witmer, L. M. (2001). Nostril position in dinosaurs and other vertebrates and its significance for nasal function. Science, 293, 850–853.

Witmer, L. M., & Ridgely, R. C. (2008). The paranasal air sinuses of predatory and armored dinosaurs (Archosauria: Theropoda and

Ankylosauria) and their contribution to cephalic structure. The Anatomical Record, 291, 1362–1388.

Witmer, L. M., & Ridgely, R. C. (2009). New insights into the brain, braincase, and ear region of tyrannosaurs (Dinosauria, Theropoda), with implications for sensory organization and behavior. The Anatomical Record, 292, 1266–1296.

Witmer, L. M., & Ridgely, R. C. (2010). The Cleveland tyrannosaur skull (*Nanotyrannus* or *Tyrannosaurus*): New findings based on CT scanning, with special reference to the braincase. Kirtlandia, 57, 61–81.

Witmer, L. M., Chatterjee, S., Franzosa, J., & Rowe, T. (2003). Neuroanatomy of flying reptiles and implications for flight, posture and behaviour. Nature, 425, 950–953.

Witton, M., & Michel, E. (2022). *The art and science of the Crystal Palace Dinosaurs*. The Crowood Press.

Witton, M. P. (2013). *Pterosaurs: Natural history, evolution, anatomy*. Princeton University Press.

Witton, M. P. (2018). *The palaeoartist's handbook*. Crowood Press.

Witton, M. P., & Habib, M. B. (2010). On the size and flight diversity of giant pterosaurs, the use of birds as pterosaur analogues and comments on pterosaur flightlessness. PLOS One, 5, e13982.

Witton, M. P., & Hone, D. W. E. (2018). Tyrannosaurid theropods: Did they ever smile like crocodiles? The 66th Symposium on Vertebrate Palaeontology and Comparative Anatomy, Universities of Manchester, September 5–8, 2018, Programme and abstracts, 67.

Witton, M. P., & Naish, D. (2008). A reappraisal of azhdarchid pterosaur functional morphology and paleoecology. PLOS One, 3, e2271.

Witton, M. P., & Naish, D. (2013). Azhdarchid pterosaurs: Water-trawling pelican mimics or "terrestrial stalkers"? Acta Palaeontologica Polonica, 60, 651–660.

Wolberg, D. L., Lozinsky, R. P., & Hunt, A. P. (1986). Late Cretaceous (Maastrichtian-Lancian) vertebrate paleontology of the McRae Formation, Elephant Butte area, Sierra County, New Mexico. New Mexico Geological Society Guidebook, 37, 227–334.

Wolff, E. D., Salisbury, S. W., Horner, J. R., & Varricchio, D. J. (2009). Common avian infection plagued the tyrant dinosaurs. PLOS One, 4, e7288.

Woodruff, D. C., Goodwin, M. B., Lyson, T. R., & Evans, D. C. (2021). Ontogeny and variation of the pachycephalosaurine dinosaur *Sphaerotholus buchholtzae*, and its systematics within the genus. Zoological Journal of the Linnean Society, 193, 563–601.

Woodward, A. S. (1910), Presidential Address to Section C, Report of the British Association for the Advancement of Science, 1909, 462–471.

Woodward, H. N., Tremaine, K., Williams, S. A., Zanno, L. E., Horner, J. R., & Myhrvold, N. (2020). Growing up *Tyrannosaurus rex*: Osteohistology refutes the pygmy "*Nanotyrannus*" and supports ontogenetic niche partitioning in juvenile *Tyrannosaurus*. Science Advances, 6, eaax6250.

Wosik, M., & Evans, D. C. (2022). Osteohistological and taphonomic life-history assessment of *Edmontosaurus annectens* (Ornithischia: Hadrosauridae) from the Late Cretaceous (Maastrichtian) Ruth Mason dinosaur quarry, South Dakota, United States, with implication for ontogenetic segregation between juvenile and adult hadrosaurids. Journal of Anatomy, 241, 272–296.

Wroe, S., Huber, D. R., Lowry, M., McHenry, C., Moreno, K., Clausen, P., ... & Summers, A. P. (2008). Three-dimensional computer analysis of white shark jaw mechanics: How hard can a great white bite? Journal of Zoology, 276, 336–342.

Wu, X. C., Shi, J. R., Dong, L. Y., Carr, T. D., Yi, J., & Xu, S. C. (2020). A new tyrannosauroid from the Upper Cretaceous of Shanxi, China. Cretaceous Research, 108, 104357.

Xu, X., Clark, J. M., Forster, C. A., Norell, M. A., Erickson, G. M., Eberth, D. A., ... & Zhao, Q. (2006). A basal tyrannosauroid dinosaur from the Late Jurassic of China. Nature, 439, 715–718.

Xu, X., Norell, M. A., Kuang, X., Wang, X., Zhao, Q., & Jia, C. (2004). Basal tyrannosauroids from China and evidence for protofeathers in tyrannosauroids. Nature, 431, 680–684.

Xu, X., Wang, K., Zhang, K., Ma, Q., Xing, L., Sullivan, C., ... & Wang, S. (2012). A gigantic feathered dinosaur from the Lower Cretaceous of China. Nature, 484, 92–95.

Ye, C. H. (1975). Jurassic system. In: Su Z. (Ed.). *Mesozoic redbeds of Yunnan,* Science Press. pp. 11–31.

Yoshida, J., Kobayashi, Y., & Norell, M. A. (2023). An ankylosaur larynx provides insights for bird-like vocalization in non-avian dinosaurs. Communications Biology, 6, 152.

Young, B. A. (1991). Morphological basis of "growling" in the king cobra, *Ophiophagus hannah*. Journal of Experimental Zoology, 260, 275–287.

Young, J. (2011). *Dino gangs: Dr Philip J. Currie's new science of dinosaurs*. Collins.

Zanno, L. E., & Makovicky, P. J. (2013). Neovenatorid theropods are apex predators in the Late Cretaceous of North America. Nature Communications, 4, 2827.

Zanno, L. E., Tucker, R. T., Canoville, A., Avrahami, H. M., Gates, T. A., & Makovicky, P. J. (2019). Diminutive fleet-footed tyrannosauroid narrows the 70-million-year gap in the North American fossil record. Communications Biology, 2, 64.

Zhai, R. J., Zheng, J. J., & Tong, Y. S. (1978). Stratigraphy of the mammal-bearing Tertiary of the Turfan Basin, Sinkiang. Memoirs of the Institute of Vertebrate Paleontology and Paleoantropology, 13, 68–81.

Zhang, L., Yin, Y., & Wang, C. (2021). High-altitude and cold habitat for the Early Cretaceous feathered dinosaurs at Sihetun, Western Liaoning, China. Geophysical Research Letters, 48, e2021GL094370.

Zheng, W., Jin, X., Xie, J., & Du, T. (2024). The first deep-snouted tyrannosaur from Upper Cretaceous Ganzhou City of southeastern China. Scientific Reports, 14, 16276.

IMAGE CREDITS

1.4A. Henry Fairfield Osborn in public domain.

1.4B. Barnum Brown in public domain.

1.8A–B. Charles R. Knight *Tyrannosaurus* artworks courtesy American Museum of Natural History Library.

1.11. AMNH fighting *T. rex* skeletons by American Museum of Natural History, public domain.

1.14. AMNH advertisement from *New York Tribune*, public domain.

1.15. *The Lost World* poster in public domain.

1.16. Carnegie *Tyrannosaurus* by Georgia Witton-Maclean.

1.18. AMNH 5027 by Georgia Witton-Maclean.

1.19. *Jurassic Park Tyrannosaurus* model by Amaury Laporte, CC BY 2.0.

2.8A. *Proceratosaurus* skull by The Trustees of the Natural History Museum, London, CC BY 4.0.

2.8B. *Eotyrannus* skull by Naish and Cau, CC BY 4.0.

2.8C. *Qianzhousaurus* skull by Junchang Lü, CC BY 4.0.

2.8D. *Gorgosaurus* skull by Georgia Witton-Maclean.

2.11. *Tarbosaurus* skull by Pavel Bochkov from Moscow, Russia, CC BY-SA 2.0.

2.14. *Nanotyrannus* skull by James St. John, CC BY 2.0 DEED.

2.15. Jane the *Tyrannosaurus* by Zissoudisctrucker, CC BY 4.0.

2.16C. *Tyrannosaurus* histology photographs by Holly N. Woodward, Katie Tremaine, Scott A. Williams, Lindsay E. Zanno, John R. Horner, & Nathan Myhrvold, CC BY-NC 4.0.

2.19. *Tyrannosaurus mcraeensis* images by Sebastian G. Dalman, Mark A. Loewen, R. Alexander Pyron, Steven E. Jasinski, D. Edward Malinzak, Spencer G. Lucas, Anthony R. Fiorillo, Philip J. Currie, & Nicholas R. Longrich, CC BY 4.0.

3.16A. *Tyrannosauripus* by Rufous-crowned Sparrow, CC BY 4.0.

3.17C. *Manospondylus* vertebra by Evolutionnumber9, CC BY 4.0.

3.18B. Barnum the *Tyrannosaurus* coprolite by Poozeum, CC BY-SA 4.0.

3.19. *Tyrannosaurus* blood vessels by Elizabeth M. Boatman, Mark B. Goodwin, Hoi-Ying N. Holman, Sirine Fakra, Wenxia Zheng, Ronald Gronsky, & Mary H. Schweitzer, CC BY 4.0.

3.29A. Komodo dragon by Yuliseperi2020, CC BY-SA 4.0.

3.29B. Lion by Bernard DUPONT from FRANCE, CC BY-SA 2.0.

3.29C. Tiger by Charles J. Sharp, CC BY-SA 4.0.

3.29D. Polar bear by Andreas Weith CC BY-SA 4.0.

3.29E. Saltwater crocodile by fvanrenterghem, CC BY-SA 2.0.

3.29F. Golden eagle by Juan Lacruz, CC BY-SA 3.0.

3.29G. Great white shark by Elias Levy, CC BY 2.0.

4.6. LACM *T. rex* growth series by Georgia Witton-Maclean.

4.9. Medullary bone of chicken and *Tyrannosaurus* by Mary Higby Schweitzer, Wenxia Zheng, Lindsay Zanno, Sarah Werning, & Toshie Sugiyama, CC BY 4.0.

4.12. Digital mass estimation by John R. Hutchinson, Karl T. Bates, Julia Molnar, Vivian Allen, Peter J. Makovicky, CC BY 4.0.

4.15. *Tyrannosaurus* jaw neurovascularity by Florian Bouabdellah, Emily Lessner, and Julien Benoit, CC BY-NC-SA 4.0.

4.17. Pigeon courting by 4028mdk09, CC BY 4.0.

4.20A–C. Floating archosaur models by Donald Henderson, CC BY 4.0.

4.22. Charles R. Knight *Leaping Laelaps* in public domain.

5.20. Lion and cape buffalo by Luca Galuzzi (Lucag), CC BY-SA 2.5.

6.2. Tyrannosaur trackways by McRea et al., CC0, public domain.

6.3. Komodo dragon feeding by Brice Lee, CC BY 2.0.

6.4. Harris's hawks by Gregory "Slobirdr" Smith, CC BY-SA 2.0.

6.8A. *Edmontosaurus* skeleton by Firsfron, CC BY 3.0.

6.8A inset. *Edmontosaurus* vertebrae detail by Kenneth Carpenter, CC BY-SA 4.0.

6.8B. *Triceratops* pelvis by Paul Gignac and Gregory M. Erickson, CC BY 4.0.

6.8C–F. Bitten dinosaur bones by Nicholas R. Longrich, John R. Horner, Gregory M. Erickson and Philip J. Currie, CC BY 4.0.

6.9. Komodo dragon juvenile and adult by Bahnfrend, CC BY-SA 4.0.

6.16. MUJA mating *Tyrannosaurus* by Mario Modesto, CC BY-SA 3.0.

7.13. AMNH Dinosaur hall by Georgia Witton-Maclean.

INDEX

Page numbers in *italics* indicate figures and tables.

abelisaurids, 53, 228
Abler, William, 236
Academy of Natural Sciences of Philadelphia, 7
Acheroraptor, 199
Acheroraptor temertyorum, *191*, 198
Acrocanthosaurus, forelimbs of, 175
Acrocanthosaurus atokensis, 21
Aerosteon riocoloradensis, 57–58
African buffalo, 212, 213, *214*
After Man (Dixon), 274
Age of Mammals, 258
Age of Reptiles, 6, 258, 259
air sacs, 47–48, 50; clavicular, 107, 159–60; in *T. rex*, 87, 105–8, 133
Alamosaurus, 210, 210–11, 215, 255, *270*; characterizing dinosaur province, *184*, 209
Alamosaurus sanjuanensis: habitats, 209, *210*; stratigraphic distribution, *194*
Alberta, Canada, 60, 184, *184*, *185*, 188, 190–91
Albertosaurinae, *54*, 60, *61*, 62; facial ornament, 118
Albertosaurus, *54*, 71, 137, 253; skin impression, *113*; survivorship rates, *254*
Albertosaurus sarcophagus, 60; bonebed, 220, 253
Alioramini, *54*, 57, 60, *61*, 62; facial ornament, 118
Alioramus, *54*
Alioramus altai, 62
Alioramus remotus, 62
alligator, *123*, 150, *152*, 153, 157, 161, 166, *167*, 242–43
alligatoroid, 196
Allosauroidea, *45*, 49, 53, 55, 59, 94; growth strategies, 137, *138*; forelimb function, 177; longevity, 255; pneumaticity, *106*
Allosaurus, 9, 22, 52, 53, 66, 156, 247, 248, 255; forelimbs of, 175; growth, 138; ribcage of, *95*; teeth of, *88*
Allosaurus fragilis, 1
alluvial plains, *T. rex* habitat, *184*, 188–91
Alvarez, Luis, 27
American Museum of Natural History (AMNH), 2, 7–10, 13–14, 18–19, 21–22, 24, 79, 278, *279*; classic *T. rex* mount in Dinosaur Hall, *16*; Department of Vertebrate Paleontology, 7, 8, 10; Hall of Fossil Reptiles, 10, 13, 18; modern *Tyrannosaurus* mount, *28*
AMNH 5027, *T. rex* specimen, 14, *16*, 17, 18, *16*, *18*, *70*, 79, 144, 184, *224*; *Jurassic Park* logo, 19, *32*
anatomy, *T. rex*: brain, 151–55; bone, internal, *75*, 110–11, 142–43, *143*; color 126–29;

cranial, 85–92; facial, 115–25; forelimb, 97–99, 175–77; gut, 108–10; hindlimb, 102–5, 163; integument, 32, 99, 105, 111–18, *119*; musculature, 87, 89–91, 93–94, *96*, 96, 99, 102, 108, 118–19, 122, 133, *134*, 140, 155–56, 164–65, 168–72, *169*, 172–73, 176, *176*, 233–34, *234*; pectoral, 97–99; pelvic, 99–102, 175–76, 242, *242*; pulmonary system, 105–8; reproductive, 244–45; ribs, 93–95; "robust" and "gracile" morphologies, 77, *77*, 78–79, 242; sensory, 155–58; size, 144–51; variation, 77–79, 241–43; vertebral column, 92–97; vocal, 158–61
Andrewsarchus mongoliensis, 5
ankylosaurids, *159*, 160, 209, *209*, 211, 213–15, *215*, 220, 251, 262
Ankylosaurus, 194, 201, 209, 211, 214, 215; abundance in Hell Creek Formation, *182*; juvenile, *215*
Ankylosaurus magniventris, 209, 209
Anning, Mary, 33
anteriorization, genetic, 173
anthropomorphism, 23
Anzu wyliei, *207*, 208
Apatosaurus, 65, 255
"Apex," *Stegosaurus* specimen, sale of, 33–34
Appalachia, 188
Appalachiosaurus montgomeriensis, *54*, 59
appearance. See life appearance of *T. rex*
Archaeopteryx, 27, 106
Archosauria, *45*, 47, 48, 53
Archosauriformes, *45*, 44–45, 53
archosaurs, 87; endothermy, 132, 133, *134*; evolution of facial anatomy, 45, 122, *123*, 157; models of floating, *167*; reproduction, 245, 248, 249; vocalization, 160–62, *161*
arctometatarsal condition, 60, *103*, 104, 163, 208. See also hindlimb
armor, 10, 13, 46, 49, 157, 209, *209*, 211; absence of, 201, *215*; armored face of *T. rex*, 115, 125; as predator deterrent, 211–14
armored dinosaurs: See ankylosaurids
armadillos, 209, 213
arms. See forelimb
Asia, origin of *T. rex*, *54*, 62–63, 65, 80
Asiatyrannus xui, 62
asteroid, Cretaceous extinction, *256*, 257, 266–74
Australovenator, 58
Australovenator wintonensis, *58*, 58
Aves. See birds
Avisaurus archibaldi, 206
Azhdarchidae, 25, 69, 199, *200*, 201, *250*

"Baby Bob," *T. rex* specimen, 33
Bakker, Robert, 27, 29, 69, 72, 162
"Barnum," *T. rex* coprolite specimen, 109, *109*
Baryonyx, 65

behavior, predatory, 213; group foraging, tyrannosaurids, 219–23; Harris's hawks, 222, *223*; influence on color, 126–29; in Lancian ecosystems, 196–201; Komodo dragons, 221–22, 232–33; prey responses to, 166, 201, 202, 204, 208, 209, *210*, 211–15; senses, 155–58; role of *T. rex* arms, 172–77, *178–79*; *T. rex*, *96*, 164–65, *166*, 168, 228–36; *T. rex* hatchlings, 248–49. *See also* toxic bite
behavior, *T. rex*: aggressive, 173, 223–28; clichéd, *37*, 38; complexity, 154–55; cooling; *134*; deducing from fossils, 217, 154–55; feeding, 237–41; gregariousness, 219–23; inferable from sensory anatomy, 118, 158; migratory, 218–19; parental, 248–50; predatory, *96*, 164–65, *166*, 168, 228–36; reproductive, 135, 243–48; vocalization, *130*, 160–62. *See also* behavior, predatory; feeding
Bell, Phillip, 111
Benson, Roger, 57
Berthoud, Edward, 10, *11*
Big Five, mass extinctions, 257
birds: brains and intelligence, 152, 153–55; bite forces, 172; color, 126, *127*; cranial kinesis, 170; evolution, *45*, 47, 49, 53, 55; eye anatomy, 87, 119; face biting, 226; feathers, 112, 114; growth, 137; hindlimb anatomy, 102, 163; in Lancian communities, 195, 196, 197, *206–7*, 249, *264*; longevity, 253, 255; as models of dinosaur appearance, 118, 121, *123*; physiology, 114, 132–33; piratical behavior, 201; predatory behavior, 221–23, *223*, 234; pulmonary system, 105, 107–8, 133, 148; reproduction, 142–43, *143*, 243, 244–45, 247–49; sexual dimorphism, 243; talon use, 175, 177; *Trichomonosis*, avian infection, 252; use as models for dinosaur mass prediction, 147–48; vision, 156; vocal anatomy, *159*, 159–60, *161*. *See also* raptor (bird)
Bistahieversor, *54*, 63
Bistahieversor sealeyi, 59
bite(s): carcharodontosaurids, 52; erythrosuchids, 44; finches, 172; as predator deterrent, 213; as predatory tactic, 234–35; tyrannosauroids, 55, 60, 63, 80, 263; greatest strength, 171, 172
bite(s), *T. rex*: changes with growth, 92, 140–41, 236, 249; during predation, 29, 172–73, 177, 234–36, *235*; fossil evidence of, 202, 223–27, 230, *231*, 232, 237–41, 245, 252; influence on skull shape, 156, 263; jaw muscles, 169; as play behavior, 154; strength, 38, 80, 92, 162, 168–72, 229, 236; toxicity, 236. *See also* face biting

INDEX

Black, Riley, 39
"Black Beauty," *T. rex* specimen, 184
Black Hills Institute, South Dakota, 29
Bluth, Don, 30
body mass: *Alamosaurus*, 211; brain size, 153–54; carcharodontosaurids, 52; ecology, 232–33, *233*; egg and clutch size, 247–48; estimation methods, extinct animals, 145, 147, 148, *149*; *Spinosaurus*, 50; thermoregulation, 112–14
body mass, *T. rex*: estimates, 144–50; hypothetical maximum, 150; impact on locomotion, 163–65; increase during growth, 137, *138*, 140–41, 144
bonebeds, 202, 218, 220, 253
"Bone Wars," 7. *See also* Cope, Edward Drinker; Marsh, Othniel Charles
Bradbury, Ray, 25, 27
brains: dinosaurs, 132, 153–55, 201; intelligence and, 20, 30, 153–55; role in extinction, 258–59; size and structure, 85, 89–90, 105, 154–55; *T. rex*, 73, 151, *152*, 155–56; tyrannosaurids, 73, 151
British Columbia, tyrannosaurid tracks, *221*
British Museum (Natural History), 28
Brochu, Christopher A., 29, 84, 91
Brontosaurus, 8, 9, 22, 23, 65, 66; mentions in literature, *20*
Brontosaurus excelsus, 1, 8
Bronx Zoological Garden, 9
Brown, Barnum, 7, 9, 10, 13, 17, 18, 22, 25, 33, 97, *109*, 172, 278
Burian, Zdeněk, 24
Burpee Museum, Illinois, 73

camouflage, 126–29, *127*
Campanian, *54*, *185*, 247, 259; dinosaur diversity, 260–62, *262*
Campione, Nicolás, 148
Carbone, Chris, 229
Carcharodontosauridae, 39, *41*, *45*, 50–52, *51*, 55, 60, 97, 278; facial sensitivity, 157–58; growth, 137, *138*; longevity, 255
Carcharodontosaurus, 21, 22, 51, 52
Carcharodontosaurus saharicus, 21, *41*, 50
Carnegie Museum, Pittsburgh, 8, 10, *11*, 24, 65; dinosaur hall, *25*; holotype *T. rex* skeleton, *42*
Carnian, 49
Carnosauria, 53
Carpenter, Kenneth, 30, 69, 73
Carr, Thomas, 34, 66, 73, 78, 79, 143
Carroll, Lewis, 211
cathedral, 3
Cenozoic Era, 186–87, 258–59, 266, 274, *276*, *277*
Ceratopsia, horned dinosaurs, 204–5, *204*, *205*, 211, *212*, *215*, 213, 254, 262, *262*, *264*; bone beds, 218; gregariousness, 218–19; as prey and food for *T. rex*, 17, 230, *231*, 232–35, *234*, *235*, 237, 238–41, *240–41*, 277; running speeds, 233–34, *234*

Ceratosauria, *45*, 53, *106*, *138*
Ceratosaurus, 53
Cerberophis, snake, 196
Champsosaurus, 196
Chatterjee, Sankar, 47
Chicxulub crater, Yucatan Peninsula. *See* K/Pg extinction
Chin, Karen, 109
China, 56, 59, 60
claws, *T. rex*: feet, *103*, 104; hands, *98*, 99; use of manual claws in taxonomy, 75, 76
cliché: dinosaurs, *41*; Mesozoic mammals, 196; *T. rex* depictions, 5, 37, 38–39, 193, 279
climate. *See* paleoenvironments of *T. rex*
cloaca, anatomy, 109, 244–45, 246
CM9380, *T. rex* holotype specimen, *12*, *14*, 24, *25*, *42*, 65, 66, 70; sale to Carnegie Museum, 24
Coelophysis, 156
Coelophysoidea, *45*, 53
Coelurosauria, *45*, 53, *54*, 55, 56, 59; brains of, 153
Coelurus fragilis, 55
cold-blooded. *See* physiology
color: in predatory animals, 126, *127*; predictions for *T. rex*, 126, 128–29; reconstruction for extinct species, 126
Colorado, 17, *185*, 267
Complete T. rex, The (Horner and Lessem), 29
Cooper, Merian C., 22
Cope, Edward Drinker, 7–8, 10, 17, 66, 67, 172, 174
coprolites, *T. rex*, 108–10
cornual process, 86, 87–88, *89*, 118. *See also* facial appearance
cranial kinesis, 170
cranium, *T. rex*: awesomeness, 2, 83; changes with growth, 135, *136*, *138*, 140–41, 144–45; differences to *Tarbosaurus*, 63, 65; facial appearance, 113, 115–25; pathologies, 223–26, *224*, 245, 252; jaw function, 168–72; musculature, 89, 90, 133, *134*; pneumaticity, *106*, 107–8; sensory anatomy, 155–58; skull and mandible, 5, 10, *11*, 13, *14*, *17*, 18, 35, *57*, *84*, 85–92, 94, 151, *152*; use in classification, 66, 71–80; use in predation, 177, 234–36. *See also* brains; dentition; face biting; headbutting
Cretaceous Period, *45*: climate, 132; environments and ecosystems, 126, *185*, 188–211, 261–65; extinction, 256–74; fossil discoveries, 9, 10, 21, 245; fossil record, 52, 259–65; *Kulindadromeus*, 112; imagined post-Cretaceous dinosaurs, 274–77; large theropods, 49–52; temporal distribution of *T. rex*, 186–87; tyrannosauroid evolution within, *54*, 55, 57–65, 67, 68
Crichton, Michael, 30
Crocodylia: bite strength, 170–71; brains, *152*, 154; digestive process, 109; evolution,

45, *45*, 46–48, 133, 157, 196; face biting, 226–27; growth, 137, 140; hunting strategies, 221–22, 234; jaw sensitivity, 157–58; as Lancian fossils, 196, 220, *264*, 265; as "living fossils," 274; as measures of ancient climates, 190; as models of dinosaur appearance, 112, 119, 121, 122, *123*, 125, 126, *127*; neck muscles, 94; physiology, 135; reproduction, 143, 244–45, 247–49, *247–48*; sexual dimorphism, 242–43; temperature regulation, 133; vocal anatomy, 159; vocal evolution, 159, 160–62
Crocodylus porosus, 127, 170. *See also* saltwater crocodile
Currie, Phillip J., 30, 140, 220

Dakotaraptor steini, 199
Dalman, Sebastian G., 80
Daspletosaurus, 54, 62, 63, 68, 137, 220; bite strength, *169*; facial skin, 115, 118; growth, *138*; survivorship rates, 254; tooth wear, *123*
Daspletosaurus horneri, 54, 62
Daspletosaurus torosus, 54, 57, 62
Daspletosaurus wilsoni, 62
"*Deinodon*," 18
Deinonychus, 27, 30, 53, 198
Deinosuchus riograndensis, 196
dentition, *Allosaurus*, 88
dentition, *T. rex*, 88, 91–92, 236; replacement rate, 92; role in feeding, 238–39; strength, 168, 229, 234; "*Trix*," 219
Denver Formation, Colorado, *184*, *185*; age, *185*, 187; environment, 191; site of oldest *T. rex* discovery, *11*, 17
Denversaurus schlessmani, 209
de Ricqlès, Armand, 27
Didelphodon vorax, 196, *197*
diet, *T. rex*: appetite, 133, 135; confirmed prey species, 109, 202, 205, 230, *231*, *234*; feeding, 237–41; hunting, 230, 232–36; isotopic insights, 219
Dilong paradoxus, *54*, 55
Dino Hunters (TV series), 35
dinosaurs: as commercialized fossils, 33–36, 40–41; cultural appeal, 2, 4–6; defining anatomical characteristics, 48; discoveries in American West, 8–10; evolution, *45*, 48–49, extinction, 257–74; longevity, 254–55; popularity, 8, 10, 13, 19–27, *20*, 29–41; variation in integument, 111–12; visualization of, 36–38, *37*; vocal anatomy, 160
Dinosaur 13 (film), 35
Dinosaur Heresies, The (Bakker), 29
Dinosaur Renaissance, *20*, 27–29, 30, 32, 228, 237
"*Dinotyrannus megagracilis*," 70, 71, 72, 80, *138*, *139*, 141, 220
Diplodocus, 8, 18, 19, *19*; auction value, 33; mentions in literature, *20*
Diplodocus carnegii, 1

disease. *See* pathologies
Disney (film company), 22–24
Disney, Walt, 24
Dixon, Dougal, 274–76, *276*
Dollo's Law, 276
Doyle, Arthur Conan, 22, 274
Dracorex, 65
Dreadnoughtus schrani, 101
Dromaeosauridae, *45*, 53, 71, *106*; as *T. rex*
    contemporaries, *191*, 198–99
Dryptosaurus, 54, 172
Dryptosaurus aquilunguis, *58*, 59
"*Dynamosaurus*," 13, 17, 28, 185
"*Dynamosaurus imperiosus*," *10*, *11*, 67
Dynamoterror, 63
Dynamoterror dynastes, 62

ears, 114, pneumaticity around, 107; in *T. rex*,
    121, 122. *See also* life appearance
ecology: predatory niches, 44, 47–49, 55, 60,
    62; of *T. rex*, 129, 133, 198, 260–62
ectothermy, 48, 131–33, 162. *See also*
    physiology
Edmontonian faunal stage, *185*, *194*, 195, 205
Edmontosaurus, 66, 156, 197, 201, 215, 232,
    263; abundance in Hell Creek Formation,
    182; musculature, 233, *234*; running
    speed, 233–34, *234*; as *T. rex* prey, 215,
    230, 232, 263
Edmontosaurus annectens, 187, 194–95, 202,
    203, 211; as *T. rex* prey, *82*, *178–79*, *231*;
    stratigraphic distribution, 187, *194*
Edmontosaurus regalis, 202
elephant, 4, 83; body mass and size, 148,
    150, 183, 247; color, 114, 128; eyeball size,
    120; longevity, 254; as prey species,
    232; running, 162–63; thermoregulation,
    114
emperor penguin, models of floating, *167*
enantiornithine, *206*. *See also* birds
encephalization quotient (EQ), 153, 155. *See
    also* brains
endocast, *T. rex*, 151–53, 155–56. *See also*
    brains
Eoneophron infernalis, 208
Eotyrannus lengi, *54*, 55, *56*, *57*
epidermal correlates, 115, *116*, *117*, 118
Erickson, Gregory M., 168, 253
erythrosuchids, 44–46
Erythrosuchus africanus, *45*, *46*
Evans, David C., 148
evolution: classification and methods,
    43–44; influenced by *T. rex*, 211–15;
    "Red Queen arms race," 211; reptile,
    43–48, *45*; success, 39; theropod, 48–52;
    tyrannosauroid, 53–59; tyrannosaurid,
    59–65
"evolutionary teratology," 173
extinction, 43, 44, 60, 211; of dinosaurs, 6,
    27, 256–74. *See also* K/Pg extinction;
    speculative evolution
extra-oral tissues: 32, 39, 115, 122–25. *See also*
    facial appearance

eye, anatomy: as restored in *T. rex*, 5, *31*,
    32, 87–88, *89*, 115, 118–20, *120*; in
    tyrannosauroid evolution, 59, 62; sizes,
    120; sockets (orbits), *86*, 87; vision in *T.
    rex*, 155–56, 229

face biting: 223, *224*, 225–27, 252; among
    living animals, 226; theropods, 226
facial appearance: evolution in
    tyrannosauroids, 55, 59, 62; relation to
    aggressive behavior, 225–26, *227*, 228; *T.
    mcraeensis*, 80, *81*; *T. rex*, 4, 32, 50, *86*,
    87–88, *89*, 115–25, 128, 144, 242–43. *See
    also* life appearance
Fantasia (movie), *20*, 22, 23–24
faunal stages, Cretaceous dinosaurs, 187, 188,
    195
feathers: digestibility, 109; evolution, 48, 49,
    111; insulative properties, 112, 114, 132;
    See also integument
feeding, *Tyrannosaurus*: as cause of arm
    reduction, 173; biting power, 168, *169*,
    170; dental adaptations, 92, 237–41
Ferris Formation, *184*, *185*, 191
fibers, skin, 48, 112, 114–15, *134*. *See also*
    feathers; protofeathers; integument
Field Museum of Natural History, Chicago, 8,
    33; Charles Knight *T. rex* mural, *15*, 22;
    as Field Columbia Museum, Chicago, 8;
    ownership of "Sue" specimen, 33, *34*. *See
    also* "Sue," *T. rex* specimen
floodplains, *T. rex* habitat, 188, 190–92, 194,
    204
foot. *See* hindlimb
footprints, *T. rex*, *104*, 105. *See also* hindlimb;
    trackways
forelimb: ornithischians, 201, 202;
    pneumatization, 107, 159; theropods, 49,
    55, 59, 177
forelimb, *T. rex*: as characterizing anatomy, 66,
    74, *75*, 76; discovery, 10, 13, 97; function,
    172–77, *176*, *178–79*, 245; musculature,
    99, 176, *176*; osteology, 77, 84, 97, *98*,
    99; pathologies, 177; range of motion of
    joints, 175, *176*; redundancy, 172–73, *174*,
    275, *276*, 276; sexual dimorphism of, 242;
    size change through growth, 74, *75*, 76,
    141–42, 175, *176*, 248, 249
forests, Cretaceous. *See* paleobotany
fossilization: biases against, 112, 150, 183, 195,
    198; as distorting process, 83, *84*, 105,
    109–10, *110*, 135, 220, 230, 238; of *T. rex*
    specimens, *84*, *182*, 198
fossil record: carcharodontosaurids, 52;
    identifying extinctions within, 257,
    259; Lancian dinosaurs, *194*, 195, 232;
    limitations of, 215; pterosaurs, 199; *T. rex*,
    34–35, 40–41, 65, 80, 141, 150, 181–83,
    187, *194*, 253; tyrannosauroids, 55, 59, 60
Frenchman Formation, Saskatchewan, 184,
    184; age, 185; environment, 190–92
frontoparietal fossae, 133, *134*
Fukuiraptor kitadaniensis, 58

Gallus domesticus, *143*
Galton, Peter, 27
"*Ganeosaurus tardus*," speculative
    tyrannosaur, 275, *276*. *See also*
    speculative evolution
gastralia, 10, *14*, *85*, 94–95
Gauthier, Jacques, 27
geographic distribution, *T. rex*: 183–86,
    184; capacity for migration, 218–19;
    contemporary plants and animals,
    194–210; paleoenvironments, 188–94
geological time: dating rocks, 186–87; dinosaur
    communities of, 194–95; distribution of
    T. rex in, *185*, 186–87, *194*; relationship to
    rock units, 183
geology, of ancient habitats, 218; of North
    American Late Cretaceous, 181–87; *T. rex*
    as an ambassador of, 36
Ghost of Slumber Mountain, The (O'Brien),
    20, 22
Giganotosaurus, *45*, 52, 128; pelvis of, 101,
    101
Giganotosaurus carolinii, 50, *51*
Gilmore, Charles W., 72
Glut, Donald, 71
Godzilla (movie), 25
golden eagle, *127*
Gorgosaurus, 54, 71, 72, 119, 187; bite force,
    169; geological range, 187; growth, *138*;
    survivorship rates, *254*
Gorgosaurus lancensis, 72. *See also*
    "*Nanotyrannus lancensis*"
Gorgosaurus libratus, *57*, 60, *61*
great white shark, *127*
gregariousness, tyrannosaurids, 219–23
growth. *See* ontogeny
Guanlong wucaii, *54*, 55
Guinness World Records, 109
gut tissues, inferred for *T. rex*, 108–10

habitats. *See* paleoenvironments of *T. rex*
hadrosaurs, *203*, 254, 262, *262*, *264*; fossil
    abundance, 197, 202, 218; footprints,
    104, 105; gregariousness, 218–19; running
    speed, 233–34, *234*; as *T. rex* prey,
    178–79, 211, 214, 230, *231*, 232–35, 238,
    247, 277. *See also Edmontosaurus*
Hall Lake Formation, New Mexico, *184*, 186;
    age, *185*, 187; environment, 192–93, *193*.
    See also Tyrannosaurus mcraeensis
Halstead, Beverly, 168
hand. *See* forelimb
Harris's hawk, 222, *223*
hatchling, *Ankylosaurus*, 215
hatchling, *Tyrannosaurus*, *137*, 140, *142*, 165,
    216, 245, *246*; ecology, 195–96, 198–201,
    248–51; survivorship, 253, *254*
Hatzegopteryx, 199. *See also* Azhdarchidae;
    pterosaurs, flying reptiles
Hay, Oliver P., 18–19; *Diplodocus*, *19*
head. *See* cranium
headbutting, 226–27, *228*
Hell Creek, Montana, 9

INDEX

Hell Creek Formation, Montana, North and South Dakota, 72, *184*; age, *185*, 187; botany, *191*, 192, 220; dinosaur abundance, 181–83; environment, 190–92; fauna, 196–99, 220, 262–63; preservation biases, 198; source of *T. rex* specimens, 184–85; stratigraphy, *78*, 181, *182*; 3 m gap, 263, *264*, 265

Henderson, Donald, 166

hindlimb: characterizing dinosaurs, 48; of theropods, 49, 58–59, 199, 208; of dinosaur herbivores, 201, 202, 233–34, *234*; pseudosuchians, 46–47

hindlimb, *T. rex*: as characterizing anatomy, 66; changes through growth, 135, *136*, 140–42, 144; discovery, 9, 10, 13, 17; functional morphology, 19, 96, *96*, 99, 102, 104–5, 145, 162–66, 177, 218, 232–34, 238, 244; medullary bone within, 243; musculature, *90*, *96*, 102; osteology, *14*, 99–105; pathologies, *250*, 252; position during copulation, 245; raptorial feet in *T. rex*, 13; role in predation, 172–73; *See also* arctometatarsal condition

holotype specimen: concept of, 65–67; of *Tyrannosaurus* species, *12*, *14*, *25*, *42*, *66*–67, *70*, *71*, 72–73, 76–77, 80. *See also* CM9380

Holtz, Thomas, Jr. 69

Hone, David W. E., 30, 145, 150

horned dinosaurs. *See* Ceratopsia

horns, 4, 59, 62, 128, 204–5, 228, 258; in *T. rex*; *37*, 88, 115, 118; use against predators, 211–15, *215*, 230

Hornaday, W. T., 9

Horner, Jack, 29, 181, 228. *See also* predator vs. scavenger controversy

Houston, David C., 229

Hoyt, Harry O., 22, 158

Hydrurga leptonyx, 1

hype, *T. rex*, 5–6, 19, 32–33. 36–41, 219, 229, 236, 258, 278

Iguanodon, 25

injuries. *See* pathologies

integument, *T. rex*, 32, 99, 105, 111–18, *119*; claw sheaths, 99, 105; facial skin, 115–18, *119*; feathers, 111, 112, 114; fossil evidence, 111–13; naked skin, 114–15; scales, 111–13; thermoregulatory concerns, 112–15

intelligence: as agent of dinosaur extinction, 258–59; dinosaur, 20, 30, 153–55, 215. *See also* brains

internal anatomy, *T. rex*, 105–11

International Commission on Zoological Nomenclature (ICZN), 65, 67

isotope(s): migration, 218–19; physiology, 132–33

"Jane," *T. rex* specimen, 73, *74*, *136*, 184, 241; age, 73, 141–42; bite strength of, 171; body mass, *147*; pathologies, *224*, 226, 253; taphonomy, *182*

Javelina Formation, Texas, 184, 186; age, *185*, 187; fauna, 196, 210

Jinbeisaurus wangi, 59

"Jordan theropod," *T. rex* specimen, 140, *141*. *See also* "*Stygivenator molnari*"

Judithian faunal stage, 195

Jurassic Period, 8, 9, *45*, 186; theropod communities, 49, 52, 260, 262; tyrannosauroid evolution in, 53, *54*, 55, 56

Jurassic Park (novel), 30

Jurassic Park (movie), 20, 71; logo, 19, 32; *T. rex*, 30–32, 122, 159, 162–63, 248. *Also see* AMNH 5027

Jurassic Park III (movie), 49

Juratyrant langhami, *54*, 55

juveniles: *Alamosaurus*, 211; *Ankylosaurus*, 215; *Edmontosaurus*, 202, Komodo dragon, 222, 232, 233; as theropod prey, 82, 109; *Triceratops*, 215

juveniles, *T. rex*: 30, 191; anatomy, 85, 87–89, 91–92, 97, *136*, 138–42; in Cretaceous ecosystems, 195, *197*, 198–99, 208, 220, 232, 236, 248, *249*, 274; bite force, 169, 171; forelimb use, 172, *175*, 176–77, 229; growth, 135–43, 248; life appearance, 114–15, *117*, 118, 129; locomotion, 164–66; mortality, 251, 253–54; private ownership and sale of specimens, 33, 40–41; scarcity of fossils, 247; taxonomic confusion, cause of, 66, *70*, *71*, 72–77; vocalization, 162

King Kong (movie): 1933, *20*, 22, 23, 274; 2005, 276

king tyrant, terminology, 39. *See also* *Tyrannosaurus rex*; tyrant king

Kirtlandian faunal stage, 195

Knight, Charles R., 13, 19, *21*, 22, 24; *Leaping Laelaps* (1897), 172, *174–75*; paintings of *T. rex*, *15*

Komodo dragon: as analogues for theropod biology, *127*, 128, 233, 238; predatory behavior, 221–22, *222*, 232, 233, 236; skull, *123*, *124*; toxic bite of, 236. *See also* lizards

Koppelhus, Eva B., 30

K/Pg extinction: asteroid impact, *256*, 257–58, 266–74; Chicxulub impact site, 266, *267*, 268, 272, 275; historic considerations, 258–59; catastrophism vs. gradualism, 27, 259–62, *261*, *262*; dating of dinosaur extinction, 186–87, 195, 263–66; evidence of spring date, 266, 268; impact geography, *267*, 268; impact effects, 268–74; victims, 257

Kulindadromeus, 112

"*Laelaps*," 172, *174–75*. *See also* *Dryptosaurus*

Lakes, Arthur, 17

Lance Formation, Wyoming, 17, 182, *184*; age, *185*, 185, environment, 190–92

Lancian dinosaur community, *191*, 194–211; Cenozoic evolution of, 276; stratigraphic distribution of, *185*, *194*, 194–95; *T. rex* as the sole predator in, 258, 260–63

Land Before Time, The (movie), 30

Laramidia, landmass, *184*, 188; dinosaur communities, 194, 232, 238, 260

Larson, Peter, 30, 73, 77

larynyx, 159, 160–61. *See also* syrinx; vocalization

Lawson, Douglas, 69

Leaping Laelaps (1897) (Knight), 172, *174–75*

legs. *See* hindlimb

Lepidosauria, *45*; vocal anatomy, 159

Leptoceratops, 188, *191*; characterizing dinosaur province, *184*, 204

Leptoceratops gracilis, 204, *204*; stratigraphic distribution, *194*

Lessem, Dom, 29

life appearance of *T. rex*, 111–29; color, 126–29; distortion in popular culture, 32, 36–37, *37*, 84; facial tissues, 115–25; of juveniles, 140, *142*. *See also* integument

lifespan. *See* longevity

lines of arrested growth (LAGs), 135, 137, *138*

Linnaeus, Carl, 43

Linnaean classification, 43–44

lion: color, *127*; cape buffalo, 214; color, *127*; predatory behavior, 232

lips. *See* facial appearance

"living fossils," 274

lizards, *45*, 112, 170, 213; Cretaceous, 195, *196*, 220, *264*, 265; facial tissues, 122–24; as models for extinct reptiles, 19, 44, 45, 126, *127*, 140, 220–22, 238; physiology, 135, 137; reproduction, 244–45, 247, 249; toxic bites, 236; vocal anatomy, *159*, 160. *See also* Komodo dragon

Lockley, Martin, 173

locomotion, dinosaurs, 48–49, 60, 132, 220, *221*; "rauisuchians," 46–48; relation to longevity, 254

locomotion, *T. rex*, 96, *96*, 142, 162–65, *166*, 218–19, 232–34; sluggish, 28, 275. *See also* footprints, *T. rex*; trackways; swimming, *T. rex*

longevity: dinosaurs, 135, 254–55; large mammals, 254; theropods, 144, 211, 253–55, *254*

Longrich, Nicholas, 73

Lost World, The (movie): 1925, 22, 23, 24, 25, 158, 274; 1960, 25

Lost World: Jurassic Park, The (Crichton), 248

Lost World: Jurassic Park (movie) 248

Lourinhanosaurus, 247, 248

Lourinhanosaurus, 105–8, 159. *See also* air sacs; pneumaticity; pulmonary system

Lythronax, *54*, 62, 63

Lythronax argestes, *54*, 62

Maastrichtian, *54*, *185*, 186–87; dinosaur diversity decline, 259–63; Lancian biota, 194–211, 247; North American geography, *184*, 188–94; predator ecology, 197–201, 232

mammals: anatomy, 83, 89, 91, 92, 105, 107, 125, 126, *127*, 153–54, 160–61, 163; as dinosaur analogues, 148; evolution, 4, 20, 24, 44, 274; evolutionary success, 39, 187; face biting, 226; longevity, 253–55; Mesozoic, 8, 195–96, *197*, *249*, 264; physiology, 27, 29, 48, 114, 132–33, 135, 137; predator defenses, 212–14; predator ecology, 5, 128, 198, 201, 213, 156, *214*, 220–23, *233*

McCrea, Richard T., 220

McRae Group, New Mexico, 80; McRae Formation, 186. *See also Tyrannosaurus mcraeensis*

Maleev, E. A., 68

Mallon, Jordan, 150

Maniraptora, 153, 156, 172; evolution, *45*, 53; feather use in display, 226; nests, 249

"Manospondylus gigas," 17, 66–67; holotype, *11*, *106*

Mapusaurus, 137, *138*

marine reptiles, 126, 188, 257

Marsh, Othniel Charles, 7–9, *11*, 17–18, 22. *See also* Osborn, Henry Fairfield; *"Ornithomimus grandis"*

masculinity, *T. rex*; 39–40, 258

mass extinctions. *See* extinction; K/Pg extinction

Matthew, William Diller, 10, *12*, 153, 162

medullary bone, 111, 142–43, 243

Megalodon. See *Otodus megalodon*

Megalosaurus, 49, 53

"Megapubis acheirus," speculative tyrannosaur, 275

Megaraptor namunhuaiquii, 58

Megaraptora, *54*, 57–59

melanosomes, 126. *See also* color

Mesozoic Era, 8, 9, *45*, 185–87, 238; birds, 159, 196–97; dinosaur evolution in, 48–49, 52, 155–56, 211–15, 257–74; ecosystems, 194–211, 260–63; flora, 188, 190, 192–93, 196, 260; marine ecosystems, 188; non-avian dinosaurs, 107, 112, 122, 137, 143, 153, 158, 160–61, 165, 167, 218, 229, 254–55; reptile evolution in, 43–47; tyrannosauroid evolution in, 53–59. *See also* K/Pg extinction

metabolism. *See* physiology

migrating behavior, dinosaurs, 218–19

Mitchell, Mary Mason, 19, *19*

model organism, *T. rex*, 30, 40, 278

Molnar, Ralph E., 30

Mongolia, 60, 63, *63*, 67–68, 204; as source of commercial fossils, 35. *See also Tarbosaurus*

monsterization, *Tyrannosaurus*, 19–24, 29, 36–39

Montana, 184, *185*; historic source of *Tyrannosaurus* fossils, 2, *7*, 9, 13, 17

Morhardt, Ashely, 155

Moros intrepidus, 59

morphodynamic compensation, 173

mountains, *T. rex* habitat, *180*, 188–90, 192–94, 205, 210. *See also* North American Cordillera

movies, *T. rex*; *20*, 22–25, 30, *31*, 32, 71, 122, 158, 159, 162, 163, 248, 274, 276

Murusraptor barrosaensis, 58

musculature, *Tyrannosaurus*: axial, 93–94, 96, *96*, 108; cranial, 87, 89–91, 118–19, 122, 133, *134*, 140, 155–56, 168–72, *169*; forelimb, 99, 172–73, 176, *176*; hindlimb, 102, 164–65, 233–34, *234*; caudofemoralis, 96, *96*, 102, 163, 233, *234*. *See also* anatomy

Museo del Jurásico de Asturias (MUJA), *244*

museums, *T. rex*: 2, *16*, *18*, *25*, *28*, *34*, *42*, *63*, 65, 73, *139*, 145, 181, *237*, 234, *244*, *275*, 278–79; as attraction, 2–4, *15*, 18, 19–22, 24, 28, 33–36, 258; collections, 40–41, 97, 253; research, 6–10, 13–14

"Nanotyrannus lancensis," 40, *70*, 72–77, 80; controversy, 73–77; holotype skull of, *71*, 140, 185; as pygmy species, 72, 197. *See also* "Jane," *T. rex* specimen; "Petey," *T. rex* specimen

Nanuqsaurus, 54, 62, 68

Nanuqsaurus hoglundi, 62

Natural History Museum, Los Angeles, 38, *63*, *139*, *275*. *See also* "Thomas," *T. rex* specimen

Natural History Museum, London, 28

Neotheropoda 53

Neovenator, 157–58

New Dinosaurs: An Alternative Evolution, *The* (Dixon), 274–76

Newman, Barney W., 28

New Mexico, 80, 183, *185*, 186, 210

Nightmare Before Christmas, The (movie), 140

North America: decline in dinosaur diversity, 259–63; impact of K/Pg event, 266–74; Lancian fauna, 194–211, 232, 247, 277; Maastrichtian geography, *184*, 188–94; paleontology, 6–19, 24; tyrannosaurids, *54*, 55, *58*, 59, 60, 62–63, 65, 68, 72, 220. *See also* Laramidia, landmass

North America, *T. rex*: fossil record, 65, 68, 69, 79; geological distribution, 183–87

North American Cordillera, *184*, 188, 190, 192, *193*, 204. *See also* alluvial plains; mountains

North Carolina Museum of Natural Sciences, 35

North Dakota, 185, *185*

North Horn Formation, Utah, *184*, 185; age, *185*, 187; environment, 192, 194; fauna, 210

nostrils, *T. rex*, 87, 115, *117*, 121, 121–22, 156. *See also* life appearance; sensory capabilities, *T. rex*

O'Brien, Willis, 22

Observer, The (newspaper), 33

Ojo Alamo Formation, New Mexico, 198

One Million Years BC (film), 24

ontogeny: study of, dinosaur, 29, 137; relation to isotopes, 218–19; complications to tyrannosaurid taxonomy, 63, 68, 135; growth rates in theropods, *138*, 137; precociality vs. altriciality, 248–49

ontogeny of *T. rex*, 85, 87–89, 91, 92, 135–44, 171, 175, 198, 233, 243, 248–51, 253–55; adult growth stage, 144; body masses of growth stages, *147*; complications to taxonomy, 70–77; juvenile growth stage, 140–41, *172*, *246*, 249–51; late juvenile– subadult growth stage, 141–44; lines of arrested growth (LAGs), 74, *75*, 135, 137, *138*; physiology of, 132; reproductive maturity, 111, 135, *138*, 142–44, 226, 253. *See also* tyrannosaurid niche assimilation

Ornithischia: classification: *45*, 48; eggs, 247; integument, 112; medullary bone in, 143; pneumaticity, 107; remains in *T. rex* coprolite, 109; vocal anatomy, 159, 160

Ornithomimidae, 104, 199, 208, 264

Ornithomimosaurs, *45*, 49, 53, *106*, 138

Ornithomimus, 17, 208; abundance in Hell Creek Formation, *182*; as prey of *T. rex*, 165, *166*; integument, 114–15, 208

"Ornithomimus grandis," 17, 185

Ornithomimus velox, 206

Ornithoscelida, 48

Osborn, Henry Fairfield, 2; classification of *T. rex*, 65, 66; early research on *T. rex*; 6–14, 17–19, 21, 22, 25, 278

osteoderm, 209, 211, 213, 215; as scutes in *Dynamosaurus*, 10, 13; in *T. rex*: 88, *89*, *117*, 118. *See also* armor

osteology, *T. rex*, 83–85; cranial, 85–92, 133, *134*; forelimb, 97–99, 175–77; hindlimb, 99–105, 163; vertebral and torso, 92–97, 105–8. *See also* anatomy

ostrich dinosaurs. *See* Ornithomimidae; ornithomimosaurs

Otodus megalodon, 171

Ostrom, John, 27

Pachycephalosauridae, 209

Pachycephalosaurus, 65; abundance in Hell Creek Formation, *182*

Pachycephalosaurus wyomingensis, *208*, 209; stratigraphic distribution, *194*

Padian, Kevin, 173

Palaeosaniwa, 196; as predator of *T. rex*, 249

paleobotany: extinction, 260, 268, 272; isotopes, 218; of *T. rex* paleoenvironments, 188, 190, 192–93, 195, 196, 220

paleoenvironments of *T. rex*, *184*, 188–94; alluvial plains, *184*, 188–91; biota, 194–211; floodplains; 188, 190–92, 194, 204; impact of Chicxulub impact on, 266–75; mountains, *180*, 188–90, 192–94, 205, 210

paleontology: commercialized, 29, 33–36, 40–41, 279; impact of world wars on, 19, 24; of Lancian North America, 180–211; popularization, 32, 36, 38; shift away from American dominance, 24; at turn of 20th century, 6–8; typological approach to taxonomy, 71; *T. rex* subculture, 2, 36–41, 181. *See also* dinosaurs: popularity

INDEX

Paleozoic Era, 186, 266
parenting, 245, 248–51
Parker, Neave, 24
Paronychodon, 198
Parrish, J. Michael, 30
pathologies: in herbivorous dinosaurs, caused by *T. rex*, 230, 231, 237–41; in *T. rex*, 85, 143, 144, 177, 223–27, 250, 251–53, 255
Paul, Gregory S., 29, 68, 69, 79, 162, 163, 230, 275–76
Peabody Museum of Natural History, Yale, 7
"Peck's rex," *T. rex* specimen, 25, 184; pathologies, 224; taphonomy, 182
Pectinodon, 198
pectoral girdle: pneumaticity, 107, 159; *T. rex*, 97, 98, 99, 141, 175–77, 252
pelvic girdle, dinosaurs, 48, 55; bitten by *T. rex*, 168, 231, 238–39; in speculative tyrannosaur descendants, 275. *See also* "*Megapubis acheirus*," speculative tyrannosaur
pelvic girdle, *T. rex*: anatomy, 83, 96, 99, 101–2, 111, 113, 145, 150, 182, 244; as characterizing anatomy, 66, 73; change through growth, 136, 140, 144; discovery of, 10, 13, 14, 16, 18; incorrectly depicted, 32; sexual dimorphism of, 242
Permian extinction, 44, 257
Peterson, Olaf A., 10
"Petey," *T. rex* specimen, 241; as purported "*Nanotyrannus*," 75, 76
Phanerozoic Eon, 186
physiology, 27, 29, 48, 128, 131–35, 144, 171, 215, 253; and growth, 135, 144; ectothermy, 48, 131–33, 162; endothermy, 48, 131–33, 218; in dinosaurs, 131–33; relevance to migration, 218. *See also* thermoregulation
Pickering, Stephen, 71
pigeon, closed-mouth vocalization, 161
Pinacosaurus, larynyx, 160
Planet Dinosaur (documentary), 32
plant life, Cretaceous. *See* paleobotany
Platytholus clemensi, 209
play behavior, *T. rex*, 154
pliosaurid, marine reptile, 171
polar bear, 5, 127
popular culture, influence on *T. rex* depictions, 32, 36–37, 37, 84, 228
population densities, *T. rex*, 133
poop. *See* coprolite
posture, *T. rex*, 15, 19, 27–29, 102, 132, 156, 166; mating, 243–46
Postosuchus kirkpatricki, 45, 46, 48
pneumaticity, 106–8, 106; in mammals, 107; functional implications, 107–8; relevance to vocalization, 159–60
predator vs. scavenger controversy, 5, 40, 228–30, 275, 277
Predatory Dinosaurs of the World (Paul), 29, 69
Prehistoric Planet (documentary), 248
Proceratosauridae, 54, 55, 56
Proceratosaurus bradleyi, 54, 55, 57
Protoceratops, 204

protofeathers, 32, 48, 55, 56, 111, 112, 114, 132. *See also* integument
provincialism, dinosaur faunas, 195, 201, 204, 209. *See also Alamosaurus; Leptoceratops; Triceratops*
Pseudosuchia, 45, 47. *See also* crocodylians
pterosaurs, flying reptiles, 25, 69, 126; evolution, 45, 44, 45, 47, 48; fossil record, 199; as heroic champions against nasty tyrannosaurs, 199, 200, 201, 250; pneumaticity, 105, 107–8
pulmonary system: adaptive properties, 107–8, 133; evolution, 47–48, 50, 159–60; in *T. rex*; 87, 105–8, 133; vocalization, 159

Qianzhousaurus sinensis, 54, 57, 61, 62
Quetzalcoatlus, 199. *See also* Azhdarchidae
Quetzalcoatlus northropi, 25, 69, 199, 200. *See also* Azhdarchidae

racial senility, extinction hypothesis, 258, 259
raptor (bird): use of talons 175, 177, 199; vision, 156
Raptorex, 54, 63, 140, 141
"Rauisuchia," 45, 46–48
reproduction: copulation, 173, 243–45; Cretaceous, 196; eggs, nests and embryos, 245, 247–48
reptiles: anatomy, 87, 89, 94, 96, 102, 104, 115, 158, 236; brains, 153–55; as dinosaur substitutes on film, 25; evolution, 41, 43–48, 53, 105; facial anatomy, 118–25; food intake, 133; growth, 73, 135, 137, 139, 140; physiology, 132–33; predatory ecology, 221–22, 232–33, 249; reproduction, 244–45, 247; survivorship rate, 253; tail regeneration, 252; vision, 155; vocal evolution, 159–61
ribs, *Allosaurus*, 95
ribs, *T. rex*, 85; cervical, 14, 93, 93; dorsal, 94–95, 95; pathologies, 250, 252
Richardoestesia, 198
Rite of Spring, The (Stravinsky), 22–24
roaring, *T. rex*, 2, 22–23, 38, 130, 158–61. *See also* cliché: *T. rex* depictions
"robust" and "gracile" morphs, *T. rex*, 77, 77, 78–80, 242
Rozhdestvensky, A. K., 68, 73
Ruxton, Graeme D., 229

Saitta, Evan T., 73
saltwater crocodile, 127, 150, 170. *See also Crocodylus porosus*
"Samson," *T. rex* specimen, 35
Saskatchewan, 184, 184, 185, 190
Saurophaganax maximus, 21
"*Saurophagus*," 21
Sauropodomorpha, 45, 48; eggs, 247
sauropods, long-necked dinosaurs; pneumaticity, 107–8; limbs and carriage, 150, 163; in Cretaceous North America, 209–11, 277; pelvic size, 101–2; as prey for *T. rex*, 211, 214, 232; longevity, 255

scales, skin covering: 22, 37, 49, 204; tyrannosaurids, 111–15, 117, 118, 119, 228. *See also* integument
scaling: body size, 50, 147–48, 150; muscle performance, 164; skeletal strength, 163–64. *See also* encephalization quotient (EQ)
scavenging: taphonomy, 182, 183; *T. rex*, 178, 201, 228–30, 241, 274, 275, 277. *See also* predator vs. scavenger controversy
Schaeffer, Joep, 219
Schoedsack, Ernest B., 22
Schweitzer, Mary, 110
scientific names, 1–2, 67, 76. *See also* holotype specimen; taxonomy
sclerotic ring(s), 85, 87; relationship to eye size, 119–20, 120
Scollard Formation, Alberta, 184, 184, 192; age, 185; environment, 188, 190
"Scotty," *T. rex* specimen, 184; body mass, 147, 148, 150; maturity, 144
Second World War, 24
sensory capabilities, *T. rex*, 155–58
sexual dimorphism: relation to facial biting, 226–27; efforts to find in *T. rex*, 77, 78, 241–43; recognition among skeletons of living animals, 243; claims to increase value of *T. rex* specimens, 40
sharks, 126, 171, 196
Short History of the World, A (Wells), 20
Sinotyrannus kazuoensis, 55
Signor-Lipps effect, 265
size, *T. rex*: adult, 5, 144–45, 146; maximum predicted, 150–51. *See also* body mass
skin. *See* integument
skull: *Andrewsarchus*, 5; erythrosuchids, 45; pachycephalosaurs, 208; "rauisuchians," 47; theropods, 21, 52; *Triceratops*, 215; tyrannosauroids, 55, 57, 60, 62, 63, 80, 81, 141, 275
skull, *T. rex*, 5, 57, 84, 116, 120, 141, 145, 150, 220; anatomy, 85–92, 94, 107–8, 113, 117, 121, 123, 124, 151, 156, 157, 169; change with growth, 138, 140–41; discovery, 10, 11, 13, 14, 17, 18, 22; iconic, 2; market value, 35; pathologies, 224, 226; strength, 170–71, 228; use in classification, 66, 70, 71–74, 76, 77, 78, 80; variation, 77, 243. *See also* brains; dentition; facial appearance
Smith, Joshua B., 91
Smithsonian National Museum of Natural History, Washington DC, 237
snakes, 45, 125, 126, 170, 257; Cretaceous, 196; reproduction, 244–45
social interactions, *T. rex*, 173, 219–28
Society of Vertebrate Paleontology, 35
soft tissues, 84, 105, 108, 126, 252; reconstruction of, *T. rex*, 90, 96, 111–25, 148, 149, 150; non-mineralized in *T. rex*, 110–11; preservation of, 183. *See also* facial reconstruction; fossilization; musculature; taphonomy

INDEX

Sound of Thunder, A (Bradbury), 25, *26*, 27
South Dakota, 29, 185, *185*
speculative evolution, 274–77, *276*, *277*
speed, *T. rex. See* locomotion, *T. rex*
Sphaerotholus buchholtzae, 209
Spinosauridae, *45*, 49–50, *51*, 53, 55, 177, 278; growth, *138*; pneumaticity, *106*
Spinosaurus, 22, 32, *45*, 49–50, *52*, 53, 128, 158; growth rate, *138*; mentions in literature, *20*; swimming, 166–67
Spinosaurus aegyptiacus, 21, 49, *51*
"Stan," *T. rex* specimen, *100*, 185, 241; body mass, *147*; pathologies, *224*, 245; maturity, 89, 144; sale, 33, 35
Stegosaurus, 9, 23, 25, 33, 34; mentions in literature, *20*. *See also* "Apex," *Stegosaurus* specimen
Stenonychosaurus, 30
Sterling, Lindsey Morris, 13
Sternberg, Charles, 33
Stevens, Kent, 155
Stokesosaurus, *54*
Stokesosaurus clevelandi, 55
Stokowski, Leopold, 23–24
Stravinsky, Igor, 22, 23
Stromer, Ernst, 21
Struthiomimus, 208
Stygimoloch, 65
Stygimoloch spinifer, *194*
"*Stygivenator molnari*," *70*, 71, 72, 80; as juvenile *T. rex*, 140, *141*
Suchomimus, 65
"Sue," *T. rex* specimen, *34*, 83, *84*, *85*, 90, *94*, *146*, 151, *157*; body mass, 145, *147*, 148, *149*, 150; description of, 84, 91; discovery, 182, 185, 219–20; forelimbs, *98*, 176; maturity, *137*, 138, 144; pathologies, *224*, *250*, 251–53; sale, *20*, 33–35; use in classification, *70*, 71, *75*, 76. *See also* paleontology
survivorship, 253–55
Suskityrannus hazelae, *54*, 59
swimming, *T. rex*, 166–68
Swinton, W. E., 20
syrinx, 159–60. *See also* larynx; vocalization

tail: clubs, ankylosaurs, 209, 213, 215; *Edmontosaurus*, 230, *231*, *234*; reptile, complication to reproduction, 244–45; *Spinosaurus*, 49; *Stegosaurus*, 24; tyrannosaurid, 28, 80
tail, *T. rex*, 22, 28, *85*, 96–97, *96*, 102, 111, *113*, 132, 164, *234*; alleged sexual dimorphism, 242–43; as counterbalance to body, 108; discovery, 13; length, 18, 84, 92, 96–97, 145, 182; pathologies, 251–53. *See also* musculature, *Tyrannosaurus*
Tanycolagreus topwilsoni, 55, 56
taphonomy, *182*, 183, 218–20. *See also* bonebeds; fossilization
Tarbosaurus, congeneric with *Tyrannosaurus*, 68–69; juvenile skulls, 140, *141*. See also *Tyrannosaurus bataar*

Tarbosaurus bataar, 35, 60, 62–65, 67 taxonomy, principles, 65–67, 68; *T. rex*, 65–81
teeth. *See* dentition
Teratophoneus, *54*, 62, 63, 68
Teratophoneus curriei, 62
territoriality, *T. rex*, 218–19
Texas, 69, 183, *185*, 186, 199, 210, 267
Thanatotheristes degrootorum, 62
thermoregulation, 112–14, 128, 131–33, *134*; thermal neutrality, 112–14
Theropoda, 2, 135; aggressive social behavior, 226, 228; anatomy, 19, 88, 91–92, *93*, 94, *95*, 99, *101*, 102, 104–5, 122, 125; brains, 153–45; color, 128; evolution, *45*, 48–55, 58–59, 132, 276; evolutionary success, 39, 277; feeding, 229, 238; forelimbs, 97, 173–74, 177; gregariousness, 219; growth, 135, 137, *138*; Lancian diversity, *191*, 197–99, 206–9, 258–60, 262, *264*; locomotion, 163–65; longevity, 211, 255; pathologies, 251; pneumaticity, *106*, 107–8; reproduction, 142, 244–45, 247–48; senses, 155–58; size, 50, 52, 144, 277; swimming, 167; vocalization, 160
Thescelosaurus, 195, 201, *264*; abundance in Hell Creek Formation, *182*
Thescelosaurus neglectus, *202*
"Thomas," *T. rex* specimen, *139*, *275*; maturity, 144
Through the Looking Glass (Carroll), 211
tiger, *127*
tongues, 122
Tornillo Group, 69
Torosaurus, 65, 195, 205, 214, *235*, *264*
"*Torosaurus*" *utahensis*, 205
Torosaurus latus, 205, *235*
Torvosaurus, 49; eggs, 247, 248
toxic bite, *T. rex* 236
trackways: swimming, 166; tyrannosaurid, 220, *221*
Triassic Period, 43–47, *45*, 49, 186, 199
Triceratops, 1, 9, 13, *15*, 24, *38*, 65, 187, 194–95, 201, 204–5, 211, *212*, 214, 215, 232, *237*, 263; abundance in Hell Creek Formation, *182*, 197, 265; characterizing dinosaur province, 204; development of frills and horns, *215*; in film, 22, 24; market value of skull, 33; mentions in literature, *20*; predated and eaten by *T. rex*, 168, *200*, 230, *231*, 232, 234–35, *237*, 238–41, *240*; running speed of, 233–34, *234*; stratigraphic distribution of 187, *194*, 265; "*Triceratops* fauna," *184*, 187
Triceratops horridus, 1, *194*, 204, 205, *264*
Triceratops prorsus, 194, 204
"Trix," *T. rex* specimen, 83, *116*, 185; dentition and diet, 219; maturity, 144; migration study, 219; pathologies, *224*, 253
Troodontidae, 53, 104, 198, 247, *264*
turtles, *45*, 132, *159*, 160, 177, 226, 274; Cretaceous, 196, 209, 220, 257, *264*, 265; reproduction, 244, 247, 248
Tyrannosaur Chronicles, The (Hone), 30

Tyrannosauridae, 44; alleged sexual dimorphism, 243; anatomy, 28, 72, 73, 90, 92, 94, 104, 107, *123*, 125, 233; bite force, 168, 169; brains, 151; distribution, 183, 186–87; ecology, 68, 128, 198, 260–62; evolution, *54*, 57, 59–65, 80, 275, 277; evolutionary success, 39, 277; facial anatomy, 118, 119–20; forelimbs, 97, 99, 172–73; gregariousness, 220, *221*, 223; growth, 76, 137, *138*, 140, *141*, 142, 150; gut content, 108; integument, 112, 113, 126; locomotion, 163; longevity, 211, 253–55; predatory behavior, 228–30; reproduction, 247, 248; research on, 30; senses, 155–56; size, 49–50; social interactions, 226, 228; taxonomy, 66–77; tracks, *221*. *See also* tyrannosaurid niche assimilation; *Tyrannosaurus rex*
tyrannosaurid niche assimilation, 129, 198, 260–61
Tyrannosaurid Paleobiology (Parrish, Molnar, Currie and Koppelhus), 30
Tyrannosaurinae, 50, *54*, 57, 62
Tyrannosaurini, *54*, 63, 65, 80
Tyrannosauripus pillmorei, track taxon, *104*, 105
Tyrannosauroidea, 172; anatomy, 55, 57, 87, 163; bite force, 168–70; evolution, *45*, 48, 53–60, 62, 133; growth, *138*; integument, 48, 55, 111, 113, 133; pneumaticity, 106; senses, 156
Tyrannosaurus: geographic origin, *54*, 62–63, 65, 80; holotypes of species, *12*, *14*, *25*, *42*, 66–67, *70*, *71*, 72–73, 76–77, 80; specimens of disputed affinity, 69–81, 186–87, 192–93; taxonomy, 65–81. *See also* "*Nanotyrannus lancensis*"; *Tyrannosaurus mcraeensis*; *Tyrannosaurus rex*; *Tyrannosaurus rex* specimens
Tyrannosaurus bataar, 67–69
"*Tyrannosaurus imperator*," *70*, 79–80; classification, 78
"*Tyrannosaurus lanpingensis*," 67
"*Tyrannosaurus luanchuanensis*," 67
Tyrannosaurus mcraeensis, 80–81, *81*, 192, *193*; geological distribution, *184*, *185*, 186–87
"*Tyrannosaurus regina*," *70*, 79–80; classification, 78
Tyrannosaurus rex, 1; abundance in Hell Creek Formation, 181–83; air sacs, 87, 105–8, 133, 159–60; anatomical variation, 77–79, 241–43; appetite, 133, 135; armored face, 115, 125; Asian origin of, *54*, 62–63, 65, 80; associated plants and animals, 194–210; brain(s), 73, 151, *152*, 155–56; bite strength, 38, 80, 92, 162, 168–72, 170–72, 229, 236; body mass, 144–50; bone, internal, *75*, 110–11; 142–43, *143*; celebrity status, 2–6, 19–24, 29–41; Charles Knight artworks of, *15*; classification, 65–67; clichéd depictions, 5, *37*, 38–39, *193*, 279; color, predictions, 126–29; confirmed prey species, 109, 202, 205, 230, *231*,

INDEX

Tyrannosaurus rex continued:
234; coprolites, 108–10; cranial anatomy, 85–92, 133, *134*; ears, *121*, 122; ecology, 129, 133, 198, 260–62; endocast, 151–53, 155–56; evolution, 43–65; extinction, 257–74; eye anatomy, restored, 5, *31*, 32, 87–88, *89*, 115, 118–20, *120*; face biting, 223, *224*, 225–27, 252; facial appearance, 4, 32, 50, *86*, 87–88, *89*, 115–25, 128, 144, 242–43; feeding, 237–41; in fiction, 22–27, 158–59, 274–77; forelimb anatomy, 97–99, 175–77; forelimb redundancy, 172–73, *174*, 275, *276*, 276; fossil evidence of biting, 202, 223–27, 230, *231*, 232, 237–41, 245, 252; fossil record, 34–35, 40–41, 65, 80, 141, 150, 181–83, 187, *194*, 253; fossilization, *84*, *182*, 198; geographic distribution, 183–86, *184*; gregariousness, 219–23; gut anatomy, 108–10; hatchlings, *137*, 140, *142*, 165, 195–96, *216*, 245, *246*, 198–201, 248–51, 253, *254*; headbutting, *226*–27, 228; hindlimb anatomy, 102–5, 163; history of discovery, 6–19; holotype, *12*, 14, 24, *25*, *42*, 65, 66, *70*; hype, 5–6, 19, 32–33, 36–41, 219, 229, 236, 258, 278; influence on dinosaur evolution, 211–15; integument, 32, 99, 105, 111–18, *119*; intelligence, 20, 30, 153–55, 215, 258–59; isotopic insights into diet, 219; *Jurassic Park* (movie) version, 30–32, 122, 159, 162–63, 248; K/Pg experience, 268–74; life appearance, 32, 36–37, *37*, *84*, 111–29; lines of arrested growth (LAGs), 74, *75*, 135, 137, *138*; locomotion, 28, *96*, *96*, 142, 162–65, *166*, 218–19, 232–34, 275; longevity, 135, 144, 211, 253–55; lungs, 105–8; masculinity, 39–40, 258; medullary bone, 111, 142–43, 243; migration, 218–19; model organism, 30, 40, 278; monsterization, 19–24, 29, 36–39; movie versions, 20, 22–25, 30, 31, 32, 71, 122, 158, 159, 162, 163, 248, 274, 276; musculature, 87, 89–91, 93–94, *96*, *96*, 99, 102, 108, 118–19, 122, 133, *134*, 140, 155–56, 164–65, 168–72, *169*, 172–73, 176, *176*, 233–34, *234*; name, 2, 10, 13, 39; niche assimilation, 129, 198, 260–61; nostrils, 87, 115, *117*, *121*, 121–22, 156; ontogeny, 85, 87–89, 91, 92, 135–44, 171, 175, 198, 233, 243, 248–51, 253–55; Osborn, early research on; 6–14, 17–19, 21, 22, 25, 278; osteoderms, 88, *89*, *117*, 118; paleoenvironments of, *184*, 188–94; parenting, 245, 248–51; pathologies, *85*, 143, 144, 177, 223–27, *225*, *250*, 251–53, 255; pectoral anatomy, 97, *98*, 99, 141, 175–77, 252; pelvic anatomy, 99–102, 175–76, 242, *242*; physiology, 29, 128, 131–35, 171; play behavior, 154; pneumaticity, 106–8; population densities, 133; posture, *15*, 19, 27–29, 102, 132, 156, 166, 243–46; predatory behavior, 164–65,

166, 168, 228–36, 248–49; pulmonary system, 105–8; reproduction, 173, 243–48; reproductive maturity, 111, 135, *138*; ribs, 14, *85*, 93–95, *95*, *250*, 252; roaring, 2, 22–23, 38, *130*, 158–61; "robust" and "gracile" morphologies, 77, 77, 78–79, 242; sale of specimens, 24, 35–36, 40–41; scales, skin covering: 111–15, *117*, 118, *119*, 228; scavenging, 229; senses, 155–58; sexual dimorphism, efforts to find, 77, 78, 241–43; size, 5, 144–45, *146*, 150–51; skull, 5, 57, *84*, 85–92, 94, 107–8, 113, *116*, *117*, *120*, 121, *123*, 124, *141*, 145, 150, 151, 156, *157*, *169*, 170–71, 220, 228; soft tissues, reconstruction of, *90*, *96*, 110–25, 148, *149*, 150; as sole predator of Lancian North America, 258, 260–63; survivorship, 253–55; tail, 22, 28, *85*, 96–97, 102, 108, 111, *113*, 132, 164, *234*, 242–43; *Tarbosaurus*, congeneric with, 68–69; taxonomy, 65–81; temporal distribution, *45*, 186–87; territoriality, 218–19; tongue, 122; toxic bite, 236; vertebral column, *14*, *15*, 28, 83, *85*, 90, 92–97, 102, 145, 182, 234, *250*, 251–53, 255; vocalization, 22–23, *130*, 158–62. See also juveniles, *T. rex*; museums, *T. rex*; *Tyrannosaurus*; *Tyrannosaurus rex* specimens
Tyrannosaurus rex (Bradbury), 27
Tyrannosaurus rex specimens: AMNH 5027, 14, *16*, 17, 18, 19, *16*, *18*, 32, 70, 79, 144, 184, *224*; "Baby Bob," 33; "Barnum," coprolite, 109, *109*; "Black Beauty," 184; CM9380, holotype, *12*, *14*, 24, *25*, *42*, 65, 66, *70*; "Jane," 73, *74*, *136*, *141*–42, *147*, 171, *182*, 184, 241, *224*, 226, 253; "Jordan theropod," 140, *141*; "Peck's rex," *25*, 184, *182*, *224*; "Petey," *75*, 76, 241; "Samson," 35; "Stan," 33, 35, *89*, *100*, 144, 147, 185, *224*, 241, 245; "Sue," *20*, 33–35, *70*, 71, *75*, 76, 83–*85*, *90*, 91, *94*, 98, *137*, 138, 144, 145–51, *157*, 176, 182, 185, 219–20, *224*, *250*, 251–53; "Thomas," *139*, 144, 275; "Trix," 83, *116*, 144, 185, 219, *224*, 253; "Wankel"/"Nation's *T. rex*," 3, 144, *147*, *149*, *182*, 184–85, *237*; "Wyrex," *98*; 111, *113*; 252
"*Tyrannosaurus stanwinstonorous*," 71
"*Tyrannosaurus turpanensis*," 67–68
"*Tyrannosaurus vannus*," 69, *70*
"*Tyrannosaurus x*," 77–79
"*Tyrannosaurus zhuchengensis*," 67–68
tyrant king, terminology, 39

United States, 47, 60, 62, 63, 65; Chicxulub impact, 266–75; commercial fossil laws, 35; dominance of early 20th century vertebrate paleontology, 6–9, 19–24; Lancian fauna, 194–211; Maastrichtian geography, 184, *184*, 188–94. See also North America

Utah, 185, *185*, 210

Valley of Gwangi (movie), 24
"*Vastatosaurus rex*," speculative tyrannosaur, 276
Velociraptor, 32, 53, 71, 198–99; mentions in literature, *20*
Velociraptor mongoliensis, 2
venom, used in predation, 236
vertebrae: adaptations to predatory lifestyles, 45; bitten by *T. rex*, 230, *231*, 238–39; hadrosaurid, *234*; posture, 47; records of Lancian fauna, 199; use in classification, 48, 55. See also tail
vertebrae, *T. rex*: anatomy, *14*, *15*, 83, *85*, 90, 92–97, 102, 145, 234; discovery, 9, 10, *11*, 13, *14*, *16*, 17; number of, 28, 145, 182; pathologies, *250*, 251–53, 255; pneumaticity, 106–8; posture, 28; scaling, 150; use in classification, 66, 67, 78
vertebrate paleontology. See paleontology
vision, in predatory animals, 60, 155–56, 157, 229; in *T. rex*, 87, 118, 155–56, 158, 275. See also eyes
vocalization: closed-mouth, *130*, 161, *161*; evolution in reptiles, 158–62; *T. rex*, 22–23, *130*, 158–62
volcanoes, 22, 186, 192–93, *193*, 260
von Ebner lines, dentition, 92
von Fuehrer, Ottmar, 24

Walking with Dinosaurs (documentary), 248
"Wankel"/ "Nation's *T. rex*," *T. rex* specimen, 3, 184–85, *237*; body mass *147*, *149*; maturity, 144; taphonomy, *182*
warm-blooded. See physiology
Wells, H. G., 20
Western Interior: geology, 183, *184*, 186–87, *194*, 259; Lancian biota, 195–211; Maastrichtian paleoenvironments, 188–94, 272
Western Interior Seaway, *184*, 188, 191–92, *267*, *271*, 272
whales, 83; eyeball size, *120*; longevity, 254
Williams, Maurice, 33
Willow Creek Formation, *184*; age, 185; paleoenvironment, 188, 190
Wilson, Maurice, 24
Woodward, Holly, 76
Wray, Fay, 22
"Wyrex," *T. rex* specimen, *98*; skin impressions, 111, *113*; tail trauma, 252
Wyoming, *185*, 185; historic *T. rex* discoveries, 10, *11*, 13, 17

Yutyrannus, *54*, 55, 111, 114, 169
Yutyrannus huali, 55, *56*

Zallinger, Rudolph, 24
Zhuchengtyrannus, *54*, 65
Zhuchengtyrannus magnus, 62, 63, 67–68